MW00913097

Dental Perspectives on Human Evolution

Vertebrate Paleobiology and Paleoanthropology

Edited by

Eric Delson
Vertebrate Paleontology, American Museum of Natural History,
New York, NY 10024, USA
delson@amnh.org

Ross D.E. MacPhee
Vertebrate Zoology, American Museum of Natural History,
New York, NY 10024, USA
macphee@amnh.org

Focal topics for volumes in the series will include systematic paleontology of all vertebrates (from agnathans to humans), phylogeny reconstruction, functional morphology, Paleolithic archaeology, taphonomy, geochronology, historical biogeography, and biostratigraphy. Other fields (e.g., paleoclimatology, paleoecology, ancient DNA, total organismal community structure) may be considered if the volume theme emphasizes paleobiology (or archaeology). Fields such as modeling of physical processes, genetic methodology, nonvertebrates, or neontology are out of our scope.

Volumes in the series may either be monographic treatments (including unpublished but fully revised dissertations) or edited collections, especially those focusing on problem-oriented issues, with multidisciplinary coverage where possible.

Editorial Advisory Board

Published and forthcoming titles in this series are listed at the end of this volume.

A Volume in the

Max Planck Institute Subseries in Human Evolution

Coordinated by

Jean-Jacques Hublin
Max Planck Institute for Evolutionary Anthropology,
Department of Human Evolution, Leipzig, Germany

Dental Perspectives on Human Evolution: State of the Art Research in Dental Paleoanthropology

Edited by

Shara E. Bailey
Max Planck Institute for Evolutionary Anthropology,
Department of Human Evolution, Leipzig, Germany
and
New York University, Department of Anthropology
New York, USA

Jean-Jacques Hublin
Max Planck Institute for Evolutionary Anthropology,
Department of Human Evolution, Leipzig, Germany

A C.I.P. Catalogue record for this book is available from the Library of Congress.

ISBN 978-1-4020-5844-8 (HB)
ISBN 978-1-4020-5845-5 (e-book)

Published by Springer,
P.O. Box 17, 3300 AA Dordrecht, The Netherlands.

www.springer.com

Printed on acid-free paper

Cover illustration: Image created by Kornelius Kupezik using VOXEL-MAN
(VOXEL-MAN Group, University Medical Center, Hamburg-Eppendorf, Germany)

Contents

Foreword

S.E. BAILEY
Department of Human Evolution
Max Planck Institute for Evolutionary Anthropology
Deutscher Platz 6
D-04103 Leipzig, Germany
and
Center for the Study of Human Origins,
Department of Anthropology, New York University,
25 Waverly Place
New York, NY 10003, USA
sbailey@nyu.edu

J.-J. HUBLIN
Department of Human Evolution
Max Planck Institute for Evolutionary Anthropology
Deutscher Platz 6
D-04103 Leipzig, Germany
hublin@eva.mpg.de

When faced with choosing a topic to be the focus of the first symposium in Human Evolution at the Max Planck Institute for Evolutionary Anthropology in Leipzig, a paleoanthropological perspective of dental anthropology was a natural choice. Teeth make up a disproportionate number of the fossils discovered. They represent strongly mineralized organs of compact shape, which allow better preservation in geological deposits and archaeological sites than any other part of the skeleton. As a result, since the discoveries of the first fossils of extinct species, vertebrate paleontology has been built primarily on analyses of teeth. The first dinosaur identified in 1825 by Gideon Mantell was actually a dinosaur tooth. Paleoanthropology is no exception to this rule, as teeth represent, by far, the most abundant material documenting different species of extinct non-human primates and hominins. As such, much of what we know about non-human primate and hominin evolution is based on teeth.

Teeth have been a focus of interest for physical anthropologists over many generations. Teeth provide a multitude of information about humans – including cultural treatment, pathology, morphological variation, and development. The presence of culturally induced wear (toothpick grooves, for example) reveals something about what humans were doing with their teeth in the past. Pathologies, such as enamel hypoplasia and dental caries, are informative for understanding the health and nutritional status of individuals and populations. Dental morphological variation among living humans has proven to be important for assessing biological relationships among recent groups. Finally, the dental sequence of calcification and eruption patterns remain, even today, the easiest way to assess the individual age of nonadult modern humans. Although dental anthropology has a long history in physical anthropology, the recent years have brought a number of new discoveries, new methods, and a renewal of interest in using the teeth to answer questions about human and nonhuman primate evolution. The goals of studies focusing on fossil humans are similar to those of recent humans noted above. In addition, of particular interest are the biological relationships among extinct species, the amount of variation one should expect in fossil species, and the polarity of dental characters.

To date, the developmental pathways of most of the skeletal features routinely used in paleoanthropological studies remain obscure. We know that the genotype interacts with the environment in a complex manner. This interaction produces a pattern that we attempt to interpret in phylogenetic and taxonomic ways, but from which we also attempt to extract other biological information. Unfortunately, the level of integration among skeletal (primarily cranial) features that are most often considered independent is, in many cases, still to be explored. Tooth size and, even more, morphology are under strong genetic control. Because dental germs are formed in an early stage of individual development while the individual is still in utero, teeth may represent organs that are widely independent from environmental influences. While sometimes considered 'less exciting' than, say, skulls, the most abundant fossils might, in fact, be some of the most meaningful for paleontological studies. In recent years, the development of 3D morphometrics and the systematic coding of non-metric traits of the dentition have opened new avenues for the assessment of dental morphological variation, which supersedes the earlier, rather disappointing, results based on simple linear measurements of length and breadth.

Another special feature of the dentition is that, in contrast to the rest of the skeleton, it is not subject to major remodeling during the course of an individual's life (aside from attrition). Enamel tissue is laid down in the early stages of life and becomes a mostly closed system interweaving mineralized prisms and organic matter. Teeth are not just an abundant fossil material that can be found in the field; they also represent fossilized stages of the life history of an individual. Because enamel is very hard and undergoes few exchanges with the surrounding environment, teeth preserve chemical signals that can be analyzed for the establishment of the geological age of specimens, as well as for the understanding of the individual biology at different ages of life. The explosive development of isotopic studies has made it possible to study dietary differences among individuals as well as environmental conditions at different stages of the life in ancient humans. In some cases, intriguing aspects of daily life, mating strategies, or landscape occupation of past populations have been revealed. It is also inside the teeth that we are more likely to find preserved fossil molecules such as proteins, which allow us to access fascinating aspects of the biology of our remote predecessors.

Soon after their eruption into the oral cavity, teeth represent a major interface between the individual and its environment, and their wear patterns and pathologies become another major source of information for both dietary and non-dietary behaviors. While the gross morphology of the tooth can tell us what an organism is capable of processing, it is the actual wear patterns on the enamel surface that tell us how this organism was actually using its teeth. Although not without problems, microwear analyses have provided a wealth of information on dietary differences in extant primates and humans. These have recently been supplemented by topographical models in three dimensions that allow angles, planes and valleys to be investigated and by three-dimensional analyses of wear planes. Both provide information on dental function and occlusion.

Once we move below gross tooth morphology a proverbial 'whole new world' opens up. New methods allow us to visualize the structures underlying tooth enamel as well as the microscopic intricacies of enamel itself. Recent research in this area has been extremely important to human paleontology, especially with regard to dental development and life history – two current and important issues in physical anthropology. The pace of development, brain maturation, length of learning period, reproductive patterns and longevity are crucial issues for understanding biological and social changes during the course of human evolution. Until recently this has been mostly a field of speculation based on the knowledge of extant apes and humans. The development of microstructural studies has revealed that dental tissues represent, by far, one of the best records of the conditions of the growth and development of individuals. The extent to which extinct hominin species are comparable to extant humans, or other large primates, has become a focus of interest for many studies. With the development of the use of new instruments such as the confocal microscope, micro-CT scanners, or the synchrotrons, researchers are now able to explore a new world inside our teeth.

One exciting aspect of publishing a volume on recent advances in dental paleoanthropological studies is the bringing together of multiple disciplines in which tooth morphology is currently being used to answer questions about human and non-human primate evolution. The different approaches mentioned here have been rapidly integrated as new methods to access biological information. Another is that here, perhaps to a greater extent than in other subfields of physical anthropology, an integration of the contributions coming from primatology and modern human variation is essential to develop meaningful interpretation of the fossil record. Dental anthropology has become a very multi-disciplinary field, by the scope of its studies as well as by the variety of techniques employed in recent analyses. In inviting the contributors of this volume to participate in its publication, we wanted to make available the state-of-the-art of our knowledge in this field. In addition, this project also illustrated the rapid progress in a variety of analytical methods that have recently emerged in the broader domain of biological anthropology.

Acknowledgments

This edited volume is based on a Dental Paleoanthropology symposium held in May 2005 at the Max Planck Institute for Evolutionary Anthropology, Leipzig, Germany. We are grateful to all the participants who attended the symposium, provided valuable feedback and discussion on the papers presented and contributed their work to this volume.

We thank the editorial staff at Springer, especially Series Editors Eric Delson and Ross MacPhee for their guidance on organizing and pulling together the edited volume. We also greatly appreciate their editorial assistance with the final version.

We are very grateful to the administrative staff of the Department of Human Evolution for their assistance with the organization of the symposium. Silke Streiber and Diana Carstens were especially helpful and patient with all the various needs of our participants who came from many different countries around the world. We also appreciate the support of Myriam Haas and the MPI Media department for their assistance with the printed matter associated with the symposium.

Special thanks goes to Allison Cleveland who coordinated the review process for this volume and whose time spent on the final editing and formatting of manuscripts, as well as compiling of the index, is greatly appreciated. It would not be an overstatement to say that without her assistance this volume would not have been possible.

List of Contributors

Juan Luis Arsuaga
Centro de Evolución y Comportamiento
Humanos Sinesio
C/Delgado 4, Pabellón 14
28029 Madrid, Spain
jlarsuaga@isciii.es

Gal Avishai
Laboratory of Bio-Anthropology and Ancient DNA
Hadassah Faculty of Dental Medicine
Hebrew University of Jerusalem
POB 12272
91120 Jerusalem, Israel
gavishai@md.huji.ac.il

Shara E. Bailey
Center for the Study of Human Origins
Department of Anthropology
New York University
25 Waverly Place
New York, NY 10003, USA, (current address),
and
Department of Human Evolution
Max Planck Institute for Evolutionary Anthropology
Deutscher Platz 6
D-04103 Leipzig, Germany
sbailey@nyu.edu

Markus Bastir
Department of Palaeobiology
Museo Nacional de Ciencias Naturales, CSIC
C/ José Gutiérrez Abascal 2
28006 Madrid, Spain
and
Hull York Medical School
The University of York
Heslington, York YO10 5DD, UK
markus.bastir@hyms.ac.uk

Jose M. Bermúdez de Castro
Centro Nacional de Investigación sobre
Evolución Humana
09004 Burgos, Spain
jmbc@mncn.csic.es

Thomas A. Bishop
Department of Statistics
The Ohio State University
Columbus, OH 43210, USA
tab@stat.ohio-state.edu

Silvia Boccone
Laboratori di Antropologia
Dipartimento di Biologia Animale e Genetica
Università di Firenze
Via del Proconsolo 12
50122 Firenze, Italy
boccone@unifi.it

Alan Boyde
Hard Tissue Research Unit, Dental Biophysics
Queen Mary University of London
London E1 1BB, UK
a.boyde@qmul.ac.uk

Jose Braga
Laboratoire d'Anthropologie Biologique
Université Paul Sabatier (Toulouse 3), FRE 2960
39 allées Jules Guesde
31000 Toulouse, France
braga@cict.fr

Timothy G. Bromage
Hard Tissue Research Unit Departments of
Biomaterials and Basic Sciences
New York University College of Dentistry
345 East 24th Street
New York, NY 10010-4086, USA
tim.bromage@nyu.edu

Alfredo Coppa
Department of Animal and Human Biology
Section of Anthropology
University of Rome "La Sapienza"
00185 Rome, Italy
alfredo.coppa@uniroma1.it

M. Christopher Dean
Department of Anatomy and Developmental Biology
University College London
Gower Street
London, WC1E 6BT, UK
ucgacrd@ucl.ac.uk

Ferran Estebaranz
Secc. Antropologia
Department of Biologia Animal
Universitat de Barcelona
08028 Barcelona, Spain
ucgacrd@ucl.ac.uk

Rebecca J. Ferrell
Department of Sociology and Anthropology
Howard University
Washington, DC 20059, USA
rferrell@howard.edu

Yankel Gabet
Bone Laboratory, Institute of Dental Sciences
Faculty of Dental Medicine
Hebrew University
91120 Jerusalem, Israel
gabet@md.huji.ac.il

Jordi Galbany
Secc. Antropologia
Department de Biologia Animal
Universitat de Barcelona
08028 Barcelona, Spain
Galbany@ub.edu

David G. Gantt
Department of Anthropology
Georgia Campus-Philadelphia College of Osteopathic Medicine
Suwanee, GA 30024, USA
davidga@pcom.edu

Laurie R. Godfrey
Department of Anthropology
University of Massachusetts
Amherst, MA 01003, USA
lgodfrey@anthro.umass.edu

Aida Gómez-Robles
Centro Nacional de Investigación sobre Evolución Humana (CENIEH)
Avda. de la Paz 28
09006 Burgos, Spain
aidagomezr@yahoo.es

Frederick E. Grine
Departments of Anthropology and Anatomical Sciences
Stony Brook University
Stony Brook, 11794-4364 NY, USA
Frederick.Grine@stonybrook.edu

Debbie Guatelli-Steinberg
Department of Anthropology
Department of Evolution, Ecology, and Organismal Biology
The Ohio State University
Columbus, OH 43210, USA
guatelli-steinbe.1@osu.edu

Yann Heuze
Laboratoire d'Anthropologie Biologique
Université Paul Sabatier (Toulouse 3), FRE 2960
39 allées Jules Guesde
31000 Toulouse, France
heuze@cict.fr

Simon Hillson
University College London
31-34 Gordon Square
London, WC1H 0PY, UK
simon.hillson@ucl.ac.uk

Leslea J. Hlusko
Department of Integrative Biology
University of California
Berkeley, CA 94720-3140, USA
hlusko@berkeley.edu

Jean-Jacques Hublin
Department of Human Evolution
Max Planck Institute for Evolutionary Anthropology
Deutscher Platz 6
D-04103 Leipzig, Germany
hublin@eva.mpg.de

Louise T. Humphrey
Department of Palaeontology
The Natural History Museum
Cromwell Road
London, SW7 5BD, UK
l.humphrey@nhm.ac.uk

Teresa E. Jeffries
Department of Mineralogy
The Natural History Museum
Cromwell Road
London SW7 5BD, UK
t.jeffries@nhm.ac.uk

John Kappelman
Department of Anthropology
The University of Texas at Austin
Austin, TX 78712-1086, USA
jkappelman@mail.utexas.edu

Richard A. Ketcham
High-Resolution X-ray CT Facility
Department of Geology
University of Texas
Austin, TX 78712-1100, USA
richk@mail.utexas.edu

Ottmar Kullmer
Research Institute Senckenberg
Department of Paleoanthropology and Quaternary Paleontology
60325 Frankfurt am Main, Germany
ottmar.kullmer@senckenberg.de

Rodrigo S. Lacruz
Center for Craniofacial Molecular Biology
School of Dentistry
University of Southern California
Los Angeles, CA 90089-0641, USA
rodrigo@usc.edu

Clark Spencer Larsen
Department of Anthropology,
Department of Evolution, Ecology,
and Organismal Biology
The Ohio State University
Columbus, OH 43210, USA
larsen.53@osu.edu

Roberto Macchiarelli
Laboratoire de Géobiologie
Biochronologie et Paléontologie humaine
Université de Poitiers
86000 Poitiers, France
roberto.macchiarelli@univ-poitiers.fr

Michael C. Mahaney
Southwest National Primate Research Center
and the Department of Genetics
Southwest Foundation for Biomedical Research
P.O. Box 760549
San Antonio, TX 78245, USA
mmahaney@sfbrgenetics.org

Patrick Mahoney
Department of Archaeology
University of Sheffield
Sheffield S1 4ET, UK
P.Mahoney@sheffield.ac.uk

Alan Mann
Department of Anthropology
Princeton University
Princeton, New Jersey 08544, USA
mann@Princeton.edu

Franz Manni
UMR 5145 – Eco-Anthropology Group
National Museum of Natural History
MNHN – Musée de l'Homme
75016 Paris, France
manni@mnhn.fr

Lawrence B. Martin
Departments of Anthropology
and Anatomical Sciences
Stony Brook University
Stony Brook, NY 11794, USA
lmartin@notes.cc.sunysb.edu

Laura M. Martínez
Secc. Antropologia, Department de Biologia Animal
Universitat de Barcelona
08028 Barcelona, Spain
lmartinez@ub.edu

Maria Martinon-Torres
Centro Nacional de Investigación sobre
Evolución Humana
09006 Burgos, Spain
mariamt@mncn.csic.es

Jacopo Moggi-Cecchi
Laboratori di Antropologia
Dipartimento di Biologia Animale e Genetica
Università di Firenze
50122 Firenze, Italy
and
Sterkfontein Research Unit Institute for Human Evolution
University of the Witwatersrand
Johannesburg 2193, South Africa
jacopo@unifi.it

Janet Monge
Department of Anthropology
Museum of Anthropology and Archaeology
University of Pennsylvania
Philadelphia, PA 19104, USA
jmonge@sas.upenn.edu

Ana Muela
Fundación Atapuerca
C/Condestable 2, 4° C
09004 Burgos, Spain
informacion@fundacionatapuerca.es

Ralph Müller
Institute for Biomedical Engineering
Swiss Federal Institute of Technology (ETH)
University of Zürich
CH-8044 Zürich, Switzerland
ralph.mueller@ethz.ch

Anthony J. Olejniczak
Department of Human Evolution
Max Planck Institute for Evolutionary Anthropology
D-04013 Leipzig, Germany
olejniczak@eva.mpg.de

Alejandro Perez-Ochoa
Department of Paleontology
Universidad Complutense de Madrid
28040 Madrid, Spain
perezochoa@msn.com

Alejandro Pérez-Pérez
Secc. Antropologia, Departament de Biologia Animal
Universitat de Barcelona
08028 Barcelona, Spain
martinez.perez-perez@ub.edu

Varsha Pilbrow
University of Melbourne
Zoology Department
Victoria 3010, Australia
vpilbrow@unimelb.edu.au

Fernando Ramirez Rozzi
UPR 2147-CNRS
44, rue de l'Amiral Mouchez
75014 Paris, France
and
Department of Human Evolution
Max Planck Institute for Evolutionary Anthropology
D-04103 Leipzig, Germany
ramrozzi@ivry.cnrs.fr

Donald J. Reid
Department of Oral Biology
School of Dental Sciences
Newcastle University
Newcastle upon Tyne NE2 4BW, UK
D.J.Reid@newcastle.ac.uk

James M. Rogér
Marquette University School of Dentistry
P.O. Box 1881
Milwaukee, WI 53201, USA
james.roger@mu.edu

Susana Sarmiento
Fundación Atapuerca, C/Condestable 2, 4° C
09004 Burgos, Spain
informacion@fundacionatapuerca.es

Friedemann Schrenk
JWG University Frankfurt
Vertebrate Paleobiology Institut for Ecology,
Evolution and Diversity
Siesmayerstrasse 70
60054 Frankfurt am Main, Germany
schrenk@bio.uni-frankfurt.de

Gary T. Schwartz
School of Human Evolution and Social
Change and The Institute of Human Origins
Arizona State University
Tempe, AZ 85287-4101, USA
garys.iho@asu.edu

Patricia Smith
Laboratory of Bio-Anthropology and Ancient DNA
Hadassah Faculty of Dental Medicine
Hebrew University of Jerusalem
POB 12272
91120 Jerusalem, Israel
pat@cc.huji.ac.il

Tanya M. Smith
Department of Human Evolution
Max Planck Institute for Evolutionary Anthropology
D-04103 Leipzig, Germany
tsmith@eva.mpg.de

Angela Stout
Temple University School of Dentistry
3223 North Broad Street
Philadelphia, PA 19140, USA
angela.stout@temple.edu

Mark F. Teaford
Center for Functional Anatomy & Evolution
Johns Hopkins University School of Medicine
Baltimore, MD 21205, USA
mteaford@jhmi.edu

Lillian Ulhaas
Research Institute Senckenberg
Department of Paleoanthropology and Quaternary Paleontology
60325 Frankfurt am Main, Germany
lilian.ulhaas@senckenberg.de

Peter S. Ungar
Department of Anthropology
University of Arkansas
Fayetteville, AR 72701, USA
pungar@uark.edu

Rita Vargiu
Department of Animal and Human Biology
Section of Anthropology
University of Rome "La Sapienza"
Piazzale Aldo Moro 5
00185 Rome, Italy
rita.vargiu@uniroma1.it

Rose Wadenya
University of Pennsylvania School of Dental Medicine
240 South 40th Street
Philadelphia, PA 19104, USA
wadenya@dolphin.upenn.edu

Bernard Wood
The George Washington University, CASHP
Department of Anthropology
2110 G Street, NW
Washington, DC 20052, USA
bernardawood@gmail.com

Introduction

S. HILLSON
University College London
31-34 Gordon Square
London WC1H 0PY, UK
simon.hillson@ucl.ac.uk

Teeth occupy a central place in the fossil evidence for human evolution. One reason for this lies in their complex biology. Strictly speaking, although teeth are preserved with the bones of the skeleton, they are biologically a separate entity; the dentition. Like bone, the three dental tissues, enamel, dentine and cement are calcium phosphate and organic composites and as the hardest parts of the body, bones and teeth are the part that remains in the fossil record, but the similarity ends there. Bone is a mineralized connective tissue, developing and remaining only within the body. It contains living cells, blood vessels and nerves. Enamel by contrast is a heavily mineralized epithelial tissue, visible on the surface of the body, containing no cells, blood or nervous supply. In effect, it is dead even in living creatures. Dentine only develops in contact with epithelium – part of the dermal armor in some ancient fish, but confined to the teeth in mammals. It also contains no complete cells, although processes from cells lining the pulp chamber inside the tooth pass through it, carried in the microscopic tubes that characterize dentine structure. Of the three dental tissues, cement is the most bone-like in composition. Both contain living cells, held in tiny chambers called lacunae, and the collagen of the organic component has a dominant role in both structures, but in primates cement has a very different organization, contains no blood supply and acts only as an attachment for the ligament that holds the tooth into its socket.

The principal difference is that dental tissues do not turn over. Bone is continually replaced throughout life by the activities of its cells, the osteoclasts and osteoblasts. The rate varies through the skeleton, but each cubic centimeter of a major long bone is probably replaced almost completely over 10 years or so. The form of a bone is actively maintained, in response to the forces acting on it. As these alter, through injury and disease, changes in posture and activity, or physiology, the overall shape of the bone is remodeled by tissue turnover. Once formed, dental tissues in primates do not do this. They therefore retain the structures put in place by their development. It is possible to see and count these in microscope preparations. In addition, teeth are formed in childhood in their final shape and size, so their morphology can be compared directly between juveniles and adults. They have an intricate form which, because it remains as originally developed, should be easier to understand in relation to the activities of the genes of development.

By contrast, bones grow in size and change in proportions through childhood, so direct comparisons cannot be made in the same way. Teeth also retain the marks left by tooth wear and disease – their main response is to line their pulp chamber with secondary dentine to maintain the covering of dental tissue over the soft tissue of the pulp. Cement is also deposited on the root throughout the life of the tooth, and there is some suggestion that it might be more rapid in heavily worn teeth but there is no close relationship. Bones respond to disease and injury by remodeling – bone is lost in some areas and deposited in others, so that the shape changes. Fractures are mended, inflammation heals and so on. The contrast between teeth and bone is clearest in the continuous eruption of teeth as a response to wear. The teeth are continually pushed into the mouth to make up for the tissue worn away, but the teeth themselves do not provide the mechanism for this. Instead, the supporting bone remodels around them so that the sockets, teeth and all, migrate up through the alveolar process. In heavy wear rate populations, the wear progresses down the root as the socket shortens and, eventually, only a tiny root fragment remains which becomes lose and is lost when it becomes too short. The teeth literally wear out with very little response from dental tissues to the changes of wear, whilst the bone responds by constant remodeling.

Teeth thus have very much their own biology. They are formed by an intricate series of processes that leave their mark in dental tissues and can be studied to give a very detailed account of development. They present a highly variable array of intricate forms. Consideration of the mammals as a whole suggests that these forms represent not only adaptation to the gathering and processing of different types of food, but also to aspects of behavior, including sexual behavior, grooming and so on. There appears to be a strong inherited component in the development of different tooth forms and, although it will probably take many years to understand how these forms are controlled, the fact that tooth formation is a single event – there is no tissue turnover – should simplify the problem in the end. Thus, there seems a real prospect of using tooth morphology to reconstruct the adaptive mechanisms involved in primate evolution. Not only this, but the teeth and jaws seem to have been a major focus of change in the evolution in the hominids. For example, one of the strongest trends in the genus *Homo* has been the reduction of tooth size, together with the prominence of the jaws in the structure of the skull.

In many anatomy texts, teeth are presented as part of the alimentary canal. They touch every particle of the food passing through the mouth and are marked by this passage. Their form is presumably an adaptation to the nature of the food, and the effect of a lifetime's food processing. They wear down but, as they spend only a small fraction of the animal's lifespan as unworn, pristine specimens, the important aspect of tooth form from a selective point of view must presumably be the worn form which it presents throughout most of its life. To put it crudely, teeth are "designed" by evolution to be worn. This means that adaptive mechanisms need to be considered in terms of not only the activities and forces producing wear, but also the developing shape of the worn tooth and the changing way in which the teeth fit together (their occlusion) to adapt to the effects of wear. This approach was first proposed in the classic paper of P.R. Begg (1954), who was interested in the heavily worn teeth of Australian aborigine people, and considered that many of the dental problems of modern urban people arose because their teeth were insufficiently worn to function in the way that they were "designed" to do.

One reason for the central place of teeth in studies of human evolution is thus their information potential (see Foreword). Even a single small tooth can sometimes yield more information than a large pile of bones. They do, however, have another important point in their favor. The tissues and forms of teeth are adapted to surviving a lifetime in the mouth, where they are subject to continuous physical and chemical attack. Teeth are very tough, durable structures. The same properties have ensured that they dominate the fossil record in mammals. Bones are less able to resist weathering. They are also more likely to have been crushed in the jaws of carnivores and scavengers because the skull is overlain by less meat. For these reasons, most vertebrate paleontology is about teeth, and primate paleontology is no exception.

All this makes the conference described in this volume a particularly important event. Dental anthropology is a small specialty within the study of human evolution. Many more studies concentrate on the skull, for example, even though complete finds of skulls are relatively speaking rather rare. The term *dental anthropology* probably has its origin in a symposium of the Society for Study of Human Biology at the Natural History Museum in London, published as an edited volume by Don Brothwell (1963). One of the developing field's distinguishing features is that it has always involved a wide range of researchers, coming from dental schools and departments of anthropology, archaeology, anatomy and biology. It therefore encompasses a variety of approaches and research questions arising from very different points of view. Some researchers may, in fact, primarily be interested in teeth from the point of view of developmental biology and find evolution an interesting application, while others see teeth as only one approach to answering their questions which focus on human evolution. This gathering of converging interests has met regularly together at various venues since 1963 and has continued to find new enthusiasts to add to its founders. It has been extraordinarily productive, perhaps because of the variation in approach. Many of its international representatives came together in 2005 for the conference on *Dental perspectives in human evolution: state of the art research in dental anthropology*, at the Max Planck Institute for Evolutionary Anthropology, in Leipzig and their papers are presented in this volume.

The papers focus on three main themes: dental morphology, dental development and methods for examining teeth. Dental morphology in this context deals with variation in the size and shape of teeth from the hominid fossil record. Living primate species to a large extent show distinctive differences in tooth form, but they also show variation within species, particularly between the sexes. It is not clear how sexually dimorphic extinct species of primates might be, and the surviving specimens in any case represent just a few glimpses at what is likely to have been a considerable range in variation. One way in which the question can be approached is by examining variation of dental features within and between species of living apes to provide a context in which variation between fossils can be interpreted, as shown in Pilbrow's chapter. A similar approach is to look for trends and variation in hominid dental morphology, within and between well defined fossil species, and assemblages which may contain a number of taxa (chapters by Bailey and Wood, and by Moggi-Cecchi and Boccone). Similarly, cladistic and phenetic analysis of dental morphology can be used to test the way in which the hominid fossil record is divided into species (chapters contributed by Martinón-Torres and colleagues and by Manni and colleagues, respectively).

The established approach to dental morphology is to classify different features of the tooth crown and roots according to a standard system. The best known is the Arizona State University dental anthropology system (ASUDAS), although this is better adapted to the study of modern *Homo sapiens* than it is either to living apes or fossil hominids (see Bailey and Wood). It is necessary to add to the system those features which show more variation between apes and between the hominins of the Pliocene. With the arrival of digital photography, it has become straightforward to take instead measurements of lengths, angles and areas of features on the tooth crown, using image analysis software. This makes it possible to assess the size and spacing of features, rather than simply to score their state of development. Another step forward has been in the measurement of enamel thickness. One of the crucial trends in hominid evolution is a change in the thickness of the enamel layer which forms the surface of the crown. In the past, this has been limited by the necessity of sectioning teeth, but recent advances in micro-computed tomography (below) have allowed non-destructive measurements to a fine resolution. This makes possible the construction of detailed three dimensional models of the enamel cap and its thickness, as discussed by Olejniczak and colleagues and also by Gantt and colleagues.

If tooth shape and size are to provide evidence for the place of different fossils in hominid evolution, then it is important to understand how different morphologies develop and the way in which different characteristics of form might be inherited. In essence, this is the question of "how to make a tooth". One approach is through experimental biology, using animals such as laboratory mice in which the actions of different genes in the developmental sequence can be investigated. Another possibility is to carry out genetic analysis of dental features in a group of individuals whose pedigree is known, as for example provided by a captive colony of baboons (Hlusko and Mahaney).

As described above, once they are erupted into the mouth, the form of the teeth is altered by wear and they are unable to respond by remodeling. Most fossil hominids show evidence of very rapid tooth wear so, until the most recent times, the teeth of all but the youngest aged individuals are heavily worn. Ungar reasons that it is the form that teeth take after wear that must be important for the adaptive mechanisms that have driven evolution. Each species has a characteristic pattern of changes with wear that defines it just as clearly as the unworn form of the teeth. It is therefore necessary to define new ways in which to describe and compare the worn surfaces, as shown by Ulhaas and colleagues. It seems logical to suggest that the pattern of wear shown on the teeth should also reflect the nature of the diet and the way in which the dentition is used to process food. If this can be assessed, then at least some of the adaptive mechanisms acting on the form of the teeth should become clearer. One established approach to this question is the microscopic study of wear, known as dental microwear. Teaford gives a concise history of microwear studies ending with the most recent measurement and analytical techniques. These have addressed a number of difficult practical and theoretical difficulties and Esteberanz and colleagues (Part IV, Chapter 6) demonstrate the use of these with a range of fossil and recent hominid and pongid specimens.

Development of the teeth has been an important theme in dental anthropology from the beginning, and a large number of the papers in this volume are concerned with the topic. Modern human children develop over a longer period than other primates, and this is seen in the dentition as well as the rest of the body. This long schedule is related to the development of cognition and the behaviors that, in effect, make us human so one of the important questions for research on human evolution is the point at which this pattern of growth appeared – which fossil forms is it associated with? The fossil record represents a relatively small number of individuals which may represent any part of a range of variation, so one of the important questions is to understand the extent to which the pattern of growth varies within one species. To investigate large numbers of children, it is necessary to use x-rays, even though there are problems in assessing the stages of growth achieved and relating these scores to the development seen by direct observation of growing teeth. Braga and Heuze discuss practical and theoretical problems with the assessment of dental x-rays, and the way in which they show patterns of development that can be used to help interpret the fossil record. Monge and colleagues describe a study of variation in development seen in a large collection of dental x-rays from modern urban children.

The bulk of the papers on dental development, however, are based on the histology of dental enamel. Enamel's heavily mineralized structure enables it to survive well in fossils, with all its microscopic features intact. Amongst these features is a pattern of layering that reflects a circadian rhythm to the secretion of the enamel matrix during development. These so-called prism cross striations can provide a daily clock beat to determine the timing of key features in the growth of the dentition. One problem with studying growth rate in fossil material is that, to estimate it, a measure of age at death is needed against which the stage of development reached by a particular specimen. All age at death estimations in children are in turn based upon the sequence of growth changes, so the researcher quickly gets into a circular argument. In addition, growth standards for recent modern humans and living primates cannot be applied to extinct hominids without making large assumptions that are difficult to test. The regularity of the enamel clock is, however, maintained throughout the formation of all the teeth in the dentition and does not appear to be affected by factors which otherwise disturb growth of the body, such as childhood infections or dietary deficiencies. It also appears to work in a similar way in all primates. So counts of cross striations are believed to provide an independent measure of age against which the rate and timing of growth can be set. Schwartz and colleagues have used this to examine the relationship between body size and timing of dental development in living and extinct primates. In a similar way, T. Smith and colleagues have compared variation in the timing of molar crown formation in chimpanzees and humans, and Ramirez-Rozzi and Lacruz have compared the development of the bonobo with that of the chimpanzee. One of the difficulties with investigating the layered structure of enamel with thin sections is the damaged this causes to specimens. Another option is to examine the surface of the crown, usually by scanning electron microscopy of high resolution casts. The surface has a pattern of coarser, but still regular lines known as perikymata. In any one individual, there is a constant number of cross striations between them, so they too represent a regular rhythm. Even if the individual's rhythm is not known, it is possible to use the constancy of the perikymata rhythm to investigate variation in the way in which a tooth crown is formed at different points in its height, as discussed by Guatelli-Steinberg and colleagues. Finally, another possibility is to combine the approaches of biochemical analysis and histology. As outlined by Humphrey and colleagues, laser ablation mass spectrometry can analyze strontium:calcium ratios in very small areas in a tooth section, allowing a comparison between different phases of development. As these ratios change with weaning, it is possible to recognize the timing of this important event in the life history of an individual.

One of the common themes through this volume is the development of new techniques. Notably, Bromage and colleagues have developed a portable confocal light microscope, designed to be packed up and carried around the world to different museums. The advantage of confocal microscopy is that it makes it possible to focus a small way under the surface of translucent specimens and provide in-focus images of structures at that depth. It is possible to focus into the undamaged enamel surface in this way, and naturally formed fracture surfaces can provide information on deeper internal structures without damaging important specimens. Gantt and colleagues, Olejniczak and colleagues, and P. Smith and colleagues, describe the use of micro-computed tomography for the non-destructive imaging of three-dimensional internal structures such as the enamel-dentine junction. Conventional computed tomography, as routinely used in hospitals, creates “slices” of 1 mm or so (some give somewhat finer resolutions) of the subject. Micro CT can reduce this to as little as 5 μm (one micrometer is one thousandth of a millimeter). This makes it possible to study a large number of specimens and allows the measurement of dental architecture and the thickness of the enamel cap in variety of ways that have never before been possible. Another new technology is the arrival of instruments that can provide high resolution scans of the three-dimensional form of surfaces, again without damaging the specimen. Some of these are based on direct mechanical contact with the specimen, others with laser depth measurement and still others are based on confocal microscopy in which a stack of images at different focal depths is used to build the three-dimensional model. The latter currently provides the highest resolution. These surface models have allowed Ungar to follow the changing form of the occlusal surface with tooth wear, and also resolve many of the problems in measuring the scratches and pits of tooth wear at a microscopic scale as shown by Teaford, and by Estebaranz and colleagues. Three-dimensional models of the surface make it possible to measure the texture of microscopic wear in a variety of new ways.

This volume can be compared with the papers of the first dental anthropology conference in 1958 (Brothwell, 1963) and the “30 years on” symposium of the American Association of Physical Anthropologists in 1988 (Kelley and Larsen, 1991). It is only another 17 years since the latter, but the field has clearly moved a considerable way, not only in the techniques available and the knowledge base that has built up, but in the research questions that can now be asked.

References

Begg, P.R., 1954. Stone Age man’s dentition. American Journal of Orthodontics 40, 298–383.

Brothwell, D.R., 1963. *Dental Anthropology*. Pergamon Press, London.

Kelley, M.A., Larsen, C.S., 1991. *Advances in Dental Anthropology*. Wiley-Liss, New York.

PART I

DENTAL EVOLUTION AND DENTAL MORPHOLOGY

1. Introduction

S.E. BAILEY
Department of Human Evolution
Max Planck Institute for Evolutionary Anthropology
Deutscher Platz 6
D-04103 Leipzig, Germany
and
Center for the Study of Human Origins,
Department of Anthropology, New York University,
25 Waverly Place
New York, NY 10003, USA
sbailey@nyu.edu

The study of the external tooth morphology can be undertaken in a non-destructive and relatively inexpensive manner. All one needs are good eyes (or a good hand lens), a decent set of calipers and a good single-lens reflex (SLR) or digital camera to keep a permanent record. As such, gross morphology (including size and shape) has long been a subject of interest to paleoanthropologists. Measurements also have a long-standing role in assessing human evolution (Wolpoff, 1971; Frayer, 1977; Brace et al., 1987; e.g., Bermúdez de Castro and Nicolás, 1996). The assessment of metric variation in living modern humans has also been applied to questions of modern populations (Hanihara and Ishida, 2005). Simple measurements of tooth length and breadth can be used to obtain broad morphological characterizations (e.g., crown indices and crown areas). However, tooth size does not discriminate well between closely related hominin species and appears to be under greater environmental influence than crown morphology (Townsend and Brown, 1978). Consequently, it may be less useful for addressing hominin evolutionary relationships.

The bumps and grooves that make up the tooth surface are more challenging to quantify than are dental metrics. Attempts to quantify morphological variation include the study of relative cusp areas (Wood and Abbott, 1983; Wood and Uytterschaut, 1987; Wood and Engleman, 1988; Bailey et al., 2004 Moggi-Cecchi and Boccone, 2007), cusp angles (Morris, 1986; Bailey, 2004), crest lengths (Pilbrow, 2003; 2007), as well as non-metric traits. Discrete, non-metric traits (e.g., shovel-shaped incisors, Carabelli's cusp) have a long history of use in characterizing modern human populations (Hrdlička, 1920; Pedersen, 1949; Turner and Scott, 1977). They have also played an important role in human evolutionary studies (Hrdlička, 1911; Weidenreich, 1937; Robinson, 1954; Grine, 1985; Suwa et al., 1996). However, the systematic study and application of dental non-metric traits to questions of human evolution has grown since standards were developed (Dahlberg,

S.E. Bailey and J.-J. Hublin (Eds.), Dental Perspectives on Human Evolution, 3–8.

1956) and made readily available to the public (Turner et al., 1991). These standards have helped to ensure that researchers are all talking about the same thing, facilitating comparative studies of contemporary (Lukacs, 1984; Turner, 1985; Sofaer et al., 1986; Hanihara, 1989; Irish and Turner, 1990; Turner, 1990; Turner, 1992; Irish, 1994; Hawkey, 1998) and fossil (Crummett, 1994; Bailey, 2002b; Irish and Guatelli-Steinberg, 2003) humans.

Some of the issues of interest in human evolutionary studies include assessing intra- and inter-specific variability, identifying and diagnosing taxa, and working out phylogenetic relationships among extinct fossil species. The application of dental morphology to these questions has come a long way since the days when the primary focus was on shovel-shaped incisors and taurodont molars. Researchers have accumulated large quantities of data on many non-metric trait frequencies in living modern humans. There is less information, however, on the morphometric variation among and within extant ape species (Uchida, 1996; Uchida, 1998a; Uchida, 1998b; Pilbrow, 2006). Assessing variation within and between species of extant apes is important if we are to successfully use dental morphology to identify the number of species represented in a particular fossil collection. Here, Pilbrow uses dental morphometrics (length, breadth and crest lengths) as a proxy for morphological variation to examine variation in extant apes. This marks a significant step forward in understanding intra- and inter-specific variation in our closest living relatives, and in making predictions about what we might expect to find in early hominins.

In the 1980s Wood and colleagues investigated dental morphological variation in early (Plio-Pleistocene) hominins using cusp areas and a limited number of morphological traits (Wood and Abbott, 1983; Wood et al., 1983; Wood and Uytterschaut, 1987; Wood and Engleman, 1988). Since then, technological advances (e.g., digital imaging and image analysis) have made the collection and manipulation of large amounts of data more feasible. In addition, many new fossils from this time period have been uncovered.

Bailey and Wood move beyond previous studies of Plio-Pleistocene dental morphology that relied solely on modern human standards (Irish, 1998; Irish and Guatelli-Steinberg, 2003), and incorporate several new traits in their analysis of dental evolution within the hominin clade. They conclude that some of the dental trends (e.g., morphological simplification) said to be characteristic of *Homo* appear relatively late in human evolution. However, a more accurate assessment of hominin dental evolution will come only after scoring standards that incorporate variation observed in Plio-Pleistocene hominins are developed.

Moggi-Cecchi and Boccone use new digital imaging and analysis techniques to assess upper molar morphometric variation of the expanded South African Pliocene fossil hominin record. Their results support trends and characterizations suggested previously by Wood and colleagues, but they also note a great deal of size variability in the expanded *A. africanus* sample. Studies in inter- and intra-specific variation in extant apes (e.g., Pilbrow, 2007) may ultimately help work out the significance of this variation.

The dentition has traditionally played a less prominent role in studies of later human evolution (Middle and Late Pleistocene) than it has in earlier human evolution. This may be because later hominins (e.g., Neandertals) have been assumed to have teeth that are indistinguishable from our own. Recent studies showing that teeth are important tools in assessing later human evolution (e.g., Bailey, 2002a, 2004, 2006; Bailey and Lynch, 2005), as well as the discovery of large samples of well-preserved European Early and Middle Pleistocene fossil hominins have led to a renewed interest in dental variation during this important time period. Here, Martinón-

Torres et al. (2007) use new dental data from Gran Dolina and Sima de los Huesos to explore questions about phylogenetic relationships between these populations, as well as to investigate possible evolutionary scenarios within Europe.

Phenetic assessments, rather than phylogenetic ones, are more commonly the focus of dental morphological study. One standard for assessing biological relationships using dental morphology is the Mean Measure of Divergence (Sjøvold, 1973; Berry, 1976; Harris, 2004). But several other methods for assessing biological distance and modeling dental relationships are available (e.g., Mahalanobis distance). Sample sizes and extent of completeness of the dentitions will play a role in choosing one's method. The Mean Measure of Divergence is used by Bailey and Wood (2007) in their assessment of Plio-Pleistocene dental relationships. Manni et al. (2007) take a novel approach and apply a method derived from artificial neural networks called Self Organizing Maps (SOMs) to assess phenetic relationships among Pleistocene hominins. One advantage of this method is allowing the use of specimens with missing data; in addition it enables one to avoid *a priori* grouping of specimens by analyzing individuals rather than trait frequencies in samples.

Each of the aforementioned studies relies on direct observation or two-dimensional representations of the crown surface. The final two papers in this section focus on new three-dimensional techniques that can be used to image and study not only the crown surface but also the morphology below. Olejniczak et al. (2007) show how micro-computer tomography (mCT) can be used to capture aspects of tooth morphology otherwise unobtainable through non-destructive means (e.g., morphology of the dentine surface, as well as enamel and dentine volumes). They present the methodological parameters relevant to mCT studies of molars, pointing out that while it may be desirable to have the highest resolution and thinnest slice sections possible in some cases (e.g., for taking accurate distance measurements), slice thickness of 10 times the minimum possible provide accurate data on enamel and dentine volumes. They also review the advantages and disadvantages to using this technique.

Gantt et al. present information on another method for producing high resolution images of internal dental structures - high resolution x-ray computer tomography (HRXCT). Like mCT, with the proper software linear measurements and volumetric data on enamel and dentine can be accurately obtained with HRXCT. According to Gantt et al., HRXCT provides higher resolution and permits the use of a greater size range of specimens (it can handle entire jaws and skulls) but otherwise there are few differences between HRXCT and mCT. Gantt et al.'s study of enamel volume using HRXCT agree with previous studies showing the relationship *Gorilla*>*Pongo*>*Homo*>*Pan* e.g., Kono, 2004; Olejniczak et al. 2007).

These new methods - HRXCT and mCT – address two major issues that have beset many paleoanthropological dental studies in the past. First, they allow one to examine morphology below the crown surface by non-destructive means. Second, they provide a more accurate representation of volumetric and linear data than can be obtained from two-dimensional representations of the crown and EDJ. As such, they open up new avenues of research on internal dental morphology of fossil hominins that were not previously possible.

References

Bailey, S.E., 2002a. A closer look at Neanderthal postcanine dental morphology. I. The mandibular dentition. The Anatomical Record (New Anat.) 269, 148–156.

Bailey, S.E., 2002b. Neandertal dental morphology: implications for modern human origins.

Ph.D. Dissertation, Arizona State University, Tempe.

Bailey, S.E., 2004. A morphometric analysis of maxillary molar crowns of Middle- Late Pleistocene hominins. Journal of Human Evolution 47, 183–198.

Bailey S.E., 2006. Diagnostic dental differences between Neandertals and Upper Paleolithic modern humans: getting to the root of the matter. In: Zadzinska, E. (Ed.), Current Trends in Dental Morphology Research. Univeristy of Lodz Press, Lodz (Poland).

Bailey, S.E., Lynch, J.M., 2005. Diagnostic differences in mandibular P4 shape between Neandertals and anatomically modern humans. American Journal of Physical Anthropology 126, 268–277.

Bailey, S.E., Pilbrow, V.C., Wood, B.A., 2004. Interobserver error involved in independent attempts to measure cusp base areas of *Pan* M1s. Journal of Anatomy 205, 323–331.

Bailey, S.E., Wood, B.A., 2007. Trends in postcanine occlusal morphology within the hominin clade: The case of *Paranathropus*. In: Bailey, S.E., Hublin, J-J. (Eds.), Dental Perspectives on Human Evolution: State of the Art Research in Dental Paleoanthropology. Springer, Dordrecht pp. 3–8.

Bermúdez de Castro, J.M., Nicolás, M.E., 1996. Changes in the lower premolar-size sequence during hominid evolution. Phylogenetic implications. Human Evolution 11, 205–215.

Berry, A., 1976. The anthropological value of minor variants of the dental crown. American Journal of Physical Anthropology 45, 257–268.

Brace, C.L., Rosenberg, K.R., Hunt, K.D., 1987. Gradual change in human tooth size in the late Pleistocene and post-Pleistocene. Evolution 41, 705–720.

Crummett, T., 1994. The evolution of shovel shaping: regional and temporal variation in human incisor morphology. Ph.D. Dissertation, University of Michigan, Ann Arbor.

Dahlberg, A., 1956. Materials for the establishment of standards for classification of tooth characteristics, attributes, and techniques in morphological studies of the dentition. Zoller Laboratory of Dental Anthropology, University of Chicago.

Frayer, D., 1977. Metric changes in the Upper Paleolithic and Mesolithic. American Journal of Physical Anthropology 46, 109–120.

Grine, F., 1985. Dental morphology and the systematic affinities of the Taung fossil hominid. In: Tobias, P. (Ed.), Hominid Evolution: Past, Present and Future. Alan R. Liss, Inc., New York, pp. 247–254.

Hanihara, T., 1989. Affinities of the Philippine Negritos as viewed from dental characters: a preliminary report. Journal of Anthropology Society Nippon 97, 327–339.

Hanihara, T., Ishida, H., 2005. Metric dental variation of major human populations. American Journal of Physical Anthropology 128, 287–298.

Harris, E., 2004. Calculation of Smith's Mean Measure of Divergence for intergroup comparisons using nonmetric data. Dental Anthropology 17, 83–93.

Hawkey, D., 1998. Out of Asia: Dental evidence for affinities and microevolution of early populations from India/Sri Lanka. Ph.D. Dissertation, Arizona State University, Tempe.

Hrdlička, A., 1911. Human dentition and teeth from the evolutionary and racial standpoint. Dominion Dentistry Journal 23, 403–417.

Hrdlička, A., 1920. Shovel-shaped teeth. American Journal of Physical Anthropology 3, 429–465.

Irish, J., 1994. The African dental complex: diagnostic morphological variants of modern sub-Saharan populations. American Journal of Physical Anthropology [Suppl] 18, 112.

Irish, J., 1998. Ancestral dental traits in recent sub-Saharan Africans and the origins of modern humans. Journal of Human Evolution 34, 81–98.

Irish, J.D., Guatelli-Steinberg, D., 2003. Ancient teeth and modern human origins: an expanded comparison of African Plio-Pleistocene and recent world dental samples. Journal of Human Evolution 45, 113–144.

Irish, J.D., Turner, C.G., II, 1990. West African dental affinity of Late Pleistocene Nubians: peopling of the Eurafrican-South Asian triangle II. Homo 41, 42–53.

Lukacs, J., 1984. Dental anthropology of South Asian populations: a review. In: Lukacs, J. (Ed.), People of South Asia. Plenum Press, New York, pp. 133–157.

Manni, F., Vargiu, R., Coppa, A., 2007. Neural Network Analysis by using self-organizing maps (SOMs) applied to human fossil dental morphology: a new methodology. In: Bailey, S.E., Hublin, J-J. (Eds.), Dental Perspectives on Human Evolution: State of the Art Research in Dental Paleoanthropology. Springer, Dordrecht pp. 3–8.

Martinon-Torres, M., Bermúdez de Castro, J.M., Gómez-Robles, A., Bastir, M., Sarmiento, S., Muela, A., Arsuaga, J.L., 2007. Gran Dolina-TD6 and Sima de los Huesos dental samples: Preliminary approach to some dental characters of interest for phylogenetic studies. In: Bailey, S.E., Hublin, J-J. (Eds.), Dental Perspectives on Human Evolution: State of the Art Research in Dental Paleoanthropology. Springer, Dordrecht pp. 3–8.

Moggi-Cecchi, J., Boccone, S., 2007. Maxillary molars cusp morphology of South African australopithecines. In: Bailey, S.E., Hublin, J-J. (Eds.), Dental Perspectives on Human Evolution: State of the Art Research in Dental Paleoanthropology. Springer, Dordrecht pp. 3–8.

Morris, D.H., 1986. Maxillary molar occlusal polygons in five human samples. American Journal of Physical Anthropology 70, 333–338.

Olejniczak, A.J., Grine, F.E., Martin, L.B., 2007. Micro-computed tomography of primate molars: methodological aspects of three-dimensional data collection. In: Bailey, S.E., Hublin, J-J. (Eds.), Dental Perspectives on Human Evolution: State of the Art Research in Dental Paleoanthropology. Springer, Dordrecht pp. 3–8.

Pedersen, P., 1949. The East Greenland Eskimo dentition. Meddelelser om Grønland 142, 1–244.

Pilbrow, V., 2003. Dental variation in African apes with implications for understanding patterns of variation in species of fossil apes. Ph.D. dissertation, New York University, New York.

Pilbrow, V., 2006. Lingual incisor traits in modern hominoids and an assessment of their utility for fossil hominoid taxonomy. American Journal of Physical Anthropology 129, 323–338.

Pilbrow, V., 2007. Patterns of molar variation in great apes and their implications for hominin taxonomy. In: Bailey, S.E., Hublin, J-J. (Eds.), Dental Perspectives on Human Evolution: State of the Art Research in Dental Paleoanthropology. Springer, Dordrecht pp. 3–8.

Robinson, J., 1954. Prehominid dentition and hominid evolution. Evolution 8, 324–334.

Sjøvold, T., 1973. The occurrence of minor nonmetrical variants in the skeleton and their quantitative treatment for population comparisons. Homo 24, 204–233.

Sofaer, J., Smith, P., Kaye, E., 1986. Affinities between contemporary and skeletal Jewish and non-Jewish groups based on tooth morphology. American Journal of Physical Anthropology 70, 265–275.

Suwa, G., White, T., Howell, F., 1996. Mandibular postcanine dentition from the Shungura Formation, Ethiopia: crown morphology, taxonomic allocations, and Plio-Pleistocene hominid evolution. American Journal of Physical Anthropology 101, 247–282.

Townsend, G.C., Brown, T., 1978. Heritability of permanent tooth size. American Journal of Physical Anthropology 49, 497–504.

Turner, C.G., II, 1985. The dental search for Native American origins. In: Kirk, R., Szathmary, E. (Eds.). Out of Asia: Peopling of the Americas and Pacific. Canberra, Journal of Pacific History, pp. 31–77.

Turner, C.G., II, 1990. Origin and affinity of the prehistoric people of Guam: a dental anthropological assessment. In: Hunter-Anderson, R. (Ed.), Recent Advances in Micronesian Archaeology, Micronesia Supplement No. 2. University of Guam Press, Mangilao.

Turner, C.G., II, 1992. Sundadonty and Sinodonty in Japan: the dental basis for a dual origin hypothesis for the peopling of the Japanese Islands. In: Hanihara, K. (Ed.), International Symposium on Japanese as a Member of the Asian and Pacific Populations. International Research Center for Japanese Studies, Kyoto, pp. 96–112.

Turner, C.G., II, Nichol, C.R., Scott, G.R., 1991 Scoring procedures for key morphological traits of the permanent dentition: The Arizona State University Dental Anthropology System. In: Kelley, M., Larsen, C. (Eds.), Advances in Dental Anthropology. Wiley Liss, New York, pp. 13–31.

Turner, C.G., II, Scott, G.R., 1977. Dentition of Easter Islanders. In: Dahlberg, A., Graber, T. (Eds.), Orofacial Growth and Development. Mouton Publishers, The Hague, pp. 229–249.

Uchida, A., 1996. *Craniodental Variation Among the Great Apes*. Harvard University, Cambridge.

Uchida, A., 1998a. Variation in tooth morphology of *Gorilla gorilla*. Journal of Human Evolution 34, 55–70.

Uchida, A., 1998b. Variation in tooth morphology of *Pongo pygmaeus*. Journal of Human Evolution 34, 71–79.

Weidenreich, F., 1937. The dentition of *Sinanthropus pekenensis*: a comparative odontography of the hominids. Paleontologia Sinica *n.s.* D, 1–180.

Wolpoff, M.H., 1971. *Metric Trends in Hominid Dental Evolution*. Press of Case Western Reserve University, Cleveland.

Wood, B.A., Abbott, S.A., 1983. Analysis of the dental morphology of Plio-Pleistocene hominids. I. Mandibular molars: crown area measurements and morphological traits. Journal of Anatomy 136, 197–219.

Wood, B.A., Abbott, S.A., Graham, S.H., 1983. Analysis of the dental morphology of Plio-Pleistocene hominids. II. Mandibular molars – study of cusp areas, fissure pattern and cross sectional shape of the crown. Journal of Anatomy 137, 287–314.

Wood, B.A., Engleman, C.A., 1988. Analysis of the dental morphology of Plio-Pleistocene hominids. V. Maxillary postcanine tooth morphology. Journal of Anatomy 161, 1–35.

Wood, B.A., Uytterschaut, H., 1987. Analysis of the dental morphology of Plio-Pleistocene hominids. III. Mandibular premolar crowns. Journal of Anatomy 154, 121–156.

2. Patterns of molar variation in great apes and their implications for hominin taxonomy

V. PILBROW
University of Melbourne
Zoology Department
Victoria 3010 Australia
vpilbrow@unimelb.edu.au.

Keywords: *Pan, Pongo, Gorilla*, models, molar metrics

Abstract

In studying the nature of variation and determining the taxonomic composition of a hominin fossil assemblage the phylogenetically closest and thus the most relevant modern comparators are *Homo* and *Pan* and following these, *Gorilla* and *Pongo*. Except for *Pan*, however, modern hominids lack taxonomic diversity, since by most accounts each one is represented by a single living species. *Pan* is the sister taxon to modern humans and it is represented by two living species. As such the species of *Pan* have greater relevance for studying interspecific variation in fossil hominin taxonomy. Despite their relatively impoverished species representations *Pan troglodytes, Gorilla gorilla* and *Pongo pygmaeus* are, nevertheless, represented by subspecies. This makes them relevant for studying the nature of intraspecific variation, in particular for addressing the question of subspecies in hominin taxonomy. The aim of this study is to examine the degree and pattern of molar variation in species and subspecies of *P. pygmaeus, G. gorilla, P. troglodytes* and *P. paniscus*. I test the hypothesis that measurements taken on the occlusal surface of molars are capable of discriminating between species and subspecies in commingled samples of great apes. The results of this study are used to draw inferences about our ability to differentiate between species and subspecies of fossil hominins. The study samples include *P. t. troglodytes* (n = 152), *P. t. verus* (n = 64), *P. t. schweinfurthii* (n = 79), *G. g. gorilla* (n = 208), *G. g. graueri* (n = 61), *G. g. beringei* (*n* = 30), *P. p. pygmaeus* (n = 140), and *P. p. abelii* (*n* = 25). Measurements taken from digital images were used to calculate squared Mahalanobis distances between subspecies pairs. Results indicate that molar metrics are successful in differentiating between the genera, species and subspecies of great apes. There was a hierarchical level of differentiation, with the greatest separation between genera, followed by that between species within the genus *Pan* and finally between subspecies within species. The patterns of molar differentiation showed excellent concordance with the patterns of molecular differentiation, which suggests that molar metrics have a reasonably strong phylogenetic signal. *Pan troglodytes troglodytes* and *P. troglodytes schweinfurthii* were separated by the least dental distance. *P. troglodytes verus* was separated by a greater distance from these two, but on the whole the distances among subspecies of *P. troglodytes* were less than among subspecies of *G. gorilla* and *P. pygmaeus*. The dental distance between *G. g. gorilla* and *G. g. graueri* was greater than that observed between *P. troglodytes* and *P. paniscus*. With size adjustment intergroup distances between gorilla subspecies were reduced, resulting in distances comparable to subspecies of *P. troglodytes*. A contrast between size-preserved and

S.E. Bailey and J.-J. Hublin (Eds.), Dental Perspectives on Human Evolution, 9–32.

size-adjusted analyses reveal that size, sexual dimorphism and shape are significant factors in the patterning of molar variation in great apes. The results of this study have several implications for hominin taxonomy, including identifying subspecies among hominins. These implications are discussed.

Introduction

Molars make up a disproportionately large part of early hominin fossil collections and figure prominently in taxonomic assessments. When determining whether the differences observed among sets of fossil hominin molars can be attributed to that of a species, or are part of the variation to be expected within a species, paleoanthropologists generally look to modern analogs. Because the fossil record does not present us with the requisite numbers of specimens, or the anatomical, behavioral and ecological details needed to gauge the nature of variation in fossil hominins, extant hominids[1], namely humans and great apes, provide the next best alternative for modeling variation. The justification behind using closely related extant taxa for models is fairly sound: through recency of common ancestry in closely related taxa are likely to have shared similar patterns and ranges of variation. Therefore, they likely provide reasonably accurate estimates of the type of variation to be expected in the fossils (see papers in Kimbel and Martin, 1993). Just as important, we have a fairly good understanding of patterns of variation in extant hominids in external morphology, breeding patterns, habitat preferences and genetic structure and we can see how these match up with variation in fossilizable attributes such as cranial and dental features. The extant hominids thus provide a comprehensive comparative model for understanding variation in the molars of fossil hominins. If we are to ultimately reconstruct the biology and lifeways of fossil forms in a manner that is consistent with living forms this modeling of variation is essential.

A consensus opinion emerging from molecular systematists (Goodman, 1962; Goodman et al., 1982; 1998; Caccone and Powell, 1989; Ruvolo, 1994; 1997), corroborated by morphological data (Begun, 1992; Gibbs et al., 2000; Guy et al, 2003; Lockwood et al., 2004), is that chimpanzees are the closest extant relatives of modern humans. Chimpanzees are therefore especially relevant for studying the nature of variation in fossil hominins. Additionally, chimpanzee patterns of variation are well documented. Two species of chimpanzees, *Pan paniscus* and *Pan troglodytes*, are recognized by a plethora of morphological, behavioral, ecological and genetic studies (Coolidge, 1933; Johanson, 1974; Shea, 1981, 1983a, b, c, 1984; Sibley and Ahlquist, 1984; Caccone and Powell, 1989; Kinzey, 1984; Shea et al., 1993; Wrangham et al., 1994; Uchida, 1996; Guy et al., 2003; Taylor and Groves, 2003; Lockwood et al., 2004; Pilbrow, 2006a, b; but see Horn, 1979). Diversity within *P. troglodytes* is also substantial. The traditional taxonomy recognizes three subspecies (Hill, 1967; 1969), but mtDNA studies have suggested that an additional one should be recognized (Gonder et al., 1997). They also suggest that the West African subspecies *P. t. verus* should be recognized as a distinct species, *P. verus* (Morin et al., 1994). Certain morphological data sets have registered the distinctiveness of *P. t. verus* compared to the other two subspecies (Braga, 1995; Uchida, 1996; Pilbrow, 2003, 2006a, b; Taylor and Groves, 2003). The substantial diversity exhibited by *Pan* at the inter- and intra-specific level, together with a close phylogenetic relationship to humans makes them particularly germane to discussions about hominin taxonomy.

Gorillas and orangutans are more distantly related to humans. According to the consensus view (above) gorillas are a sister taxon to the chimp-human clade and orangutans are the closest outgroup (for a contrary viewpoint see Schwartz, 1984). Regardless, both are members of the family Hominidae making them phylogenetically appropriate taxa for studying fossil hominin variation. Gorillas and orangutans differ from chimpanzees in their taxonomic diversity. According to the traditional taxonomy, the diversity within *Gorilla gorilla* is no more than can be accommodated within a single species with three recognized subspecies (Coolidge, 1929; Groves, 1970). The same is true for *Pongo pygmaeus*, where two subspecies are traditionally recognized (Courtney et al., 1988). There have been several recent attempts to revise the conventional taxonomy, particularly by molecular systematists. They advocate that the east and west African gorillas best represent two distinct species, and likewise that the Bornean and Sumatran orangutans should be distinguished at the species level (Ruvolo et al., 1994; Garner and Ryder, 1996; Xu and Arnason, 1996; Zhi et al., 1996; Saltonstall et al., 1998; Jensen-Seaman and Kidd, 2001). Several studies also propose that one or more additional subspecies of *G. gorilla* be recognized (Sarmiento and Butynski, 1996; Sarmiento and Oates, 2000; Stumpf et al., 2003). The proposal of two species within *Gorilla* and *Pongo* (Groves, 2001) has not gained wide acceptance, however, because several molecular and morphological studies show that relative variation within the traditional subspecies of *Gorilla gorilla* and *Pongo pygmaeus* is high relative to that between the subspecies (Courtenay et al., 1988; Gagneux et al., 1999; Muir et al., 2000; Jensen-Seaman et al., 2003; Leigh et al., 2003).

Even if we adhere to a conventional taxonomy (e.g., Jenkins, 1990) while awaiting final word on alternative taxonomic scenarios, the diversity within and among great ape species is impressive. If considered to be a single species, both *G. gorilla* and *P. pygmaeus* exhibit considerable variation at the infraspecific level. Together with the species of *Pan* and subspecies of *P. troglodytes*, they provide comprehensive data for models of inter- and intraspecific variation, which can be applied to understanding variation in fossil hominins.

Great ape patterns of diversity are especially relevant to discussions about identifying subspecies from the hominin fossil record. Subspecies are an unresolved quandary in systematics and taxonomy. Defined as geographically circumscribed, phenotypically distinct units (Futuyma, 1986; Smith, et al., 1997), or the point at which we no longer lump populations (Groves, 1986), subspecies are more difficult to identify than species (Tattersall, 1986; Kimbel, 1991; Shea et al., 1993; Templeton, 1999; Leigh et al., 2003). As explained by Templeton (1999), this is because the criterion of reproductive isolation, which gives ontological strength to the concept of a species, is lacking for the subspecies. A subspecies, can encompass everything from a population to a species. Kimbel (1991), citing Mayr's (1982) view that subspecies cannot be treated as incipient species, reasoned that subspecies are an arbitrary tool of taxonomy and are best ignored in paleoanthropology, being a hindrance to the task of determining the phylogeny of fossil hominins. Tattersall (1986, 1991, 1993) demonstrated that only a few subtle morphological features differ between closely related species of *Lemur*. He proposed that when morphological distinctions are observed in the fossil record they should be considered as evidence for species, rather than lower-order taxonomic units. Thus Neandertals and modern human are likely to have been distinct species.

Evolutionary biologists who study the nature of variation in extant primates argue, to the contrary, that subspecies provide important information about the structure of the gene pool, and patterns of genetic contact and evolutionary divergence among populations. From a neontologist's perspective these are meaningful aspects of the population biology and history of a species, vital enough to advocate conservation status for endangered taxa (Templeton, 1999; Leigh et al., 2003). Several researchers have called for a need to identify subspecies among extinct hominins so as to gain a better appreciation of their population dynamics (Shea et al., 1993; Jolly, 1993, 2001; Leigh et al, 2003). There is a strong contingent of paleoanthropologists who believe that Neandertals constitute an extinct subspecies of modern humans (Wolpoff et al., 2001). Disagreeing with Tattersall (1991), Jolly (1993) argued that Tattersall's criteria for recognizing species based on morphological distinction cannot be applied to baboons because baboon populations achieve morphological distinctiveness without reproductive isolation, and such populations may never become extinct in the same sense as phylogenetic species. According to Jolly if craniodentally distinct baboon taxa were identified as species in the paleontological context, significant attributes of their genetic structure, or zygostructure, as he calls it, would be obscured. Lemurs and baboons clearly have contrasting signatures of genetic isolation relative to morphological distinctiveness. Using baboons as models Jolly (2001), in his turn, has suggested that the interactions between hominins like Neandertals and modern humans could have involved a certain amount of interbreeding. Although this implies that Neandertals are a subspecies of modern humans, at least according to the dominant biological species concept, Jolly prefers to use the term allotaxa (Grubb, 1999) to describe them, to avoid the distinction between the biological and phylogenetic species concepts, and to place the emphasis on population history rather than on naming names.

As phylogenetic kin, great ape patterns of taxonomic diversity should have greater relevance than lemur or baboon patterns for addressing the question of infraspecific diversity in extinct hominins. The purpose of this paper is to document the apportionment of molar variation among species and subspecies of extant great apes. Phenetic distances separating the traditional subspecies of *P. troglodytes* are compared with those separating *P. troglodytes* from *P. paniscus*. These in turn are compared with the distances between subspecies within *G. gorilla* and *P. pygmaeus*. The patterns of dental divergence are then compared with the patterns derived from previous cranial, dental and genetic studies to evaluate the relevance of this material for understanding diversity and as models for understanding variation in fossil hominin molars. A close match with the patterns revealed by selectively neutral molecular data is particularly important because this helps to determine whether dental data can reveal patterns of genetic divergence (Collard and Wood, 2000, 2001; Guy et al., 2003; Lockwood et al., 2004).

Because great apes range considerably in size and because *G. gorilla* and *P. pygmaeus* also display marked sexual dimorphism, both raw and size-adjusted measurements, and sex-pooled and sex-regregated samples were used in the analysis. The following specific questions were addressed:

(1) Are molar metrics able to differentiate between the four species of great apes?
(2) How successful are molar metrics in classifying subspecies within each species?
(3) How do inter- and intra-species dental distances compare within and across species?
(4) In what way do the phenetic distances between taxa change when adjusted for size and sex?
(5) How do dental patterns of divergence compare with molecular patterns of divergence?

Several studies have previously examined the nature of inter- and intraspecific morphological variation in the great apes. Shea et al. (1993) were able to differentiate between *P. paniscus* and *P. troglodytes* and between subspecies of *P. troglodytes* using craniometric data. Braga (1995) demonstrated the differences between these same groups using a larger data set of discrete cranial traits. Groves (1967, 1970) and more recently, Stumpf et al. (2003) used craniometric data to differentiate between gorilla subspecies. Taylor and Groves (2003) found that patterns of diversity based on molecular data in the African apes were discernible using mandibular measurements. Guy et al. (2003) and Lockwood et al. (2004) used a geometric morphometric analysis of the craniofacial complex and the temporal bone to differentiate among great ape subspecies.

Dental data have also been used to address the question of variation in the great apes (Mahler, 1973; Johanson, 1974; Swindler, 1976, 2002; Kinzey, 1984; Scott and Lockwood, 2004). The most significant study from the perspective of the present one is Uchida's (1992, 1996). Uchida measured molar cusp base areas on photographs of the occlusal surface to examine patterns of variation within and among great ape species and subspecies. She demonstrated that molar cusp areas successfully discriminate between subspecies within great ape species. Her work provided significant insight into the nature of dental diversity in great apes.

This study borrows from Uchida's in its technique for taking measurements from the occlusal surface of molars. It differs from it and the others in several important respects. Molar crest lengths, which have not been used by previous studies, were used to test whether other aspects of molar morphology are as effective as cusp base areas or length/breadth dimensions in differentiating taxa. In addition, this study seeks to determine whether molar metrics are capable of differentiating between species and subspecies of great apes when these are commingled. Consequently, nine taxa, including *P. paniscus*, and the subspecies of *P. troglodytes*, *G. gorilla*, and *P. pygmaeus* were included in a single analysis. Most importantly, in this study samples were drawn from populations representing the entire range of distribution for great apes. This provides an understanding of how variation is partitioned at infraspecific levels of the species, rather than in select populations representing the species, as has often been done. This provides a comprehensive hierarchical model from which to approach fossil hominin variation (Albrecht et al., 2003; Miller et al., 2004).

Materials and Methods

The analysis is based on the unworn dentitions of 804 adult individuals, including 341 chimpanzees, 298 gorillas and 165 orangutans (Table 1). Only individuals with third molars erupted or erupting were selected. Samples were obtained from major museums in the USA and Europe taking care to select specimens from the known geographic distribution of the great apes (for locality information and museum listings, see Pilbrow, 2003). Locality data from museum records were verified against the United States Geographic Names Database and compared with previous museum based studies (Groves, 1970; Röhrer-Ertl, 1984; Shea et al., 1993; Braga, 1995). Localities

Table 1. Sample sizes and sex proportions of taxa used in this study

Taxon	*N*	*M/F %*
P. t. verus	64	47/53
P. t. troglodytes	152	46/54
P. t. schweinfurthii	79	52/48
P. paniscus	46	39/61
G. g. gorilla	208	66/34
G. g. graueri	60	67/33
G. g. beringei	30	52/48
P. p. pygmaeus	140	44/56
P. p. abelii	25	44/56
Total	**804**	**53/47**

were then aggregated into the traditionally recognized species and subspecies (Jenkins, 1990): *P. paniscus, P. t. verus, P. t. troglodytes, P. t. schweinfurthii, G. g. gorilla, G. g. graueri, G. g. beringei, P. p. pygmaeus, P. p. abelii.*

The overall sex ratios for the study samples are fairly well balanced (Table 1). However, within individual subgroups the sexes are not distributed equally. There is also an imbalance in the sample sizes representing each subspecies. This is because museum collections are biased towards certain locales and sexes. To provide an estimate of overall variation I carried out separate analyses on pooled and segregated sexes to evaluate if and how variation differs between them.

Molar dimensions were measured on a digital image of the occlusal surface of the molar. Measurements consisted of mesiodistal and buccolingual dimensions and the length of molar crests. The technique for taking photographs is described in detail elsewhere (Pilbrow, 2003; Bailey et al., 2004) and will not be elaborated here. The mesiodistal dimension was identified as the longest dimension across the tooth crown. Two buccolingual dimensions were taken at the mesial and distal cusps and identified as the widest dimensions across the tooth crown at these points. To measure crest (or cristid) lengths I first identified cusp boundaries using the longitudinal, lingual and buccal development grooves for the upper molar, and the longitudinal, lingual, mesiobuccal and distobuccal developmental grooves for the lower molar (Figure 1). I then measured crest lengths from cusp boundary to cusp tip. If accessory cuspules were encountered they were not included in the crest length measurement. A total of 13 dimensions were taken on the lower molar and 11 on the upper. Figure 1 illustrates the measurements taken. NIH Image, a public domain image analysis program was used to take measurements (http://rsb.info.nih.gov/nih-image/).

In an intra-observer error study using 23 gorilla molars and premolars, I found that the average error in measuring the length of the tooth on the actual specimen versus measuring it on a digitized image was 1.36% (SD = 0.53%, range = 0.12–2.76%). An inter-observer study (Bailey et al., 2004) was also undertaken, comparing cusp base area measurements, on images obtained using slightly different photographic techniques, photo equipment and measurement software. There were no statistically significant differences in the measurements taken by two observers.

The dental measurements were size adjusted by indexing each measurement against the geometric mean of all measurements for that tooth (Mosimann and James, 1979; Darroch and Mosimann, 1985; James and McCulloch, 1990; Falsetti et al., 1993). Separate analyses were performed using raw and size-adjusted measurements. This helped to evaluate how molar size contributes to subspecies differences. Because of missing teeth and differential wear patterns the sample sizes differed for the molars. To maximize sample sizes the data set was subdivided according to molar type and separate analyses were performed for each molar. The results from six molars were then averaged to get an overall pattern of molar differentiation. However, the role of each molar in contributing to the differences was also evaluated.

A step-wise discriminant analysis (SPSS 12.0) was used to see how accurately molar morphometrics differentiate the nine taxa. The percentage accuracy by which individuals were classified helped to verify the preconceived separation of the taxa. The loading of the variables on the discriminant functions helped determine which variables influenced the differentiation. Group centroids were used to calculate Mahalanobis distances or squared generalized distances (D^2) between taxonomic pairs. This provided a phenetic distance between groups. The F statistic was

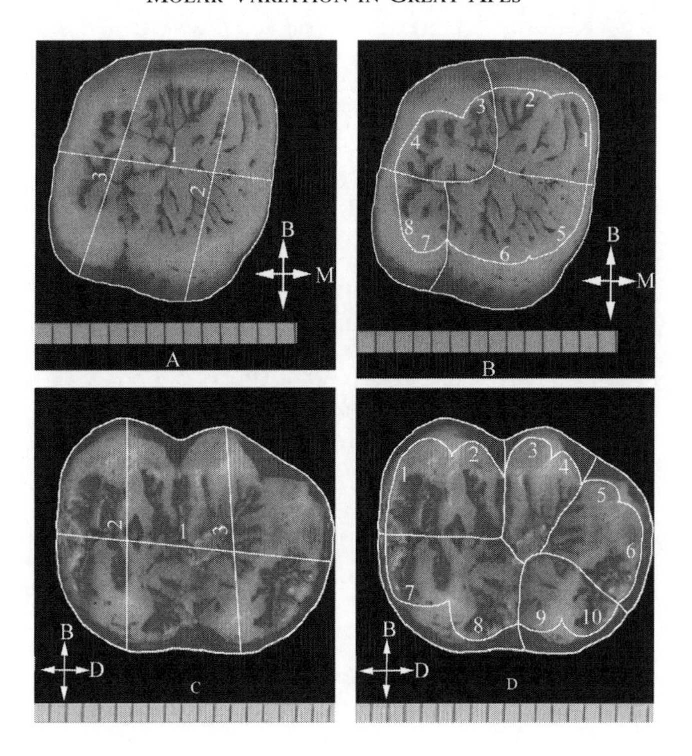

Figure 1. Measurements taken on digital image of upper molar (A, B) and lower molar (C, D). A, C: 1, Length; 2, breadth across mesial cusps; 3, breadth across distal cusps. B: 1, Preparacrista; 2, Postparacrista; 3, Premetacrista; 4, Postmetacrista; 5, Preprotocrista; 6, Postprotocrista; 7, Prehypocrista; 8, Posthypocrista. D: 1, Preprotocristid; 2, Postprotocristid; 3, Prehypoconidcristid; 4, Posthypoconidcristid; 5, Prehypconulidcristid; 6, Posthypoconulidcristid; 7, Premetaconidcristid; 8, Postmetaconidcristid; 9, Preentoconidcristid; 10, Postentoconidcristid. Both molars from the collections of the Zoologische Staatssaammlung, Munich.

used to test for the significance of pairwise distances. The first two discriminant functions, which most often accounted for a large proportion of the variance, were used in two-dimensional scatter-plots to show within-group variance, as well as between-group separation. Finally, the geometric mean was used as a generalized size factor and Pearson's correlations between the scores on discriminant functions and the geometric mean were used to determine the role of size or allometry in discriminating groups.

Results

Raw Variables

When raw variables were used in the analysis the classification accuracy for the nine taxa, averaged over six molars, was around 70% (Table 2). Classification accuracy was low for the UM3 (50%), but higher for the other molars (63% to 73%, not shown in table). Although the percentages of cases correctly assigned did not differ much between the sexes, classification accuracy was lower when the sexes were combined. Mayr (1942) suggested that if at least 75% of individuals within populations can be accurately differentiated from other populations within the species, these intraspecific groups may be described as subspecies. Classification accuracy was highest for *P. paniscus*, the only taxon not differentiated at the subspecies level, indicating that this taxon is most distinct from the others. Accuracy of classification was lowest for *P. t. troglodytes* and *P. t. schweinfurthii*. The

Table 2. Average classification accuracy using raw variables

	Ptv	*Ptt*	*Pts*	*Pp*	*Ggg*	*Gggr*	*Ggb*	*Ppp*	*Ppa*
Sex combined									
P. t. verus	**70**	10	16	3	0	0	0	0	1
P. t. troglodytes	16	**47**	24	10	0	0	0	1	2
P. t. schweinfurthii	17	16	**56**	9	0	0	0	0	1
P. paniscus	3	3	3	**90**	0	0	0	0	1
G. g. gorilla	0	0	0	0	**72**	9	16	2	1
G. g. graueri	0	0	0	0	9	**71**	19	0	0
G. g. beringei	0	0	0	0	19	20	**59**	1	0
P. p. pygmaeus	3	2	2	0	3	0	1	**66**	23
P. p. abelii	0	4	1	1	1	0	1	16	**76**
Average classification accuracy: 67%									
Males									
P. t. verus	**72**	11	15	1	0	0	0	0	1
P. t. troglodytes	18	**48**	24	9	0	0	0	0	1
P. t. schweinfurthii	17	17	**56**	9	0	0	0	0	1
P. paniscus	3	4	3	**90**	0	0	0	0	0
G. g. gorilla	0	0	0	0	**78**	7	10	3	2
G. g. graueri	0	0	0	0	5	**74**	20	1	0
G. g. beringei	0	0	0	0	16	22	**62**	0	0
P. p. pygmaeus	1	2	0	0	2	0	1	**72**	22
P. p. abelii	0	4	0	0	0	0	0	25	**71**
Average classification accuracy: 69%									
Females									
P. t. verus	**70**	14	12	2	0	0	0	1	1
P. t. troglodytes	17	**52**	18	11	0	0	0	1	1
P. t. schweinfurthii	12	16	**59**	11	0	0	0	2	0
P. paniscus	3	5	1	**91**	0	0	0	0	0
G. g. gorilla	0	0	0	0	**77**	6	14	2	1
G. g. graueri	0	0	0	0	5	**83**	12	0	0
G. g. beringei	0	0	0	0	14	21	**65**	0	0
P. p. pygmaeus	3	1	1	0	2	0	1	**70**	22
P. p. abelii	1	3	0	1	0	0	0	20	**75**
Average classification accuracy: 71%									

Cross-matrix shows percentage of cases of taxa from column one classified into taxa from row one. Correct classifications.

rate of misclassifications, which can be determined by noting the percentage of cases from each taxon in the row, classified into one or the other taxa from each column, demonstrates that misclassified subspecies are likely to be assigned to other subspecies within the same species. Average classification accuracy for the four great ape species, calculated by summing the percentage accuracy for the subspecies within each species, was close to 95%. This demonstrates that there is greater overlap among subspecies than species, which fits with the understanding that the gene pools of subspecies are not as completely segregated as that of species, allowing greater genetic exchange. It should be remembered, of course, that apart from *P. paniscus* and *P. troglodytes*, the great ape species are considered to be distinct genera.

Mahalanobis distances (Table 3) show that the greatest distance is between the smallest (*P. paniscus)* and the largest (*G. g. graueri)* of the great apes. The distances between subspecies of *P. troglodytes* and those of *G. gorilla* are only marginally lower. Intermediate distances separate the subspecies of

Table 3. Average mahalanobis distances using raw variables

	Ptv	*Ptt*	*Pts*	*Pp*	*Ggg*	*Gggr*	*Ggb*	*Ppp*	*Ppa*
Sex combined									
P. t. verus	0.00		*	*	*	*	*	*	*
P. t. troglodytes	1.82	0.00		*	*	*	*	*	*
P. t. schweinfurthii	1.59	0.81	0.00	*	*	*	*	*	*
P. paniscus	5.56	3.06	4.22	0.00	*	*	*	*	*
G. g. gorilla	42.51	44.94	42.06	63.04	0.00	*	*	*	*
G. g. graueri	69.56	72.90	69.70	95.88	7.39	0.00	*	*	*
G. g. beringei	55.98	59.34	56.44	80.62	3.92	2.93	0.00	*	*
P. p. pygmaeus	13.24	13.11	13.60	23.82	19.12	38.16	26.73	0.00	*
P. p. abelii	15.09	14.40	14.39	24.77	18.83	40.11	28.34	2.99	0.00
Males									
P. t. verus	0.00			*	*	*	*	*	*
P. t. troglodytes	1.61	0.00		*	*	*	*	*	*
P. t. schweinfurthii	1.52	0.81	0.00	*	*	*	*	*	*
P. paniscus	6.23	3.86	4.52	0.00	*	*	*	*	*
G. g. gorilla	46.34	48.69	47.41	70.53	0.00	*	*	*	*
G. g. graueri	79.86	82.83	81.73	111.20	8.58	0.00		*	*
G. g. beringei	67.87	71.28	69.96	98.02	5.36	2.92	0.00	*	*
P. p. pygmaeus	16.76	17.14	18.33	31.11	17.10	38.05	28.86	0.00	
P. p. abelii	17.51	17.19	17.59	29.44	17.52	41.47	31.21	3.29	0.00
Females									
P. t. verus	0.00	*	*	*	*	*	*	*	*
P. t. troglodytes	2.46	0.00		*	*	*	*	*	*
P. t. schweinfurthii	2.21	1.24	0.00	*	*	*	*	*	*
P. paniscus	7.49	3.92	5.71	0.00	*	*	*	*	*
G. g. gorilla	54.05	59.77	53.94	84.69	0.00	*	*	*	*
G. g. graueri	87.99	96.65	89.86	128.86	9.61	0.00	*	*	*
G. g. beringei	66.80	74.03	68.32	103.02	5.04	4.12	0.00	*	*
P. p. pygmaeus	14.33	15.00	14.75	29.35	25.91	50.70	33.10	0.00	*
P. p. abelii	15.60	15.96	15.09	29.49	27.14	55.35	37.40	3.55	0.00

Asterisk above the diagonal indicates that pair wise distances are significant ($p < 0.05$).

P. pygmaeus from other groups. Subspecies within a species display the lowest Mahalanobis distances, which are not always statistically significant (Table 3).

Distances between subspecies within species differ markedly among the great apes. The distances separating subspecies of *P. troglodytes* are remarkably low. *Pan troglodytes verus* is separated by a greater distance from *P. t. troglodytes* and *P. t. schweinfurthii* than either of these are from each other. Distances that are three to four times greater separate *Pan paniscus* from the subspecies of *P. troglodytes*. Of the *P. troglodytes* subspecies, *P. t. troglodytes* is closest to *P. paniscus*. The two subspecies of *P. pygmaeus* are separated by a greater distance than are the subspecies of *P. troglodytes*. However, the distance is less than that separating *P. paniscus* from the subspecies of *P. troglodytes*. Relative to the other great apes, the subspecies of *G. gorilla* are separated by large distances. The Mahalanobis distances between the western gorilla (*G. g. gorilla*) and the eastern gorillas (*G. g. graueri* and *G. g. beringei*) are even greater than that separating *P. paniscus* from *P. troglodytes*.

Using the lower second molar (LM2) as an example, Figure 2 shows the distribution

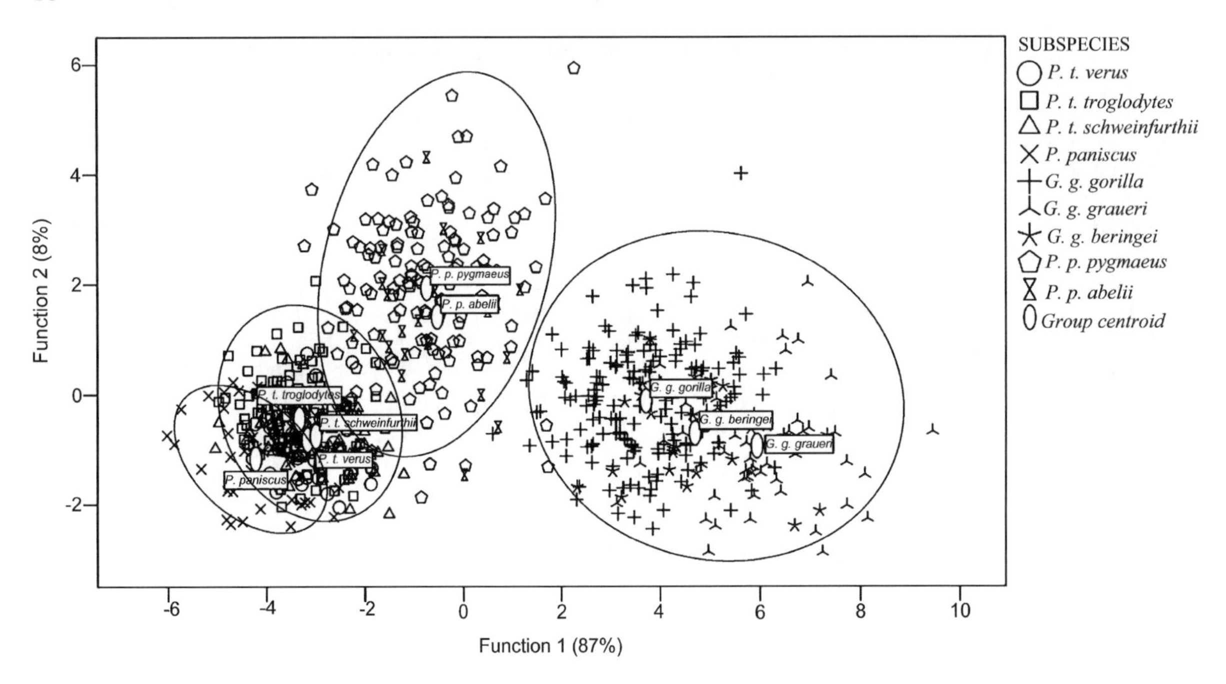

Figure 2. Dispersion of great apes along first two discriminant functions in the analysis of raw measurements on the LM2. Sexes combined.

of the taxa along the first two discriminant functions of sex-combined samples. The LM2 lies in the middle of the range for classificatory result for the molars, and therefore, for consistency, the LM2 is used to illustrate the spread of data in scatter plots throughout this paper. The four species are enclosed by 95% confidence ellipsoids, which helps to illustrate the distribution of the species and the relative variance within each. The major separation along function 1, with non-overlapping distributions, is between gorillas at one end of the axis and chimpanzees, including bonobos, at the other end. *P. pygmaeus* is separated from the others along function 1 and 2 although it shows slight overlap with *P. troglodytes*. *P. paniscus* is tightly clustered with *P. troglodytes* showing greater overlap. The relative size of the confidence ellipsoids indicates that *P. paniscus* and *P. troglodytes* are characterized by lower overall variance in raw molar metrics than *P. pygmaeus* and *G. gorilla*. Gorilla molars display comparatively higher variance.

Size strongly influences the segregation when raw molar metrics are used. The first discriminant function accounts for 85% to 90% of the variance, depending on molar position. Pearson's correlations between the discriminant scores for this function and the geometric mean, a proxy for size, are between 0.89 and 0.95, $p < 0.01$. The second function accounts for 5–10% of the variance, but has lower correlations with the geometric mean (0.01 to 0.26, not always significant).

Figure 3 presents the same data as Figure 2 but males and females of the four species are separated and encircled by 95% confidence ellipsoids. The close clusters formed by *P. paniscus* and *P. troglodytes*, which are not dimorphic in molar dimensions, contrast strongly with the larger cloud of data points formed by the dimorphic orangutans and gorillas. The larger ellipses surrounding gorilla and orangutan males suggests that the variance in males is greater than in females. Even among the dimorphic apes though, there is greater overlap between male and female data points of the same species than between

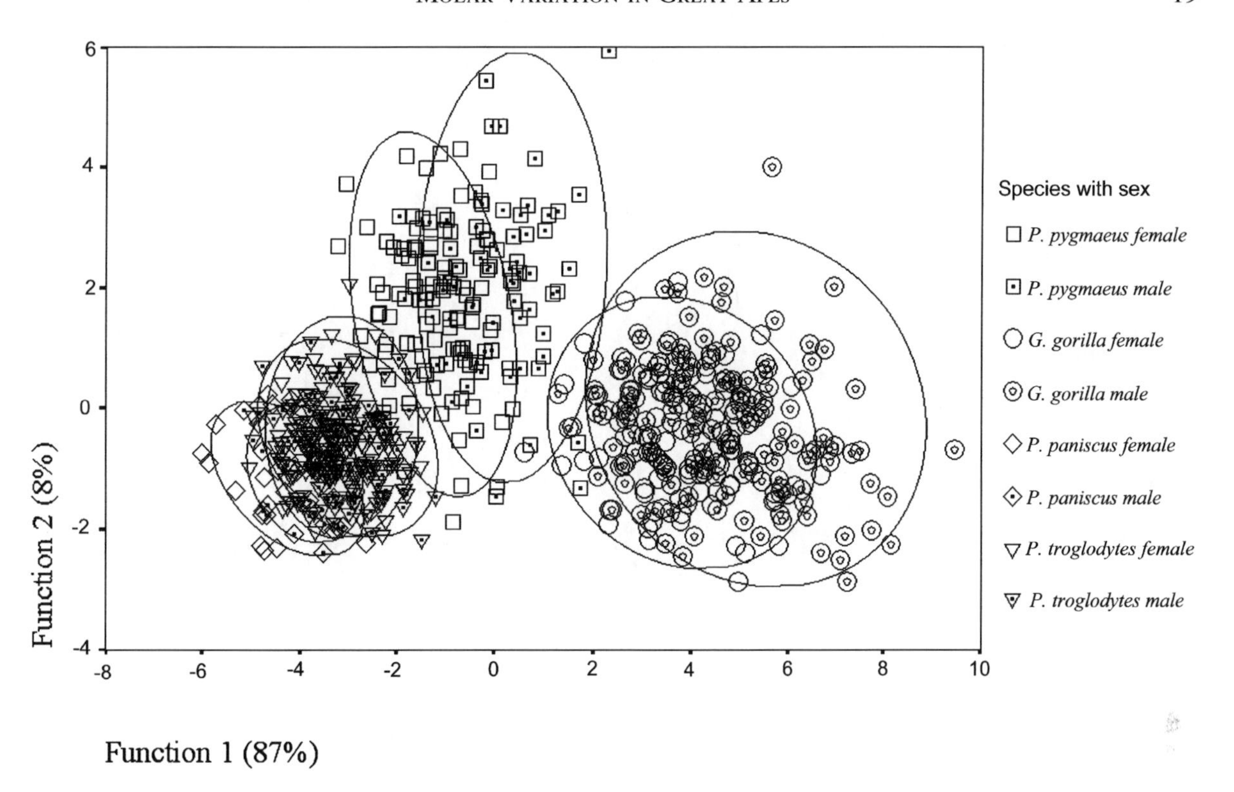

Figure 3. Spread of data along functions 1 and 2 of the LM2 using raw variables. 95% confidence ellipsoids surrounding males and females of *P. paniscus, P. troglodytes, P. pygmaeus* and *G. gorilla.*

females of a larger species and males of a smaller one. At all molar positions the variables most strongly affecting the segregation are the length and the breadth at mesial and distal cusps.

Size Adjusted Variables

Since size appears to be a strong factor affecting separation, the results of the size-adjusted discriminant analyses are instructive. Table 4 shows the percentage of cases correctly classified using size-adjusted dimensions averaged over all molars. Once again, accuracy is higher with sex-segregated molars but only slightly. Accuracy is lower for the UM3 (44%) than the other molars (50% to 61%, not shown). The correct classification is reduced by about 14% after size adjustment. Of all groups the decrease in accuracy is greatest for *P. paniscus* (Table 4). In this taxon, which had the highest accuracy using raw variables, classification accuracy is about 35% lower when scaled variables are employed. Misclassified individuals are likely to be assigned to other subspecies within the same species, with *P. paniscus* individuals assigned to subspecies of *P. troglodytes*. The percentage accuracy for species groups is about 80%; 15% lower than in the previous analysis.

Table 5 shows the size-adjusted Mahalanobis distances averaged over all molars for combined and sex-separated samples. After size adjustment, the distance between *Pan* and *Gorilla* is reduced considerably. However, in this analysis the greatest overall distance is between *P. t. verus* and *G. g. graueri*. Distances among subspecies within species also differ with size-scaled data. All inter-group distances are lower, except for the distances between subspecies

Table 4. Average classificatory accuracy using shape variables

	Ptv	*Ptt*	*Pts*	*Pp*	*Ggg*	*Gggr*	*Ggb*	*Ppp*	*Ppa*
Sex combined									
P. t. verus	**61**	9	15	10	0	0	1	3	1
P. t. troglodytes	17	**34**	22	13	1	2	2	7	3
P. t. schweinfurthii	14	13	**54**	10	2	2	2	2	2
P. paniscus	13	12	10	**56**	2	0	1	4	1
G. g. gorilla	1	1	2	1	**61**	14	14	3	4
G. g. graueri	1	1	1	2	11	**63**	19	2	1
G. g. beringei	0	2	3	1	14	23	**52**	2	3
P. p. pygmaeus	3	7	4	4	3	3	3	**53**	20
P. p. abelii	0	4	1	3	9	2	4	15	**63**
Average classification accuracy: 55%									
Males									
P. t. verus	**65**	8	13	9	0	1	2	1	1
P. t. troglodytes	16	**37**	25	11	0	2	2	6	1
P. t. schweinfurthii	14	14	**54**	7	2	2	1	3	3
P. paniscus	12	11	11	**54**	4	1	1	5	1
G. g. gorilla	1	0	2	2	**57**	13	16	3	6
G. g. graueri	0	0	0	3	10	**60**	20	3	4
G. g. beringei	1	0	2	1	13	17	**59**	3	4
P. p. pygmaeus	2	7	4	5	2	4	2	**55**	19
P. p. abelii	0	2	0	3	3	3	6	23	**60**
Average classification accuracy: 56%									
Females									
P. t. verus	**63**	10	12	10	0	0	1	3	1
P. t. troglodytes	12	**36**	21	15	1	2	2	7	4
P. t. schweinfurthii	14	13	**50**	12	2	3	1	3	2
P. paniscus	12	17	9	**56**	0	0	3	3	0
G. g. gorilla	1	0	2	1	**69**	12	11	2	2
G. g. graueri	0	1	1	0	16	**65**	15	1	1
G. g. beringei	0	3	3	2	13	20	**53**	3	3
P. p. pygmaeus	3	6	3	4	3	4	4	**54**	19
P. p. abelii	1	3	3	3	5	1	4	10	**70**
Average classification accuracy: 57%									

Cross-matrix shows the percentage of cases for taxa in the first column classified into taxa in the first row. Accurate classifications in bold.

of *P. troglodytes*, which are *higher* following size adjustment. The overall distance between *P. t. verus* and the other two subspecies is similar to their distance from *P. paniscus*. In lower molars *P. t. verus* was placed further away from the central and east African chimpanzees than *P. paniscus*. The distances between subspecies of *P. troglodytes* and *P. paniscus* although reduced are statistically significant for the analyses with combined sexes and females, but not with males.

As was the case with *Pan*, distances among subspecies of *G. gorilla* are also reduced, so that the distances between east and west African gorillas are now lower than the distances separating *P. paniscus* from *P. troglodytes* subspecies. However, they are still higher than the distances separating *P. troglodytes* subspecies. Similarly, the relative distance among gorilla subspecies does not change after size adjustment. Eastern gorilla subspecies are closely clustered, well separated from the western subspecies.

Table 5. Mahalanobis distances using shape variables

	Ptv	*Ptt*	*Pts*	*Pp*	*Ggg*	*Gggr*	*Ggb*	*Ppp*	*Ppa*
Sex combined									
P. t. verus	0.00	*	*	*	*	*	*	*	*
P. t. troglodytes	2.14	0.00	*	*	*	*	*	*	*
P. t. schweinfurthii	2.25	1.10	0.00	*	*	*	*	*	*
P. paniscus	3.24	2.09	2.53	0.00	*	*	*	*	*
G. g. gorilla	13.26	10.76	10.22	12.65	0.00	*		*	*
G. g. graueri	13.60	10.60	10.63	11.99	2.31	0.00		*	*
G. g. beringei	12.44	9.99	9.99	12.05	1.94	1.38	0.00	*	*
P. p. pygmaeus	9.15	6.28	8.28	8.34	9.07	10.57	8.36	0.00	*
P. p. abelii	11.57	8.60	9.72	11.22	7.87	10.93	8.48	2.65	0.00
Males									
P. t. verus	0.00	*	*		*	*	*	*	*
P. t. troglodytes	2.44	0.00	*		*	*	*	*	*
P. t. schweinfurthii	2.41	1.22	0.00		*	*	*	*	*
P. paniscus	4.07	2.34	3.13	0.00	*	*	*	*	*
G. g. gorilla	13.78	11.48	10.74	12.39	0.00	*		*	*
G. g. graueri	15.13	11.83	11.82	12.34	2.61	*		*	*
G. g. beringei	14.00	11.71	11.32	12.70	2.43	1.97	*	*	*
P. p. pygmaeus	10.42	7.43	9.35	8.90	9.10	10.95	9.46	*	
P. p. abelii	14.10	10.94	11.83	12.84	7.32	10.18	8.15	3.64	*
Females									
P. t. verus	0.00			*	*	*	*	*	*
P. t. troglodytes	2.75	0.00		*	*	*	*	*	*
P. t. schweinfurthii	2.99	1.30	0.00	*	*	*	*	*	*
P. paniscus	3.46	2.40	2.68	0.00	*	*	*	*	*
G. g. gorilla	14.56	11.35	10.73	13.62	0.00	*		*	*
G. g. graueri	13.08	10.22	10.28	11.91	2.71	0.00		*	*
G. g. beringei	12.35	9.57	9.78	11.98	2.58	1.55	0.00	*	*
P. p. pygmaeus	9.20	6.02	8.10	8.67	10.04	10.47	8.31	0.00	
P. p. abelii	11.17	7.90	9.28	11.10	9.55	12.09	9.94	3.27	0.00

Asterisk above diagonal shows statistically significant ($p < 0.05$) pair-wise distances.

D^2 values between *P. p. pygmaeus* and *P. p. abelii* do not change substantially after size adjustment. Mahalanobis distances are higher between Bornean and Sumatran orangutans than the intraspecific distances within *P. troglodytes* and *G. gorilla*.

The first discriminant function after size adjustment accounts for a lower proportion of variance than before. The variance explained by the first function is between 55% to 73% in sex-pooled samples, depending on molar position. In contrast, the second function accounts for a higher proportion of the variance, increasing from 20% to 28%, depending on the molar position. With this change in the distribution of variance, the separation of taxa is not as distinct on the first axis as it was before. Figure 4 illustrates the spread of data on the first two functions of the LM2 using size-adjusted measures on sex-pooled samples. Compared to the non-size-adjusted data (Figure 2) the first axis now separates *P. troglodytes* from *G. gorilla*, with *P. paniscus* occupying an intermediate position. Whereas in the analysis of raw data *P. troglodytes* and *P. paniscus* formed small, tight clusters, size-adjusted data reveals that the ellipsoid surrounding *P. troglodytes* is similar in size to that of *G. gorilla*. *P. paniscus* is virtually confined within *P. troglodytes*,

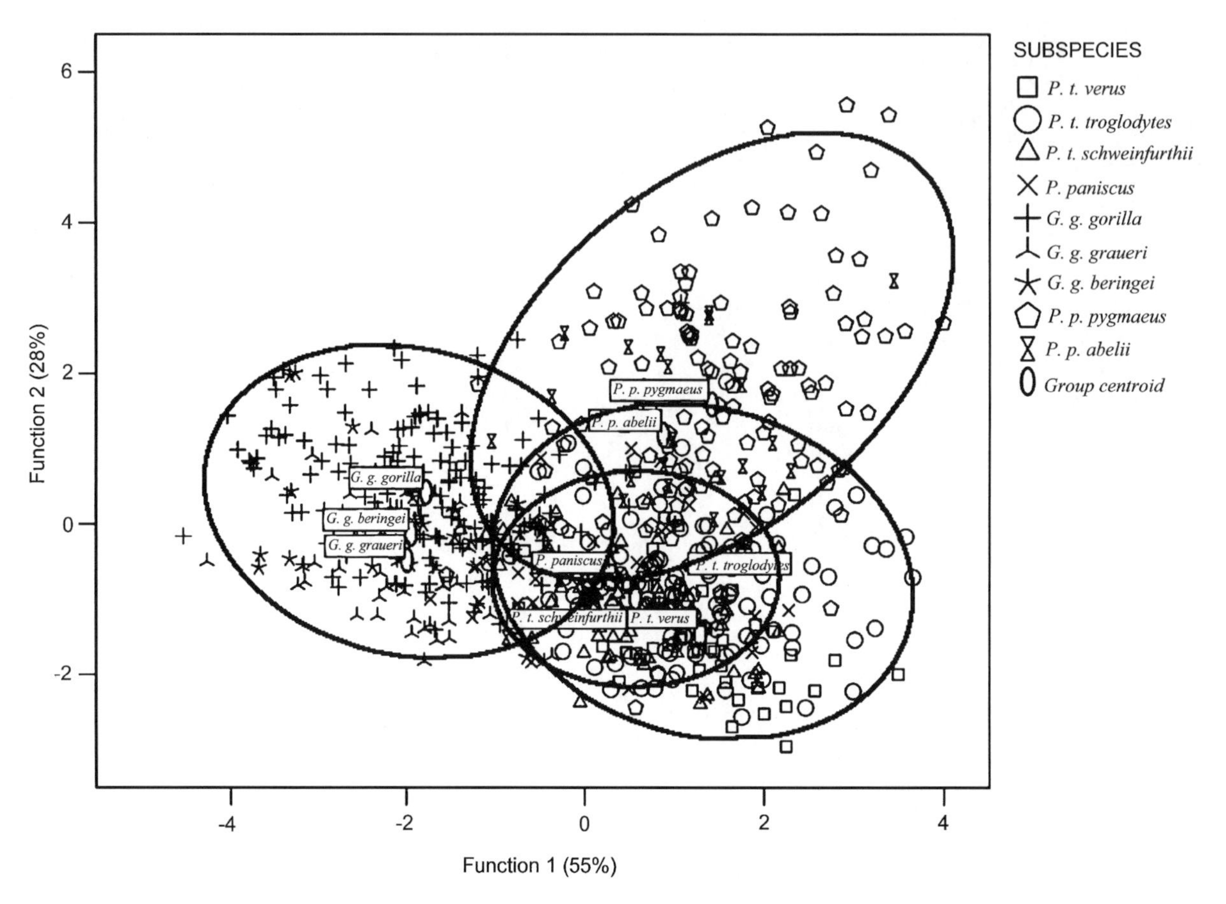

Figure 4. Distribution of great ape subspecies along first two discriminant functions in the analysis using shape variables on the LM2. Sexes combined.

which is also reflected in the low classification accuracy. The second function separates *G. g. gorilla* from *G. g. graueri* and *G. g. beringei* and the subspecies of *P. pygmaeus* from the other great apes, although there is considerable overlap. Size-adjusted *P. pygmaeus* data display greater variance than the other groups.

Pearson's correlations of the discriminant scores on function 1 with the geometric mean are negative but quite high on all size-adjusted molars (−0.63 to −0.74, $p < 0.01$), except the LM1 and LM2 (−0.37 and −0.49, respectively, $p < 0.01$). The variance explained by the first function is also lower in these two molars (57% and 55%), although the classification accuracy is relatively high (56% and 58%, respectively). The second function has a fairly strong correlation with tooth size in the LM1 and LM2 (0.43 and 0.38, respectively, $p < 0.01$), but not in other molars. This suggests that a size component is still preserved in great ape taxa even after the effect of isometric size is reduced; for the LM1 and LM2 this effect is seen in the first two functions.

In Figure 5 the four great ape species are identified by sex in a scatter plot of the first two discriminant functions of size-scaled variables in the LM2. In contrast with the scatter using raw data (Figure 3), the sexes overlap considerably in their distributions. This is because the size difference between male and female orangutans and gorillas is adjusted, contributing to a reduction in overall variance in these taxa compared to *Pan*.

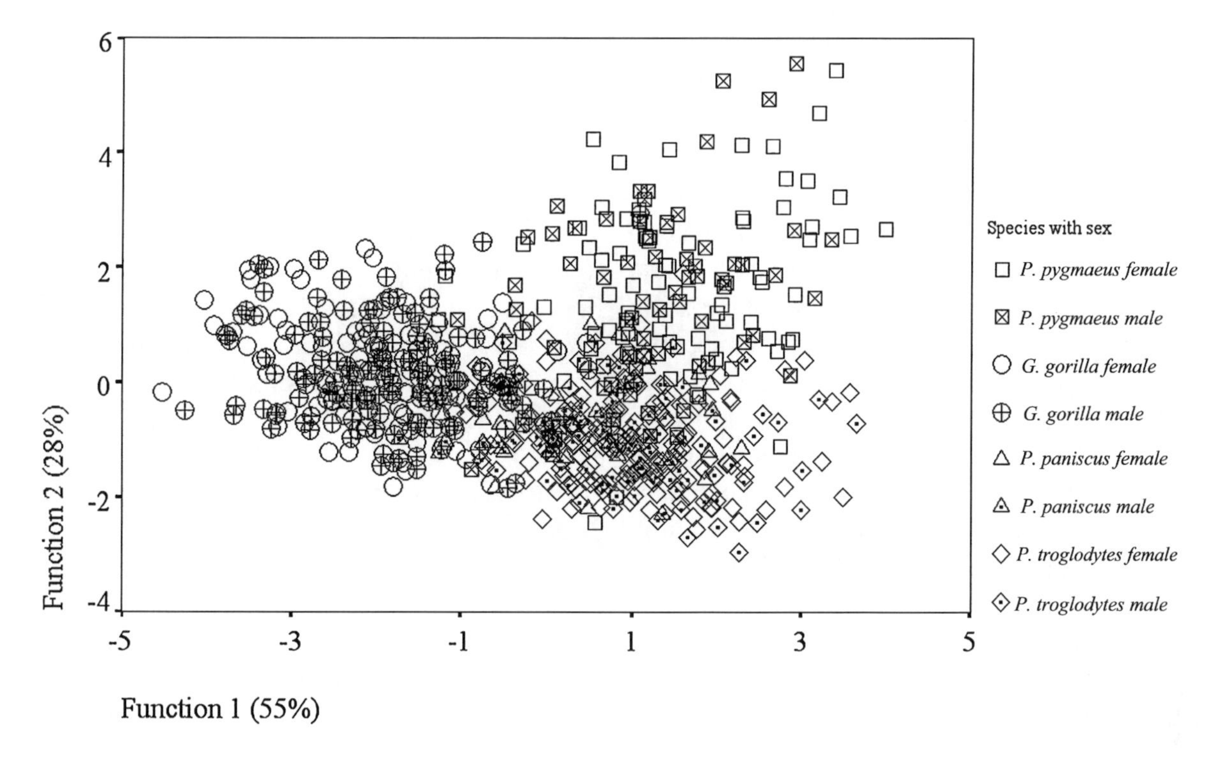

Figure 5. Distribution of sexes of great ape species along factor 1 and 2 of the LM2 using shape data. Ellipsoids not displayed because of extensive overlap between sexes.

After size adjustment, the variables most responsible for causing separation among groups are the length of the preprotocrista on the upper molars and the lengths of postprotocristid and prehypocristid on the LM1 and LM2. The mesiodistal length of the molars also has a fairly strong correlation with function 1 in all analyses.

Discussion

The results of this study open up the following questions for further discussion: how reliable are molar metrics for recognizing patterns of diversity and taxonomy in the great apes? How do size differences among the great apes affect molar discrimination and patterns of variation? How does sexual dimorphism and shape affect the nature of molar variation among great apes? Can great ape models be used to understand the taxonomy of fossil hominins? What inferences can we draw from great ape patterns of intraspecific variation for recognizing subspecies among fossil hominins? Each of these is addressed below.

Concordance Between Dental and Molecular Patterns of Divergence

A primary conclusion of this study is that molar metrics are able to discriminate between *Pan, Gorilla* and *Pongo*, and their constituent species, *P. paniscus, P. troglodytes, G. gorilla* and *P. pygmaeus*. These are well-established taxa with very little debate regarding their taxonomic status. Regarding subspecies, *P. t. verus* is clearly distinguishable from *P. t. troglodytes* and *P. t. schweinfurthii*, but the latter two are not as easily differentiated from each other. All three *G. gorilla* subspecies are easily recognizable and *G. g. gorilla* is well differentiated from *G. g. graueri* and *G. g. beringei.*

Finally, *P. p. pygmaeus* is well differentiated from *P. p. abelii*. These patterns of differentiation deviate slightly from the traditionally understood taxonomy, in the separation of *P. t. verus* from the other subspecies of *P. troglodytes*, and in the clear distinction between east and west African gorillas. However, they are consistent with the conclusions of several recent molecular studies (Morin et al., 1994; Ruvolo et al., 1994; Garner and Ryder, 1996; Saltonstall et al., 1998; Gagneux et al., 1999; Jensen-Seaman and Kidd, 2001). Since molecular data ideally provide accurate information regarding patterns of divergence (Collard and Wood, 2000, 2001), a close match between the infraspecific measures of divergence revealed by this study and those of molecular studies suggests that molar metrics are successful at revealing taxonomically relevant patterns of divergence in great apes. Similar results obtained by several morphological studies (Braga, 1995; Uchida, 1996; Guy et al., 2003; Stumpf et al., 2003; Taylor and Groves, 2003; Lockwood et al.; 2004) support this conclusion.

The Role of Size

Molar size plays a significant role in classifying great ape taxa in this study. This is seen in the high classification accuracy when raw variables are used and a strong correlation between the most significant discriminant functions and the geometric mean in these analyses. The importance of molar size in driving dispersion is most evident in the loss of the distinctive status of *P. paniscus* following size adjustment. The distance between gorilla subspecies is also reduced, so that the distance between the eastern and western gorillas is no longer greater than the distance between *P. paniscus* and *P. troglodytes*. Furthermore, following size adjustment gorilla and orangutan male and female molars do not form discrete clusters. Previous multivariate studies of the cranium (Shea et al., 1993) and mandible (Taylor and Groves, 2003) have reported a reduction in classification accuracy for *P. paniscus* and subspecies of *G. gorilla* following size correction. To explain their finding, Shea et al. (1993) suggested that overall size difference is an important criterion by which morphological differentiation is achieved. They argued (Shea et al., 1993: 278) that "...size and allometric effects are not merely troublesome obfuscation to be removed to yield a clearer view of "true" biological or phylogenetic distinctions; indeed size differences have so many pervasive effects on myriad biological systems and levels (e.g., Lindstedt and Calder, 1981) that they should be central to systematic conclusions in and of themselves."

Indeed, size appears to be an important element in the niche partitioning and adaptive divergence of the African apes. Sympatric chimpanzees and gorillas differ in their fallback food preferences (fruits versus leaves), habitat preferences (greater versus lesser arboreality), and social systems (multi-male versus single-male groups), all of which are associated with size differences in primates. Size related differences between *P. pansicus* and *P. troglodytes* are achieved and maintained through ontogenetic scaling (Shea 1981; 1983a, b, c; 1984). However, the differences are not proportional in all body systems: the face and teeth of *P. paniscus* are relatively smaller whereas the hindlimbs are relatively longer than *P. troglodytes* (Jungers and Susman, 1984; Morbeck and Zihlman, 1989). Also, as pointed out by Shea et al. (1993), size differences do not exhaust the differences between African apes; they differ quite significantly in size-adjusted morphological systems (Taylor and Groves, 2003; Guy et al., 2003; Lockwood et al., 2004), was as found in this study.

The Nature and Pattern of Molar Variation

The contrast between size-preserved and size-adjusted analyses provides important insights into the nature and pattern of molar variation in great apes. For example, size-preserved molars demonstrate how sexual dimorphism affects dispersion. It was seen that the highly dimorphic gorillas and orangutans, especially the males in these taxa, display higher molar size variance, which increased the divergence between the species. When molar size was adjusted by indexing all variables against the geometric mean, the size difference between males and females was reduced, as was the overall variance in gorillas. This resulted in similar levels of variance among chimpanzees and gorillas.

It is noteworthy, however, that despite these high levels of molar sex-dimorphism, the males and females of each species clustered together in size-preserved analyses (Figure 3). Even in the dimorphic gorillas and orangutans there was greater overlap between males and females than between members of distinct species. This suggests that in fossil hominin molars where sex attribution and level of dimorphism is unknown, variation due to sexual dimorphism is likely to produce a pattern of two overlapping distributions with a bimodal peak rather than two mostly non-overlapping distributions.

Size-adjusted analyses also help to highlight the finding that subspecies of *P. troglodytes* show greater distinction when shape data are used. The distances among subspecies of *P. troglodytes* were greater, comparable to those among subspecies of *G. gorilla* and *P. pygmaeus*. In particular, the separation of *P. t. verus* from *P. t. troglodytes* and *P. t. schweinfurthii* was heightened, so that in some analyses it was more divergent than *P. paniscus*.

A comparison of raw and shape analyses also revealed that molar variance in orangutans was affected by shape as well as size. Although shape data reduced the size difference between sexes it did not result in a substantial change in inter-subspecies distance or overall variance in *P. pygmaeus*.

A final point worth noting in this regard is that even after size conversion there was clear separation between east and west African gorillas, lowland and mountain gorillas, west African chimpanzees and central and east African chimpanzees, and Bornean and Sumatran orangutans. Strong correlations with the geometric mean even in the size-adjusted analyses indicates that allometric size is important in distinguishing between these taxa.

Great Apes as Models for Hominin Variation

A major finding that emerges from this study is that no single pattern can be used to characterize great ape molar variation. Isometric size, sexual dimorphism and allometric size all contribute to variable extents. Several previous studies have underscored the disparity in great apes in patterns of variation and the signals revealed by different body systems (Shea and Coolidge, 1988; Ruvolo et al., 1994; Gagneux et al., 1999; Ackermann, 2002; Taylor and Groves, 2003; Guy et al., 2003; Schaefer et al., 2004; Scott and Lockwood, 2004; Taylor, 2006).

A related finding, that chimpanzees are characterized by low variance in dental metrics, so that the distances between *P. paniscus* and *P. troglodytes* are similar or lower than the distances between subspecies of gorillas and orangutans, also fits with existing studies. Shea and Coolidge, 1988; Ruvolo et al., 1994; Guy et al., 2003; Taylor and Groves, 2003; Lockwood et al., 2004. Ruvolo et al. (1994) used their findings to advocate for a species level distinction between east and west African gorillas. Jolly et al. (1995), however, pointed out that it is problematic to use the distance from one taxon to judge the distance from another since no consistent standards of distances can be established among vertebrates. This study reveals that chimpanzees and gorillas are characterized by fundamental differences in patterns of variation, in particular in sexual dimorphism, which affects the magnitude of variance within and among subgroups (see also, Leigh et al., 2003; Pilbrow and Bailey, 2005). Previous studies have alluded to the different biogeographic and evolutionary histories of the great apes resulting in differential patterns of ontogenetic development, mating behaviors and migratory patterns (Shea and Coolidge, 1988; Kingdon, 1989; Goldberg, 1998; Leigh and Shea, 1995; Gagneux et al., 1999; Leigh et al., 2003).

Corroboration with other studies helps to strengthen the conclusions of this study, that great ape patterns of dental variation cannot easily be applied to understanding patterns of molar variation in extinct hominin groups (see also Ackermann, 2002; Taylor, 2006). My results suggest that chimpanzees will provide a different interpretation of hominin patterns of molar variation compared to gorillas and orangutans. Specifically, chimpanzee molars could falsify a single species hypothesis where gorilla and orangutan molars may not. Greater Mahalanobis distance between two subspecies of *G. gorilla* than between *P. troglodytes* and *P. paniscus* in this study suggests that even if both chimpanzee species are combined, so as to model combined-species ranges of variation, they could falsify a single-species hypothesis in a situation where gorillas may uphold it. The right choice of model could be a difficult question here. A greater reliance on the chimpanzee interpretation may be justified by the closer phylogenetic affinity between chimpanzees and hominins. However, this must be based on the assumption that the pattern and magnitude of variance in chimpanzees and hominins is similar, and is part of their shared evolutionary history. Given that closely related great apes differ so remarkably in their patterns of molar variation, this may not be a valid assumption.

On the other hand, chimpanzee molars may not falsify the converse null hypothesis, namely that variation in a fossil hominin assemblage is more than can be accommodated within a single species, when gorilla and orangutan molars may. If combined chimpanzee species are still unable to falsify this hypothesis, the likelihood that more than one species is present is likely to become more robust.

The different interpretations for hominin taxonomy provided by chimpanzees and gorillas are illustrated by the analysis of Harvati et al. (2004). In their study pair-wise distances between Neandertal and modern human populations were compared with distances between subspecies in twelve species of Old World monkeys and apes. The distances between *G. g. gorilla* and *G. g. beringei* were not significantly smaller than the distances between Neandertals and modern humans, but the distances between *P. troglodytes* subspecies and *P. paniscus* were, as were the distances between the subspecies of Old World monkeys. Harvati et al. (2004) concluded that Neandertals and modern humans constitute distinct species, and suggested that the gorilla subspecies might be distinct species as also advocated by Groves (2001). However, based on the present study, it appears that their results were influenced by the differing patterns of variation in the African apes, whereby low variation within and among chimpanzee species could falsify the single species hypothesis but higher variation among gorilla subspecies could not (see also Shea and Coolidge, (1988). Their reliance on a wider comparative sample of Old World monkeys upholds their conclusion, nonetheless, and suggests that a wider comparative database from which to model hominin variation is a likely solution to the use of phylogenetically affiliated great ape species. Thus baboons may be used as ecological models along with the great apes (Jolly, 2001).

Implications for Differentiating Subspecies Among Fossil Hominins

Finally, the hierarchical levels of differentiation evident among great ape molars in this study has several implications for the ability to recognize subspecies among fossil hominins. Discrimination among great ape genera is better than among species, which in turn is better than among subspecies (Tables 2 and 4). Subspecies have the lowest classification accuracy, but they are distinguishable even when shape data are used. The distances among subspecies within *P. troglodytes, G. gorilla*, and *P. pygmaeus* are not comparable, but that is to be expected because subspecies, by nature, are transitory units, documenting the fluid nature of population dynamics. Nevertheless, they provide information about historical processes underlying diversification. These results suggest a potential for similar levels of diversity to be recognized among extinct hominins. In light of the foregoing discussion, however, where subspecies of one great ape group are equivalent to species of another, no firm taxonomic standards can be set for identifying subspecies, or species, of fossil hominins. It might be preferable then to use terms such as operational taxonomic units or allotaxa (Grubb, 1999) to describe distinct units from the hominin fossil context, Jolly (2001). Especially in the paleoanthropological context, however, direct corroborating evidence for population-level interactions is lacking. Our interpretations of these interactions, including diversification, reticulation, genetic exchange and isolation are necessarily based on how we designate populations, in taxonomic terms based on criteria we establish from the comparative context. I would suggest, based on the results of this study, that paleoanthropologists are likely to have greater success in identifying and appreciating intraspecific variation if the comparative taxa are sorted into infraspecific units such as subspecies. This will enable them to contrast interspecific patterns of variation with those that are observed within the species. The present study provides such a dataset.

Conclusions

Great apes vary substantially in the magnitude and patterns of variation in molar occlusal morphology. Size, shape, and sexual dimorphism are among the factors that affect the nature of variation. Molar dimensions and crest lengths provide clear signals of diversification among subspecies within species, among species and among genera of great apes. These hierarchical levels can be differentiated. The patterns of diversification match those of molecular studies. In general, sex-segregated samples provide clearer discrimination than sex-pooled samples, but the difference between the two is negligible. Despite high sexual dimorphism sexes within species cluster together. This indicates that in mixed species samples of fossil hominin molars of unknown sex variation due to sexual dimorphism is unlikely to be confused with interspecific variation. The difference between great apes in the patterns of molar variation signals caution when applying these models for studying alpha taxonomy in fossil hominins. The choice of model will affect interpretations regarding the number of species and patterns of variation proposed for the fossils. Keeping this caveat in mind, the great apes provide an exhaustive database of patterns and ranges of molar variation in taxa closely related to the fossil hominins. When used in combination with other ecological models they provide an excellent understanding of species-level diversification, but more importantly of patterns of intraspecific variation, which provide an understanding of the processes of interaction among hominins

Acknowledgments

Heartfelt thanks to Shara Bailey and Jean-Jacques Hublin for inviting me to attend the dental anthropology conference in Leipzig. I really enjoyed the conference, especially the efficiency and hospitality of the Max Planck Institute. For this I should also thank Silke Streiber, Diana Carstens, Allison Cleveland, Michelle Hänel and Jörn Scheller. An extra thanks goes to Shara for going beyond the call of duty to organize my consular visit to Berlin while I was there.

For help with this paper I would like to thank the museums that made their collections available for study: American Museum of Natural History, NY; Anthropologisches Institüt und Museum der Universität Zürich-Irchel, Zürich; British Museum of Natural History, London; Field Museum of Natural History, Chicago; Museum of Comparative Zoology, Harvard; Muséum National d'Histoire Naturelle, Paris; Powell-Cotton Museum, Kent; Peabody Museum of Anthropology, Harvard; United States National Museum, Washington, D.C.; Musée Royal de l'Afrique Centrale, Tervuren; Zoologisches Museum, Berlin; Anthropologische und Zoologische Staassammlung, Münich. The project was funded by grants from the LSB Leakey Foundation, National Science Foundation (SBR-9815546), and the Wenner-Gren Foundation. The paper benefited greatly from comments by an anonymous reviewer, Leslea Hlusko, and Shara Bailey. I thank them for that.

Note

1. The terms hominin and hominid used here follow Wood and Richmond (2000). However, when referring to chimpanzees, gorillas and orangutans together, the conventional term great apes is used.

References

Ackermann, R.R., 2002. Patterns of covariation in the hominoid craniofacial skeleton: implications for paleoanthropological models. Journal of Human Evolution 42, 167–187.

Albrecht, G.H., Gelvin, B.R., Miller, J.M.A., 2003. The hierarchy of intraspecific craniometric variation in gorillas: a population-thinking approach with implications for fossil species recognition studies. In: Taylor, A.B., Goldsmith, M.L. (Eds.), Gorilla Biology. Cambridge University Press, Cambridge, pp. 62–103.

Bailey, S.E., Pilbrow, V.C., Wood, B.A., 2004. Inter-observer error involved in independent attempts to measure cusp base areas of Pan M^1s. Journal of Anatomy 205, 323–331.

Begun, D.R., 1992. Miocene fossil hominids and the chimp-human clade. Science 257, 1929–1933.

Braga, J.C., 1995. Définition de certains caractères discrets crâniens chez *Pongo, Gorilla, et Pan*. Perspectives taxonomiques et phylogénétiques. Ph.D. Dissertation, University of Bordeaux.

Caccone, A., Powell, J.R., 1989. DNA divergence among hominoids. Evolution 43, 925–942.

Collard, M., Wood, B., 2000. How reliable are human phylogenetic hypothesis? Proceedings of the National Academy of Sciences of the USA 97, 5003–5006.

Collard, M., Wood, B., 2001. Homoplasy and the early hominid masticatory system: inferences from analyses of extant hominoids and papionins. Journal of Human Evolution 41, 167–94.

Coolidge, H.J., 1929. A revision of the genus *Gorilla*. Memoirs of the Museum of Comparative Zoology Harvard 50, 293–381.

Coolidge, H.J., 1933. *Pan paniscus*: Pygmy chimpanzee from south of the Congo River. American Journal of Physical Anthropology 18, 1–57.

Courtenay, J., Groves, C.P., Andrews, P., 1988. Inter– or intra-island variation? An assessment of the differences between Bornean and Sumatran orang-utans. In: Schwartz, J.H. (Ed.), Orangutan Biology. Oxford University Press, Oxford, pp. 19–29.

Darroch, J.N., Mosimann, J.E., 1985. Canonical and principal components of shape. Biometrika 72, 241–252.

Falsetti, A.B., Jungers, W.L., Cole III, T.M., 1993. Morphometrics of the callitrichid forelimb: a case study in size and shape. International Journal of Primatology 14, 551–572.

Futuyma, D.J., 1986. *Evolutionary Biology*. Sinauer Associates, Inc., Sunderland, MA

Gagneux, P., Wills, C., Gerloff, U., Tautz, D., Morin, P.A., Boesch, C., Fruth, B., Hohmann, G., Ryder, O., Woodruff, D.S., 1999. Mitochondrial sequences show diverse evolutionary histories of African hominoids. Proceedings of the National Academy of Sciences, USA 96, 5077–5082.

Garner, K.J., Ryder, O.A., 1996. Mitochondrial DNA diversity in gorillas. Molecular Phylogenetics and Evolution 6, 39–48.

Gibbs, S., Collard, M., Wood, B., 2000. Soft-tissue characters in higher primate phylogenetics. Proceedings of the National Academy of Sciences, USA 97, 11130–11132.

Goldberg, T.L., 1998. Biogeographic predictors of genetic diversity in populations of eastern African chimpanzees (*Pan troglodytes schweinfurthii*). International Journal of Primatology 19, 237–254.

Gonder, M.K., Oates, J.E., Disotell, T.R., Forstner, M.R.J., Morales, J.C., Melnick, D.J., 1997. A new west African chimpanzee subspecies? Nature 388, 337.

Goodman, M., 1962. Evolution of the immunologic species specificity of human serum proteins. Human Biology 34, 104–150.

Goodman, M., Romero-Herrera, A.E., Dene, H., Czelusniak, J., Tashian, R.E., 1982. Macromolecular sequences in systematic and evolutionary biology. In: Goodman, M. (Ed.), Macromolecular Sequences in Systematic and Evolutionary Biology. Plenum Press, New York, pp. 115–191.

Goodman, M., Porter, C.A., Czelusniak, J., Page, S.L., Schneider, H., Shoshani, J., Gunnell, G., Groves, C.P., 1998. Toward a phylogenetic classification of primates based on DNA evidence complemented by fossil evidence. Molecular Phylogenetics and Evolution, 585–598.

Groves, C.P., 1967. Ecology and taxonomy of the *Gorilla*. Nature 213, 890–893.

Groves, C.P., 1970. Population systematics of the *Gorilla*. Journal of Zoology. London. 161, 287–300.

Groves, C.P., 1986. Primate systematics. In: Swindler, D.R., Erwin, J. (Eds.), Comparative Primate Biology, Volume 1. Systematics, Evolution, and Anatomy. Alan R. Liss, New York, pp. 187–218.

Groves, C.P., 2001. *Primate Taxonomy*. Smithsonian University Press, Washington, D.C.

Grubb, P., 1999. Evolutionary processes implicit in distribution patterns of modern African mammals. In: Bromage, T.J., Schrenk, F. (Eds.), African Biogeography, Climate Change, and Human Evolution. Oxford University Press, New York, pp. 150–164.

Guy, F., Brunet, M., Schmittbuhl, M., Viriot, L., 2003. New approaches in hominoid taxonomy: morphometrics. American Journal of Physical Anthropology 121, 198–218.

Harvati, K., Frost, S.R., McNulty, K.P., 2004. Neanderthal taxonomy reconsidered: implications of 3D primate models of intra- and interspecific differences. Proceedings of the National Academy of Sciences, USA 101, 1147–1152.

Hill, W.C.O., 1967. The taxonomy of the genus *Pan*. In: Starck, D. Schneider, R., Kuhn, H.-J. (Eds.), Neue Ergebnisse der Primatologie. Fischer, Stuttgart, pp. 47–54.

Hill, W.C.O., 1969. The nomenclature, taxonomy and distribution of Chimpanzees. In: Bourne, G.H. (Ed.). The Chimpanzee. Vol. 1. Karger, Basel, pp. 22–49.

Horn, A.D., 1979. The taxonomic status of the bonobo chimpanzee. American Journal of Physical Anthropology 51, 273–282.

James, F.C., McCullough, C.E., 1990. Multivariate statistical methods in ecology and systematics: Panacea or Pandora's Box. Annual Review of Ecology and Systematics 211, 129–166.

Jenkins, P., 1990. *Catalogue of Primates in the British Museum (Natural History). Part V.* Natural History Museum Publications, London.

Jensen-Seaman, M.I., Kidd, K.K., 2001. Mitochondrial DNA variability and biogeography of eastern gorillas. Molecular Ecology 10, 2241–2247.

Jensen-Seaman, M.I., Deinard, A.S., Kidd, K.K., 2003. Mitochondrial and nuclear DNA estimates of divergence between western and eastern gorillas. In: Taylor, A.B., Goldsmith, M.L. (Eds.), Gorilla Biology. Cambridge University Press, Cambridge, pp. 247–268.

Johanson, D.C., 1974. An odontological study of the chimpanzee with some implications for hominoid evolution. Ph.D. Dissertation, University of Chicago.

Jolly, C.J., 1993. Species, subspecies, and baboon systematics. In: Kimbel, W.H., Martin, L.B. (Eds.) Species, Species Concepts, and Primate Evolution. Plenum Press, New York, pp.67–107.

Jolly, C.J., 2001. A proper study for mankind: analogies from the papionin monkeys and their implications for human evolution. Yearbook of Physical Anthropology 44, 177–204.

Jolly, C.J., Oates, J.F., Disotell, T.R., 1995. Chimpanzee kinship. Science 268, 185–187.

Jungers, W.L. Susman, R.S., 1984. Body size and skeletal allometry in African apes. In: Susman, R.L. (Ed.), Evolutionary Morphology and Behavior of the Pygmy Chimpanzee. Plenum Press, New York. pp. 131–178.

Kimbel, W.H., 1991. Species, species concepts and hominid evolution. Journal of Human Evolution 20, 355–372.

Kimbel, W.H., Martin, L.B., 1993. (Eds). Species, Species Concepts and Primate Evolution. Plenum Press, New York.

Kingdon, J., 1989. Island Africa. Princeton University Press, Princeton.

Kinzey, W.G., 1984. The dentition of the pygmy chimpanzee, *Pan paniscus*. In: Susman, R.L. (Ed.), The Pygmy Chimpanzee: Evolutionary Biology and Behavior. Plenum Press, New York, pp. 65–88.

Leigh, S.R., Shea, B.T., 1995. Ontogeny and the evolution of adult body size dimorphism in apes. American Journal of Primatology 36, 37–60.

Leigh, S.R., Relethford, J.H., Park, P.B., Konigsberg, L.W., 2003. Morphological differentiation of *Gorilla* subspecies. In: Taylor, A.B., Goldsmith, M.L. (Eds.), Gorilla Biology. Cambridge University Press, Cambridge, pp. 104–131.

Lindstedt, S.L., Calder, W.A.III., 1981. Body size, physiological time, and longevity of homeothermic animals. Quaterly Review of Biology 56, 1–16.

Lockwood, C.A., Kimbel, W.H., Lynch, J.M., 2004. Morphometrics and hominoid phylogeny: Support for a chimpanzee-human clade and differentiation among great ape subspecies. Proceedings of the National Academy of Sciences, USA101, 4356–4360.

Mahler, P.E., 1973. Metric variation in the pongid dentition. Ph.D. Dissertation, University of Michigan.

Mayr, E., 1942. *Systematics and the Origin of Species*. Columbia University Press, New York.

Mayr, E., 1982. Of what use are subspecies? The Auk 99, 593–595.

Miller, J.M.A., Albrecht, G.H., Gelvin, B.R., 2004. Craniometric variation in early *Homo* compared to modern gorillas: a population-thinking approach. In: Anapol, F., German, R.Z., Jablonski, N.G. (Eds.), Shaping Primate Evolution. Cambridge University Press, Cambridge, pp. 66–96.

Morbeck, M.E., Zihlman, A.L., 1989. Body size and proportions in chimpanzees, with special reference to *Pan troglodytes schweinfurthii* from Gombe National Park, Tanzania. Primates 30, 369–382.

Morin, P.A., Moore, J.J., Chakraborthy, R., Jin, L., Goodall, J., Woodruff, D.S., 1994. Kin selection, social structure, gene flow, and the evolution of chimpanzees. Science 265, 1193–1201.

Mosimann, J.E., James, F. C., 1979. New statistical methods for allometry with application to Florida red-winged blackbirds. Evolution 33, 444–459.

Muir C.C., Galdikas B.M., Beckenbach A.T., 2000. MtDNA sequence diversity of orangutans from the islands of Borneo and Sumatra. Journal of Molecular Evolution 51, 471–80.

Pilbrow, V.C., 2003. Dental variation in African apes with implications for understanding patterns of variation in species of fossil apes. Ph.D. Dissertation, New York University.

Pilbrow, V.C., 2006a. Lingual incisor waits in the modern hominoids and their utility for fossil hominoid taxonomy. American Journal of Physical Anthropology 129, 323-338.

Pilbrow, V.C., 2006b. Population systematics of chimpanzees using molar morphometrics. Journal of Human Evolution 51, 646-662.

Pilbrow, V.C. Bailey, S.E., 2005. Patterns of intraspecific variation in dental metrics of *Gorilla gorilla, Pan troglodytes* and *Homo sapiens*: a comparison using Wright's (1969) *Fst* statistic. PaleoAnthropology 3, 13. http://pa.pennpress.org/

Röhrer-Ertl, O., 1984. Research history, nomenclature and taxonomy of the orang-utan. In: Schwartz, J.H., (Ed.), Orang-utan Biology. Oxford University Press, Oxford, pp.7–18.

Ruvolo, M., 1994. Molecular evolutionary processes and conflicting gene trees: the hominoid case. American Journal of Physical Anthropology 94, 89–113.

Ruvolo, M., 1997. Molecular phylogeny of the hominoids: inferences from multiple

independent DNA sequence data sets. Molecular Biology and Evolution 14, 248–265.
Ruvolo, M., Pan, D., Zehr, S., Goldberg, T., Disotell, T.R., 1994. Gene trees and hominoid phylogeny. Proceedings of the National Academy of Sciences of the USA 91, 8900–8904.
Saltonstall, K., Amato, G., Powell, J., 1998. Mitochondrial DNA variability in Grauer's gorillas of Kahuzi-Biega National Park. Journal of Heredity 89, 129–135.
Sarmiento, E.E., Butynski, T., 1996. Present problems in gorilla taxonomy. Gorilla Journal 19, 5–7.
Sarmiento, E.E., Oates, J.F., 2000. The Cross River gorillas: a distinct subspecies, *Gorilla gorilla diehli* Matschie 1904. Am. Mus. Novitates 3250, 56 pp.
Schaefer, K., Mitteroecker, P., Gunz, P., Bernhard, M., Bookstein, F.L., 2004. Craniofacial sexual dimorphism patterns and allometry among extant hominids. Annals of Anatomy 186, 471–478.
Schwartz, J.H., 1984. The evolutionary relationships of man and orang-utans. Nature 308, 501–505.
Scott, J.E., Lockwood C.A., 2004. Patterns of tooth crown size and shape variation in great apes and humans and species recognition in the hominid fossil record. American Journal of Physical Anthropology 125, 303–319.
Shea, B.T., 1981. Relative growth of the limbs and trunk in the African apes. American Journal of Physical Anthropology 56, 179–201.
Shea, B.T., 1983a. Size and diet in the evolution of African ape craniodental form. Folia Primatologica 40, 32–68.
Shea, B.T., 1983b. Allometry and heterochrony in the African apes. American Journal of Physical Anthropology 62, 275–289.
Shea, B.T., 1983c. Paedomorphy and neoteny in the pygmy chimpanzee. Science 222, 521–522.
Shea, B.T., 1984. An allometric perspective on the morphological and evolutionary relationships between pygmy (*Pan paniscus*) and common (*Pan troglodytes*) chimpanzees. In: Susman, R.L. (Ed.), The Pygmy Chimpanzee: Evolutionary Biology and Behavior. Plenum Press, New York, pp. 89–130.
Shea, B.T., Coolidge, H.J. Jr., 1988. Craniometric differences and systematics in the genus *Pan*. Journal of Human Evolution 17, 671–685.
Shea, B.T., Leigh, S.R., Groves, C.P., 1993. Multivariate craniometric variation in chimpanzees: implications for species identification in paleoanthropology. In: Kimbel, W.H., Martin, L.B. (Eds.), Species, Species Concepts, and Primate Evolution. Plenum Press, New York, pp. 265–296.
Sibley, C., Ahlquist, J., 1984. The phylogeny of the hominoid primates, as indicated by DNA-DNA hybridization. Journal of Molecular Evolution 20, 1–15.
Smith, H., Chiszar, M.D., Montanucci, R.R., 1997. Subspecies and classification. Herpetological Review 28, 13–16.
Stumpf, R.M., Polk, J.D., Oates, J.F., Jungers, W.L., Heesy, C.P., Groves, C.P., Fleagle, J. G., 2003. Patterns of diversity in gorilla cranial morphology. In: Taylor, A.B., Goldsmith, M.L. (Eds.), Gorilla Biology. Cambridge University Press, Cambridge, pp. 35–61.
Swindler, D.R., 1976. *Dentition of Living Primates.* Academic Press, London.
Swindler, D.R., 2002. *Primate Dentition: An Introduction to the Teeth of Non-Human Primates.* Cambridge University Press, Cambridge.
Taylor, A.B., 2006. Size and shape dimorphism in great ape mandibles and implications for fossil species recognition. American Journal of Physical Anthropology 129, 82–98.
Taylor, A.B., Groves, C.P., 2003. Patterns of mandibular variation in *Pan* and *Gorilla* and implications for African ape taxonomy. Journal of Human Evolution 44, 529–561.
Tattersall, I., 1986. Species recognition in human paleontology. Journal of Human Evolution 15, 165–175.
Tattersall. I., 1991. What was the human revolution? Journal of Human Evolution 20, 77–83.
Tattersall, I., 1993. Speciation and morphological differentiation in the genus *Lemur*. In: Kimbel, W.H., Martin, L.B. (Eds.), Species, Species Concepts, and Primate Evolution. Plenum Press, New York, pp. 163–176.
Templeton, A.R., 1999. Human Races: A genetic and evolutionary perspective. American Anthropologist 100, 632–650.
Uchida, A., 1992. Intra-species variation among the great apes: Implications for taxonomy of fossil hominoids. Ph.D. Dissertation, Harvard University.
Uchida, A., 1996. Craniodental variation among the great apes. Peabody Museum Bulletin, 4. Harvard University.
Wolpoff, M.H., Hawks, J., Frayer, D.W., Hunley, K. 2001. Modern human ancestry at the

peripheries: a test of the replacement theory. Science. 291, 293–297.

Wood, B., Richmond, B.G., 2000. Human evolution: taxonomy and paleobiology. Journal of Anatomy 197, 19–60.

Wrangham, R.W., de Waal, F.B.M., McGrew, W.C., 1994. The challenge of behavioral diversity. In: Wrangham, R.W., McGrew, W.C., de Waal, F.B.M., Heltne, P.G. (Eds.), Chimpanzee Cultures. Harvard University Press, Cambridge, Mass., pp. 1–18.

Xu, X., Arnason, U., 1996. The mitochondrial DNA molecule of Sumatran orangutan and a molecular proposal for two (Bornean and Sumatran) species of orangutan. Journal of Molecular Evolution 43, 431–437.

Zhi, L., Karesh, W.B., Janczewski, D.N., Frazier-Taylor, H., Sajuthi, D., Gombek, F., Andau, M., Martenson, J.S., O'Brien, S.J., 1996. Genomic differentiation among natural populations of orang-utan (*Pongo pygmaeus*). Current Biology 6, 1326–1336.

3. Trends in postcanine occlusal morphology within the hominin clade: The case of *Paranthropus*

S.E. BAILEY
Department of Human Evolution
Max Planck Institute for Evolutionary Anthropology
Deutscher Platz 6
D-04103 Leipzig, Germany
and
Center for the Study of Human Origins,
Department of Anthropology, New York University,
25 Waverly Place
New York, NY 10003, USA
and
The George Washington University
CASHP
Department of Anthropology
2110 G Street, NW
Washington, DC 20052, USA
sbailey@nyu.edu

B.A. WOOD
The George Washington University
CASHP
Department of Anthropology
2110 G Street, NW
Washington, DC 20052, USA
bernardawood@gmail.com

Keywords: dental morphology, Plio-Pleistocene hominins, *Australopithecus*, early *Homo*

Abstract

We have examined the crowns of chimpanzee, australopith, and *Paranthropus* species and early *Homo* in order to investigate two different, widely recognized, dental trends in Plio-Pleistocene hominin evolution. They are a reduction in crown size and morphological complexity in *Homo*, and an increase in crown size and morphological complexity in *Paranthropus*. A phenetic assessment of maxillary and mandibular molar crown non-metrical traits revealed that two australopith species (*Au. africanus* and *Au. afarensis*) are much more similar to each other than either is to *Paranthropus*, and together all hominins are distinctively different from chimpanzees (*P. troglodytes* and *P. paniscus*). The difference between *Paranthropus* and australopith postcanine teeth was 20–30 times greater than that between the australopith species and the difference between the two australopith species was about half the difference between the two extant chimpanzee species. The characters that contribute to the increase in crown complexity seen in *Paranthropus* do not appear to be primitive retentions from a great ape ancestor, and there is

S.E. Bailey and J.-J. Hublin (Eds.), Dental Perspectives on Human Evolution, 33–52.

some evidence that the same, or a very similar, trend towards trait intensification is already present in australopiths. These traits include additional cusps on the maxillary and mandibular molars, and the expanded P_4 talonid. Early *Homo* exhibits the primitive condition for many of the molar traits, but it has also lost many other primitive traits (upper molar anterior and posterior foveae, for example) that are present in the australopiths. Relative to *Pan*, and similar to the australopiths, early *Homo* possesses a larger P_4 with a somewhat expanded talonid, but this trend is subsequently reversed in later *Homo*. Our study reveals that some of the dental trends said to be characteristic of *Homo* actually appear relatively late in human evolution.

Introduction

Half a century ago Robinson (1954a, b) proposed that an adaptive distinction be made between *Paranthropus* and *Australopithecus*. Robinson interpreted what he judged to be the distinctive morphology of *Paranthropus* from the southern African cave sites as evidence that members of this taxon were dietary specialists. His interpretation of the morphology of *Australopithecus* was that it had adopted a more generalist strategy. Initially, the attention of researchers was focused on the hominin fossil evidence from the southern African caves sites, the only relevant evidence then available. Subsequently, researchers have turned their attention to the distinction between *Paranthropus* and *Homo* instead of the distinction between *Paranthropus* and *Australopithecus* and they have focused more on the fossil hominin evidence recovered from East African sites (e.g., Tobias, 1967; Suwa, 1990; Tobias 1991; Wood, 1991).

Dental morphology has always been a major component of the morphological evidence for the distinctiveness of *Paranthropus* (e.g., Robinson, 1956; Wood, 1991). Robinson (1956) particularly emphasized the discrepancy between the size of the anterior and the postcanine teeth as well as differences in occlusal morphology (e.g., a molarized first mandibular deciduous molar, molarized mandibular premolars, and upper molars with a trigon and hypocone). Although some researchers have subsequently addressed the differences in relative tooth size (e.g., Wood and Stack, 1980), most attention has been paid to documenting the non-metrical and metrical differences in the occlusal morphology of the deciduous and permanent postcanine teeth of *Paranthropus* and *Homo* (e.g., Wood and Abbott, 1983; Grine, 1985; Wood and Uytterschaut, 1987; Wood and Engleman, 1988).

It has been widely assumed that it is the occlusal morphology of *Paranthropus* and not *Homo* that is derived with respect to the symplesiomorphic condition for the hominin clade. Initially this was based on the explicit or implicit assumption that *Au. afarensis* was an appropriate model for the primitive condition for the hominin clade. This was not an unreasonable assumption at a time when (i) the morphology of *Au. afarensis* was interpreted as being almost entirely primitive (de Bonis et al., 1981; White et al., 1981); (ii) when it was assumed that *Au. afarensis* was similar to, if not actually, the common ancestor of all later hominins; and (iii) when the relationships among the extant higher primates were regarded as unresolved or unresolvable. However, discoveries and advances made over the course of the last decade or so have changed our interpretation of *Au. afarensis*. First, fossil evidence has been recovered of presumed hominins that are evidently more primitive than *Au. afarensis* (White et al., 1994; 1995; Senut et al., 2001; Brunet et al., 2002; Haile-Selassie et al., 2004). Second, molecular evidence now points very strongly to a sister group relationship between *Homo* and *Pan* (i.e., (((*Homo*, *Pan*) *Gorilla*) *Pongo*)) (Ruvolo, 1997). Thus, we now have the opportunity to test the hypothesis that *Paranthropus* morphology, particularly *P. boisei* morphology, is derived within a wider context that includes extant taxa.

The standard scoring system used to assess dental morphology in modern humans is the Arizona State University Dental Anthropology System (ASUDAS). This system, consisting of more than 50 crown, root and osseous traits scored using a combination of rank-scale reference plaques and written procedures (Turner et al., 1991), is based on nearly a century of careful comparative studies (Gregory, 1916; Hrdlička, 1920; Weidenreich, 1937; Dahlberg, 1956; Hanihara, 1961; Scott, 1973; Morris, 1975; Harris, 1977; Morris et al., 1978; Harris and Bailit, 1980; Nichol et al., 1984).

The publication and dissemination of the ASUDAS enabled studies of modern human variation aimed at recovering information about biological relationships and prehistoric population movements (Haeussler and Turner, 1992; Turner, 1992a, b and c; Irish, 1993; Haeussler, 1995; 1996; Lipschultz, 1997; Hawkey, 1998). It was only a matter of time before researchers would attempt to apply this system to the hominin fossil record. In the past decade, several researchers have established that the ASUDAS can be used to study morphological variation in relatively recent fossil hominin taxa (Crummett, 1994; Irish, 1998; Bailey, 2000; Coppa et al., 2001; Bailey, 2002b). However, several researchers have questioned whether the ASUDAS, a scoring system developed for use on modern human populations, is appropriate for capturing the differences between *Paranthropus* and early *Homo* (Reid and Van Reenen, 1995; Van Reenen and Reid, 1995; Irish and Guatelli-Steinberg, 2003; Hlusko, 2004). There are at least two reasons why the application of a modern human scoring system to fossil hominin and extant hominoid data sets may result in biased results. First, it makes all samples look more like modern humans than they are. Second, it will not distinguish groups that do not vary with respect to modern human traits but which may be distinctive with respect to other traits.

Given these difficulties, finding traits that would be comparable among early hominins and *Pan* proved to be a more difficult task than first imagined. The final trait list used in this study excluded many ASUDAS traits that would have been uninformative (e.g., UM hypocone reduction, molar and premolar enamel extensions, congenital absence of UM3, premolar odontome, M_2 Y-pattern, LM anterior fovea, four cusped M_2, LM deflecting wrinkle) and included some traits that are not currently part of the ASUDAS. However, there are other traits that have not been systematically collected for hominins that would likely provide for a more complete analysis (e.g., UM anterior and posterior fovea).

Whether or not a particular scoring system or scoring criteria is appropriate depends on the question being addressed. If the question focuses on variation within a species' – as in examining the biological relationships among different geographic populations of modern humans – then a system based on that species' variation may be sufficient to answer the question. However, when we begin to make interspecific comparisons, as when attempting to undertake cladistic analyses, it is important that the system used encompasses variation that is relevant to the particular groups being analyzed. For example, many of the ASUDAS traits are invariant in extant hominoids and Pliocene hominins (Irish and Guatelli-Steinberg, 2003). Therefore, they are not useful for ascertaining interspecific relationships among these taxa. Although attempts have been made to establish new traits and devise scoring criteria for certain fossil hominin taxa (Suwa, 1990; Crummett, 1994; Van Reenen and Reid, 1995; Bailey, 2002a, b; 2004; Hlusko, 2004; Bailey and Lynell, 2005) and non-human higher primates (Hlusko, 2002; Pilbrow, 2003), to date few of these new traits have been standardized as they have for modern human traits.

This study focuses on the metrical and non-metrical aspects of occlusal morphology and addresses the following questions. First, do metrical and non-metrical data support a distinction between *Paranthropus* and *Homo*? Second, are the tools that have been developed to use non-metrical data to distinguish the dentition of regional samples of modern humans appropriate for assessing the extent of phenetic differences among early hominins? Third, do the metrical and non-metrical trait data support the hypothesis that the distinctive occlusal morphology of *Paranthropus* is derived? Fourth, do non-metrical trait data carry a phylogenetic signal? In other words, is there evidence that non-metrical traits are more effective than metrical data (Hartman, 1988; Collard and Wood, 2000) for recovering the relationships among extant higher primates that are supported by molecular evidence?

Materials

The fossil hominin dental data included in this analysis come from two australopith species (*Au. afarensis, Au. africanus*), two *Paranthropus* species (*P. robustus, P. boisei*), and two early *Homo* species (*H. habilis, H. rudolfensis*) (Table 1). The *Au. afarensis* sample includes specimens from both Hadar and Laetoli. Data were collected from high-resolution casts, made available by W. Kimbel of the Institute of Human Origins, Arizona State University. The *Au. africanus* sample includes specimens from Sterkfontein; the *P. robustus* sample includes specimens from Swartkrans; the *P. boisei* sample includes specimens from Koobi Fora, West Turkana, Olduvai and Peninj; and the early *Homo* sample includes specimens from Olduvai and Koobi Fora. Morphological data for *Au. africanus* and *P. robustus* were collected from high-resolution casts provided by P. Ungar (University Arkansas, Fayetteville) and F. Grine (Stony Brook University, New York). The remaining fossil data were collected directly from the fossils (see Methods below).

The extant data for this study are confined to two chimpanzee species (*Pan troglodytes troglodytes* and *Pan paniscus*). *Pan* data were gathered from skeletal collections at the Powell Cotton Museum, Birchington Kent (*P. t. troglodytes*) and the Musée Royal d'Afrique Central, Tervuren (*P. paniscus*).

Methods

In the present study we focused on premolar and molar non-metrical traits that contribute to the complex crown morphology suggested to be derived in *Paranthropus*. Some of these traits (UM Carabelli's trait, UM Cusp 5, LM Cusp 6, and LM Cusp 7) are part of the ASUDAS and were scored (or converted to scores) according to that system. Other traits that contribute to crown complexity but are not included in the ASUDAS include UM "Cusp 6" (Figure 1) and LM "Double Cusp 6" (Figure 2). Carabelli's trait was scored as present for a Y-shaped groove and/or ridge on the protocone and stronger (cusp or shelf) expressions [i.e., grade 3 and above Turner et al. (1991) and grade 2 and above Wood and Abbott (1983)]. Cusp 7 was scored as present only when a well-defined cusp or cuspule was present [i.e., grade 2 and above for Turner et al. (1991) and grade 2 and above for Wood and Abbott (1983)]. For all other traits, any degree of cusp expression (grade 1 and above) was scored as present. Protostylid expression in *Pan* and fossil hominins is often quite different from that observed in modern humans, and it is difficult to score using the ASUDAS. Because we were unsure which features on the protoconid were homologous with the human protostylid and because we lacked an appropriate way to score this variation in *Pan*, we did not include it in our analysis.

Both authors contributed to the data. All *Pan* and *Au. afarensis* data (metrical and morphological) were collected by SEB. Metrical crown component data for the P_4 of

Table 1. List of specimens used in the various analyses

	M_1	M_2	M_3	M^1	M^2	M^3	P_4
Au. afarensis	AL 200-1b	AL 128-23	AL 207-13	AL 200-1a	AL 200-1a	AL 200-1a	AL 128-23
	AL 128-23	Al 145-35	AL 266-1	AL 333-86	AL 200-12-13	AL 333x-1	AL 176-35
	AL 145-35	AL 207-13	Al 288-1	AL 436-1	AL 436-1	AL 436-1	AL 207-13
	AL 330-5	AL 241-14	AL 330-5	LH 3-h	LH 11		AL 266-1
	AL 333s-1	AL 266-1	AL 400	LH 6-e			AL 330-5
	AL 417	AL 288-1	AL 462-7	LH 21-a			AL 333w-1
	LH 2	AL 330-5	LH 15				Al 400-1a
	LH 3-t	AL 333-1	LH 17				AL 417
	LH 15	AL 333w-48					LH 3r
		AL 400					
		AL 417					
		AL 443-1					
Au. africanus	Stw 106	Stw 61	Stw 212	Sts 8	Sts 8	Sts 8	Sts 52b
	Stw 123	Stw 213	Stw 404	Sts 1	Sts 1	Sts 28/37	STW/H 14
	Stw 151a	Stw 308	Stw 487b	Sts 56	Sts 22	Stw 6	STW/H 56
	Stw 145	Stw 327	Stw 14	Sts 57	Sts 32	Stw 179	TM 1523
	Stw 309	Stw 404	Sts 55	Stw 151	Stw 73	Stw 189	
	Stw 327	Stw 14	TM 1518	Stw 183	Stw 151*	Stw 252	
	Stw 404	Stw 424		Stw 402	Stw 183	TM 1511	
	Sts 9	Stw 412		Stw 450	Stw 188	TM 1561	
	Sts 24	Stw 234		Stw 252	Stw 204		
	Stw 246			STS 24	Stw 252		
	Stw 421				TM 1511		
P. robustus	SK 6	SK 1	SK 6	SK 52	SK 47	SKW 29	SK 7
	SK 23	SK 6	SK 23	SK 102	SK 48	SK 31	SK 9
	SK 61	SK 23	SK 55	SK 829	SK 49	SK 36	SK 23
	SK 63	SK 25	SKW5	SK 826/828	SKW 11	SK 48	SK 34
	SK 104	SK 37		SK 89		SK 52	SK 74
	SK 828	SK 55		SKW 33		SK 49	SK 88
						SKW 11	SK 826
							SK 827
							TM 1517*
							TM 1601

(Continued)

Table 1 (Continued)

	M_1	M_2	M_3	M^1	M^2	M^3	P_4
P. boisei	KNM-ER 729	KNM-ER 729	KNM-ER 729*	KNM-ER 733E	KNM-ER 1171H	OH 30*	KNM-ER 729*
	KNM-ER 802	KNM-ER 801	KNM-ER 802	OH 5*	KNM-ER 1171G	CH 1	KNM-ER 802*
	KNM-ER 1171	KNM-ER 802	KNM-ER 810	OH 6	CH 1		KNM-ER 818
	KNM-ER 1509	KNM-ER 1171*	KNM-ER 1509	OH 30*			KNM-ER 1171
	KNM-ER 3230	KNM-ER 1816	KNM-ER 3230*	CH 1			KNM-ER 1816*
	KNM-ER 3890	KNM-ER 3230	Peninj*				KNM-ER 3229*
	Peninj*	Peninj*					KNM-ER 3230*
							KNM-ER 3885
							Peninj*
Early Homo	OH 7*	OH 7*	OH 4	KNM-ER 1590L*	OH 13*	OH 13*	KNM-ER 992*
(H. habilis	OH 13	OH 13	OH 13*	OH 13*	OH 16*	OH 16*	KNM-ER 3734
H. rudolfensis)	OH 16	OH 16	OH 16*	OH 16	KNM-ER 1590L	OH 24	OH 7*
	KNM-ER 806	KNM-ER 806	KNM-ER 730	OH 24*			OH 13*
	KNM-ER 809	KNM-ER 992*	KNM-ER 806				OH 16*
	KNM-ER 820*		KNM-ER 992*				
	KNM-ER 992*						
	KNM-ER 1502						
	KNM-ER 1507						
	KNM-ER 3734						

*indicates both antimeres were used in the analysis

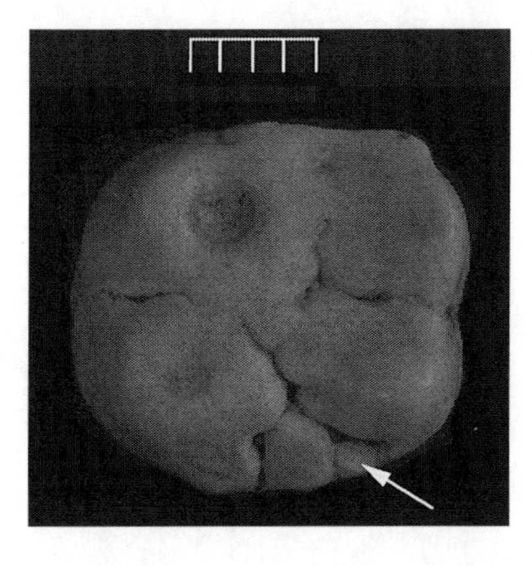

Figure 1. Maxillary molar showing expression of Cusp 6 in *Au. afarensis* (AL 200-1a).

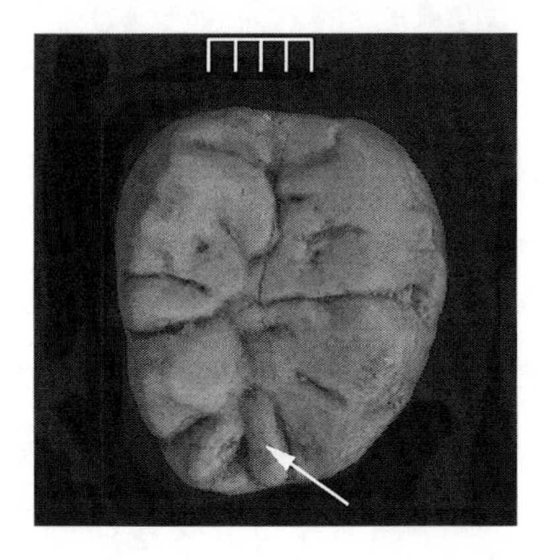

Figure 2. Mandibular molar showing expression of "double" Cusp 6 in *Au. afarensis* (AL 400).

Au. africanus, *P. robustus*, *P. boisei*, and early *Homo* had been collected previously (Wood and Uytterschaut, 1987). Morphological data for *P. boisei* and early *Homo* (*H. habilis* and *H. rudolfensis*) were also collected previously (Wood and Abbott, 1983), while morphological data for *Au. africanus* and *P. robustus* were collected by SEB.

The combination of published data with that collected for this study by SEB naturally introduces some degree of interobserver error. However, we believe that our choice of traits limited the potential for error to some extent. Traits like Cusp number (e.g., Cusp 6, Cusp 7) appear to have a higher scorability index than traits like incisor shoveling where the distinction between presence and absence may be more subjective (Nichol and Turner, 1986). With the exception of Carabelli's trait, all the traits used in this study are of this type. Sofaer et al. (1972) found presence/absence concordance between observers for Carabelli's trait to be more problematic. Still, we believe the detailed descriptions outlining how morphology was scored in Wood and Abbott (1983) and Wood and Engleman (1988), allowed us to generate the appropriate ASUDAS scores. Regarding the interobserver error for the metrical portion of this study, we are fairly certain that it would have be low. A recent study suggests that interobserver error of relative cusp area measurements does not differ significantly from intraobserver error, even when methods for image capture differ (Bailey et al., 2004).

Inter-Trait Correlations

Although Scott and Turner (1997) suggest using only one tooth from each tooth district in statistical analyses, in this study we use traits expressed on all three molars (with the exception of UM Carabelli's trait and UM Cusp 5 – see Table 2). Therefore, a brief discussion of inter-trait correlations is warranted here. Inter-trait correlations can occur between different teeth in the same tooth district (e.g., Carabelli's trait on M^1 and M^2) and between different traits on the same or different teeth (e.g., Carabelli's trait on M^1 and hypocone on M^1). Scott and Turner (1997) have found that the former produces most of the significant inter-trait correlations in modern populations. Of the traits used in this study, the highest inter-class correlations (between the same tooth district, e.g., M^1–M^2) appear to be for Carabelli's trait ($r = 0.31$–0.60) and UM Cusp 5 ($r = 0.41$–0.59) (Scott and Turner, 1997). And lower inter-class correlations are found for Cusp 6 (0.35, Sofaer et al., 1972) and for Cusp 7 (0.23–0.32, Scott and Turner, 1997). Inter-class correlations for the two other traits

Table 2. Trait frequencies

Traits	*Au. afarensis*		*Au. africanus*		*Early Homo*		*P. robustus*		*P. boisei*		*P. troglodytes*		*P. paniscus*	
	Present	*%*	*Present*	*%*	*Present*	*%*	*Present*	*%*	*Present*	*%*	*Present*	*%*	*Present*	*%*
Maxilla														
M^1 Carabelli's trait	4/6	66.7	7/7	100	0/4	0.0	5/5	100	3/3	100	33/35	94.3	19/24	79.2
M^1 Cusp 5	3/7	42.8	6/9	66.7	0/3	0.0	5/6	83.3	3/3	100	4/33	12.1	2/30	6.7
M^1 Cusp 6 +	0/7	0.0	0/8	0.0	0/3	0.0	1/6	16.7	0/2	0.0	0/33	0.0	0/30	0.0
M^2 Carabelli's trait +	1/4	25.0	11/12	91.7	2/3	66.7	2/4	50.0	2/2	100	27/30	90.0	18/26	69.2
M^2 Cusp 5	2/4	50.0	11/12	91.7	3/3	100	3/4	75.0	3/3	100	6/28	21.4	3/25	12.0
M^2 Cusp 6 +	1/4	25.0	3/12	25.0	0/3	0.0	0/4	0.0	0/2	0.0	1/27	3.7	0/25	0.0
M^3 Carabelli's trait +	1/3	33.3	7/8	87.5	1/3	33.3	6/6	100	2/2	100	11/13	84.6	5/11	45.5
M^3 Cusp 5 +	3/3	100	8/8	100	2/2	0.0	5/6	83.3	3/3	100	6/14	42.9	3/12	25.0
M^3 Cusp 6 +	3/3	100	3/8	37.5	0/2	0.0	2/5	40.0	0/2	0.0	0/14	0.0	0/12	0.0
Mandible														
M_1 Cusp 6	2/8	25	2/9	22.2	0/11	0.0	6/6	100	5/5	100	6/37	16.1	0/33	0.0
M_1 'double' Cusp 6+	0/9	0.0	0/9	0.0	0/11	0.0	0/6	0.0	1/2	50.0	1/29	3.4	0/28	0.0
M_1 Cusp 7	0/8	0.0	2/10	20.0	3/12	25.0	1/8	12.5	0/4	0.0	0/37	0.0	3/32	9.3
M_2 Cusp 6	5/12	41.7	2/8	25.0	2/6	33.3	6/7	85.7	9/9	100	3/31	9.7	0/29	0.0
M_2 'double' Cusp 6	0/12	0.0	0/8	0.0	0/6	0.0	1/7	14.3	1/6	16.7	1/29	3.4	0/28	0.0
M_2 Cusp 7	1/12	8.3	3/8	37.5	3/6	50.0	1/8	12.5	0/5	0.0	0/32	0.0	1/29	3.4
M_3 Cusp 6	6/6	100	6/6	100	5/9	55.6	4/4	100	8/8	100	3/18	16.7	1/12	8.3
M_3 'double' Cusp 6	3/6	50.0	2/6	33.3	0/9	0.0	0/4	0.0	4/9	44.4	0/18	0.0	0/12	0.0
M_3 Cusp 7	3/6	50.0	3/5	60.0	2/9	22.2	0/4	0.0	4/9	44.4	0/16	0.0	0/13	0.0

+ indicates trait was not used in MMD analysis because sample size for one of the groups < 3. Relative premolar area not used in the MMD analysis but are used in a later comparison.

used in this study – UM Cusp 6, LM double Cusp 6 – are not known. We were unable to test whether the same inter-class correlations in modern humans apply to our sample of Pliocene hominins because (i) few individuals were represented by all three molars, (ii) most teeth were found in an isolated context (not in their associated jaw); and (iii) because with samples of 3 to 7 individuals any correlations that do exist are not likely to be significant.

Regarding the second type of inter-trait correlation (different traits on the same or different teeth), data suggest that only two inter-trait correlations are significant in modern populations (Scott and Turner, 1997). These occur between M^1 Carabelli's trait and M^1, M^2 hypocone and between M^1, M^2 Carabelli's trait and M_1, M_2 protostylid. Neither hypocone size nor protostylid expression were used in the study. Whether the other traits (UM Cusp 6 and LM double Cusp 6) are significantly correlated with one another or other traits awaits further study with larger sample sizes.

In the end, we decided to use all three molars in the analysis for several reasons. First, the presence or absence of traits on different teeth in the same tooth district may be informative. For example, in some groups certain traits are present only on the M3 (e.g., LM Cusp 7 in *P. boisei*). If we exclude the M3 from the analysis we may miss out on potentially taxonomically important variation. Second, we lack knowledge about inter-correlations between dental traits in *Pan* and Plio-Pleistocene hominins and should not assume they are the same as in contemporary humans. Finally, we suspect that different patterns in inter-district correlations may carry taxonomic information.

Data Collection

In addition to the discrete molar crown traits mentioned above, we assessed talonid expansion in the P_4. Measurements of the P_4 crown area and its component parts (protoconid, metaconid, talonid) in *Au. africanus*, *P. boisei* and early *Homo* were taken from occlusal photographs as described in Wood and Uytterschaut (1987). The method for image capture is described in Wood and Abbott (1983). Briefly, teeth were positioned so that their cervical plane was perpendicular to the optical axis of the camera. A Nikkor Microlens system was used to take 1 to 1 images of each tooth. Measurements were then taken from 7x enlarged photographic prints. SEB used a Nikon CoolPix 950 camera to take digital images of *Au. afarensis* and *Pan* P_4s. The teeth were oriented in the same manner as described above, and a scale placed at the level of the cusp tips was included in each image. Downloaded images were analyzed using SigmaScan Pro 5.0® software. The scale was used to calibrate the software and cusp areas were measured directly from the digital images in the same way that is outlined by Wood and Uytterschaut (1987) (see Figure 3).

Males and females were pooled in the analyses primarily because it is not possible to dichotomize the fossil data by sex. This was not deemed to be a problem since studies have shown that for most crown traits there is little

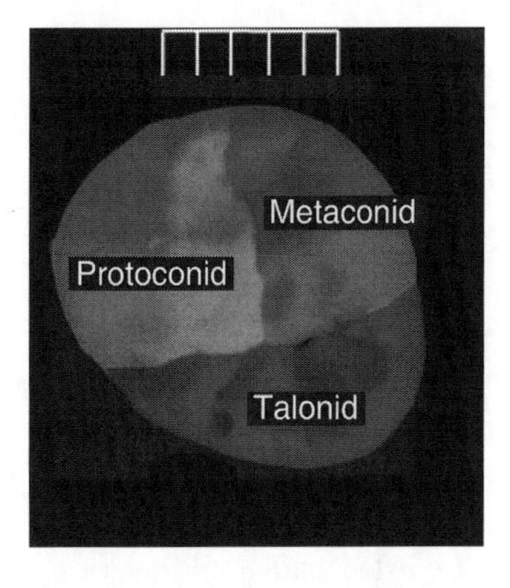

Figure 3. Figure showing how crown component areas were divided up and measured in the analysis.

sex dimorphism in trait expression (Scott and Turner, 1997). For data collected by SEB, if both sides of the dentition were present, both were scored but only the side showing the strongest expression for a given trait was used in the analysis. This approach assumes that the tooth with the highest expression reflects that individual's genetic potential for a given trait (Turner and Scott, 1977). However, the fossil samples comprised many isolated teeth. Therefore, when antimeres were not available, data were scored on whichever tooth was preserved. For data collected from the literature (*Paranthropus* and early *Homo*), if both antimeres were present both were used in the analysis (Wood and Abbott, 1983; Wood and Engleman, 1988). A similar procedure applies to the P4 crown component analysis. For data collected by SEB, if both antimeres were preserved, only one side (the left) was used in the analysis. If the left side was missing the right side was used instead. The *Paranthropus* and early *Homo* morphology collected from the literature (Wood and Uytterschaut, 1987) include both antimeres, when present. Table 1 provides a list of specimens for which antimeres were used. Although this may artificially inflate sample sizes, considering the marked differences we observed between samples, in the end we believe it did not distort the larger picture (see Discussion section).

Statistical Analysis

We used the Mean Measure of Divergence statistic to assess dental phenetic similarity. The Mean Measure of Divergence (MMD) is a multivariate statistic that utilizes multiple traits to provide a relative dissimilarity among groups. It is important, however, to remember that the results (distances) apply only to the present trait list because adding or excluding traits in the analysis may change the resulting MMD value (e.g., 26- vs. 12-trait study in Bailey, 2002b). Potential problems arise when using small sample sizes in MMD analysis (see Harris, 2004 for a review). One of these is the risk of obtaining MMD values that are 0 or negative because the correction factor is larger than the MMD; the smaller the sample size the larger the correction factor. We attempted to control for this by only including traits in the analysis where at least two groups differed by a minimum of 20%, and by using samples consisting of a minimum of three individuals. Traits on several teeth, therefore, were not included in the analysis (Carabelli's trait on M^2 and M^3, UM Cusp 6, and Cusp 5 on M^3, see Table 2). In addition, although the MMD program utilizes the Freeman and Tukey angular transformation to correct for small sample sizes (Berry and Berry, 1967; Sjøvold, 1973; Green and Suchey, 1976), our sample sizes of 3 to 7 individuals are likely too small even for this correction. Therefore, the statistical significance of the differences among groups should be interpreted cautiously.

Typically, when phenetic analyses such as the MMD are applied to the dentition, only one tooth in a series (e.g., incisors, premolars, molars) is used in the analysis. This is done to avoid redundancy caused by inter-tooth trait correlations. Often, but not always, this tooth is the "key tooth" of a series (Dahlberg, 1945) i.e., the most stable according to Butler's (1939) field concept. However, as mentioned earlier, in this study the analysis of molar occlusal morphology is not limited to a single tooth in the molar district (e.g., M1). Although the MMD statistic does not "deal well" with intercorrelated traits, the small sample sizes – which preclude the construction of a correlation matrix – prevented our use of an analytical method that may have dealt with inter-trait correlations better (e.g., Mahalanobis D^2). In sum, given the potential problems with sample sizes and possible inter-district trait correlations, the results should not be over-interpreted.

Clustering analyses using Ward's method and multidimensional scaling were used to visualize the phenetic distances among the samples. Ward's method is the clustering algorithm generally preferred by dental anthropologists because it has been shown that the clusters produced conform to known population relationships based on other (e.g., genetic) data. This method bases cluster membership on the total sum of squared deviations from the mean of the cluster. It joins the two clusters that will result in the smallest increase in the pooled within-cluster variation (Ward, 1963).

Results

Mean Measure of Divergence

In the MMD analysis the resulting distances provide a measure of difference in molar crown morphology. The results of this analysis revealed that species within the same genus are similar to one another. In fact, none of the intrageneric comparisons are statistically significant and MMD values are relatively low (Table 3). Moreover, the intergeneric differences between *Paranthropus* and *Australopithecus* are between 20 and 60 times greater than those within *Australopithecus* and *Paranthropus* genera, respectively. The cluster analysis (Figure 4) and multidimensional scaling results (Figure 5) illustrate these close intrageneric and more distant intergeneric relationships. The MMD values are slightly lower for the *Au. afarensis – Paranthropus* comparisons than for the *Au. africanus – Paranthropus* comparisons (Table 3). However, the implications of this should be interpreted with caution since an examination of trait frequencies (Table 2) shows that *Au. africanus* is more similar to *Paranthropus* species in upper molar trait frequencies but *Au. afarensis* is more similar in lower molar trait frequencies. Because of sample size issues, the MMD analysis is dominated by lower molar traits, which may explain why the MMD value for the *Au. afarensis – Paranthropus* comparison is lower.

Early *Homo* differs significantly from all non-*Homo* samples. The MMD values for early *Homo–Australopithecus* and *Paranthropus–Australopithecus* comparisons are similar (but not identical) in magnitude (Table 3); however, early *Homo* is more than twice the distance from *Paranthropus* species as it is from the australopiths. The cluster analysis (Figure 4) shows early *Homo* linking more closely with the australopiths and more distantly with *Paranthropus*. When the MMD data are visualized using multidimensional scaling in three dimensions (Figure 5) this relationship is also supported, although it is clear that early *Homo* is still quite different from the australopiths.

All early hominin taxa are quite different from *Pan* but the MMD values between *Paranthropus* and *Pan* are two to three

Table 3. MMD values for multiple comparisons based on 11 mandibular tooth crown traits

	Au. afarensis	*Au. africanus*	*P. robustus*	*P. boisei*	*Early Homo*	*P. t. troglodytes*	*P. paniscus*
Au. afarensis	0	0.007	**0.423**	**0.427**	**0.524**	**0.683**	**0.888**
Au. africanus		0	**0.324**	**0.416**	**0.668**	**0.981**	**1.192**
P. robustus			0	0.011	**1.323**	**1.102**	**1.679**
P. boisei				0	**1.698**	**1.723**	**2.431**
H. habilis					0	**0.902**	**0.764**
P. t. troglodytes						0	0.045
P. paniscus							0

Values in bold are statistically significant at $p < .025$.

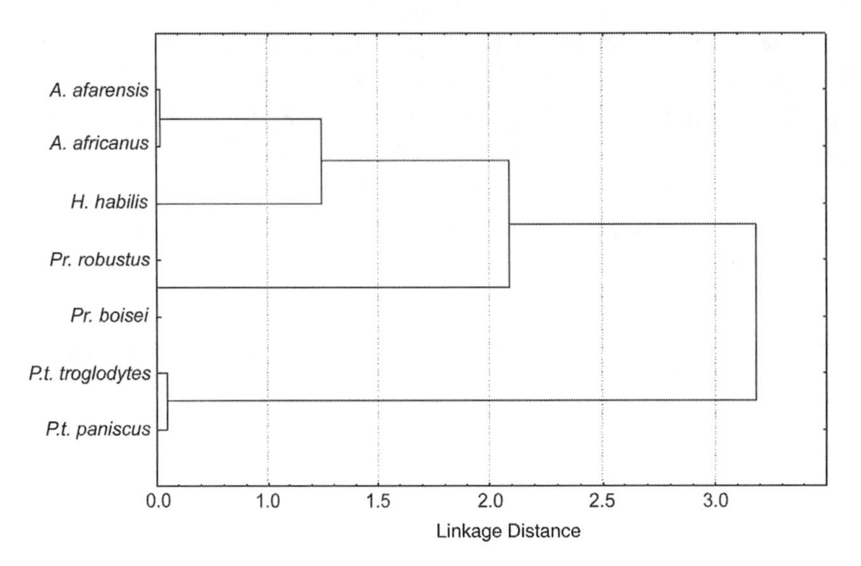

Figure 4. Cluster analysis of MMD distances using Ward's method.

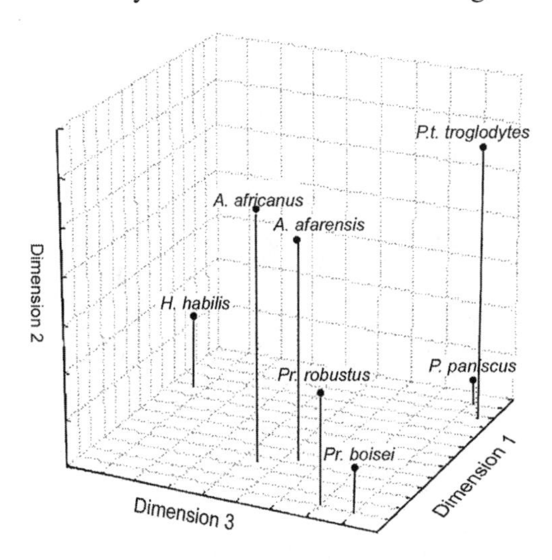

Figure 5. Multidimensional scaling scatter plot showing the results of the MMD analysis of seven groups in three dimensional space.

times those observed between *Pan* and the other hominin taxa and five to seven times those observed between *Paranthropus* and the two *Australopithecus* species. The interspecific differences between the two *Pan* species are approximately two to four times greater than the interspecific differences within the australopiths and between the *Paranthropus* taxa.

Because of small sample size issues outlined above, we conducted a second MMD analysis using pooled samples of *Australopithecus* and *Paranthropus*. The results of this analysis are similar to the first (Tables 4 and 5), and the larger picture does not change. All intergeneric distances are large and significant. The distance between *Australopithecus* and the two *Pan* species is two to two-and-a-half times larger than the distance between the two hominin species. The distance between *Paranthropus* and the two *Pan* species is even larger – three-and-a-half to five times that between the two hominin species. A cluster analysis visualizing these differences illustrates

Table 4. Pooled trait frequencies for Australopithecus and Paranthropus

Traits	*Australopithecus*		*Paranthropus*	
	Present	*%*	*Present*	*%*
Maxilla				
M^1 Carabelli's trait	11/13	84.6	8/8	100
M^1 Cusp 5	9/16	56.3	8/9	83.3
M^1 Cusp 6	0/15	0.0	1/8	16.7
M^2 Carabelli's trait	12/16	75.0	4/6	50.0
M^2 Cusp 5	13/16	81.3	6/7	75.0
M^2 Cusp 6	4/16	25.0	0/6	0.0
M^3 Carabelli's trait	8/11	72.7	8/8	100
M^3 Cusp 5	11/11	100	8/9	83.3
M^3 Cusp 6	6/11	54.5	2/7	40.0
Mandible				
M_1 Cusp 6	4/17	23.5	11/11	100
M_1 'double' Cusp 6	0/18	0.0	0/8	0.0
M_1 Cusp 7	2/18	11.1	1/12	12.5
M_2 Cusp 6	7/20	35.0	15/16	85.7
M_2 'double' Cusp 6	0/20	0.0	2/13	14.3
M_2 Cusp 7	4/20	20.0	1/13	12.5
M_3 Cusp 6	12/12	100	12/12	100
M_3 'double' Cusp 6	5/12	41.7	4/13	0.0
M_3 Cusp 7	6/11	54.5	4/13	0.0

Table 5. MMD values for multiple comparisons of pooled Australopithecus and Paranthropus samples based on 11 mandibular tooth crown traits

	Australopithecus	*Paranthropus*	*P. t. troglodytes*	*P. paniscus*
Australopithecus	0	**0.425**	**0.916**	**1.121**
Paranthropus		0	**1.530**	**2.128**
P. t. troglodytes			0	0.045
P. paniscus				0

Values in bold are statistically significant at $p < .025$.

the marked differences between hominins and *Pan* for the molar traits (Figure 6).

Premolar Morphology

The expanded talonid said to characterize *Paranthropus* taxa is difficult to assess in terms of presence or absence, as is necessary for MMD analysis. Instead, we assessed talonid expansion by measuring the contribution of the two main cusps and the talonid to total crown area. Wood and Uytterschaut (1987) previously analyzed premolar crown areas in Plio-Pleistocene hominins, but did not include *Pan* in their analysis. Measuring the P_3 talonid in a homologous way in *Australopithecus* and *Pan* proved to be impossible, so we limited our analysis to the P_4.

Results indicate that all of the early hominin P_4s sampled possess significantly larger measured crown and talonid areas than the two *Pan* species (Table 6). Moreover, the relative talonid size in *Au. afarensis*, *P. boisei*, *P. robustus*, but not in early *Homo*, was

Table 6. Descriptive statistics for absolute and relative P_4 crown and talonid area[1]

Taxa	*n*	*Mean crown area (range)*	*Mean measured talonid area (range)*	*Mean relative talonid area (range)*
P.t. troglodytes[2]	26	54.5 (42.7–66.9)	14.0 (7.3–20.1)	25.5 (17.2–34.1)
P. paniscus	28	42.7 (34.3–53.8)	12.1 (6.6–17.9)	28.1 (18.3–34.8)
Au. afarensis	10	82.8* (60–101.5)	24.3* (19.1–34.9)*	30.6* (27.2–36.4)
Au. africanus [3]	2	106.7 (100.7, 112.8)	31.4 (27.2, 35.6)	29.3 (27.0, 31.6)
P. robustus	6	123.7* (102–135.9)	43.7* (33.6–50.5) *	36.0* (26.9–40.4)
P. boisei	8	169.4* (136.1–190.8)	68.4* (56.2–76.7)*	40.5* (37.5–44.1)
H. habilis	7	78.3* (55.1–88.9)	22.8* (14.7–28.5)*	27.8 (19.9–33.1)

*designates value is significantly different from *Pan* after using Bonferroni correction for multiple tests.
[1]Hominin (except *Au. afarensis*) data from Wood and Uytterschaut (1987), *Pan* and *Au. afarensis* data collected by SEB.
[2]*P. paniscus* and *P.t. troglodytes* are significantly different (after correction) in total crown area only.
[3]*Au. africanus* sample too small to test for significance.

significantly larger than *Pan* (Table 6). Small sample size prevented us from including *Au. africanus* in this part of the study .

Discussion

Although they may not agree about the alpha taxonomy of the hominin fossil record, or about the relative importance of anagenesis and cladogenesis in hominin evolution, most paleoanthropologists agree on one aspect of macroevolution within the hominin clade. This is that between three and two million years ago a cladogenetic event resulted in at least two subclades within the hominin clade. One of these subclades includes modern humans and subsumes the taxa most researchers include in the genus *Homo*. The other subsumes taxa informally referred to as "robust" australopiths, which we include in the genus *Paranthropus*. This contribution uses some of the detailed dental evidence for the relevant taxa in order to address four questions involving their affinities and relationships.

The first question we set out to address was "Do metrical and non-metrical dental data support a distinction between *Paranthropus* and early *Homo*?" The results of this study confirm that there are clear metrical and non-metrical dental differences between *Paranthropus* and early *Homo*. These differences are manifest in relative expansion of the P_4 talonid, in different trait frequencies and the patterns in which they are expressed.

While *Australopithecus*, *Paranthropus* and early *Homo* are all quite different from *Pan*, they differ in distinctive ways and *Paranthropus* is by far the most distinctive of the hominins. *Paranthropus* (and to some degree the australopiths) departs from *Pan* molar morphology in having high frequencies of crown-additive traits such as accessory molar cusps and expanded premolar talonids. Early *Homo*, on the other hand, departs from the *Pan* morphological pattern by lacking several primitive traits, such as UM anterior fovea and LM posterior fovea (Bailey, unpublished data). In contrast to *Paranthropus*, *Pan* and early *Homo* both have relatively low frequencies for traits that add to crown complexity. In addition, while the overall size of the P_4 crown is larger in early *Homo* than in *Pan*, its relative cusp areas are similar to those seen in *Pan*. It is only later in the *Homo* clade that the relative size of the P_4 talonid reduces. (for example, the mean relative talonid area for a sample of five *Homo ergaster/erectus* specimens is 20.7[1] see also (Martinon-Torres et al., 2006). And it is much later e.g., in *Homo sapiens*) that that talonid is sometimes missing completely (Bailey, unpublished data). It would appear

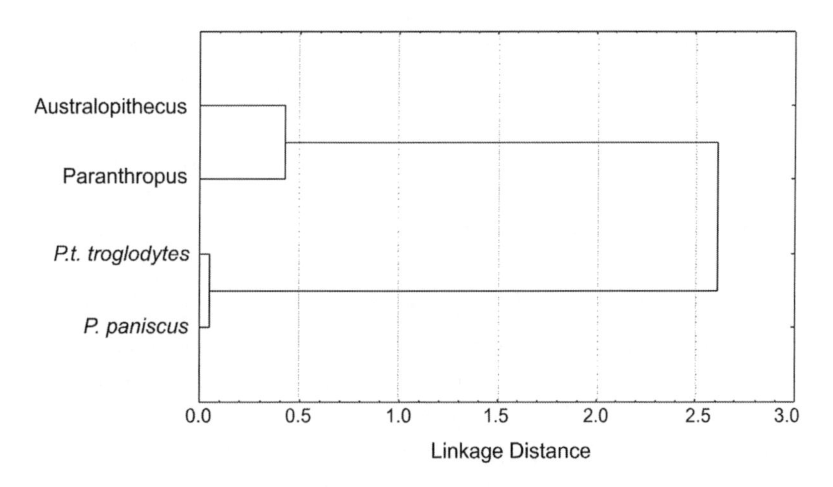

Figure 6. Cluster analysis of MMD distances using Ward's method (pooled samples).

that in some senses early *Homo* is derived in that it has lost some traits (e.g., foveae), but it is more primitive (that is, more *Pan*-like) in its non-metrical molar morphology than is either *Paranthropus* or the taxa within the australopiths. In its absence of a shelf-like Carabelli's structure early *Homo* upper molars also differ from both those of *Pan* and the australopiths, which both tend towards a shelf-like Carabelli's structure rather than the groove and cusp-like form seen in later *Homo*.

Our second question addressed the appropriateness of using a modern human system for scoring dental non-metrical traits to assess the extent of phenetic differences among early hominin taxa. Our results confirm the conclusion of Irish and Guatelli-Steinberg (2003), that using morphological standards based on modern humans to assess fossil hominin variation can lead to unreliable results. For example, in their study of the plesiomorphic nature of sub-Saharan African dental morphology, Irish and Guatelli-Steinberg (2003) found there was relatively little difference between australopiths and *Paranthropus* – less difference, in fact, than between two geographically proximate and presumably closely related modern human populations (S.E. Asia and Polynesia). This finding clearly contrasts with our own results. As possible explanations, they suggested that (i) the ASUDAS may not be able to capture differences among early hominins in which modern human traits are fixed at or near 0% and 100%; and (ii) there may be other traits that distinguish between *Paranthropus* on the one hand, and the australopiths and early *Homo*, on the other, that were not used in their study (Irish and Guatelli-Steinberg, 2003: 134–135). A later study focused only on the M^1, thereby eliminating many traits that are invariable in early hominins. It confirmed that there were significant differences between *Paranthropus* and the australopiths/early *Homo* grouping that were not apparent in their earlier study (Guatelli-Steinberg and Irish, 2005).

In the present study we addressed Irish and Guatelli-Steinberg (2003) concerns by eliminating traits that are invariant in early hominins and by including other (i.e., non-ASUDAS) traits that are known to, or are suspected of being able to, differentiate the early hominin taxa. As a result, we found marked and significant differences in molar crown morphology between the australopiths/early *Homo* and *Paranthropus*. This confirms the principle that in order to be able to uncover the dental morphological differences among early hominins, one cannot rely on traits developed for differentiating among regional populations of modern humans; researchers must use traits that are relevant to the target taxa.

More can be done, however. In addition to the traits included in this study, other molar traits we believe may have potential for assessing phenetic relationships among early hominins are: UM anterior and posterior fovea, UM mesial accessory cuspules, LM posterior fovea, LM mesial marginal cuspules and LM double cusp 7. Many of these are present and variable in extant hominoids (Bailey, unpublished data) and may be particularly useful as characters for cladistic analysis.

In addition to eliminating some traits and adding others, there is a need to reassess and develop new scoring criteria for several traits relevant to the study of early hominins. This need has been already been recognized and to some extent met by Hlusko (2004) and Van Reenen and Reid (1995) who have developed new standards for protostylid and Carabelli's structure variation, respectively. However, other traits need to be revised in the same fashion. For example, the full range of parastyle variation observed in early hominins is not represented by the current scoring system. Neither is much of the morphological variation in *Pan*, which would be important if one wanted to conduct cladistic analyses.

Our third question referred to the apparently derived nature of the *Paranthropus* dentition. Although we did not conduct a formal cladistic analysis the marked differences in molar crown morphology between *Paranthropus* and *Pan* (about twice that between *Paranthropus* and *Au. afarensis*) suggest that the morphological complexity that makes *P. robustus* and *P. boisei* distinct is unlikely to be a primitive retention. The postcanine dentition of *Paranthropus* is no more similar to *Pan* than any of the australopith taxa. The australopith taxa, while still quite different from *Paranthropus*, are more similar to them than they are to *Pan*. This, together with an examination of trait frequencies, suggests that the trend for increasing crown complexity is already present in *Au. afarensis* and *Au. africanus*, although it appears to be manifest differently in the two species.

Without conducting a formal cladistic study using a large suite of dental traits and without including other extant hominoids in the dataset, it is difficult to say for sure whether the non-metrical dental data will correctly reconstruct the known molecular tree for extant higher primates. Phenetic analysis using dental non-metrical traits do correctly reconstruct known molecular relationships among chimpanzee species (Bailey, 2007 and unpublished data), which suggests the potential for such analyses. The biggest challenge is finding dental crown traits that are relevant to assessing the full range of early hominin taxa. In order to do this we need to continue to collect data on all hominoids (not just *Pan*), which will require developing scoring standards that are applicable to all groups. Assessing the extent of homoplasy within metrical and non-metrical measures of occlusal morphology will be a large part of this research agenda, as will be the determination of the primitive and derived nature of important dental traits.

Summary

This study, like many that rely on fossil data, is limited by small sample sizes and our lack of knowledge regarding inter-trait correlations. None-the-less, we have identified a number of interesting results and trends that are worthy of additional study. In addition, we believe that even with its potential weaknesses our analysis shows that dental morphological variation has the potential to address important questions about early hominin evolution when appropriate traits and scoring methods are utilized. A scoring system developed for use on recent modern humans is not sufficient in this regard. We suggest that the development of a new or supplemental system for early hominins would go a long way to helping to answer these questions. This is currently underway.

Acknowledgments

Thanks to W. Kimbel for providing access to casts of *Au. afarensis*, and P. Ungar and F. Grine for access to casts of *A. africanus*, *Paranthropus* and *Homo* specimens. SEB is grateful to W. Van Neer and W. Wendelen from Musée Royal d'Afrique Central and M. Harman from the Powell Cotton Museum for their assistance with the chimpanzee collections. The authors also thank the two anonymous reviewers for their helpful suggestions. Data collection and travel for SEB were funded by The George Washington University CASHP and the Max Planck Institute for Evolutionary Anthropology, Department of Human Evolution.

Note

1. *H.ergaster/erectus* sample includes: Zhoukoudian (Nos. 90, 29 and 89), OH22 and WT15000.

References

Bailey, S.E., 2000. Dental morphological affinities among late Pleistocene and recent humans. Dental Anthropology 14, 1–8.

Bailey, S.E., 2002a. A closer look at Neanderthal postcanine dental morphology. I. The mandibular dentition. New Anatomist 269, 148–156.

Bailey, S.E., 2002b. Neandertal dental morphology: implications for modern human origins. Ph.D. Dissertation, Arizona State University.

Bailey, S.E., 2004. A morphometric analysis of maxillary molar crowns of Middle-Late Pleistocene hominins. Journal of Human Evolution 47, 183–198.

Bailey, S. in press. Inter and intraspecific variation in *Pan* tooth crown morphology: implications for Neandertal taxonomy. In: Irish, J. and Nelson, G. (Eds.), Technique and Application in Dental Anthropology. Cambridge University Press: Cambridge.

Bailey, S.E., Lynch, J.M., 2005. Diagnostic differences in mandibular P4 shape between Neandertals and anatomically modern humans. American Journal of Physical Anthropology 126, 268–277.

Bailey, S.E., Pilbrow, V.C., Wood, B.A., 2004. Interobserver error involved in independent attempts to measure cusp base areas of *Pan* M1s. Journal of Anatomy 205, 323–331.

Berry, A., Berry, R., 1967. Epigenetic variation in the human cranium. Journal of Anatomy 101, 361–379.

Brunet, M., Guy, F., Pilbeam, D., Mackaye, H., Likius, A., Ahounta, D., Beauvilain, A., Blondel, C., Bocherens, H., Boisserie, J., De Bonis, L., Coppens, Y., Dejax, J., Denys, C., Duringer, P., Eisenmann, V., Fanone, G., Fronty, P., Geraads, D., Lehmann, T., Lihoreau, F., Louchart, A., Mahamat, A., Merceron, G., Mouchelin, G., Otero, O., Campomanes, P., De Leon, M., Rage, J., Sapanet, M., Schuster, M., Sudre, J., Tassy, P., Valentin, X., Vignaud, P., Viriot, L., Zazze, A., Zellikefer, C., 2002. A new hominid from the Upper Miocene of Chad, Central Africa. Nature 418, 145–151.

Butler, P.M., 1939. Studies of the mammalian dentition. Differentiation of the post-canine dentition. Proceedings of the Zoological Society of London. B. 109, 1–36.

Collard, M., Wood, B., 2000. How reliable are human phylogenetic hypotheses? Proceedings of the National Academy of Sciences of the USA 97, 5003–5006.

Coppa, A., Dicintip, F., Vargiu, R., Lucci, M., Cucina, A., 2001. Morphological dental traits to reconstruct phenetic relationships between Late Pleistocene-Ancient Holocene human groups from Eurasia and North Africa (Abstract). American Journal of Physical Anthropology 32, 54.

Crummett, T., 1994. The evolution of shovel shaping: regional and remporal variation in human incisor morphology. Ph.D. Dissertation, University of Michigan, Ann Arbor.

Dahlberg, A., 1945. The changing dentition of man. Journal of the American Dental Association 32, 676–680.

Dahlberg, A., 1956. *Materials for the Eestablishment of Standards for Classification of Tooth*

Characteristics, Attributes, and Techniques in Morphological Studies of the Dentition. University of Chicago, Mimeo.

de Bonis, L., Johanson, D., Melentis, J., White, T., 1981. Variations métriques de la denture chez les Hominidés primitifs: comparison entre *Australopithecus afarensis* et *Ouranopithecus macedoniensis*. Comptes Rendus de l'Academie des Sciences, Paris, D 292, 373–376.

Green, R.F., Suchey, J.M., 1976. The use of inverse sign transformations in the analysis of non-metric cranial data. American Journal of Physical Anthropology 50, 629–634.

Gregory, W., 1916. Studies on the evolution of the Primates. I. The Cope-Osborn "theory of trituberculy' and the ancestral molar patterns of the Primates. Bulletin of the American Museum of Natural History 35, 239–257.

Grine, F., 1985. Australopithecine evolution: the deciduous dental evidence. In: Delson, E. (Ed.), Ancestors: The Hard Evidence. Alan R. Liss, Inc., New York, pp. 153–167.

Guatelli-Steinberg, D., Irish, J.D., 2005. Early hominin variability in first molar dental trait frequencies. American Journal of Physical Anthropology 128, 477–484.

Haeussler, A., 1995. Upper Paleolithic teeth from the Kostenki sites on the Don River, Russia. In: Moggi-Cecchi, J. (Ed.), Aspects of Dental Biology: Palaeontology, Anthropology and Evolution. International Institute for the Study of Man., Florence, pp. 315–332.

Haeussler, A., 1996. Biological relationships of Late Pleistocene and Holocene Eurasian and American peoples: the dental anthropological evidence. Ph.D. Dissertation, Arizona State University.

Haeussler, A.M., Turner, C.G., II, 1992. The dentition of Soviet Central Asians and the quest for New World ancestors. Journal of Human Ecology Special Issue 2, 273–297.

Haile-Selassie, Y., Asfaw, B., White, T.D., 2004. Hominid cranial remains from Upper Pleistocene deposits at Aduma, Middle Awash, Ethiopia. American Journal of Physical Anthropology 123, 1–10.

Hanihara, K., 1961. Criteria for classification of crown characters of the human deciduous dentition. Zinriugaku Zassi 69, 27–45.

Harris, E., 1977. Anthropologic and genetic aspects of the dental morphology of Solomon Islanders, Melanesia. Ph.D. Dissertation, Arizona State University.

Harris, E., 2004. Calculation of Smith's Mean Measure of Divergence for intergroup comparisons using nonmetric data. Dental Anthropology 17, 83–93.

Harris, E., Bailit, H., 1980. The metaconule: a morphologic and familial analysis of a molar cusp in humans. American Journal of Physical Anthropology 53, 349–358.

Hartman, S., 1988. A cladistic analysis of hominoid molars. Journal of Human Evolution 17, 489–502.

Hawkey, D., 1998. Out of Asia: dental evidence for affinities and microevolution of early populations from India/Sri Lanka. Ph.D. Dissertation, Arizona State University.

Hlusko, L.J., 2002. Expression types for two cercopithecoid dental traits (interconulus and interconulid) and their variation in a modern baboon population. International Journal of Primatology 23, 1309–1318.

Hlusko, L.J., 2004. Protostylid variation in *Australopithecus*. Journal of Human Evolution 46, 579–594.

Hrdlička, A., 1920. Shovel-shaped teeth. American Journal of Physical Anthropology 3, 429–465.

Irish, J., 1993. Biological affinities of Late Pleistocene through modern African Aboriginal populations: the dental evidence. Ph.D. Dissertation, Arizona State University.

Irish, J., 1998. Ancestral dental traits in recent sub-Saharan Africans and the origins of modern humans. Journal of Human Evolution 34, 81–98.

Irish, J.D., Guatelli-Steinberg, D., 2003. Ancient teeth and modern human origins: an expanded comparison of African Plio-Pleistocene and recent world dental samples. Journal of Human Evolution 45, 113–144.

Lipschultz, J.G., 1997. Who were the Natufians? A dental assessment of their population affinities. Master's Thesis, Arizona State University.

Martinón-Torres, M., Bastir, M., Bermudez de Castro, J.M., Gomez, A., Sarmiento, A., Muela, A., Arsuaga, J.L., 2006. Hominin lower second premolar morphology: evolutionary inferences through geometric morphometric analysis. Journal of Human Evolution 50, 523-533.

Morris, D.H., 1975. Bushmen maxillary canine polymorphism. South African Journal of Science 71, 333–335.

Morris, D.H., Glasstone Hughes., S., Dahlberg, A.A., 1978. Uto-Aztecan premolar: the anthropology of a dental trait. In: Butler, P., Joysey, K. (Eds.),

Development, Function and Evolution of Teeth. Academic Press, New York, pp. 69–79.

Nichol, C.R., Turner, C.G., II, 1986. Intra-and inter-observer concordance in classifying dental morphology. American Journal of Physical Anthropology 69, 299–315.

Nichol, C.R., Turner, C.G., II, Dahlberg, A.A., 1984. Variation in the convexity of the human maxillary incisor labial surface. American Journal of Physical Anthropology 63, 361–370.

Pilbrow, V., 2003. Dental variation in African apes with implications for understanding patterns of variation in species of fossil apes. Ph.D. Dissertation, New York University.

Reid, C., Van Reenen, J., 1995. The Carabelli trait in early South African hominids: a morphometrics study. In: Moggi-Cecchi, J. (Ed.), Aspects of Dental Biology: Paleontology, Anthropology and Evolution. International Institute for the Study of Man, Florence, pp. 299–304.

Robinson, J., 1954a. The genera and species of the Australopithecinae. In: Howells, W. (Ed.), Ideas on Human Evolution. Harvard University Press, Cambridge, pp. 268–278.

Robinson, J., 1954b. Prehominid dentition and hominid evolution. Evolution 8, 324–334.

Robinson, J., 1956. *The Dentition of the Australopithecinae*. Transvaal Museum, Pretoria.

Ruvolo, M., 1997. Molecular phylogeny of the hominoids: inferences from multiple independent DNA sequence data sets. Molecular and Biological Evolution 14, 248–265.

Scott, G.R., 1973. Dental morphology: a genetic study of American White families and variation in living Southwest Indians. Ph.D. Dissertation, Arizona State University.

Senut, B., Pickford, M., Gommery, D., Mein, P., Cheboi, K., Coppens, Y., 2001. First hominid from the Miocene (Lukeino Formation, Kenya). Comptes Rendus de l'Academie des Sciences 332, 137–144.

Sjøvold, T., 1973. The occurrence of minor non-metrical variants in the skeleton and their quantitative treatment for population comparisons. Homo 24, 204–233.

Sofaer, J.A., Niswander, J., MacLean, C.J., Workman,P., 1972. Population studies on Southwestern Indian tribes. V. Tooth morphology as an indicator of biological distance. American Journal of Physical Anthropology 7, 357–366.

Suwa, G., 1990. A comparative analysis of hominid dental remains from the Shungura and Usno Formations, Omo Valley, Ethiopia. Ph.D. Dissertation, University of California at Berkeley.

Tobias, P. (Ed.). 1967. The Cranium and Maxillary Dentition of *Australopithecus* (*Zinjanthropus*) *boisei*. Cambridge University Press, Cambridge.

Tobias, P., 1991. *Olduvai Gorge, Volume IV. The Skulls, Endocasts and Teeth of Homo habilis*. Cambridge University Press, Cambridge.

Turner, C.G., II, 1992a. The dental bridge between Australia and Asia: following Macintosh into the East Asian hearth of humanity. Perspectives in Human Biology 2/Archaeology in Oceania 27, 143–152.

Turner, C.G., II, 1992b. Microevolution of East Asian and European populations: a dental perspective. In: Akazawa, T., Aoki, K., Kimura, T. (Eds.), The Evolution and Dispersal of Modern Humans in Asia. Hokusen-Sha Pub. Co., Tokyo, pp. 415–438.

Turner, C.G., II, 1992c. Sundadonty and Sinodonty in Japan: the dental basis for a dual origin hypothesis for the peopling of the Japanese Islands. In: Hanihara, K. (Ed.), International Symposium on Japanese as a Member of the Asian and Pacific Populations. International Research Center for Japanese Studies, Kyoto, pp. 96–112.

Turner, C.G., II, Nichol, C.R., Scott, G.R., 1991. Scoring procedures for key morphological traits of the permanent dentition: The Arizona State University Dental Anthropology System. In: Kelley, M. and Larsen, C. (Eds.), Advances in Dental Anthropology. Wiley Liss, New York, pp. 13–31.

Turner, C.G., II, Scott, G.R., 1977. Dentition of Easter Islanders. In: Dahlberg, A., Graber, T. (Eds.), Orofacial Growth and Development. Mouton Publishers, The Hague, pp. 229–249.

Van Reenen, J., Reid, C., 1995. The Carabelli trait in early South African hominids: a morphological study. In: Moggi-Cecchi, J. (Ed.), Aspects of Dental Biology: Palaeontology, Anthropology and Evolution. International Institute for the Study of Man, Florence, pp. 291–298.

Ward, J., 1963. Hierarchial groupings to optimize an objective function. Journal of the American Statistical Association 58, 236–244.

Weidenreich, F., 1937. The dentition of *Sinanthropus pekenensis*: a comparative odontography of the hominids. Paleontologia Sinica. n.s. D, 1–180.

White, T., Johanson, D., Kimbel, W., 1981. *Australopithecus africanus*: its phyletic position reconsidered. South African Journal of Science 77, 445–470.

White, T., Suwa, G., Asfaw, B., 1994. *Australopithecus ramidus*, a new species of early hominid from Aramis, Ethiopia. Nature 371, 306–312.

White, T., Suwa, G., Asfaw, B., 1995. *Australopithecus ramidus*, a new species of early hominid from Aramis, Ethiopia. Corrigendum. Nature 375, 88.

Wood, B.A., 1991. Koobi Fora Research Project Volume 4: Hominid Cranial Remains. Clarendon Press, Oxford.

Wood, B.A., Abbott, S.A., 1983. Analysis of the dental morphology of Plio-Pleistocene hominids. I. Mandibular molars: crown area measurements and morphological traits. Journal of Anatomy 136, 197–219.

Wood, B.A., Abbott, S.A., Graham, S.H., 1983. Analysis of the dental morphology of Plio-Pleistocene hominids. II. Mandibular molars – study of cusp areas, fissure pattern and cross sectional shape of the crown. Journal of Anatomy 137, 287–314.

Wood, B.A., Engleman, C.A., 1988. Analysis of the dental morphology of Plio-Pleistocene hominids. V. Maxillary postcanine tooth morphology. Journal of Anatomy 161, 1–35.

Wood, B.A., Stack, C.G., 1980. Does allometry explain the differences between "gracile" and "robust" australopithecines? American Journal of Physical Anthropology 52, 55–62.

Wood, B.A., Uytterschaut, H., 1987. Analysis of the dental morphology of Plio-Pleistocene hominids. III. Mandibular premolar crowns. Journal of Anatomy 154, 121–156.

4. Maxillary molars cusp morphology of South African australopithecines

J. MOGGI-CECCHI
Laboratori di Antropologia
Dipartimento di Biologia Animale e Genetica
Università di Firenze
Via del Proconsolo, 12
50122 Firenze
Italy
jacopo@unifi.it
and
Institute for Human Evolution
University of the Witwatersrand
Johannesburg, South Africa

S. BOCCONE
Laboratori di Antropologia
Dipartimento di Biologia Animale e Genetica
Università di Firenze
Via del Proconsolo, 12
50122 Firenze
Italy
boccone@unifi.it

Keywords: maxillary molars, cusp areas, *A. africanus, A. robustus*, Australopithecinae, protocone, paracone, metacone, hypocone

Abstract

The South African Plio-Pleistocene sites where large numbers of fossil hominid specimens have been discovered in the last 20 years are Sterkfontein, Swartkrans and, most recently, Drimolen. Hominid specimens recovered from these sites have usually been attributed to *A. africanus* (from Sterkfontein), *A. robustus* (Swartkrans, Drimolen and Sterkfontein) and South African early *Homo* (Swartkrans, Drimolen and Sterkfontein). We recently started a research project aimed at characterizing cheek teeth cusp morphology of South African Australopithecinae employing digital photographs of their occlusal surfaces. In this paper an analysis of the basic metrical features of maxillary molar cusp areas and proportions of *A. africanus* and *A. robustus* is presented. We analyzed 92 permanent maxillary molar teeth of South African Australopithecinae. The main results suggest that: a) crown base areas of the three molars are broadly similar in *A. africanus* and *A. robustus*; b) significant differences between the two species in relative cusp areas are evident for the protocone of M^1 (with *A. africanus* larger than *A. robustus*), the paracone of

S.E. Bailey and J.-J. Hublin (Eds.), Dental Perspectives on Human Evolution, 53–64.

M^1, and the protocone of M^2 and M^3 (with *A. robustus* larger than *A. africanus*); c) in the total crown area *A. robustus* shows the sequence $M^1 < M^2 < M^3$ as previously described; d) in *A. africanus* the sequence observed is $M^1 < M^2 > M^3$, as in living apes. This different sequence between *A. africanus* and *A. robustus* appears to be related mostly to differences in mesial cusp size, which in *A. robustus* shows a marked relative expansion from M^1 to M^3. Also, the variability in absolute cusp areas of the *A. africanus* sample seems to be related to the presence of specimens with notably large teeth.

Introduction

The number of fossil hominid specimens recovered from southern African Plio-Pleistocene sites has dramatically increased in the last 20 years. The sites where large numbers of fossil hominid specimens have been discovered include Sterkfontein (e.g., Lockwood and Tobias, 2002; Moggi-Cecchi et al., 2006), Swartkrans (e.g., Brain, 1993) and, most recently, Drimolen (Keyser et al., 2000). Hominid specimens recovered from these sites have usually been attributed to *Australopithecus africanus* (Sterkfontein), *Australopithecus robustus* (Swartkrans, Drimolen and Sterkfontein) and southern African early *Homo* (Swartkrans, Drimolen and Sterkfontein).

Among these, the fossils recovered from the Sterkfontein Formation represent the largest collection of early hominid specimens from a single locality. Hominids from Sterkfontein Member 4 have, with few exceptions, been assigned to *A. africanus*. In recent years, some authors have suggested that, on the basis of the analysis of the cranial anatomy, a few specimens from Sterkfontein Member 4 may represent another taxon (e.g. Clarke, 1988, 1994; Lockwood and Tobias, 2002). However, different studies based on dental metrics have found no evidence for substantial heterogeneity within the Sterkfontein Member 4 hominid dental sample (Suwa, 1990; Wood, 1991a; Calcagno et al., 1999; Moggi-Cecchi, 2003).

It is becoming apparent that analytical studies of the dentition employing traditional linear measurements (mesio-distal and bucco-lingual diameters) may not be the appropriate approach for addressing the issue of morphological variability within the Sterkfontein Member 4 hominid dental sample (Moggi-Cecchi, 2003). For this reason we recently started a research project aiming to characterize the dental morphology of the Sterkfontein hominid sample (in comparison with the other South African fossil hominid species) employing digital photographs of the occlusal surface of the cheek teeth. Although 2D images are just a crude approximation of the complex shape of the tooth crown, they are relatively easy and quick to collect using digital photographs, and they are more informative than the traditional linear measurements. This is because, among other things, they allow measurements of the absolute areas of the individual cusps.

Little work has been carried out on the analysis of cusp areas of the teeth of Plio-Pleistocene hominids since the series of papers by Wood and colleagues two decades ago (Wood and Abbott, 1983; Wood et al., 1983; Wood and Uytterschaut, 1987; Wood and Engleman, 1988). The number of South African dental specimens has vastly increased since then, thus allowing a more detailed analysis of issues pertaining to the intra- and interspecific variability in the fossil samples.

In this paper, we focus on the basic metrical features of maxillary molars cusp areas and proportions of *A. africanus* and *A. robustus*.

Materials and Methods

We analyzed 92 permanent maxillary molar teeth of South African Australopithecinae representing 20 individuals and 15 isolated teeth of *A. africanus* from the sites of Sterkfontein and Makapansgat, and 9 individuals and 30 isolated teeth of *A. robustus* from the sites of Swartkrans, Kromdraai and Drimolen (Table 1). Taxonomic allocation of the specimens for *A. robustus* follows previous studies (e.g., Grine, 1989; Keyser et al, 2000), whereas for *A. africanus*, the specimens recently described from Sterkfontein Member 4 (Moggi-Cecchi et al., 2006), were provisionally considered as belonging to the species *A. africanus*, as all specimens recovered before 1966.

Heavily worn teeth were excluded from the analysis. Teeth in which the fissures between the main cusps were not evident were excluded as well. Each of the selected teeth was positioned with the cusp tips in their approximate anatomical position or, for the worn teeth, with the mesial and the buccal cervical enamel line parallel to the camera lens. A graduated scale was placed next to it, half way between the cusp tips and the cervix. Photographs were taken with a Nikon Coolpix 885 digital camera with a 2048 X 1536 pixel resolution. The images were then stored on a PC and measured with image analysis software (NIH Image J free software) (Boccone, 2004). The intra-observer error was about 2%.

Table 1. Number of maxillary molars of A. africanus *and* A. robustus *analyzed in this study*

	M^1	M^2	M^3
A. africanus			
Sterkfontein	13	16	9
Makapansgat	1	1	1
A. robustus			
Swartkrans	12	11	16
Kromdraai	3	1	2
Drimolen	2	2	2

The four main cusps (paracone, protocone, metacone and hypocone) were defined following Wood and Engleman, (1988), and their absolute areas were measured; the total measured area (TMA) of the crown was computed from the individual cusp areas. One of us (S.B.) measured the cusp areas three times over a six month period, and the average of the three readings was used in the analysis.

A series of univariate non-parametric statistical comparisons, in the form of Mann-Whitney tests, were performed to test for differences in cusp size between *A. africanus* and *A. robustus*.

Results

Tables 2 and 3 present the data on the absolute and relative cusp areas of the teeth assigned to *A. africanus* and *A. robustus*. Mean values for the TMA are also presented in Table 2.

The mean value of TMA of M^1 in *A. africanus* is slightly smaller than *A. robustus*. In M^2 the TMA of *A. africanus* is larger than *A. robustus*, whereas in M^3 the opposite is true. None of the pair-wise differences between *A. africanus* and *A. robustus* (TMA M^1 *A. africanus* vs. TMA M^1 *A. robustus*, etc.) is statistically significant.

In *A. africanus*, when the mean values of the total area are examined, we found that the M^2 is the largest tooth, followed by M^3 and M^1, respectively. In *A. robustus*, the M^3 is the largest tooth, followed by M^2 and M^1. This condition results in a molar size sequence $M^1 < M^2 > M^3$ in *A. africanus*, whereas in *A. robustus* it is $M^1 < M^2 < M^3$ (Figure 1).

Inspection of the mean absolute values of the individual cusps showed only minor differences between the two species in the three molars. In M^1, both species have similar values for the lingual cusps (protocone and the hypocone), whereas *A. robustus* shows slightly larger buccal cusps; in M^2, *A. africanus* has slightly larger distal cusps

Table 2. Absolute cusp areas and Total Measured Area (TMA) (in mm^2) of maxillary molars in A. africanus *and* A. robustus

	A. africanus						A. robustus					
	N	x	*Min*	*Max*	*s.d.*	*CV*	*N*	x	*Min*	*Max*	*s.d.*	*CV*
M^1												
Total measured area	14	159.7	115.6	247.7	33.1	20.7	14	163.0	113.4	240.1	32.8	20.1
Protocone	14	51.7	38.2	85.5	11.8	22.9	17	50.8	37.5	66.0	8.6	16.9
Paracone	14	33.8	22.5	51.0	8.1	23.8	16	36.9	25.1	52.7	6.6	17.8
Metacone	14	36.3	26.4	49.9	7.5	20.8	16	39.7	27.4	59.4	8.3	21.0
Hypocone	14	37.4	28.6	62.0	8.4	22.6	15	36.6	20.7	62.2	10.0	27.2
M^2												
Total measured area	14	194.2	130.5	272.8	43.0	22.1	11	183.8	147.3	215.6	22.4	12.2
Protocone	15	64.6	45.5	83.7	12.2	18.9	13	65.6	47.5	82.6	11.5	17.5
Paracone	17	45.9	25.5	68.8	11.5	25.0	13	44.2	31.3	55.2	7.2	16.3
Metacone	16	40.6	23.9	59.5	10.7	26.2	13	35.2	22.2	45.1	8.4	24.0
Hypocone	14	42.9	21.1	68.4	13.2	30.7	13	38.1	23.8	46.7	5.7	15.0
M^3												
Total measured area	9	180.0	137.8	242.0	32.1	17.8	18	195.6	154.3	264.4	31.9	16.3
Protocone	9	61.2	48.2	83.5	10.4	17.0	20	71.3	51.0	130.2	18.3	25.7
Paracone	10	45.6	30.2	68.4	10.4	22.8	19	49.6	38.0	72.1	10.8	21.8
Metacone	10	35.1	21.8	47.1	9.4	26.8	19	35.1	20.2	60.0	10.2	29.2
Hypocone	10	37.7	21.2	53.7	9.0	23.9	19	37.2	18.3	68.3	11.0	29.5

Table 3. Relative cusp areas of maxillary molars in A. africanus *and* A. robustus. (*) *indicates significant differences. See text*

	A. africanus						A. robustus					
	N	x	*Min*	*Max*	*s.d.*	*CV*	*N*	x	*Min*	*Max*	*s.d.*	*CV*
M^1												
Protocone	14	32.4*	27.5	37.0	2.4	7.4	14	30.9	27.5	34.0	1.7	5.5
Paracone	14	21.1	17.8	25.4	2.0	9.6	14	22.9*	20.1	28.0	1.9	8.4
Metacone	14	22.8	19.8	25.5	1.9	8.3	14	24.0	20.2	27.6	2.2	9.1
Hypocone	14	23.5	18.7	25.7	1.9	8.3	14	22.1	17.7	25.9	2.5	11.3
M^2												
Protocone	14	32.9	28.3	37.1	2.6	7.9	11	35.0*	27.5	40.2	3.3	9.5
Paracone	14	24.1	19.7	28.8	2.7	11.2	11	24.8	21.2	29.4	2.8	11.4
Metacone	14	21.3	16.7	25.0	2.0	9.4	11	19.5	11.2	24.3	3.6	18.7
Hypocone	14	21.8	16.2	26.0	2.9	13.3	11	20.7	15.8	26.1	2.5	12.0
M^3												
Protocone	9	34.1	30.6	38.7	2.5	7.4	18	36.9	25.3	49.3	5.7	15.5
Paracone	9	24.8	19.9	28.3	2.5	10.2	18	25.7	19.0	35.8	4.3	16.7
Metacone	9	20.1	15.5	25.7	4.2	20.9	18	18.3	13.1	31.6	4.4	24.2
Hypocone	9	20.6	15.4	26.6	3.4	16.6	18	19.1	9.5	26.4	4.7	24.6

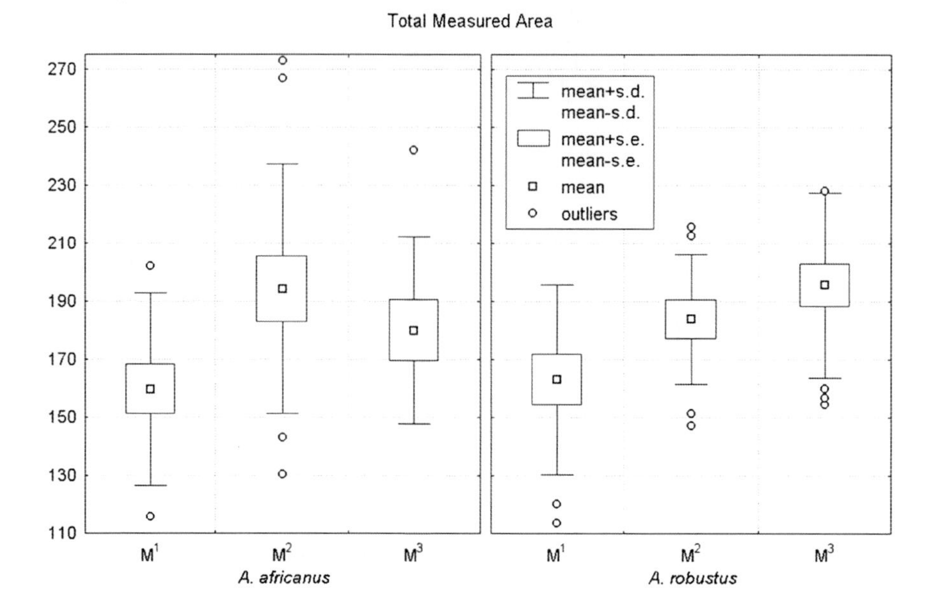

Figure 1. Total crown base area (in mm^2) of maxillary molars of *A. africanus* and *A. robustus*.

than *A. robustus* and similarly sized mesial cusps. In M^3, *A. robustus* shows larger mesial cusps, on average, than *A. africanus*, and distal cusps of the same size. As was the case for TMA, none of the pair-wise differences between *A. africanus* and *A. robustus* is statistically significant.

Both species showed remarkably high levels of variability in absolute cusp areas as expressed by the coefficient of variation (CV), with values always larger than 15. *A. africanus* shows higher CV values than *A. robustus* in the mesial cusps of M^1, and in all four cusps of M^2. However, in M^3 *A. robustus* has higher CV values than *A. africanus* in the protocone, metacone and hypocone. With respect to the TMA, the CV values for *A. africanus* were always larger than for *A. robustus*, especially in M^2.

When absolute mean cusp size is examined along the tooth row (Figure 2) in *A. africanus*, in each cusp the sequence is $M^1 < M^2 > M^3$, as was the case for the TMA. Specifically, there is a marked increase ($>12\,mm^2$) from M^1 to M^2 for both mesial cusps, less evident for the distal cusps (~ 4–$5\,mm^2$). On the other hand, the difference in individual cusp size between M^2 to M^3 is less marked and not statistically significant.

The picture is different in *A. robustus*. The cusp size sequence is $M^1 < M^2 < M^3$ for the protocone and the paracone, with a notable increase both from M^1 to M^2 ($>14\,mm^2$ for the protocone and $>7mm^2$ for the paracone) and less from M^2 to M^3 ($> 5\,mm^2$). The cusp sequence is $M^1 > M^2 = M^3$ for the metacone and $M^1 < M^2 > M^3$ for the hypocone, although in the latter case the differences between the three teeth are minimal and not statistically significant.

Analysis of the relative cusp areas showed differences between the two species that were not apparent in the analysis of absolute values. In M^1, the protocone is relatively large in *A. africanus* compared to *A. robustus*, and the difference is statistically significant ($p < 0.05$). A significant difference also emerges in paracone size, with *A. robustus* being significantly larger than *A. africanus* ($p < 0.05$). In M^2 the protocone size is significantly larger in *A. robustus* than in *A. africanus* ($p < 0.05$). In M^3 the protocone size is also larger in *A. robustus*, with a p value close to significant ($p = 0.057$).

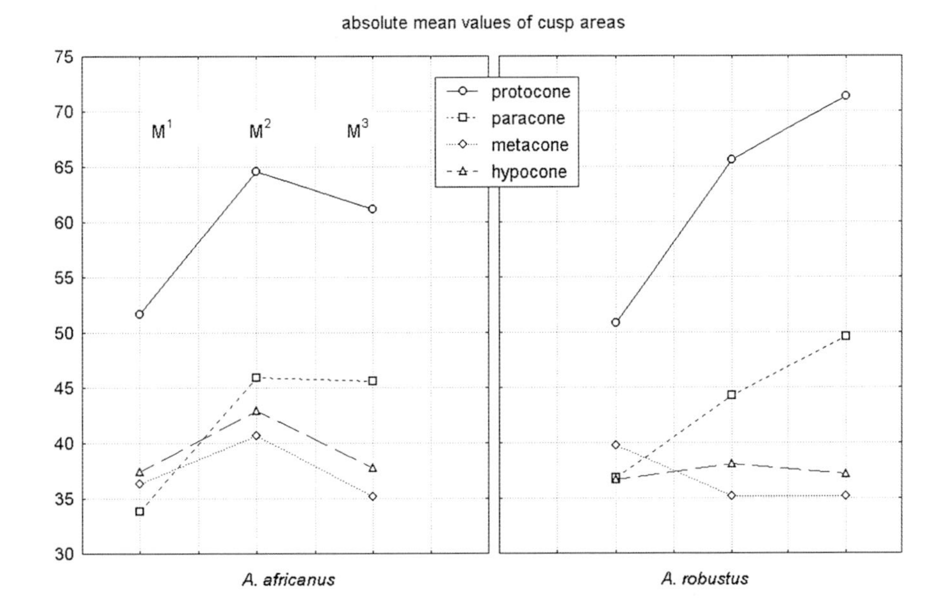

Figure 2. Mean values of absolute cusp areas (in mm^2) of maxillary molars of *A. africanus* and *A. robustus*.

It is also interesting to observe the relative contribution of each cusp to the total area in each molar of both species. For both *A. africanus* and *A. robustus*, in M^2 and M^3 the protocone is the largest cusp, followed by the paracone, hypocone and metacone. In M^1, for *A. africanus* the protocone is the largest cusp, followed by the hypocone, metacone and paracone, whereas in *A. robustus* the sequence is protocone>metacone>paracone>hypocone.

Relative cusp size examined along the tooth row adds complementary information to the analysis of absolute areas (Figure 3). In *A. africanus*, there is an increase in the relative size of the mesial cusps from M^1 to M^2, with a proportional reduction of both distal cusps. From M^2 to M^3, there is also a small increase of the relative size of the mesial cusps, and a decrease in relative size of the distal cusps. In *A. robustus*, on the other hand, the trend for relative cusp area is the same as for absolute cusp area. Moreover, the enlargement of both mesial relative to the distal cusps from M^1 to M^3 becomes more evident.

Discussion

The morphological and metrical characterization of the dental features of early hominids is crucial to the interpretation of their phylogenetic position (e.g., Robinson, 1956; Johanson et al, 1982; Grine, 1989; Tobias, 1991; Wood, 1991b; Ward et al., 2001). In the case of South African Plio-Pleistocene hominids, because of the large number of specimens recovered from long-term excavations, we are now in a position to get a fairly good idea of the characteristics of the dentition of the two most abundant hominid species (*A. africanus* and *A. robustus*). In particular, the large samples allow us to address important issues such as intra- and interspecific variability in the features considered.

The issue of metrical and morphological variability in the dental record of South African hominids is, at the present time, a matter of debate. The existence of different species or different 'morphs' in the fossil sample from Sterkfontein has been suggested by several authors (e.g. Clarke, 1988, 1994; Kimbel and White, 1988, Lockwood

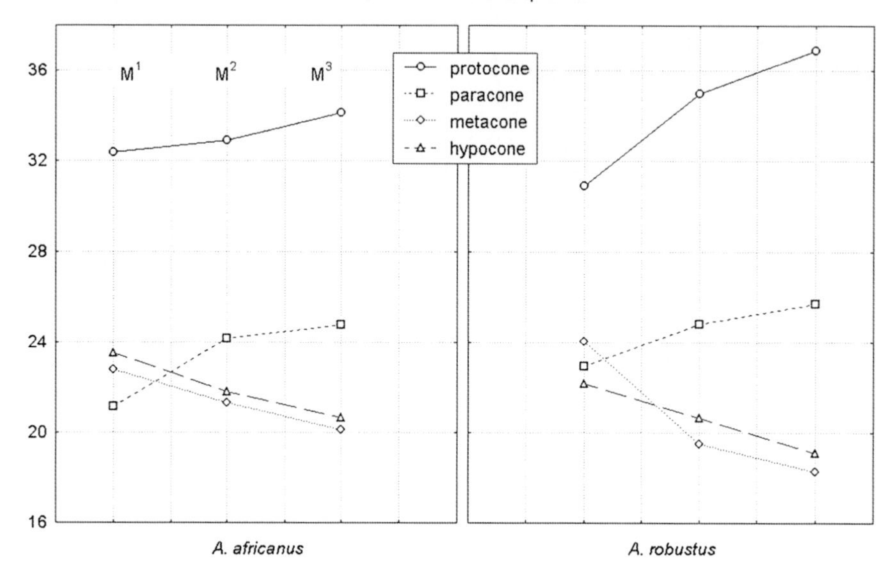

Figure 3. Mean values of relative cusp areas of maxillary molars of *A. africanus* and *A. robustus*.

and Tobias, 2002; Schwartz and Tattersall, 2005), but without reliable information about intraspecific variability these proposals cannot be assessed. On the other hand, previous studies based on dental metrics found no evidence for substantial heterogeneity within the Sterkfontein Member 4 hominid dental sample (Suwa, 1990; Wood, 1991a; Calcagno et al., 1999; Moggi-Cecchi, 2003).

Thus, the detailed metrical analysis of cusp morphology of *A. africanus* and *A. robustus* analyzed in terms of cusp areas of maxillary molars provides a framework within which the existence of a second hominid taxon within Sterkfontein Member 4 can be assessed. Detailed analysis of molar morphology may provide additional information, possibly employing a method of analysis that has been successfully employed to discriminate late Pleistocene taxa (Bailey, 2004).

Results show that broad similarities exist between *A. africanus* and *A. robustus* in terms of TMA of the upper molars. This result confirms observations from previous studies (Robinson, 1956; Sperber, 1973; Wood and Engleman, 1988). In terms of variability in crown size, as expressed by the CV, this is similar for M^1 and M^3 in *A. africanus* and *A. robustus*. However, in *A. africanus* the CV of M^2 is almost twice that in *A. robustus*. This finding will require further investigation in future studies.

In terms of molar size sequence based on TMA values, the uniquely derived condition seen in *A. robustus* described in previous studies (Robinson, 1956; Sperber, 1973; Wood and Engleman, 1988) with $M^1 < M^2 < M^3$ is confirmed. On the other hand, in *A. africanus*, the sequence observed is $M^1 < M^2 > M^3$, which differs from that described by Wood and Engleman (1988), who reported that in their "SAFGRA" sample M^2 and M^3 were subequal in size. Our results indicate the existence of the primitive condition in *A. africanus*, as observed in living apes (Wood and Engleman, 1988).

This different crown size sequence between *A. africanus* and *A. robustus* appears to be related primarily to differences in the size of the mesial cusps. In fact, in *A. africanus* the $M^1 < M^2 > M^3$ sequence of the TMA is also present when the individual cusps are considered in both absolute and in relative terms. On the other hand, in *A. robustus*, the mesial cusps show a marked increase in size from M^1 to M^3, whereas the size of the distal cusps remains the same (in absolute terms) along the molar row (or decreases slightly – in relative terms). This result suggests that the overall increase of the total crown base area from M^1 to M^3 in *A. robustus* is the result of a selective enlargement of the mesial cusps.

This result is interesting, since it differs from the pattern described for mandibular molars of robust australopithecines, where there is a relative expansion of the distal portion of the tooth (the talonid), and relative reduction of the mesial portion (Wood et al., 1983). It will be important to explore this finding in more detail in future studies to discern whether or not a difference between upper and lower molars will be confirmed.

The relative contribution of the different cusps to the total crown area varies between *A. africanus* and *A. robustus*, and from M^1 to M^3. In general, the protocone is the largest cusp. Robinson (1956, p. 97) noted that, in *A. africanus*, the protocone provides a greater contribution to the total area than it does in *A. robustus*. On the basis of the present analysis this statement appears to be confirmed only for the M^1, whereas the opposite is true in M^2 and M^3.

It is interesting to compare cusp area mean values obtained in this study with those published by Wood and Engleman (1988). There are obvious differences in sample size and composition between the two studies (and also there are slightly different measuring techniques), but their comparative analysis provides relevant information.

In *A. africanus*, the mean values of absolute cusp areas obtained in this study are consistently larger that those reported by Wood and Engleman (1988), with the only exception being the hypocone of M^3. In *A. robustus*, the protocone always shows larger mean values in our sample, whereas the distal cusps always have larger mean values in the Wood and Engleman sample. No consistent trend in differences along the tooth row is evident for the metacone. The picture is different when the mean values of relative cusp areas are examined. Differences between the two samples are always minor, with most around 1% and a few less than 3%. Two exceptions are the relative area of the hypocone of M^3 in *A. africanus*, which is 5% larger in the Wood and Engleman sample, and relative area of the protocone of M^2 in *A. robustus*, which is 4.3% larger in our extended sample.

Taken together, this evidence suggests that the large sample used in the present analysis confirms the relative contribution of the main cusps to the total area in *A. africanus* and in *A. robustus* noted by Wood and Engleman (1988). At the same time, analysis of the absolute values suggests that our expanded sample of *A. africanus* includes more individuals with large teeth than were in the Wood and Engleman sample.

The sample of South African fossil hominids used by Wood and Engleman (1988) in their study included almost exclusively specimens recovered from the excavations by Broom and by Robison at Swartkrans and at Sterkfontein (except for Taung and Makapansgat). No specimens included in their study derived from the later excavations conducted by Brain at Swartkrans (labeled SKX and SKW) and by Tobias, Hughes and Clarke at Sterkfontein (labeled Stw – except Stw 6 and Stw 19, used by Wood and Engleman), and, for obvious reasons, the specimens from Drimolen (Keyser et al., 2000).

Among the hominid specimens more recently recovered from Sterkfontein Member 4 (labeled Stw), a few have notably large teeth, as is shown in Figure 4 (comparing the M^2 of Sts 22 and of Stw 183). Some of these Stw specimens are those identified by Clarke (1994) as being distinct from *A. africanus*. (It will be thus essential to perform a more detailed analysis of cusp areas in order to evaluate potential

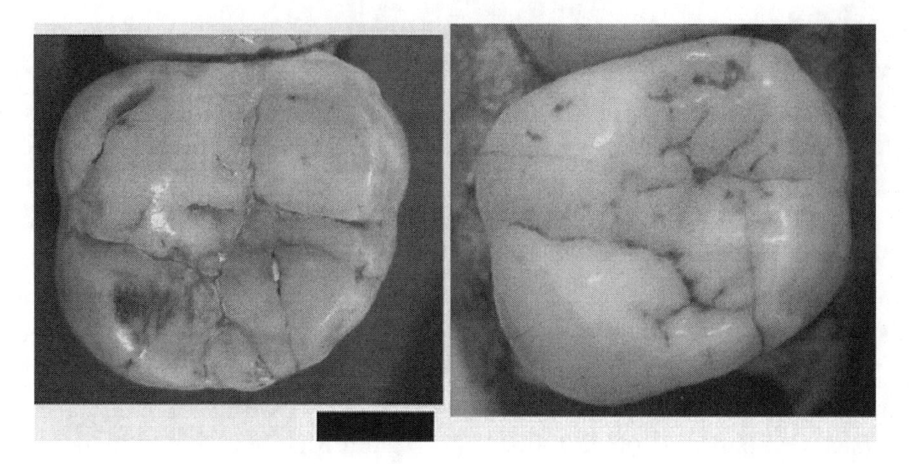

Figure 4. Occlusal photographs of M^2 of Sts 22 (on the left) and Stw 183 (on the right) at the same scale. Note differences in overall crown morphology. Scale = 1 cm.

differences in cusp base areas within the *A. africanus* extended sample.) Further, these larger specimens appear to have a broad based crown relative to the occlusal basin. Such a complex occlusal morphology cannot be captured by standard cusp area analysis and will require a different approach to be quantified (e.g., Bailey, 2004).

Conclusions

Analysis of the maxillary molar cusp areas and proportions of *A. africanus* and *A. robustus* provides the following main results:

1. Crown base areas for the three molars are broadly similar in *A. africanus* and *A. robustus*.
2. Significant differences between the two species in relative cusp areas are evident for the protocone of M^1 (with *A. africanus* larger than *A. robustus*), the paracone of M^1 (with *A. robustus* larger than *A. africanus*), and for the protocone of M^2 and M^3 (with *A. robustus* larger than *A. africanus*).
3. The total crown area sequence for *A. robustus* is $M^1 < M^2 < M^3$ as previously described (Wood and Engleman, 1988).
4. In *A. africanus* the total crown area sequence is $M^1 < M^2 > M^3$, as in living apes (Wood and Engleman, 1988).
5. The different sequences of *A. africanus* and *A. robustus* appear to be related primarily to differences in mesial cusp size, which in *A. robustus* shows a marked expansion from M^1 to M^3 relative to the distal cusps.
6. Differences in the absolute cusp areas of the present expanded *A. africanus* sample and the smaller sample of Wood and Engleman (1988) are likely related to the presence of specimens with notably large teeth in the expanded sample.

Further analysis will specifically address the issue of variability within the *A. africanus* dental sample, with special reference to the specimens from Sterkfontein Member 4.

Acknowledgments

We would like to thank Shara Bailey, Jean-Jacques Hublin and all the staff of the Dept. of Human Evolution of the Max Planck Institute for Evolutionary Anthropology for organizing such a remarkable conference and for inviting JMC to participate. In South Africa, we would like to thank

L. Backwell, L. Berger, R. Clarke, H. Fourie, A. Keyser, B. Kramer, K. Kuykendall, C. Menter, E. Mofokeng, S. Potze, M. Raath, J. F. Thackeray, P. V. Tobias and H. White for permission to study the fossils and for assistance at the various stages of the project. Part of this work was carried out within the Agreement on Scientific and Technological Co-operation between Italy and South Africa (2001–2004), program "Knowledge and promotion of the South African palaeoanthropological heritage", whose referring institutions, the Italian Ministero degli Affari Esteri and the South African National Research Foundation, are gratefully acknowledged. Financial support from the Italian Ministero degli Affari Esteri (project 'Conservazione di reperti paleontologici sudafricani' 2004) and University of Florence (Finanziamenti di Ateneo 2004) are also acknowledged.

References

Bailey, S.E., 2004. A morphometric analysis of maxillary molar crowns of Middle-Late Pleistocene hominins. Journal of Human Evolution 47, 183–198.

Boccone, S., 2004. *Alcuni aspetti dello sviluppo dentale nelle Australopitecine Sudafricane. Tesi di Dottorato di Ricerca in Scienze Antropologiche*. Università di Firenze, Florence.

Brain, C.K., 1993. *Swartkrans: A Cave's Chronicle of Early Man. Pretoria*. Transvaal Museum Monograph No. 8.

Calcagno, J.M., Cope, D.A., Lacy, M.G., Moggi-Cecchi, J., Tobias, P.V., 1999. Reinvestigating the number of hominid species in Sterkfontein Member 4. American Journal of Physical Anthropology Suppl. 28, 101.

Clarke, R.J., 1994. On some new interpretations of Sterkfontein stratigraphy. South African Journal of Science 90, 211–214.

Clarke, R.J., 1988. A new *Australopithecus* cranium from Sterkfontein and its bearing on the ancestry of *Paranthropus*. In: Grine F.E. (Ed.), Evolutionary History of the "Robust" Australopithecines. Aldine de Gruyter, New York, pp. 285–292.

Grine, F.E., 1989. New hominid fossils from the Swartkrans Formation (1979–1986 excavations): craniodental specimens. American Journal of Physical Anthropology 79, 409–449.

Johanson, D.C., White, T.D., Coppens, Y., 1982. Dental remains from the Hadar Formation, Ethiopia: 1974–1977 collections. American Journal of Physical Anthropology 57, 545–603.

Keyser, A.W., Menter, C.G., Moggi-Cecchi, J., Pickering, T.R., Berger, L.R., 2000. Drimolen: a new hominid-bearing site in Gauteng, South Africa. South African Journal of Science 96, 193–197.

Kimbel, W.H., White, T.D., 1988. Variation, sexual dimorphism, and taxonomy of *Australopithecus*. In: Grine F.E. (Ed.), Evolutionary History of the "Robust" Australopithecines. Aldine de Gruyter, New York, pp. 175–192.

Lockwood, C.A., Tobias, P.V., 2002. Morphology and affinities of new hominin cranial remains from Member 4 of the Sterkfontein Formation, Gauteng Province, South Africa. Journal of Human Evolution 42, 389–450.

Moggi-Cecchi, J., 2003. The elusive 'second species' in Sterkfontein Member 4: the dental metrical evidence. South African Journal of Science 96, 268–271.

Moggi-Cecchi, J., Grine, F.E., Tobias, P.V., 2006. Early hominid dental remains from Members 4 and 5 of the Sterkfontein Formation (1966–1996 excavations): catalogue, individual associations, morphological descriptions and initial metrical analysis. Journal of Human Evolution 50, 239–328.

Robinson, J.T., 1956. The dentition of the Australopithecinae. Transvaal Museum Memoirs 9, 1–179.

Schwartz, J.H., Tattersall, I., 2005. *The Human Fossil Record. Volume 4. Craniodental morphology of early Hominids (Genera Australopithecus, Paranthropus, Orrorin), an Overview*. Wiley-Liss, Hoboken, NJ.

Sperber, G.H., 1973. Morphology of the cheek teeth of early South African Hominids. Ph.D. Dissertation, University of the Witwatersrand, Johannesburg.

Suwa, G., 1990. A comparative analysis of hominid dental remains from the Shungura and Usno Formations, Omo Valley, Ethiopia. Ph.D. Dissertation, University of MI, Ann Arbor.

Tobias, P.V., 1991. *Olduvai Gorge, Volume IV. The Skulls, Endocasts and Teeth of Homo habilis*. Cambridge University Press, Cambridge.

Ward, C.V., Leakey, M.G., Walker, A., 2001. Morphology of *Australopithecus anamensis* from Kanapoi and Allia Bay, Kenya. Journal of Human Evolution 41, 255–368.

Wood, B.A., 1991a. A paleoanthropological model for determining the limits of early hominid taxonomic variability. Palaeontologia Africana 28, 71–77.

Wood, B.A., 1991b. *Koobi Fora Research Project IV: Hominid Cranial Remains from Koobi Fora*. Clarendon Press, Oxford.

Wood, B.A., Abbott, S.A., 1983. Analysis of the dental morphology of Plio-Pleistocene hominids. I. Mandibular molars: crown area measurements and morphological traits. Journal of Anatomy 136, 197–219.

Wood, B.A., Abbott, S.A., Graham, S.H., 1983. Analysis of the dental morphology of Plio-Pleistocene hominids. II. Mandibular molars – study of cusp areas, fissure pattern and cross sectional shape of the crown. Journal of Anatomy 137, 287–314.

Wood, B.A., Engleman, C.A., 1988. Analysis of the dental morphology of Plio-Pleistocene hominids. V. Maxillary postcanine tooth morphology. Journal of Anatomy 161, 1–35.

Wood, B.A., Uytterschaut, H., 1987. Analysis of the dental morphology of Plio-Pleistocene hominids. III. Mandibular premolar crowns. Journal of Anatomy 154, 121–156.

5. Gran Dolina-TD6 and Sima de los Huesos dental samples: Preliminary approach to some dental characters of interest for phylogenetic studies

M. MARTINÓN-TORRES
Centro Nacional de Investigación sobre Evolución Humana (CENIEH)
Avda. de la Paz 28, 09006 Burgos, Spain
maria.martinon.torres@gmail.com

J.M. BERMÚDEZ DE CASTRO
Centro Nacional de Investigación sobre Evolución Humana (CENIEH)
C/Toledo 4-5ª, 09004 Burgos, Spain
mcnbc54@mncn.csic.es

A. GÓMEZ-ROBLES
Centro Nacional de Investigación sobre Evolución Humana (CENIEH)
Avda. de la Paz 28, 09006 Burgos, Spain
aidagomezr@yahoo.es

M. BASTIR
Department of Palaeobiology, Museo Nacional de Ciencias Naturales
CSIC, C/José Gutiérrez Abascal 2
28006 Madrid, Spain
and
Hull York Medical School
The University of York; Heslington
York YO10 5DD, United Kingdom
markus.bustir@hyms.ac.uk

S. SARMIENTO
Fundación Atapuerca
C/Condestable 2, 4°C, 09004 Burgos, Spain
informacion@fundacionatapuerca.es

A. MUELA
Fundación Atapuerca.
C/Condestable 2, 4°C, 09004 Burgos, Spain
informacion@fundacionatapuerca.es

J.L. ARSUAGA
Centro de Evolución y Comportamiento Humanos
Sinesio Delgado 4, Pabellón 14, 28029 Madrid, Spain
jlarsuaga@isciii.es

Keywords: Atapuerca, *Homo antecessor, Homo heidelbergensis*, Pleistocene, Europe, teeth

S.E. Bailey and J.-J. Hublin (Eds.), Dental Perspectives on Human Evolution, 65–79.

Abstract

The Sima de los Huesos (SH) and Gran Dolina-TD6 sites in Sierra de Atapuerca (Spain) have each yielded an impressive fossil hominin sample representing Middle Pleistocene and Late Lower Pleistocene European populations, respectively. Paleontological evidence, paleomagnetic analyses, and radiometric dates (U/Th) suggest an interval of 400 to 500 ky for the SH hominins. At Gran Dolina, radiometric dates (ESR and U-series) combined with paleomagnetic analyses and fossil evidence indicate an age range between 780 and 860 ky for the Aurora Stratum of the TD6 level where the fossil hominins were found. We have assigned the SH hominins to the *Homo heidelbergensis* species, whereas the TD6 hominins are representative of *Homo antecessor*, the species named in 1997 (Bermúdez de Castro et al., 1997) to accommodate the variability observed in the TD6 fossil human assemblage. Dental collections of the SH and TD6 sites include more than five hundred deciduous and permanent teeth. The detailed description and morphological comparison of the Atapuerca dental samples will be published elsewhere in a near future, but the examination of an extensive human fossil record, has already revealed some dental characters we consider crucial for phylogenetic studies. We describe those characters and provide an overview of their distribution across the hominin fossil dental record. On the basis of these traits we explore some questions about the phylogenetic relationship between TD6 and SH hominins as well as the evolutionary scenario of these two populations.

Introduction

The Sima de los Huesos and Gran Dolina sites in Sierra de Atapuerca (Spain) have each yielded an impressive fossil hominin sample representing Middle Pleistocene and Late Lower Pleistocene European populations, respectively (Arsuaga et al., 1997b; Bermúdez de Castro et al., 1999a).

The Sima de los Huesos (SH) site is placed well inside the Cueva Mayor-Cueva del Silo Karst System of the Sierra de Atapuerca. At present, the Sima de los Huesos site has provided more than five thousand fossil human remains. All of them come from the same level and belong to a minimum of 28 individuals (Bermúdez de Castro et al., 2004a, b). The human assemblage was assigned to the *Homo heidelbergensis* species (Arsuaga et al., 1997a, 1999). Paleontological evidence, paleomagnetic analyses, and radiometric dates (U/Th) suggest an interval of 400,000 to 500,000 years for the hominin level (Bischoff et al., 2003).

Gran Dolina (TD) is one of the filled caves located at the right wall of the Mining Railway Trench, opened at the end of the nineteenth century. The infilling of the Gran Dolina is 20 meters high and its stratigraphic sequence consists of 11 levels (Parés and Pérez-González, 1995) deposited from the late Early Pleistocene to the end of the Middle Pleistocene. Seven levels (TD4, 5, 6, 7, 8, 10, and 11) are rich in fossils and stone tools. The Aurora Stratum of the TD6 level has yielded a rich archaeological-paleontological assemblage, including approximately one hundred human fossil remains (Carbonell et al., 1995). Radiometric dates (ESR and U-series) combined with paleomagnetic analyses and fossil evidence indicate an age range between 780 and 860 kyr for the Aurora Stratum of the TD6 level (Falguères et al., 1999).

The TD6 hominins are representative of *Homo antecessor*, the species named in 1997 to accommodate the variability observed in the TD6 fossil human assemblage with its unique combination of a modern face with a primitive dentition (Bermúdez de Castro et al., 1997). Different skeletal parts are represented in the TD6 hypodigm, which correspond to a minimum of seven individuals. We have proposed that *H. antecessor* could represent the last common ancestor for Neandertals and modern humans (Bermúdez de Castro et al., 1997). This is an alternative hypothesis to the one that considers the Afro-Eurasian Middle Pleistocene species *H. heidelbergensis* the last common ancestor for both the Neandertal and "sapiens" lineages (Rightmire, 1996; Tattersall, 1996).

The recent study of a well-preserved mandible recovered in 2003 from the Aurora Stratum of the TD6 level suggested that the TD6 hominins might have an Asian origin, thus entering into conflict with the previous hypothesis of an African origin of *H. antecessor* and a possible relationship between this species and *H. ergaster* (Carbonell et al., 2005a).

The finding in 1994 of the TD6 human fossil assemblage associated with nearly 300 lithic artifacts of the Mode 1 technology led to the collapse of the so-called "short chronology hypothesis" for the earliest occupation of Europe, which asserted that all human occupation in Europe postdated 500,000 years BP (Roebroeks and van Kolfschoten, 1994). Nevertheless, this hypothesis could still be applied for the Central and Northern European regions. In fact, the presence of hominins in Western Europe during the Lower Pleistocene can only be demonstrated, at the moment, in the Iberian and Italian Peninsulas (Carbonell et al., 1999). The earliest evidence for a human settlement in Western Europe are the lithic assemblages recovered from Barranco León and Fuentenueva 3, in southern Spain, which are placed in the biozones of *Allophaiomys pliocaenicus* and *Allophaiomys burgondiae*, dated between 1.3 and 1.2 Ma (Turq et al., 1996) and Sima del Elefante site in Atapuerca (Parés et al., 2006). The lithic assemblages of all European Lower Pleistocene sites belong to the Mode 1 technologies. It is very probable that the first Europeans occupied mainly the more temperate Mediterranean regions. In contrast, the European Middle Pleistocene sites are generally younger than 500,000 years (except for the French site of Carrière Carpentier in Abbeville and the level F from Notarchirico in Italy) and the lithic assemblages of most belong to the Mode 2 technology (Carbonell et al., 1999). In summary, we affirm that the distribution of these Middle Pleistocene sites does not cross the parallel 53.

H. antecessor shares two apomorphic cranial traits with Neandertals and modern humans not seen in *H. ergaster*: a convex superior border of the temporal squama and an anterior position of the incisive canal, which is nearly vertical (Martínez and Arsuaga, 1997; Bermúdez de Castro et al., 1997). Moreover, in the TD6 hominins there is a marked prominence of the nasal bones, a trait that is also present in Neandertals and modern humans (Arsuaga et al., 1999). For these reasons, the TD6 hominins could represent the root of the European Pleistocene lineage that led to the appearance of the "classic" Neandertals. However, some morphometric differences between the TD6 and SH dental samples led us to explore the possible discontinuity in these European Pleistocene populations (Bermúdez de Castro et al., 2003). We suggested that the early Europeans might have been replaced or genetically absorbed by a new arrival of African emigrants during the Middle Pleistocene, carrying the Mode 2 technology. We have also observed that the Arago teeth share some dental traits with the TD6 hominins such as the particular premolars' root morphology, which, in contrast, are absent in the SH teeth sample (Bermúdez de Castro et al., 2003). Since the site of Arago is approximately contemporaneous with the SH site, we suggest that the model of colonization of the European continent during the Middle Pleistocene was probably very complex.

The aim of this paper is to present a preliminary approach to the dental variation of these two Atapuerca populations through the selection of some characters we consider tentatively useful for future cladistic studies. On this basis, we approach some questions about the phylogenetic relationship of the SH and TD6 hominins and their evolutionary scenario in Europe.

The dental evidence

Dental collections at the SH and TD6 sites include more than 500 deciduous and permanent teeth. Sima de los Huesos comprises 519 teeth (511 permanent and 8 deciduous) assigned to a minimum number of 28 individuals (Bermúdez de Castro et al., 2004a, b). The Gran Dolina-TD6 collection includes 40 teeth (37 permanent and 3 deciduous) belonging to at least six individuals (Bermúdez de Castro et al., 1999a). We are currently performing an extensive morphometric study, which records the morphological variability in both Atapuerca samples, and compares them with hominins from other European, African and Asian sites. The analysis we are presenting here is a preliminary approach to the evolutionary scenario of the first human settlement in Europe in light of dental variability between these two European Lower and Middle Pleistocene populations. After a detailed examination of the hominin dental evidence from the Pleistocene, we particularly focus on a set of eleven dental characters that have been selected because of their variability and relatively clear polarity across the fossil record. We consider these characters to be potentially useful for future cladistic studies, given the high degree of polymorphism existing in the dental characters and the difficulty in finding characters that can be appropriately employed in studies of this type. Most of these characters are usually recorded in morphological analysis but some of them have been "redefined" to better cover the observed variability across the hominin fossil record. The majority of the standards developed to assess dental variability among hominins, like the Arizona State University Dental Anthropology System (Turner et al., 1991) are based on modern human populations so, despite their enormous aid in comparative dental studies, they fail to cover hominin variability (Bailey, 2002a; b). This is why modifications in the description and graded scale of expression in certain traits we find are important for accurately describing some extinct hominin species.

Upper Lateral Incisor Shovel Shape

The first of the characters is the shovel shape of the upper lateral incisors. Incisor shovel shape is a plesiomorphic trait displayed by australopithecines and early *Homo* species, where marginal ridges are thickened to a variable degree (Hrdlička, 1920; Dahlberg, 1956; Mizoguchi, 1985). In *H. sapiens* this trait can range from complete absence (exclusive of this species) to the most pronounced grades, where marginal ridges define a relatively closed lingual space, also called "barrel shovel shape" (Turner et al., 1991: 15). We have found that some Neandertals from Hortus and Krapina, as well as those from Monsempron and Le Moustier, display a particular shovel shape wherein marginal ridges are very conspicuous, quite massive and invade the palatal face, defining a deep longitudinal lingual fossa at their meeting point. This particular morphology is very common in the SH specimens (Figure 1). The combination of this lingual morphology plus a strong convexity of the labial surface gives a very characteristic morphology to the occlusal section which we have named "triangular shovel shape" based on de Lumley's descriptions (1973) of some Hortus specimens. Crummett (1994) and Mizoguchi (1985) also described the Neandertal shovel shape as morphologically different from *H. sapiens* shovel shape. In the typical Neandertal shovel shape, the marginal borders run parallel from the cervix to the incisal edge. Interestingly, ATD6-69 also shows a less pronounced form of this triangular shovel shape with a still more spacious lingual fossa, but already a triangular section with labial curvature (Figure 1). The marginal ridges are narrower at the neck

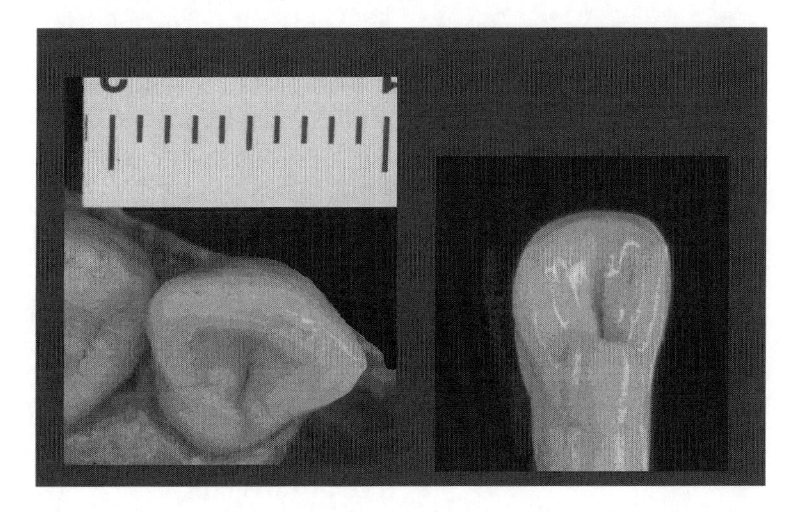

Figure 1. Gran Dolina-TD6 (left) and Sima de los Huesos (right) upper lateral incisors.

of the teeth than at the incisal edge, which is a primitive conformation when compared to the typical Neandertal lateral incisors. We have personally noticed similar marginal ridge development in Sangiran and Zhoukoudian *H. erectus* specimens (Weidenreich, 1937), which agrees with observations on these two groups by Mizoguchi (1985: 44–49).

Upper Canine Morphology

We suggest that the primitive shape of the permanent upper canines can be generally defined by a sharp cusp, defined by steeply inclined cutting edges that cross the occlusal plane of the rest of the teeth. The mesial and distal cutting edges are approximately the same length comprising more than a third. The marginal ridges are flared, thick, and end before reaching the occlusal margin in the shape of free tubercles or cusplets. Both mesial and distal marginal ridges are continuous with the cingulum and are also delimitated on the buccal surface by grooves or depressions. We consider this canine morphotype to be the primitive state since it is typical of the *Australopithecus* species as well as some early *Homo* specimens like *H. habilis* (Tobias, 1991) and the specimens from the Dmanisi site (personal observations on originals by the 1st and 2nd authors). SH and TD6 hominins display more derived canines, with a less prominent cutting edge as in later *Homo*. The mesial and distal cutting edges are shorter and less sharply angled. The mesial edge is shorter and less sloped than the distal edge. The marginal ridges are longer and run in a straighter course towards the incisal edge. These marginal ridges do not end in tubercles; neither are they separated by buccal grooves. There is no buccal *cingulum* or barely a vestige of it. It is interesting to note, however, that deciduous canines in *H. antecessor* retain the "flared" primitive state.

Lower Canine Morphology

In primitive specimens such as *Australopithecus* and *H. georgicus* (1st and 2nd authors' personal observation), the mesial part of the cutting edge is considerably shorter than the distal one. The mesial half slopes gently upwards to the cusp tip, while the distal counterpart slopes steeply downwards to the distolabial-incisal point, where an accessory distal cuspule tends to be present. SH and TD6 canines are more derived in their conformation. The cutting edges of the cusps are of similar length and present a much gentler slope. The marginal borders are longer and run in a straighter course to the cutting edges. However, *H. antecessor* lower canines (and premolars as well) differ from those of the SH hominins in their expression of a buccal *cingulum* (Bermúdez de Castro et al., 1999b).

Transverse Crest in Lower Second Premolars

The transverse crest in second lower premolars has been recorded in fossil hominins by many dental specialists (e.g. Genet-Varcin, 1962; Patté, 1962; Wood and Uytterschaut, 1987), although its utility in taxonomic and phylogenetic studies has been particularly emphasized by Bailey and coworkers (Bailey, 2002a, b, Bailey and Lynch, 2005). The transverse crest develops as an enamel bridge that connects protoconid and metaconid, and obscures the sagittal groove in that portion. It is typical of Neandertals – although not exclusive to this taxon – and it is rare in modern humans (Figure 2) (Bailey, 2002a; b). One of the *H. antecessor* lower second premolars expresses the transverse crest while the other does not, displaying an uninterrupted sagittal fissure instead. This trait is frequent in the SH hominins, although its expression is polymorphic and not all the specimens display it. We have noticed it also in some of the *H. erectus* specimens, particulary those from Sangiran (see Sangiran 6, Figure 3).

Lower Second Premolar Morphology

Several studies have shown that the morphology of mandibular premolars provides a source for taxon-specific diagnostic analysis in human evolution (Ludwig, 1957; Patte, 1962; Biggerstaff, 1969; Wood and Uytterschaut, 1987; Uytterschaut and Wood, 1989; Bailey, 2002a, b; Bailey and Lynch, 2005).

We suggest that the primitive morphology of P4s is defined by an asymmetrical contour with a mesially displaced metaconid, development of a bulging talonid, and wide occlusal polygon (Martinón-Torres et al., 2006). This is the shape displayed by the majority of *A. afarensis, A. africanus*, and early *Homo* specimens from Africa and Asia (Wood and Uytterschaut, 1987; Martinón-Torres et al., 2006). This talonid comprises a considerable buccal part, sometimes delimited by a buccal furrow in the external outline (Figure 3). The trend for dental reduction operating from the beginning of the Pleistocene (Wolpoff, 1971; Wood and Abbott, 1983; Bermúdez de Castro, 1993; Bermúdez de Castro and Nicolás, 1995) could have led to different morphological variants with a reduced occlusal polygon and decreased lingual occlusal surface in later *Homo* species. Based on our observations, we propose that *H. heidelbergensis/neanderthalensis* would have fixed plesiomorphic traits in high percentages, while modern humans would have derived into a symmetrical outline with centered metaconid and talonid reduction.

H. heidelbergensis from SH, *H. neanderthalensis* and *H. antecessor* appear to have retained the asymmetry but display a relatively smaller occlusal polygon relative to earlier hominins. The small occlusal polygon area could be considered derived in these species (Figure 2). The buccal portion of the talonid is reduced relative to earlier hominins. Likewise, generally speaking, the lingual half of the tooth is reduced, although extra lingual cusps still develop. In contrast, a centered metaconid in lower P4s is commonly associated with a more regular and circular outline, and seems to be a derived state in *H. sapiens* (Figure 4).

Lower First Premolar Shape

We suggest that the primitive morphology of the P3 comprises a strongly asymmetrical contour and large talonid, as observed in *Australopithecus* and early *Homo* species. *H.*

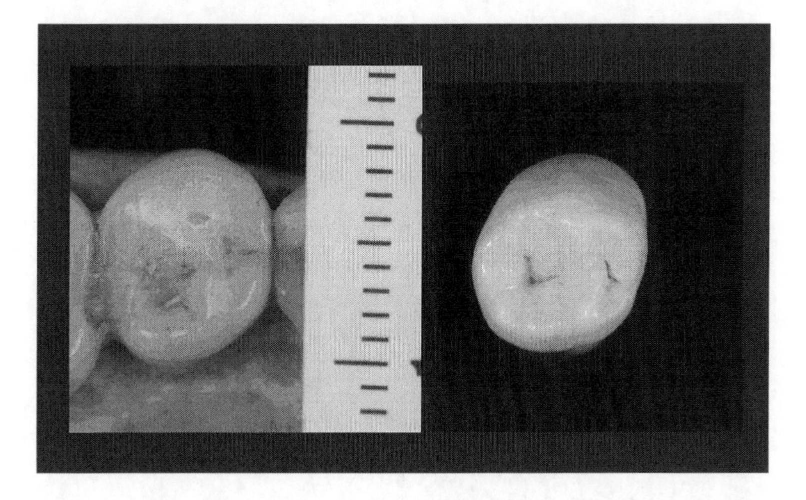

Figure 2. Gran Dolina-TD6 (left) and Sima de los Huesos (right) lower second premolar.

antecessor displays this primitive morphology (Figure 4). The buccolingual axis is perceptibly oblique relative to the mesiodistal axis and the occlusal polygon appears wide and centered. In comparison, the SH hominins present a derived shape with a much more symmetrical contour – close to a round shape – where the talonid is extremely reduced or even absent. The occlusal polygon is very small and located near the lingual border of the crown, resembling a canine in its conformation (Figure 5). The morphological variation of this tooth across the hominin fossil record is currently being evaluated by the third author using geometric morphometric methods. Thus far, results confirm the differences between the TD6 and SH hominin samples.

Lower First Premolar Root Morphology

At this point it is also interesting to evaluate premolar root morphology and the way it covaries with crown morphology. Dental roots have an important function in supporting the tooth crown, so it is reasonable to expect parallel changes in their morphology. *Paranthropus* and *Australopithecus* P3s are characterized by the high incidence of distinct mesial and distal roots, each with a buccal and a lingual interradicular process (Sperber, 1974; Wood et al., 1988; Bermúdez de Castro et al., 1999b). The great crown area attributed to "robust" australopithecine taxa is mainly related to the expansion of the talonid (Wood et al., 1988) so that a reinforced distal root structure is needed. In early representatives of

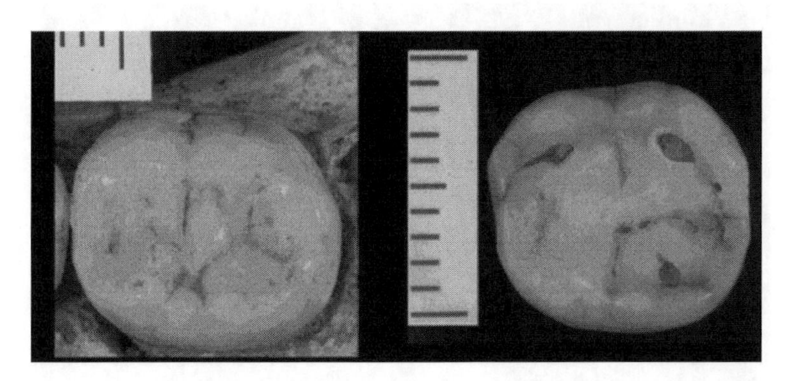

Figure 3. Gran Dolina-TD6 (left) and Sima de los Huesos (right) lower second molar.

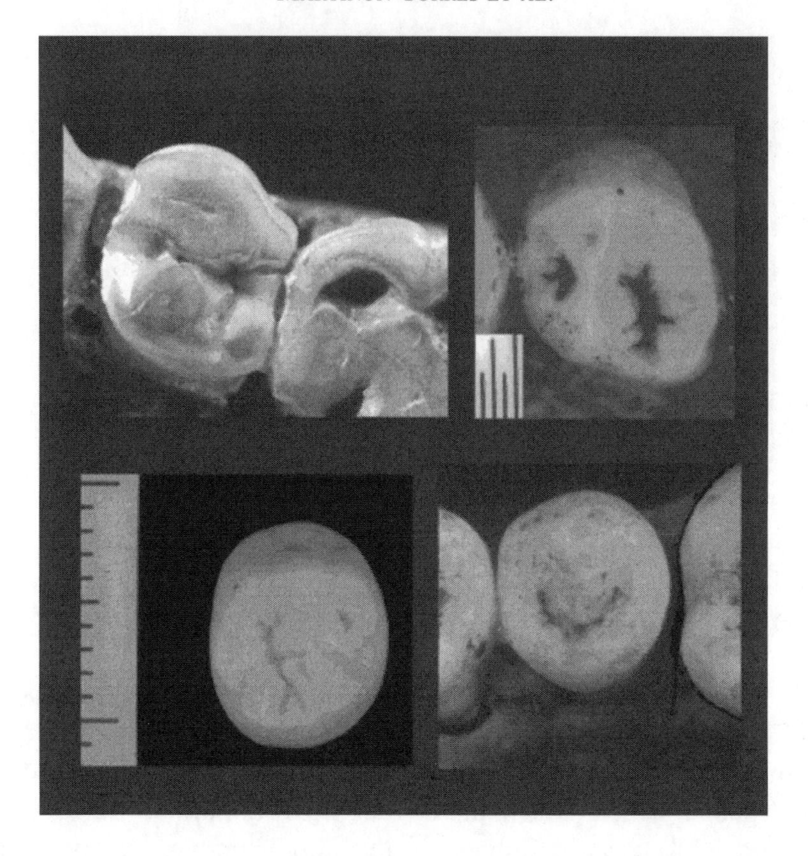

Figure 4. Some examples illustrating the lower second premolar's variability. A: *Homo mauritanicus*; B: *Homo erectus*; C: *Homo neanderthalensis*; D: *Homo sapiens*. Note the buccal indentation in the external contour of A (arrow) delimiting the buccal portion of the talonid. Also note the expression of transverse crest in B and C.

Homo (e.g., *H. habilis, H. ergaster* and *H. erectus*), the tendency is for root morphology to be simplified.

One of the TD6 specimens shows a particular derived root morphology, which could be a novelty in the fossil record (Bermúdez de Castro et al., 1999b). It shows a distolingual single root – supporting a well-developed distolingual talonid – and a mesiobuccal root that has buccal and mesial components. This particular root morphology could result from the suppression of the

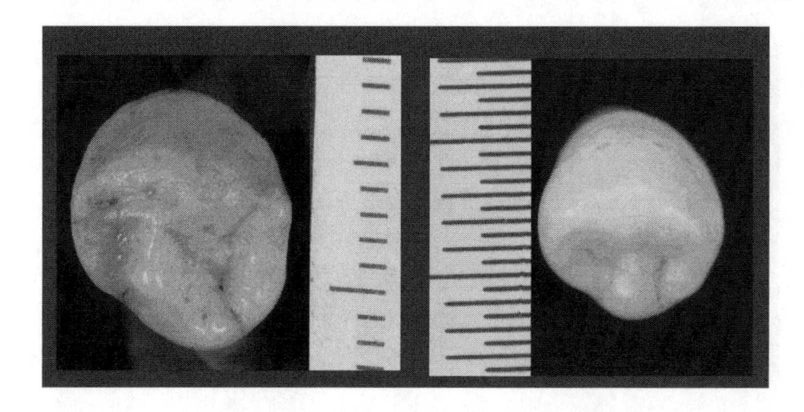

Figure 5. Gran Dolina-TD6 (left) and Sima de los Huesos (right) lower first premolar.

distobuccal interradicular process found in the more primitive forms. This radicular reduction might be related to the reduction of the crown. During human evolution the talonid would start decreasing in area by first a reduction in the buccal portion (Genet-Varcin, 1962), so the distobuccal interradicular process is suppressed. The second root would move from a distal position to distolingual location, where it would provide support for the talonid of the asymmetric premolar crown (Wood et al., 1988). The simplification in the number and morphology of the root is a process that is first observed in the earlier *Homo* specimens (Wood et al., 1988) but reaches its maximum expression in modern humans (Scott and Turner, 1997) as well as in the *H. heidelbergensis/neanderthalensis* lineage (Bermúdez de Castro, 1993, 1988; Bermúdez de Castro et al., 2003). Out of 74 lower premolars from SH, there is only one fragment of a P3 root *in situ* with mesial and distolingal roots. This could be a variant of the *H. antecessor* root but it is difficult to assess the evolutionary significance of this type of exception. Arago 13 also presents two-rooted lower first premolars, with mesiobuccal and buccodistal components (Bermúdez de Castro et al., 2003).

Mid-Trigonid Crest in M1 and/or M2

The mid-trigonid crest is an enamel crest that connects the mesial portions of the protoconid and the metaconid. This trait was added to the ASU classification in 1993 (Wu and Turner, 1993). According to Bailey (2002a) and Zubov (1992a,b), the presence of a continuous crest, that is uninterrupted by the central longitudinal fissure is very rare in modern humans, but very frequent in Neandertals. A strongly developed mid-trigonid crest is also uncommon in early *Homo* individuals and, in general, in the *Australopithecus* genus although it is possible to observe it in some Sangiran *H. erectus* molars (Kaifu et al., 2005). This trait is developed in almost all of the SH individuals (including M1, M2 and M3) and in one of the two TD6 lower M2s. Although, in the TD6 M2 it is slightly less pronounced than in the typical Neanderthal and *H. heidelbergensis* molars (Figure 5).

Four Cusped M1 and/or M2

The hypoconulid or Cusp 5 is a distal cusp situated between the hypoconid and the entoconid in a five-cusped tooth or between the hypoconid and the *tuberculum sextum* (Cusp 6) in a six-cusped tooth. Lower first molars with four cusps are very uncommon in primates (Swindler, 1976), while in the later stages of human evolution the trend in lower molar cusp reduction has mainly affected the hypoconulid (Bermúdez de Castro and Nicolás, 1995). The majority of *H. sapiens* M1s express five cusps, while M2s are more prone to display inter-population variability (Bath-Balogh and Selma, 1997; Scott and Turner, 1997). Although four-cusped M1s are less frequent than four-cusped M2s, four-cusped M2s pattern of geographic variation shows interesting results (Scott and Turner, 1997).

The absence of Cusp 5 can be found in both the M1 and M2 of some *H. sapiens* individuals. The absence of Cusp 5, although exceptional, can be found in the M2 of certain *H. neanderthalensis* specimens such us Hortus V-J9, which lacks a hypoconulid although it expresses C6 (this condition is also seen in some SH individuals (Bermúdez de Castro, 1987 and personal observation) and *H. heidelbergensis* (Bermúdez de Castro, 1987, 1988; and personal observation). *H. antecessor*, however, does not present hypoconulid reduction in M1 or M2 (Figure 5) (Bermúdez de Castro, 1993, and personal observation).

Upper First Molar Shape

A recent study has explored the possible distinctive morphology of upper first molars in Neandertals (Bailey, 2004), which may be helpful to distinguish this taxon from modern human specimens. We are currently exploring this tooth by geometric morphometric analysis and we have also found significant differences between Neanderthals and the rest of hominins (Gómez-Robles, in press). As outlined previously by Bailey (2004), modern humans tend to present more squared outlines while Neanderthals are characterized by rhomboidal shapes with bulging hypocones in their distolingual corners. There is a distal displacement of the lingual cusps relative to the buccal ones, particularly at the expense of the hypocone, which becomes closer to the lingual border of the crown and protrudes in the external outline. We have only observed this morphology in the majority of the SH and Neandertal specimens, as well as in both of the TD6 specimens (Figure 6). In constrast, modern humans maxillary M1 morphology would align better with early *Homo* specimens.

Discussion

Bermúdez de Castro et al. (2003) found that the dental morphological differences between the SH and TD6 samples support the hypothesis that they belong to different paleospecies. In agreement with previous studies, (Arsuaga et al., 1993, 1997a, b; Martínez and Arsuaga, 1997; Rosas and Bermúdez de Castro, 1998; Bermúdez de Castro et al., 2004b), the dental characters included in this study suggest that the dental morphology of the *H. heidelbergensis* specimens from SH are already derived toward that observed in the *H. neanderthalensis* lineage. For this reason, we believe that *H. heidelbergensis* Schoetensack 1908 (including African and European specimens), cannot represent the last common ancestor of both Neandertal and modern human lineages like some authors have suggested (Stringer, 1985; Rightmire, 1996; Stringer and McKie, 1996; Tattersall, 1996). Instead, we conclude that this taxon should only be assigned only to the Middle Pleistocene specimens from Europe (Arsuaga et al., 1993, 1997a, b; Martínez and Arsuaga, 1997; Rosas and Bermúdez de Castro, 1998; Bailey, 2002a; Bermúdez de Castro et al., 2004b). The African Middle Pleistocene specimens lack some of the Neanderthal apomorphic traits present in their European counterparts. Therefore, we support their classification as *H. rhodesiensis* Woodward, 1921, a chronospecies of the African lineage *H. rhodesiensis-H. sapiens*.

Our observations indicate that *H. antecessor* dental morphology presents certain similarities

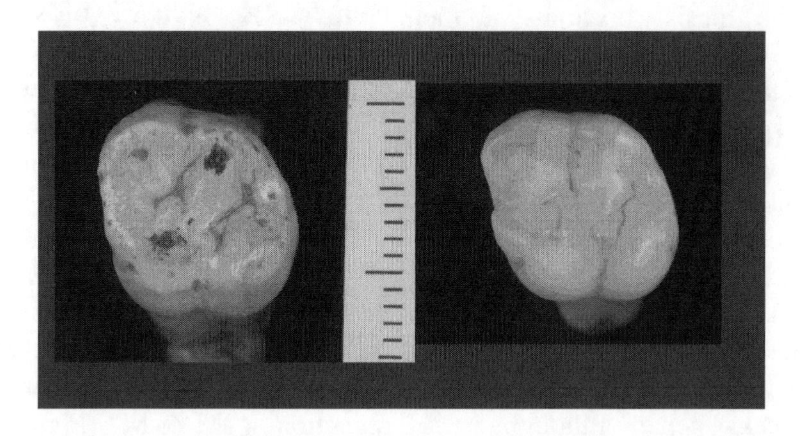

Figure 6. Gran Dolina-TD6 (left) and Sima de los Huesos (right) upper first molar.

with *H. erectus* populations such as their upper lateral incisor morphology, which might suggest a close relationship of this taxon to Asian populations. This evidence would be consistent with the mandibular traits found in the ATD6-69 specimen recently recovered from TD6 (Bermúdez de Castro et al., 2005; Carbonell et al., 2005a). The modern midfacial morphology as well as other cranial traits described in the *H. antecessor* hypodigm (Arsuaga et al., 1999) could point to some similarities shared with Nanjing I and Zhoukoudian specimens (Wang and Tobias, 2000). Moreover, the morphology of *H. antecessor* upper first molars and lower second premolars, differentiates this species from the Ternifine specimens, despite that some authors have suggested they belong to the same taxon (Hublin, 2001; Stringer 2003).

In 2003, we explored the discontinuity scenario (Bermúdez de Castro et al., 2003) as a working hypothesis for the first human settlement of Europe. The identification of a set of dental characters that we believe are useful in characterizing the Middle and Lower Pleistocene hominin species has led us to reassess this scenario under a new light.

The primitive morphology of first lower premolar crown and root, the size sequence of lower molar series (M1 > M2), a well-developed hypoconulid always present in M1 and M2, the expression of a buccal cingulum in their mandibular canines and premolars, and the extensively crenulated enamel, along with the primitive nature of other characters, support that morphological differences exist between the TD6 and the SH populations. However, some of the characters considered in this study would suggest continuity between the European Lower and Middle Pleistocene populations. These include similarities in the shovel shape of their upper lateral incisors, a rhomboidal first upper molar with protruding hypocone, and an asymmetrical P4 with a reduced lingual half, extra lingual cusps and frequent expression of a transverse crest (Martinón-Torres et al., 2006). Under a continuity scenario *H. antecessor* would have evolved in Europe culminating with the appearance of the *H. heidelbergensis/neanderthalensis* lineage.

The continuity scenario apparently presents some problems in reconciling the archaeological evidence (Bermúdez de Castro et al., 2003) relative to the transition of Mode 1 to Mode 2 lithic industries in Europe (Carbonell et al., 1999). The lithic artifacts found associated with *H. antecessor* fossils belong to Mode 1 (Carbonell et al., 1995). Other European Lower Pleistocene lithic collections such as Monte Poggiolo (Peretto and Ferrari, 1995: Gagnepain et al., 2000), Campo Grande (Ceprano), Barranco León, Fuentenueva 3 (Tixier et al., 1995, Turq et al., 1996) and the TD4 and TD5 levels of Gran Dolina (Carbonell and Rodríguez, 1994) have also been assigned to the Mode 1 typology. Some authors do not consider the appearance of Mode 2 in Europe as an evolution from the Mode 1 technology present in this continent during the Lower Pleistocene (Carbonell et al., 1999). According to them, the oldest Mode 2 technology appeared in Europe between 0.6 and 0.5 myr ago, and had all the features of the well-developed African Mode 2, lacking the archaic characteristics of the early African Mode 2. They proposed that Mode 2 could have been introduced into Europe by a new migration of hominins that originated in socially complex populations of the African Rift System, with Mode 2 technology at that time (Manzi, 2004).

However, if continuity between Lower and Middle Pleistocene populations is the case, other alternatives have to be considered. Mode 2 technology could also have been introduced into the continent by cultural diffusion (Carbonell et al., 1999). The new technique may have been socialized and assimilated by the European populations provoking a "social reorganization" (Carbonell et al., 2005b) of these groups.

The acquisition of the new mode would have had major effects on the complexity of the groups leading to their demographic explosion towards the north.

The colonization of the European continent remains as a complex scenario that still needs to be clarified. This issue will undoubtedly be advanced by an understanding of the archaeological context, especially from the more dynamic "operative chain" frame (Geneste, 1985, 1991; Karlin et al., 1991; Carbonell and Sala, 1993; Otte, 1998, Turq, 1999), which considers the lithic artifacts beyond mere typological classification. We believe that a real coordination between the analysis of biological and cultural evidence will lead to important advances in our knowledge of human evolution.

Acknowledgments

We are sincerely grateful to J.-J. Hublin and S. Bailey for their kind invitation to participate in the Dental Conference held at the Max Planck Institute (Leipzig) and in this volume, as well as to all the personnel involved in the meeting organization. We are grateful to all members of Atapuerca research team. We also want to thank to C. Bernis and J. Rascón from Universidad Autónoma de Madrid; A. Pérez-Pérez, J. Galbany, L. M. Martínez and F. Estebaranz from the Universitat de Barcelona; D. Lordkipanidze and A. Vekua from the Georgian National Museum; J. Svoboda and M. Oliva from the Institute of Archaeology-Paleolithic and Paleoethnology Research Center, Dolní Vestonice, Czech Republic; H. and M. A. de Lumley from Institut de Paléontologie Humaine, Paris; and I. Tattersall, K. Mowbray and G. Sawyer from the American Museum of Natural History, New York; for providing access to the comparative sample and their helpful assistance when examining it. We also thank J. Baena from the Universidad Autónoma de Madrid for his helpful comments. This research was supported by funding from the Dirección General de Investigación of the Spanish M. E. C., Project N° BOS2003-08938-C03-01, 02, 03, Spanish Ministry of Science and Education, Fundación Atapuerca, Fundación Duques de Soria, and Fundación Caja Madrid. This research was partially carried out under the Cooperation Treaty between Spain and the Republic of Georgia, hosted by the Fundación Duques de Soria and the Georgian National Museum. Fieldwork at Atapuerca is supported by Consejería de Cultura y Turismo of the Junta de Castilla y León.

References

Arsuaga J.L., Bermúdez de Castro J.M., Carbonell E., 1997a. The Sima de los Huesos Hominid Site. Journal of Human Evolution Special Issue, 33.

Arsuaga, J.L., Martínez, I., Gracia, A., Carretero, J.M., Carbonell, E., 1993. Three new human skulls from the Sima de los Huesos site in Sierra de Atapuerca, Spain. Nature 362, 534–537.

Arsuaga, J.L., Martínez, I., Gracia, A., Lorenzo, C., 1997b. The Sima de los Huesos crania (Sierra de Atapuerca, Spain). A comparative study. Journal of Human Evolution 33, 219–281.

Arsuaga, J.L., Martínez, I., Lorenzo, C., Gracia, A., Muñoz, A., Alonso, O, Gallego, J., 1999. The human cranial remains from Gran Dolina Lower Pleistocene site (Sierra de Atapuerca, Spain). Journal of Human Evolution 37, 431–457.

Bailey, S.E., 2002a. A closer look at Neanderthal postcanine dental morphology: the mandibular dentition. Anatomical Record (New Anat.) 269, 148–156.

Bailey, S.E., 2002b. Neandertal dental morphology: implications for modern human origins. Ph.D. Dissertation, Arizona State University, Tempe.

Bailey, S.E., 2004. A morphometric analysis of maxillary molar crowns of Middle-late Pleistocene hominins. Journal of Human Evolution 47, 183–198.

Bailey, S.E., Lynch, J.M., 2005. Diagnostic differences in mandibular P4 shape between Neandertals and anatomically modern humans. American Journal of Physical Anthropology 126(3), 268–277.

Bath-Balogh, M., Selma, K., 1997. *Illustrated Dental Embriology, Histology and Anatomy*. Elsevier Science Health Science.

Bermúdez de Castro, J.M., 1987. Morfología comparada de los dientes humanos fósiles de Ibeas (Sierra de Atapuerca, Burgos). Estudios geologia 43, 309–333.

Bermúdez de Castro, J.M., 1988. Dental remains from Atapuerca/Ibeas (Spain) II. Morphology. Journal of Human Evolution 17, 279–304.

Bermúdez de Castro, J. M., 1993. The Atapuerca dental remains: new evidence (1987–1991 excavations) and interpretations. Journal of Human Evolution 24, 339–371.

Bermúdez de Castro, J.M., Arsuaga, J.L., Carbonell, E., Rosas, A., Martínez, I., Mosquera, M., 1997. A hominid from the Lower Pleistocene of Atapuerca, Spain: possible ancestor to Neandertals and modern humans. Science 276, 1392–1395.

Bermúdez de Castro, J.M., Carbonell, E., Arsuaga, J.L., 1999a. Gran Dolina Site: TD6 Aurora Stratum (Burgos, Spain). Journal of Human Evolution, Special Issue, 37.

Bermúdez de Castro, J.M., Martinón-Torres, M., Carbonell, E., Lozano, M., Gómez, A., Sarmiento, S., 2005. Origen y filogenia de los primeros homínidos de Europa. Munibe (Antropología-Arkeologia) 57/3, 279–287. Homenaje al Prof. Jesús Altuna.

Bermúdez de Castro, J.M., Martinón-Torres, M., Lozano, M., Sarmiento, S., 2004a. Palaeodemography of the Atapuerca-SH hominin sample: a revision and new approaches to the palaeodemography of the European Middle Pleistocene Population. Journal of Anthropological Research 60, 5–26.

Bermúdez de Castro, J.M., Martinón-Torres, M., Carbonell, E., Sarmiento, S., Rosas, A., Lozano, M., 2004b. The Atapuerca sites and their contribution to the knowledge of human evolution in Europe. Evolutionary Anthropology 13, 25–41.

Bermúdez de Castro, J.M., Martinón-Torres, M., Sarmiento, S., Lozano, M., 2003. Gran Dolina-TD6 versus Sima de los Huesos dental samples from Atapuerca: evidence of discontinuity in the European Pleistocene population? Journal of Archaeological Science 30, 1421–1428.

Bermúdez de Castro, J.M., Nicolás, M.E., 1995. Posterior dental size reduction in hominids: the Atapuerca Evidence. American Journal of Physical Anthropology 96, 335–356.

Bermúdez de Castro, J.M, Rosas, A., Nicolás, M.E., 1999b. Dental remains from Atapuerca-TD6 (Gran Dolina site, Burgos, Spain). Journal of Human Evolution 37, 523–566.

Biggerstaff, R.H., 1969. The basal area of posterior tooth crown components: the assessment of within tooth variation of premolars and molars. American Journal of Physical Anthropology 31, 163–170.

Bischoff, J.L., Shampa, D.D., Aramburu, A., Arsuaga, J.L., Carbonell, E., Bermúdez de Castro, J.M., 2003. The Sima de los Huesos hominids date to beyond U/Th equilibrium (> 350 kyrs) and perhaps to 400–600 kyrs: new radiometric dates. Journal of Archaeological Science 30, 275–280.

Carbonell, E., Bermúdez de Castro, J.M., Arsuaga, J.L., Allué, E., Bastir, M., Benito, A., Cáceres, I., Canals, T., Díez, J.C., van der Made, J., Mosquera, M., Ollé, A., Pérez-González, A., Rodríguez, J., Rodríguez, X.P., Rosas, A., Rosell, J., Sala, R., Vallverdú, J., Vergés, J.M., 2005a. An Early Pleistocene hominin mandible from Atapuerca-TD6, Spain. Proceedings of the National Academy of Sciences of the USA 102, 5674–5678.

Carbonell, E., Bermúdez de Castro, J.M., Arsuaga, J.L., Díez, J.C., Rosas, A., Cuenca-Bescós G., Sala, R., Mosquera, M., Rodríguez, X.P., 1995. Lower Pleistocene hominids and artifacts from Atapuerca-TD6 (Spain). Science 269, 826–830.

Carbonell, E., Mosquera, M., Rodríguez, X.P., Sala, R., Ollé, A., Vergés, J.M., Martínez-Navarro, B., van der Made, J., Bermúdez de Castro, J.M., 2005b. Hominid dispersals: The Out of Africa Technological Hypothesis. Human Nature, submitted.

Carbonell, E., Mosquera, M., Rodríguez, X.P., Sala, R., van der Made, J., 1999. Out of Africa: the dispersal of the earliest technical systems reconsidered. Journal of Anthropological Archaeology 18, 119–136.

Carbonell, E., Rodríguez, X.P., 1994. Early Middle Pleistocene deposits and artefacts in the Gran Dolina site (TD4) of the "Sierra de Atapuerca" (Burgos, Spain). Journal of Human Evolution 26, 292–311.

Carbonell, E., Sala, R., 1993. Cadena opertiva i "transfer" en els objectes d'ús humà i llur context. Empúries 48–50(I), 176–183.

Crummett T., 1994. The evolution of shovel shaping: Regional and temporal variation in human incisor morphology. Ph.D. dissertation, University of Michigan, Ann Arbor.

Dahlberg, A.A., 1956. *Materials for the Establishment of Standards for Classification of Tooth Characteristics*. Zoller Laboratory of Dental Anthropology, University of Chicago, Chicago.

Falguères, C., Bahain, J-J., Yokoyama, Y., Arsuaga, J.L., Bermúdez de Castro, J.M., Carbonell, E., Bischoff, J.L., Dolo, J-M., 1999. Earliest humans in Europe: the age of TD6 Gran Dolina, Atapuerca, Spain. Journal of Human Evolution 37, 343–352.

Gagnepain, J., Hedley, I., Bahain, J.-J., Wagner, J.-J., 2000. Ètude magnétostratigraphique du site de Ca'Belvedere di Monte Poggiolo (Forli, Italie), et de son contexte stratigraphique. Prémiers ŕsultats. In: Pretto, C. (Ed.), I Primi Abitanti della Valle Padana: Monte Poggiolo Nel Quadro delle Conoscenze Europee. Jaka Book, Milan, pp. 319–335.

Geneste, J.M., 1985. Analyse lithique d'industries Mousteriennes du Périgord: Une approche technologique du comportament des Groupes Humaines au Paleolithique Moyen. Ph.D. dissertaion. University of Bourdeaux.

Geneste, J.M., 1991. L'Approvisionnement en matières premières dans les systemes de production lithique: la dimension spatiale de la technologie·Tecnología y Cadenas Operativas Líticas. Treballs d'Arqueología 1, 1–36.

Genet-Varcin, E., 1962. Évolution de la couronne de la seconde prémolaire inférieure chez les hominidés. Annals Paléo (Vert) XLVIII, 59–81.

Gómez-Robles, A., Martinón-Torres, M., Bermúdez de Castro, J.M., Margvelashvili, A., Bastir, M., Arsuaga, J.L., Pérez-Pérez, A., Estebaranz, F., Martínez, L. A geometric morphometric análisis of hominin upper first molar shape. Journal of Human Evolution (in press).

Hrdlička, A., 1920. Shovel-shaped teeth. American Journal of Physical Anthropology 3, 429–465.

Hublin, J.-J., 2001. Northwestern African Middle Pleistocene hominids and their bearing on the emergence of *Homo sapiens*. In: Barham, L., Robson-Brown, K. (Eds.), Human Roots: Africa and Asia in the Middle Pleistocene. Western Academic and Specialist Press, Bristol, pp. 99–121.

Kaifu, Y., Aziz, F., Baba, H., 2005. Hominid mandibular remains from Sangiran: 1952–1986. American Journal of Physical Anthropology 128, 497–519.

Karlin, C., Boud, P., Pelegrin, J., 1991. Processus techiques et chaînes opératoires. Comment les préhistoriense s'appropient un concept élaboré par les ethnologues». In: Balfet, H. (Ed.), Observer l'action techniques des chaînes opératoires, pour quoi faire?. CNRS editions, pp. 101–117.

Ludwig, F.J., 1957. The mandibular second premolars: morphologic variation and inheritance. Journal of Dental Research 36(2), 263–273.

Lumley, M.A., 1973. Anténéandertaliens et Néandertaliens du bassin Méditerranéen Occidental Européen. Études quaternaires 2. Université de Provence.

Manzi, G., 2004. Human evolution at the Matuyama-Brunhes boundary. Evolutionary Anthropology 13(1), 11–24.

Martínez, I., Arsuaga, J.L., 1997. The temporal bones from Sima de los Huesos Middle Pleistocene site (Sierra de Atapuerca, Spain). A phylogenetic approach. Journal of Human Evolution 33, 283–318.

Martinón-Torres, M., Bastir, M., Bermúdez de Castro, J.M., Gómez, A., Sarmiento, S., Muela, A., Arsuaga, J.L., 2006. Hominin lower second premolars morphology: evolutionary inferences through geometric morphometric analysis. Journal of Human Evolution 50, 523–533.

Mizoguchi, Y., 1985. Shovelling: a statistical analysis of its morphology. The University Museum, The University of Tokyo – Bulletin 26, 1–52.

Otte, M., 1998. *Industrie lithique de la Couche 5. In Recherches aux Grottes de Sclayn* Vol. 2 L'Archeologie E. R. A. U. L. 79. Lieja, pp. 277–278.

Parés, J.M., Pérez-González, A., 1995. Paleomagnetic age for hominid fossils at Atapuerca site, Spain. Science 269, 830–832.

Parés, J. M., Pérez-González, A., Rosas, A., Benito, A., BermúdezdeCastro,J.M.,Carbonell,E.&Huguet, R., 2006. Matuyama-age lithic tools from the SimadelElefantesite,Atapuerca(northernSpain). JournalofHumanEvolution50,163–169.

Patte, É., 1962. La dentition des Néanderthaliens. Masson et Cie, Éditerus, Paris, pp. 237–242.

Peretto, C., Ferrari, M., 1995. Techno-typological consideration on the industry from Ca'Belvedere di Monte Poggiolo (Forly, Italy), Cahier Noir 7, 3–16.

Rightmire, G.P., 1996. The human cranium from Bodo, Ethiopia: evidence for speciation in the Middle Pleistocene? Journal of Human Evolution 31, 21–39.

Roebroeks, W., van Kolfschoten, T., 1994. The earliest occupation of Europe: a short chronology. Antiquity 68, 489–503.

Rohlf, F.J., Slice, D., 1990. Extensions of the Procrustes Method for the optimal superimposition of landmarks. Systematic Zoology 39, 40–59.

Rosas, A., Bermúdez de Castro, J.M., 1998. The Mauer mandible and the evolutionary significance of *Homo heidelbergensis*. Geobios 31, 687–697.

Schoetensack, O. 1908. *Der Unterkiefer des* Homo heidelbergensis *aus den Sanden von Mauer bel Heidelberg: ein Beitrag zur Palaöntologie des Menschen*. Engelmann, Leipzig.

Scott, G.R., Turner II, C.G., 1997. *The Anthropology of Modern Human Teeth*. Cambridge University Press, Cambridge.

Sperber, G., 1974. Morphology of the check teeth of early South African hominids. Ph.D. Thesis, University of Witwatersrand.

Stringer, C.B., 1985. Middle Pleistocene variability and the origin of Late Pleistocene humans. In: Delson, E. (Ed.), Ancestors: The Hard Evidence. Alan R Liss, New York, pp. 289–295.

Stringer, C., 2003. Out of Ethiopia. Nature 423, 692–694.

Stringer, C.B., McKie, R., 1996. *African Exodus. The Origin of Modern Humanity*. Jonathan Cape, London.

Swindler, D.R., 1976. *Dentition of Living Primates*. Academic Press, London.

Tattersall, I., 1996. *The Last Neanderthal. Macmillan*, New York.

Tixier, J., Roe, D., Turq, A., Gibert, J., Martínez, B., Arribas, A., Gibert, L., Gaete, R., Myillo, A., Iglesias, A., 1995. Présence d'industries lithiques dans le Plésitocène inférieur de la region d'Orce (Grenade, Espagne): Quel est l'état de la question? Comptes Rendus de l'Academie des Sciences Paris 321, 71–78.

Tobias, P.V., 1991. *Olduvai Gorge. The Skulls, Endocasts and Teeth of Homo habilis. Vol 4: Parts V–IX*. Cambridge University Press, Cambridge.

Turner II, C.G., Nichol, C.R., Scott, G.R., 1991. Scoring Procedures for Key Morphological Traits of the permanent dentition: The Arizona State University Dental Anthropology System. Advances in Dental Anthropology. In: Kelley, M., Larsen, C. (Eds.), Advances in Dental Anthropology. Wiley Liss, New York, pp. 13–31.

Turq, A., 1999. Reflections on the Middle Palaeolithic or Aquitania Basin. In: Roebroeks, W., Gamble, C. (Eds.), The Middle Paleolithic Occupation of Europe, University of Leiden, Leiden, pp. 107–120.

Turq, A., Martínez-Navarro, B., Palmqvist, P., Arribas, J., Agusti, J., Rodríguez Vidal, J., 1996. Le Plio-Pleistocene de la région d'Orce, Province de Granade, espagne: bilan et perspectives de reserche. Paléo 8, 161–204.

Uytterschaut, H.T., Wood, B.A., 1989. Dental morphology: characterization and identification of Australopithecines and *Homo habilis*. Hominidae. Proceedings of the 2nd International Congress of Human Paleontology. Editoriale Jaca Book, Milan, pp. 183–188.

Wang, Q., Tobias, P.V., 2000. Review of the phylogenetic position of Chinese *Homo erectus* in light of midfacial morphology. Acta Anthropologica Sinica 19 (Supplement), pp. 26–36.

Weidenreich, F., 1937. The dentition of *Sinanthropus pekinensis*: a comparative odontography of the Hominids. Palaeontologica Sinica. N. S. D n°1. T, The Geological Survey of China.

Wolpoff, M.H., 1971. Metric trends in hominid dental evolution. Case Western Reserve Universty, Studies in Anthropology 2, 1–244.

Wood, B.A., Abbot, S.A., 1983. Analysis of the dental morphology of Plio-Pleistocene hominids. I. Mandibular molars: crown area measurements and morphological traits. Journal of Anatomy 136(1), 197–219.

Wood, B.A., Abbott, S.A., Uytterschaut, H., 1988. Analysis of the dental morphology of Plio-Pleistocene hominids. IV. Mandibular postcanine root morphology. Journal of Anatomy 156, 107–139.

Wood, B.A., Uytterschaut, H., 1987. Analysis of the dental morphology of Plio-Pleistocene hominins. III. Mandibular premolar crowns. Journal of Anatomy 154, 121–156.

Woodward, A. S. 1921. A new cave man from Rhodesia, South Africa. Nature 108, 371–372.

Wu, L., Turner II, C. G., 1993. Variation in the frequency and form of the lower permanent molar middle trigonid crest. American Journal of Physical Anthropology 91, 245–258.

Zubov, A. A., 1992a. The epicristid or middle trigonid crest defined. Dental Anthropology Newsletter 6, 9–10.

Zubov, A. A., 1992b. Some dental traits in different evolutionary lines leading to modern man. Dental Anthropology Newsletter 6, 4–8.

6. Neural network analysis by using the Self-Organizing Maps (SOMs) applied to human fossil dental morphology: A new methodology

F. MANNI
UMR 5145 – Eco-Anthropology Group
National Museum of Natural History
MNHN – Musée de l'Homme
17, Place du Trocadéro, 75016 Paris – France
manni@mnhn.fr

R.VARGIU
Department of Animal and Human Biology
Section of Anthropology
University of Rome "La Sapienza"
Piazzale Aldo Moro, 5, 00185 Rome, Italy
rita.vargiu@uniroma1.it

A. COPPA
Department of Animal and Human Biology
Section of Anthropology
University of Rome "La Sapienza"
Piazzale Aldo Moro, 5, 00185 Rome, Italy
alfredo.coppa@uniroma1.it

Keywords: neural network analysis, Self-Organizing Maps (SOMs), multidimensional scaling, dental morphology, Arizona State University Dental Anthropology System (ASUDAS), Lower Pleistocene specimen, Middle Pleistocene specimen, Late Pleistocene specimen

Abstract

Recent studies focusing on dental morphology of extinct and extant human populations have shown, on a global scale, the considerable potential of dental traits as a tool to understand the phenetic relations existing between populations. The aim of this paper is to analyze the dental morphologic relationships between archaic *Homo* and anatomically modern *Homo sapiens* by means of a new methodology derived from artificial neural networks called Self Organizing Maps (SOMs). The graph obtained by SOMs to some extent recalls a classical Multidimensional Scaling (MDS) or a Principal Component Analysis (PCA) plot. The most important advantages of SOMs is that they can handle vectors with missing components without interpolating missing data. The analyzed database consisted of 1055 Lower-Middle and (Early) Late Pleistocene specimens, which were scored by using dental morphological traits of the Arizona State University Dental Anthropology System (ASUDAS). The principal result indicates a close relationship between the *Homo erectus s.l.* and Middle Pleistocene specimens and the

S.E. Bailey and J.-J. Hublin (Eds.), Dental Perspectives on Human Evolution, 81–101.

later Neandertal groups. Furthermore, the dental models of anatomically modern *Homo sapiens* are particularly different compared to the more archaic populations. Thus, SOMs can be considered a valuable tool in the field of dental morphological studies since they enable the analysis of samples at an individual level without any need *i)* to interpolate missing data or *ii)* place individuals in predetermined groups.

Introduction

Studies focusing on dental morphology of anatomically modern human populations have shown, on a global scale, the considerable potential of dental traits as a tool to understand the phenetic relationships between populations (e.g., Coppa et al., 1997, 1999a; Scott and Turner, 1997). Based on these encouraging results some researchers, as a second step, have addressed the question regarding the emergence of anatomically modern humans using the same approach (e.g., Stringer et al., 1997; Irish 1997, 1998, 2000; Bailey and Turner, 1999; Bailey 2002a, 2002b, 2004; Irish and Guattelli-Steinberg, 2002; Bailey and Lynch, 2005).

An extension of this research has been to focus on the frequencies of specific dental traits and how they change over time in a specific area of interest. In this context, the study of Upper Palaeolithic Italian populations has led to the identification of a specific "Italian Upper Palaeolithic dental complex" that enables a clear discrimination between such Palaeolithic populations and more recent Italian groups (Coppa et al., 1999c, 2000a). The identification of this dental complex also makes it possible to distinguish between the different populations that inhabited the Italian peninsula over time. Furthermore, the "Italian Upper Palaeolithic dental complex" has been shown to be a common trait of all the other Upper Palaeolithic European populations in what can now be defined a "European Upper Palaeolithic dental complex" (Coppa et al., 2000b).

Similarly, the comparison of a large number of European, Asian and North-African Upper Pleistocene remains has shown that Neandertals have their own distinct pattern of trait frequencies or "Neandertal dental complex" (Bailey, 2002b; Coppa et al., 1999b, c, 2004). A more refined analysis suggested there is a clear subdivision of the "Neandertal dental complex" into three well-defined groups: "European pre-Würmian" and "European Würmian and Middle Eastern Würmian" subgroups that are distinct from the dental morphologies of anatomically modern *Homo sapiens* remains (AMHS) (Coppa et al., 2001).

While some researchers have pointed out that the Arizona State University Dental Anthropology System (ASUDAS) does not account for all fossil hominid dental morphology (Bailey, 2002a, b), others have applied it to Pliocene specimens (Irish and Guattelli-Steinberg, 2002). While relying only on ASUDAS traits may introduce some bias to this study, our main objective here is to explore a new method for assessing phenetic relationships among groups and not to make broad conclusions about the mode or pattern of human evolution. Because many, if not most, of the ASUDAS traits can be found in fossil hominins, we believe it is, at least a good starting point for investigating relationships among members of the genus *Homo*. For this reason we decided to extend our investigations to more ancient specimens including *Homo erectus s.l.*, non-Neandertal Middle Pleistocene *Homo*, Neandertals, anatomically modern *Homo sapiens* of the Middle Paleolithic from Israel and later Upper Paleolithic AMHS

Table 1. Number of individuals used in MDS and SOMs analyses

Label	*Group*	*Sample size*
HER	*Homo erectus s.l.*	45
PLE	Middle Pleistocene	46
PNE	Prewürmian Neandertal Europe	119
NEA	Würmian Neandertal Europe	100
NME	Würmian Neandertal Middle East	19
AMH	Levantine AMHS Israel	28
PAT	Early Upper Paleolithic Europe	141
PRT	Late Upper Paleolithic Europe	197
NAT	Natufian (Israel)	228
IBE	Iberomaurusian (North Africa)	132
	Total	**1055**

from (Eurasia and North-Africa) (Table 1). The aim of the study is to apply a new methodology derived from Artificial Neural Networks—Self Organizing Maps (SOMs)—to highlight the dental morphologic relationships between Early-Archaic *Homo*, Neandertal and AMHS and compare it to results based on multivariate analysis (e.g., Multidimensional Scaling or MDS).

Materials and Methods

In the last ten years the use of Artificial Neural Networks (ANN) has become widespread in a number of different disciplines (engineering, economics, ecological and environmental studies, biology, medicine, etc.). ANN have usually been used to solve problems of classification, prediction, categorization, optimisation and data-mining since they represent a method to approximate systems that cannot be effectively modelled by classic statistical methods. Their use seems more successful where there is no linear relation between the predicted variable and the data used for the prediction itself.

In anthropological sciences predictions are commonplace, but the relations between the different variables are seldom linear. This is the case for bone and dental measurements, which change through time and space under the influence of genetic and environmental factors that interact in a complex manner (Schmitt et al., 2001). For this reason we think that ANN might be a suitable tool to address the classification of dental remains from palaeoanthropological samples.

Self Organizing Maps (SOMs)

Self Organizing Maps (SOMs) constitute a technique to visualize data, derived from research on ANN, invented by Professor Teuvo Kohonen (Kohonen, 1982, 1984, 2001). SOMs reduce the dimensions of data by grouping similar items together in a two-dimensional representation.

Historically, they were developed to simulate the functioning of nets of neurons in the brain; therefore each cell is called a "neuron" since, according to the self-organizing algorithm, they are able to exchange information with neighboring neurons, to an extent that is specified by the user. A main difference between SOMs and 'classical' ANN is that the former technique does not need a training sample to obtain a stable map that will, afterwards, be used to map new data-items. In fact, SOMs use the complete dataset as a training sample, which means that each change in the dataset leads to a new map that cannot be compared to a map computed on a subset of the data or with a larger dataset. This difference is essential to fully understand SOMs.

The Philopsophy

The map consists of a regular grid of processing units called "neurons". A model of some multidimensional observations, being a vector which consists of features (e.g., different dental traits and measures) is associated with each unit. The map attempts to represent all the available observations with optimal accuracy using a restricted set of models. At the same time, the models are ordered on the grid so that similar models are close to each other and dissimilar models are far from each other. Therefore, the map is a general statement of relationships where the positioning of items suggests their similarity. It differs from a Cartesian graph in which two-dimensional data relate in accordance with the variation of only two variables (X and Y). Even if the visual aspect of data-representation obtained by SOMs is somewhat similar to a classical Multidimensional Scaling (MDS) (Seber, 1984; Torgerson, 1958) or to a Principal Component Analysis (PCA) (Gabriel, 1968), SOMs describe items on a surface that is not proportional to real distances between the items in the multidimensional space. But, when compared to such multivariate methods, SOMs describe the neighborhood of items more accurately (Kaski, 1997) since their main virtue is to be topology-orientated. As a consequence, SOMs can be preferred to MDS or PCA when all the various inputs (vectors) differ slightly from one another, as is often the case with palaeontological data.

It is worthy of note that, besides SOMs, there are other classification methods able to handle missing values such as support-vector-machines (SVM) variants based on Vapnik's statistical learning theory (Ben-Hur et al., 2001), nearest neighbor variants, and Bayesian classifiers. Nevertheless, the use of such prediction and classification methods, developed in the frame of machine learning approaches, can be very problematic. Moreover, they cannot be regarded as a standard technique based on reference software that is easy to use, as SOMs are.

The Functioning

Mathematically, neurons consist of vectors with the number of components corresponding to the traits to be analyzed (e.g., 1,0,1,0,0,0,1,0,1,0 for data consisting in 10 traits=components). The main idea is that different neurons specialize to represent different types of inputs resulting in a sort of specialization of the different areas of the map. In fact, at the beginning of the analysis, the map is not specialized, meaning that each of the components of the neurons have random values. Then inputs (e.g., real data vectors consisting of dental trait presence/absence scores) start to be analyzed, meaning that they interact with the neurons on the map, that is, with the randomly initialized vectors associated to them. By chance there will be one of the neurons of the map that will be more similar to the data-vector than the others. This neuron-vector is then called the "winner" because it is the one that best represents the input. Such a neuron will undergo a change in its components and will acquire a different state in order to better represent the input datum. Maintaining the metaphor of the brain, this phase has been called "learning". The learning phase is not only local but also applies to neighbouring vectors according to a radius that depends on the user's settings. In such a way, neurons very close to the winning one will be strongly influenced by the input and will change their components in a similar way, whereas distant neurons of the map will be less influenced.

Similarly, when another input-vector enters the map, there will be other winning vectors that will, by chance, best describe it. A new learning process will take place and the specialization of the map will become apparent. This specialization results in areas of the map where similar items are mapped together. During the learning-phase the map is rearranged because contradictory data items (ex: 1,0,1,0,0,0,1,0,1,0 and 0,1,0,1,1,1,0,1,0,1) will influence other neurons of the map in completely different ways. This procedure will result in the differentiation of the map-space: *a*) identical vectors will be mapped at the same position of the map *b*) slightly different ones close to each other, while *c*) very different vectors will be mapped far from each other. The degree of specialization is enhanced by the competition among cells, when an input arrives the neuron that is best able to represent it "wins" the competition and can continue the learning process (Figure 1).

For statistical accuracy the learning process requires an appreciable number of steps

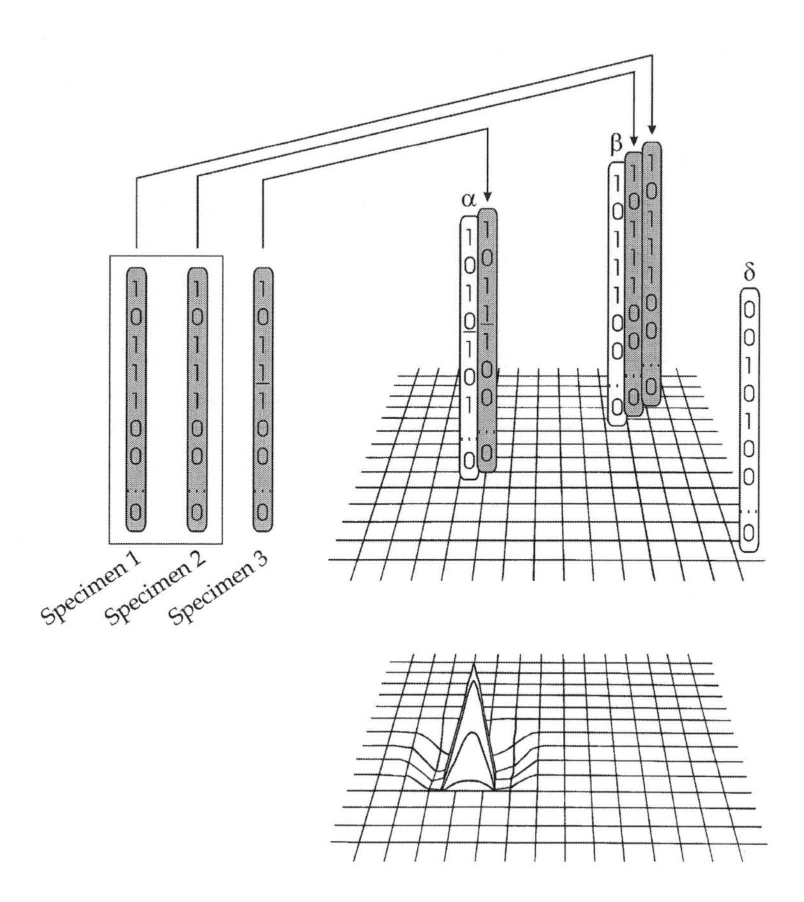

Figure 1. Mapping specimens with Self Organizing Maps (SOMs). In the example we have three tooth specimens (specimens 1; 2; 3 in gray) that are represented as vectors, meaning that their components correspond to the presence or absence of some morphological traits (1 = presence; 0 = absence). When a SOM is created, each cell (neuron) is associated with a "reference vector" (in white) randomly generated, meaning that each reference vector of the map has a number of components that exactly corresponds to those of the inputs (here specimens 1, 2 and 3) without constraints concerning the value of each component (that can be '0' or '1'). In this example specimens 1 and 2 (which are identical) will be mapped to the same reference vector β. Differently, specimen 3 is mapped to reference vector.α which, between shown reference vectors (α, β, γ) is the most similar to it. In the text, we have mentioned the "learning function" which is visualized in the bottom grid. In this grid the adaptation/learning-process that applies to neuron α, as well as to its neighbourhood, is visible. This example has been included to visualize some properties of SOMs such as a) the mapping of similar (identical) items on a same neuron, and b) the fact that some neurons may be "empty" since their components do not correspond to any of the inputs (as is the case of γ). The size of the map can be varied (here we show a 15 × 15 neuron map): a small map provides few neurons ("reference vectors") and, therefore, a poor mapping and vice versa. More details about the method can be found in the Materials and Methods section of the text.

(e.g., 100,000) each consisting in the mapping of a single data-item. The data-item enters the map, the map is modified according to it and the learning takes place. Since the number of available samples is usually smaller, it follows that data-items must be used reiteratively in training. In this case we adopted cyclic iterations of randomly permuted data-items in order to achieve several thousands of steps.

The Size of the Map

The Self Organizing Maps are divided into an arbitrary number of cells (5 × 5; 6 × 6; 7 × 7; etc.) according to user specifications. The correct size of the map depends on the number of data-items and can be effectively found by empirical trials. If the map is too small, many samples will be mapped together; whereas a majority of the cells of a map will be empty if it is too large. The empirical trials are intended to achieve an acceptable visualization compromise. Given that the stringency of the classification depends on the size of the map, a larger map will provide a clearer depiction of the patterns in the data than a smaller map. In a very small map (e.g., a 2 × 2 cell map) there will be only four neurons to which the data-items can be associated; whereas in a 10 × 10 map there will be 100 possible neurons to choose from. If we have one hundred data-items, the representation will not be more informative by using a 100 × 100 map since the patterns found will be similar to those of the 10 × 10 map. The only difference will result in a large number of unused neurons.

Missing Data

The SOM algorithm has proved to be very robust by handling vectors that have missing components (Kaski, 1997). This is probably one of the most important advantages of Self Organizing Maps. Furthermore, it has been demonstrated that better results can be obtained by keeping individuals with missing data instead of discarding those in which certain components (i.e., traits) are lacking (Samad and Harp, 1992). When including individuals with missing data, the only problem is that the normalization of data-items cannot be straightforward since such normalization has to be done separately for each component (e.g., trait).

When vectors present a missing descriptor, only the available values will contribute to the learning process of the map. For an effective learning process, it is obviously advisable that vectors where no descriptors are missing represent a large proportion of the data. Nevertheless, missing data are commonplace in paleoanthropological data. Therefore, to further validate the use of SOMs with such special kinds of datasets, we applied the SOM algorithm to a same dataset containing 0%, 10%, 20% and 30% of missing descriptors. By randomly deleting some vector components we wanted to verify if the mapping of data was still accurate when data items are incomplete. The results (not shown because of space limitations) highly support the clustering obtained with the complete dataset, thus reinforcing the interest of SOMs in paleoanthropology.

Identification of Clusters

In the sections above we mentioned the discrete nature of SOMs, which is related to the discrete number of neurons that compose a map regardless of its size. As was discussed in the section concerning the map size, it becomes apparent that a small map can be adopted if the user wants to 'force' the data into a limited number of categories with the advantage of obtaining an unsupervised clustering and the disadvantage of loosing topological detail in the mapping of items.

Nevertheless, we also mentioned that neighboring neurons on a map could be very similar or, even when quite close, considerably different. This ambiguity is only apparent because the representation of data items on a SOM is enhanced by a gray-scale that helps the user to identify the similarity/difference between neighboring neurons. In this sense SOMs are easy to understand: if neurons are close together and there is light gray connecting them, then they are similar. If there is a black ravine between them, then they are different. The degree of differentiation is proportional to the intensity of the gray that is computed by the software. Full descriptions of this step can be found in the original publication by Kohonen (1982) and the manual of the SOM-PAK vs. 3.1 software.

Outliers and Borders

An advantage of SOMs is that outliers do not affect the results of the analysis. With SOMs each outlier affects only one map unit and its neighborhood, while the rest of the map still provides a good representation of data. Furthermore, the outliers can be easily detected since the map-space is, by definition, very sparsely populated near outliers (Kaski, 1997).

The Algorithm and the Software

A short mathematical description of this non-analytic method is provided here since, to our knowledge, it has never been applied to anthropological data before. The fitting of the model vectors is usually carried out by a sequential regression process, where $t = 1, 2, \ldots$ is the step index: For each sample $\mathbf{x}(t)$, first the winner index c (best match) is identified by the condition

$$\forall i, ||\mathbf{x}(t) - \mathbf{m}_c(t)|| \leq ||\mathbf{x}(t) - \mathbf{m}_i(t)||. \quad (1)$$

After that, all model vectors or a subset of them that belong to nodes centered around node $c = c(\mathbf{x})$ are updated as

$$\mathbf{m}_i(t+1) = \mathbf{m}_i(t) + h_{c(\mathbf{x}),i}(\mathbf{x}(t) - \mathbf{m}_i(t)). \quad (2)$$

Here $h_{c(\mathbf{x}),i}$ is the "neighborhood function", a decreasing function of the distance between the ith and cth nodes on the map grid. This regression is usually reiterated over the available samples (Kohonen, 1993; Kaski & Kohonen, 1994). By virtue of its learning algorithm, the SOM forms a nonlinear regression of the ordered set of reference vectors onto the input surface. The reference vectors form a two-dimensional "elastic network" that follows the distribution of data.

Data, Clustering, and Software Used

Starting from the original database comprised by 1055 individuals (Table 1) we scored 79 dental morphological traits following the Arizona State University Dental Anthropology System (Turner et al., 1991; Scott and Turner, 1997). First, we selected 45 traits to be used in the Multidimensional Scaling analysis (Table 2 in bold and Figure 2). These traits included only those that were shared by all the groups, and which had sample sizes that exceeded five individuals.

Second, in order to build the Kohonen's Maps, we utilized three databases (see below) composed of different sets of traits from those used in the multivariate analysis. We applied this procedure because a large number of human fossil dentitions are incomplete, and because only a minority of individuals present both jaws. Furthermore, since the Kohonen's maps need as many data as possible, and because fossil dentitions are rarely complete, we chose all traits scored in the specimens and the same traits present on multiple teeth in order to maximize data collection. Therefore, we analyzed separately the upper and lower dentitions as well as

Table 2. Frequencies of the 79 ASUDAS traits grouped in 10 samples. The 45 traits used in the MDS analysis (Fig. 2) are in bold type

		Homo erectus s.l		*Middle Pleistocene*		*Pre-würmian Neandertal Europe*		*Würmian Neandertal Europe*		*Würmian Neandertal Middle East*		*Levantine AMHS Israel*		*Early Upper Palae-olithic Europe*		*Late Upper Palae-olithic Europe*		*Natufian (Israel)*		*Iberomaurusian (North Africa)*	
		HER		*PLE*		*PNE*		*NEA*		*NME*		*AMH*		*PAT*		*PRT*		*NAT*		*IBE*	
	Dichotomy	N	%	N	%	N	%	N	%	N	%	N	%	N	%	N	%	N	%	N	%
I^1 Curvature	**2-4/0-4**	**9**	**77.8**	**5**	**100.0**	**18**	**94.4**	**11**	**100.0**	**6**	**83.3**	**13**	**84.6**	**23**	**43.5**	**19**	**42.1**	**37**	**24.3**	**9**	**33.3**
I^1 Shoveling	**3-6/0-6**	**8**	**62.5**	**3**	**66.7**	**21**	**66.7**	**8**	**87.5**	**6**	**100.0**	**13**	**7.7**	**20**	**5.0**	**19**	**0.0**	**37**	**0.0**	**11**	**18.2**
I^2 Shoveling	**3-7/0-7**	**4**	**50.0**	**7**	**71.4**	**20**	**75.0**	**12**	**91.7**	**9**	**100.0**	**14**	**21.4**	**18**	**5.6**	**27**	**0.0**	**49**	**18.4**	**27**	**29.6**
I^1 Double Shoveling	**2-6/0-6**	**8**	**12.5**	**5**	**0.0**	**18**	**0.0**	**11**	**0.0**	**6**	**0.0**	**13**	**0.0**	**25**	**4.0**	**21**	**0.0**	**37**	**5.4**	**14**	**7.1**
I^2 Double Shoveling	**2-6/0-6**	**4**	**0.0**	**6**	**0.0**	**20**	**0.0**	**12**	**0.0**	**9**	**0.0**	**13**	**7.7**	**19**	**5.3**	**29**	**0.0**	**49**	**2.0**	**26**	**11.5**
I^1 1 Interruption. Groove	*1/0-1*	*8*	*0.0*	*1*	*0.0*	*15*	*6.7*	*5*	*0.0*	*3*	*33.3*	*7*	*14.3*	*4*	*20.0*	*16*	*18.8*	*37*	*10.8*	*11*	*36.4*
I^2 Interruption Groove	**1/0-1**	**4**	**0.0**	**3**	**0.0**	**15**	**26.7**	**9**	**33.3**	**7**	**42.9**	**8**	**50.0**	**15**	**60.0**	**29**	**41.4**	**47**	**12.8**	**24**	**20.8**
I^2 Mesial Bending	**1/0-1**	**3**	**0.0**	**5**	**0.0**	**20**	**5.0**	**11**	**0.0**	**5**	**0.0**	**13**	**0.0**	**17**	**23.5**	**29**	**37.9**	**44**	**22.7**	**27**	**14.8**
I^1 Tuberculum Dentale	*2-6/0-6*	*7*	*28.6*	*2*	*0.0*	*17*	*88.2*	*5*	*80.0*	*4*	*75.0*	*12*	*25.0*	*4*	*31.2*	*16*	*37.5*	*36*	*22.2*	*9*	*22.2*
I^2 Tuberculum Dentale	**2-6/0-6**	**4**	**75.0**	**4**	**50.0**	**17**	**100.0**	**8**	**75.0**	**6**	**100.0**	**10**	**90.0**	**14**	**64.3**	**26**	**65.4**	**46**	**80.4**	**26**	**96.2**
I^2 Peg Shaped	**2/0-2**	**7**	**0.0**	**11**	**0.0**	**25**	**0.0**	**21**	**0.0**	**11**	**0.0**	**14**	**0.0**	**29**	**0.0**	**46**	**2.2**	**68**	**0.0**	**62**	**0.0**
C′ Tuberculum Dentale	**2-6/0-6**	**6**	**100.0**	**2**	**100.0**	**20**	**100.0**	**7**	**100.0**	**4**	**100.0**	**6**	**66.7**	**10**	**90.0**	**24**	**58.3**	**40**	**67.5**	**10**	**100.0**
C Distal Accessory Ridge	*2-5/0-5*	*7*	*14.3*	*1*	*100.0*	*12*	*41.7*	*5*	*80.0*	*1*	*100.0*	*3*	*66.7*	*5*	*40.0*	*20*	*50.0*	*33*	*15.2*	*4*	*50.0*
C′ Mesial Rridge	*1-3/0-3*	*7*	*28.6*	*2*	*100.0*	*16*	*25.0*	*7*	*57.1*	*2*	*100.0*	*7*	*0.0*	*7*	*0.0*	*22*	*0.0*	*31*	*12.9*	*4*	*0.0*
P^3 Cusps Number	*1/0-1*	*11*	*0.0*	*5*	*20.0*	*11*	*9.1*	*7*	*28.6*	*2*	*50.0*	*6*	*0.0*	*7*	*25.0*	*19*	*10.5*	*38*	*5.3*	*7*	*0.0*
P^4 Cusps Number	*1/0-1*	*5*	*20.0*	*7*	*14.3*	*11*	*27.3*	*7*	*28.6*	*3*	*33.3*	*2*	*50.0*	*4*	*25.0*	*22*	*4.5*	*31*	*9.7*	*7*	*28.6*
M^1 Metacone	**4-5/0-5**	**18**	**100.0**	**12**	**100.0**	**23**	**100.0**	**23**	**100.0**	**10**	**90.0**	**16**	**100.0**	**37**	**97.3**	**50**	**96.0**	**90**	**100.0**	**68**	**100.0**
M^2Metacone	**3.5-5/0-5**	**12**	**100.0**	**10**	**100.0**	**25**	**92.0**	**14**	**100.0**	**8**	**87.5**	**10**	**100.0**	**42**	**97.6**	**50**	**94.0**	**86**	**100.0**	**66**	**100.0**
M^3Metacone	**3.5-5/0-5**	**11**	**63.6**	**6**	**100.0**	**21**	**81.0**	**11**	**90.9**	**9**	**100.0**	**8**	**100.0**	**28**	**92.9**	**44**	**93.2**	**72**	**86.1**	**69**	**92.8**
M^1 Hypocone	**4-5/0-5**	**18**	**100.0**	**12**	**91.7**	**23**	**100.0**	**23**	**100.0**	**10**	**100.0**	**16**	**93.8**	**32**	**100.0**	**47**	**93.6**	**85**	**100.0**	**58**	**96.6**
M^2 Hypocone	*3.5-5/0-5*	*12*	*91.7*	*10*	*100.0*	*21*	*66.7*	*13*	*92.3*	*7*	*100.0*	*9*	*100.0*	*17*	*60.0*	*45*	*66.7*	*68*	*70.6*	*58*	*86.2*
M^3 Hypocone	*3-5/0-5*	*8*	*87.5*	*4*	*50.0*	*16*	*62.5*	*9*	*88.9*	*3*	*100.0*	*6*	*100.0*	*11*	*50.0*	*39*	*56.4*	*61*	*55.7*	*58*	*81.0*
M^1 Cusp 5	*1-5/0-5*	*8*	*62.5*	*6*	*16.7*	*13*	*69.2*	*15*	*80.0*	*2*	*100.0*	*6*	*50.0*	*11*	*23.8*	*22*	*36.4*	*33*	*27.3*	*22*	*40.9*
M^2 Cusp 5	**1-5/0-5**	**8**	**55.6**	**6**	**62.5**	**13**	**83.3**	**15**	**62.5**	**2**	**100.0**	**6**	**57.1**	**21**	**52.4**	**22**	**63.9**	**33**	**34.8**	**22**	**40.7**
M^3 Cusp 5	**1-5/0-5**	**9**	**80.0**	**8**	**50.0**	**12**	**54.5**	**8**	**55.6**	**3**	**100.0**	**7**	**83.3**	**21**	**73.9**	**36**	**61.1**	**46**	**74.2**	**27**	**67.9**

M^1 Carabelli's Cusp	2-7/0-7	7	42.9	8	62.5	20	75.0	15	73.3	5	100.0	13	84.6	21	42.9	37	56.8	48	58.3	26	76.9
M^2 Carabelli's Cusp	2-7/0-7	8	25.0	7	42.9	16	56.2	9	33.3	1	0.0	7	14.3	10	17.4	40	7.5	66	15.2	32	25.0
M^3 Carabelli's Cusp	2-7/0-7	8	25.0	3	0.0	14	21.4	5	80.0	–	–	5	0.0	10	19.0	29	44.8	49	22.4	55	45.5
M^1 Parastyle	1-5/0-5	9	44.4	8	37.5	14	71.4	18	55.6	4	50.0	12	0.0	18	5.6	43	2.3	71	5.6	42	7.1
M^2 Parastyle	1-5/0-5	7	28.6	5	20.0	18	38.9	12	8.3	6	16.7	8	12.5	29	13.8	44	4.5	81	2.5	56	1.8
M^3 Parastyle	1-5/0-5	4	0.0	3	33.3	15	33.3	7	14.3	6	33.3	8	12.5	20	15.0	40	2.5	66	6.1	61	9.8
M^1 Enamel Extension	2-3/0-3	8	0.0	7	14.3	15	6.7	16	25.0	4	0.0	12	0.0	30	6.7	45	17.8	72	9.7	56	5.4
M^2 Enamel Extension	2-3/0-3	3	0.0	5	0.0	17	5.9	11	9.1	5	0.0	7	0.0	32	12.5	45	20.0	76	1.3	57	0.0
M^3 Enamel Extension	2-3/0-3	3	0.0	3	0.0	12	0.0	10	0.0	6	0.0	7	0.0	9	0.0	35	11.4	55	0.0	60	1.7
C' Root Number	1/1-2	6	100.0	4	100.0	20	100.0	6	100.0	2	100.0	3	100.0	6	100.0	29	100.0	30	100.0	30	100.0
P^3 Root Number	1/1-2	9	44.4	2	0.0	12	66.7	7	85.7	1	100.0	1	100.0	8	40.0	20	25.0	23	21.7	23	60.9
P^4 Root Number	1/1-2	3	100.0	3	66.7	13	100.0	9	88.9	1	100.0	3	100.0	4	90.9	23	91.3	23	82.6	18	94.4
M^1 Root Number	3/1-3	2	50.0	3	66.7	9	44.4	6	66.7	1	0.0	–	–	2	80.0	23	78.3	16	81.2	14	100.0
M^2 Root Number	3/1-3	3	33.3	2	50.0	12	0.0	5	40.0	1	0.0	1	100.0	4	66.7	19	78.9	13	92.3	16	81.2
M^3 Root Number	3/1-3	1	0.0	2	0.0	9	11.1	6	33.3	3	0.0	1	0.0	7	10.0	19	21.1	14	28.6	13	53.8
M^3 Peg Shaped	2/0-2	10	0.0	6	0.0	18	0.0	14	0.0	9	11.1	9	0.0	31	9.7	45	2.2	73	1.4	71	2.8
M^3 Congenital Absence	1/0-1	10	0.0	7	0.0	13	0.0	12	0.0	6	0.0	9	0.0	33	0.0	40	10.0	88	9.1	75	0.0
I_1 Shoveling	2-3/0-3	7	14.3	6	50.0	9	44.4	11	27.3	2	50.0	8	0.0	12	9.1	31	0.0	37	0.0	36	0.0
I_2 Shoveling	2-3/0-3	11	81.8	5	40.0	15	80.0	14	57.1	6	50.0	11	0.0	24	8.3	38	2.6	53	0.0	44	0.0
C, Distal Accessory Ridge	2-5/0-5	3	66.7	5	60.0	15	73.3	13	76.9	2	100.0	7	85.7	15	53.3	26	30.8	36	5.6	16	12.5
P_3 Cusp Number	2-9/0-9	6	83.3	9	22.2	23	56.5	19	52.6	5	20.0	7	14.3	15	0.0	40	20.0	42	26.2	16	6.2
P_4 Cusp Number	2-9/0-9	9	100.0	7	85.7	23	100.0	19	100.0	4	100.0	8	75.0	9	44.4	37	43.2	33	60.6	17	88.2
M_1 Anterior Fovea	2-4/0-4	5	100.0	3	100.0	8	62.5	8	100.0	1	100.0	4	25.0	2	77.8	12	50.0	12	0.0	9	22.2
M_1 Groove Pattern	Y/X.+.Y	13	92.3	7	71.4	20	95.0	23	100.0	4	100.0	11	100.0	26	96.2	37	83.8	39	97.4	22	90.9
M_2 Groove Pattern	Y/X.+.Y	13	69.2	11	36.4	26	50.0	20	80.0	6	83.3	5	60.0	25	40.0	44	25.0	62	33.9	36	50.0
M_3 Groove Pattern	Y/X.+.Y	7	71.4	8	37.5	22	13.6	14	42.9	6	50.0	5	60.0	27	33.3	47	27.7	56	32.1	45	20.0
M_1 Cusp Number	6/4-6	7	42.9	7	14.3	21	14.3	12	25.0	–	–	11	0.0	19	6.1	44	6.8	38	2.6	31	9.7
M_2 Cusp Number	4/4-6	14	0.0	13	7.7	33	27.3	22	4.5	6	16.7	6	66.7	37	81.1	49	81.6	66	90.9	45	60.0
M_3 Cusp Number	3-4/3-6	9	0.0	8	0.0	26	11.5	15	0.0	7	0.0	8	12.5	33	21.2	48	45.8	59	52.5	51	25.5
M_1 Deflecting Wrinkle	2-3/0-3	6	83.3	1	100.0	5	0.0	8	0.0	–	–	5	20.0	2	50.0	13	15.4	10	0.0	8	50.0
M_2 Deflecting Wrinkle	2-3/0-3	2	100.0	2	0.0	6	0.0	5	20.0	–	–	1	0.0	4	0.0	14	0.0	20	0.0	5	0.0
M_3 Deflecting Wrinkle	2-3/0-3	5	80.0	2	50.0	6	0.0	4	0.0	1	0.0	1	0.0	5	9.1	29	10.3	16	0.0	17	17.6
M_1 Middle Trigonid Crest	1/0-1	11	54.5	8	75.0	21	90.5	20	80.0	4	100.0	9	33.3	17	23.5	22	31.8	27	48.1	12	58.3
M_2 Middle Trigonid Crest	1/0-1	12	33.3	8	62.5	22	100.0	21	90.5	4	100.0	3	0.0	17	29.4	30	26.7	51	37.3	26	34.6

(Continued)

Table 2. (Continued)

		Homo erectus s.l		Middle Pleistocene		Pre-würmian Neandertal Europe		Würmian Neandertal Europe		Würmian Neandertal Middle East		Levantine AMHS Israel		Early Upper Palaeolithic Europe		Late Upper Palaeolithic Europe		Natufian (Israel)		Iberomaurusian (North Africa)	
		HER		PLE		PNE		NEA		NME		AMH		PAT		PRT		NAT		IBE	
	Dichotomy	N	%	N	%	N	%	N	%	N	%	N	%	N	%	N	%	N	%	N	%
M_3 Middle Trigonid Crest	*1/0-1*	*5*	*40.0*	*3*	*33.3*	*19*	*94.7*	*14*	*92.9*	*7*	*85.7*	*3*	*33.3*	*11*	*40.0*	*40*	*35.0*	*37*	*37.8*	*27*	*33.3*
M_1 Protostylid	*2-7/0-7*	*12*	*83.3*	*3*	*0.0*	*16*	*25.0*	*14*	*78.6*	–	–	*7*	*57.1*	*10*	*23.8*	*35*	*25.7*	*33*	*6.1*	*17*	*35.3*
M_2 Protostylid	**2-7/0-7**	**11**	**90.9**	**7**	**42.9**	**20**	**75.0**	**13**	**76.9**	**3**	**100.0**	**4**	**75.0**	**19**	**57.9**	**28**	**46.4**	**35**	**42.9**	**30**	**50.0**
M_3 Protostylid	*2-7/0-7*	*9*	*100.0*	*1*	*100.0*	*17*	*88.2*	*8*	*100.0*	*4*	*100.0*	*4*	*100.0*	*9*	*95.5*	*42*	*88.1*	*37*	*86.5*	*38*	*92.1*
M_1 Cusp 5	**3-5/0-5**	**10**	**100.0**	**7**	**100.0**	**20**	**90.0**	**21**	**95.2**	**2**	**100.0**	**11**	**100.0**	**28**	**96.4**	**42**	**92.9**	**38**	**92.1**	**30**	**100.0**
M_2 Cusp 5	**3-5/0-5**	**8**	**100.0**	**9**	**88.9**	**26**	**50.0**	**21**	**76.2**	**5**	**80.0**	**6**	**33.3**	**30**	**20.0**	**47**	**17.0**	**64**	**9.4**	**43**	**37.2**
M_3 Cusp 5	**3-5/0-5**	**7**	**85.7**	**6**	**83.3**	**18**	**44.4**	**11**	**90.9**	**3**	**100.0**	**7**	**71.4**	**29**	**72.4**	**40**	**42.5**	**54**	**31.5**	**43**	**62.8**
M_1 Cusp 6	*2-5/0-5*	*7*	*14.3*	*7*	*0.0*	*20*	*5.0*	*12*	*25.0*	–	–	*11*	*0.0*	*15*	*3.4*	*44*	*0.0*	*38*	*0.0*	*30*	*3.3*
M_2 Cusp 6	**2-5/0-5**	**8**	**62.5**	**8**	**12.5**	**27**	**14.8**	**19**	**31.6**	**4**	**25.0**	**6**	**0.0**	**31**	**9.7**	**48**	**10.4**	**64**	**1.6**	**44**	**0.0**
M_3 Cusp 6	**2-5/0-5**	**6**	**16.7**	**6**	**83.3**	**21**	**19.0**	**8**	**75.0**	**2**	**0.0**	**5**	**0.0**	**27**	**7.4**	**40**	**12.5**	**53**	**1.9**	**41**	**9.8**
M_1 Cusp 7	**1-4/0-4**	**11**	**36.4**	**7**	**28.6**	**25**	**40.0**	**19**	**42.1**	**3**	**66.7**	**11**	**27.3**	**26**	**3.8**	**45**	**2.2**	**56**	**3.6**	**39**	**28.2**
M_2 Cusp 7	**1-4/0-4**	**14**	**50.0**	**7**	**28.6**	**28**	**25.0**	**18**	**33.3**	**3**	**0.0**	**6**	**0.0**	**21**	**4.8**	**44**	**2.3**	**61**	**1.6**	**39**	**23.1**
M_3 Cusp 7	**1-4/0-4**	**8**	**87.5**	**6**	**33.3**	**27**	**37.0**	**10**	**80.0**	**5**	**60.0**	**5**	**0.0**	**23**	**0.0**	**46**	**4.3**	**56**	**0.0**	**39**	**20.5**
C, Root Number	*1/1-2*	*6*	*100.0*	*5*	*100.0*	*14*	*100.0*	*15*	*100.0*	*1*	*100.0*	*2*	*100.0*	*6*	*100.0*	*37*	*97.3*	*40*	*97.5*	*25*	*100.0*
P_3 Root Number	*1/1-2*	*2*	*100.0*	*2*	*100.0*	*13*	*100.0*	*8*	*100.0*	–	–	–	–	*2*	*100.0*	*35*	*100.0*	*39*	*100.0*	*24*	*100.0*
P_4 Root Number	*1/1-2*	*1*	*100.0*	*1*	*100.0*	*14*	*100.0*	*16*	*100.0*	–	–	*1*	*100.0*	*1*	*100.0*	*43*	*100.0*	*31*	*96.8*	*19*	*100.0*
M_1 Root Number	*3/1-3*	*4*	*0.0*	*4*	*0.0*	*12*	*0.0*	*14*	*0.0*	*1*	*0.0*	*1*	*0.0*	*4*	*0.0*	*42*	*0.0*	*37*	*0.0*	*19*	*15.8*
M_2 Root Number	*1/1-3*	*2*	*50.0*	*3*	*66.7*	*15*	*86.7*	*17*	*58.8*	*3*	*100.0*	*3*	*100.0*	*4*	*40.0*	*29*	*31.0*	*32*	*6.2*	*12*	*8.3*
M_3 Root Number	*1/1-3*	–	–	*3*	*33.3*	*8*	*100.0*	*11*	*81.8*	*1*	*100.0*	*1*	*0.0*	*9*	*47.8*	*29*	*34.5*	*29*	*24.1*	*10*	*20.0*
M_3 Congenital Absence	**1/0-1**	**12**	**0.0**	**12**	**8.3**	**28**	**0.0**	**24**	**0.0**	**7**	**0.0**	**10**	**0.0**	**44**	**11.4**	**67**	**9.0**	**107**	**9.3**	**61**	**4.9**

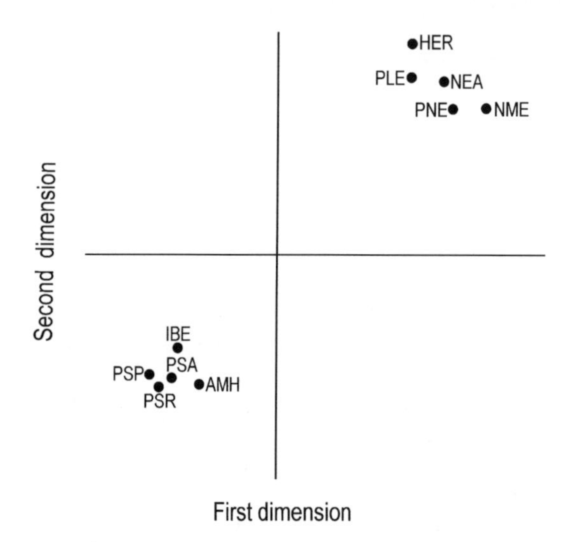

Figure 2. Multidimensional Scaling plot of fossil specimens. The plot corresponds to a distance matrix based on traits frequencies distributed in culturally and geographically defined populations. Populations are labeled according to Table 1. Stress: 0.1408.

individuals with complete dentitions (maxilla plus mandible) to maximize the number of samples versus number-of-traits-ratio and to reduce missing data as much as possible. Table 3 shows the list of the traits used (divided into maxillary, mandibular and maxillary plus mandibular) and the dichotomy used for trait presence (1) and absence (0). In Figures 3, 4, 5 we show the Kohonen maps respectively applied to:

- a maxillary database (Table 3), including 133 specimens and 17 traits, with 18.7% missing data.
- a mandibular database including 125 specimens and 28 traits (Table 3), with 28.3% missing data.
- a maxillary plus mandibular dentition database, including 89 specimens and 46 traits (Table 3), with 20.8% missing data. Please note that in this third data matrix the lower number of specimens corresponds to a higher number of measures, since individuals with complete dentitions represented a fraction of all scored specimens.

When choosing which traits to use in this analysis we limited ourselves to those without a great deal of missing data and those that discriminated amongst groups (i.e., those for which trait frequencies were not 100% present or 100% absent). For example, traits as the M^1 hypocone and M^1 metacone presence were not used in the analysis because they are present in all the specimens and are therefore not useful for this analysis.

In Table 4 we present the list of the Lower-Middle – (Early) Late Pleistocene specimens of Early-Archaic *Homo*, Neandertal and Middle Paleolithic AMHS from Israel analyzed by using the SOMs (Figures 3–5). The samples labeled as Sangiran, B, C, D and Zhoukoudian A were composites of specimens and isolated teeth as indicated in the note of Table 4. All the AMHS teeth analyzed are original, whereas those belonging to the Early-Archaic *Homo* and Neandertal specimen are both original and casts (Table 4). The considerable decrease in sample sizes in the SOM analysis compared to the original table of 1055 specimens (used in the Multidimensional Scaling analysis, Table 1), is the result of our effort to minimize the amount of missing data.

The three aforementioned databases, consisting in binary matrices coded as ‘0’ (trait absence) and ‘1’ (trait presence), were displayed on three different 6 × 9 Kohonen maps (consisting in 54 reference vectors) whose size was chosen by empirical trials for a better visualization of differences and clusters. As mentioned above, the frequency of missing descriptors in data-vectors was indicated in Table 3 and, therefore, the corresponding traits did not play a role in the ‘learning’ of the SOM. SOMs were generated by a MS-DOS compiled version of the package SOM_PAK vs. 3.1written by Kohonen and co-workers (1996). The software is freely available at the following

Table 3. Traits used in the SOMs analysis and absence (0)/presence (1) breakpoints

Maxillary	*Mandible*	*Individuals with complete dentition (maxilla+mandible)*	*absence (0)*	*presence(1)*
17 traits	28 traits	46 traits		
133 Specimens	125 Specimens	89 Specimens		
23 Archaic *Homo*	45 Archaic *Homo*	14 Archaic *Homo*		
18.7% missing data	28.3% missing data	20.8% missing data		
I1 Curvature		Upper I1 Curvature	0–1	2–4
I1 Shoveling		Upper I1 Shoveling	0–2	3–6
I2 Shoveling		Upper I2 Shoveling	0–2	3–7
I1 Double Shoveling			0–1	2–6
I2 Double Shoveling		Upper I2 Double Shoveling	0–1	2–6
		Upper I1 Interruption Groove	0	1
		Upper I2 Interruption Groove	0	1
		Upper I2 Mesial Bending	0	1
		Upper I1 *Tuberculum Dentale*	0–1	2–6
I2 *Tuberculum Dentale*		Upper I2 *Tuberculum Dentale*	0–1	2–6
M1 Metacone			0–3,5	4–5
M2 Metacone			0–3	3,5–5
M3 Metacone		Upper M3 Metacone	0–3	3,5–5
M1 Hypocone			0–3,5	4–5
M2 Hypocone		Upper M2 Hypocone	0–3	3,5–5
M3 Hypocone		Upper M3 Hypocone	0–2	3–5
		Upper M2 Cusp 5	0	1–5
M3 Cusp 5		Upper M3 Cusp 5	0	1–5
		Upper M1 Carabelli's Cusp	0–1	2–7
		Upper M2 Carabelli's Cusp	0–1	2–7
		Upper M3 Carabelli's Cusp	0–1	2–7
		Upper M1 Parastyle	0	1–5
M2 Parastyle		Upper M2 Parastyle	0	1–5
M3 Parastyle		Upper M3 Parastyle	0	1–5
M1 Enamel Extension			0–1	2–3
M2 Enamel Extension		Upper M2 Enamel Extension	0–1	2–3
		Upper M3 Peg Shape	0–1	2
	I1 Shoveling	Lower I1 Shoveling	0–1	2–3
	I2 Shoveling	Lower I2 Shoveling	0–1	2–3
	C Distal Accessory Ridge	Lower C Distal Accessory Ridge	0–1	2–5
	P3 Cusps Number	Lower P3 Cusps Number	0–2	2–9
	P4 Cusps Number	Lower P4 Cusps Number	0–2	2–9
	M1 Groove Pattern	Lower M1 Groove Pattern	X−+	Y
	M2 Groove Pattern	Lower M2 Groove Pattern	X−+	Y
	M3 Groove Pattern	Lower M3 Groove Pattern	X−+	Y
	M1 Cusps Number		4–5	6
	M2 Cusps Number	Lower M2 Cusps Number	5–6	4
	M3 Cusps Number	Lower M2 Cusps Number	5–6	3–4
	M3 Deflecting Wrinkle		0–1	2–3
	M1 Trigonid Crest		0	1
	M2 Trigonid Crest	Lower M2 Middle Trigonid Crest	0	1
	M3 Trigonid Crest	Lower M3 Middle Trigonid Crest	0	1
	M1 Protostylid	Lower M1 Protostylid	0–1	2–7
	M2 Protostylid	Lower M2 Protostylid	0–1	2–7
	M3 Protostylid	Lower M3 Protostylid	0–1	2–7
	M1 Cusp 5	Lower M1 Cusp 5	0–2	3–5
	M2 Cusp 5	Lower M2 Cusp 5	0–2	3–5
	M3 Cusp 5	Lower M3 Cusp 5	0–2	3–5

(Continued)

Table 3. (Continued)

M1 Cusp 6		0–1	2–5
M2 Cusp 6	Lower M2 Cusp 6	0–1	2–5
M3 Cusp 6	Lower M3 Cusp 6	0–1	2–5
M1 Cusp 7	Lower M1 Cusp 7	0	1–4
M2 Cusp 7	Lower M2 Cusp 7	0	1–4
M3 Cusp 7	Lower M3 Cusp 7	0	1–4
M3 Congenital Absence	Lower M3 Congenital Absence	0	1

web address: http://www.cis.hut.fi/research/som_lvq_pak.shtml.

Results and Discussion

Multidimensional Scaling

First, we computed a distance matrix of the 79 ASUDAS trait frequencies in populations defined in Table 2. The matrix was computed from trait frequencies. Each frequency is computed on a different number of individuals, complete dentitions often being unavailable. Only 45 traits (in bold type in Table 2) were used in the Multidimensional Scaling analysis, which were selected among those shared by all the groups and for which samples sizes exceeded 5 individuals. The Multidimensional scaling plot (Figure 2) shows a clear clustering of predefined groups of specimens in two clouds of points. The cluster in the upper-right portion of the plot is exclusively composed by Early-Archaic *Homo* and Neandertal specimens, while the remaining cluster (in the lower-left part of the plot) is composed by AMHS from Middle- and Upper-Paleolithic. Such classical multidimensional representation, based on ASUDAS traits, suggests a quite striking phenetic similarity between Early-Archaic *Homo* and Neandertal samples during all Pleistocene periods. Of course it is important to keep in mind that non-ASUDAS traits (c.f. Bailey, 2002) were not part of this analysis and their inclusion could show distinctions between these groups not recognized here.

Self Organizing Maps

The analysis of maxillary traits by SOMs (Figure 3) shows all Early-Archaic *Homo* and Neandertal specimens cluster in seven specific neurons (x=8; y=1, x=9; y=1, x=8; y=4, x=9; y=6, x=7; y=1, x=8; y=2, x=8; y=3), regardless of their geographic location or chronological age (Figure 3). The only exception to this general pattern is represented by the specimen "PLE8" that clusters together with later AMHS on the neuron x=9; y=6. As a result, Figure 3 indicates that the patterns of variability of dental features of archaic samples are quite stable, since a limited number of descriptors (neurons of the map) can effectively represent them. A deeper analysis of such mapping, reveals that that the seven Lower-Middle- Pleistocene specimens ("HER" and "PLE") are grouped in four specific neurons (x=8; y=1, x=9; y=1, x=8; y=4, x=9; y=6)), while Neandertal and "pre-Neandertal" specimens ("NME", "NEA" and "PNE") map together with the seven Lower-Middle Pleistocene specimens x=7; y=1, x=9; y=1. On the one hand, all the Lower- and Middle-Pleistocene specimens are clustered together, rarely being grouped with later samples (the notable exception being the previously mentioned "PLE8"), whereas on the other, Neandertal specimens are sometimes mapped together with later samples (for example x=8; y=3 e x=9; y=4). In conclusion, the mapping of such Early-Archaic *Homo* and Neandertal samples in a specific part of the SOM, where the colors connecting these cells are quite dark

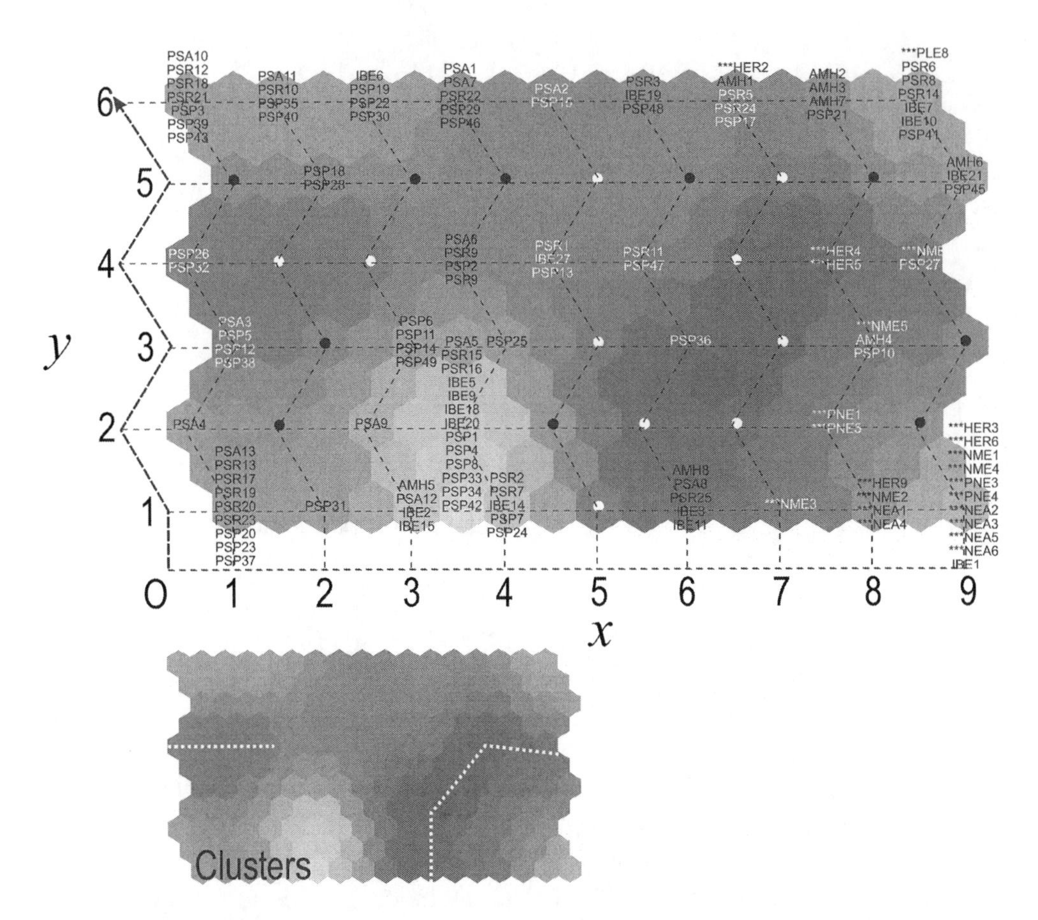

Figure 3. Self organizing map based on maxillary dentitions. This 6 × 9 neuron map visualizes 133 specimens scored for 17 traits (18.7% of missing data). When individuals in the maxillary dataset were mapped to one of the 54 neurons their label appears on it. To provide an easier cross-reference with the text, we provide 'x' and 'y' axes that unambiguously define neurons. All Middle Pleistocene sample are mapped on close neurons (x=7; y=1, x=8; y=1, x=9; y=1, x=8; y=2, x=8; y=3, x=8; y=4 and x=9; y=3) the only exception being PLE8 which is mapped on the neuron x=9; y=6. The fact that Middle Pleistocene samples lie inside an area surrounded by a major barrier (bottom of the figure) indicates that they form a cluster and that they are quite different from the other samples. To increase readability of the SOM, the labels of analyzed Middle Pleistocene samples are preceded by three asterisks, therefore in the neuron x=8; y=4 the specimens labeled as HER4 and HER5 are reported as ***HER4 and ***HER5. Close neurons can be very different or quite similar and it is possible to compute this similarity and visualize it by a gray-scale. A light gray (as between x=3; y=1 and x=4; y=1) means that neurons are quite similar (therefore "AMH5, PSA12, IBE2, IBE15" are similar to "PSR2, PSR7, IBE14, PSP7, PSP24"). Dark gray between neurons indicates important differences. Such gray shades define clusters that can be compared to a conventional PCA, MDS or dendrogram. An intuitive mapping of major barriers (clusters) is provided at the bottom of the figure in a small map. Differently from the example of Figure 1, we adopted an hexagonal neighborhood between neurons; therefore the 'y' axis is represented as zigzagging. Empty neurons are reported as dots.

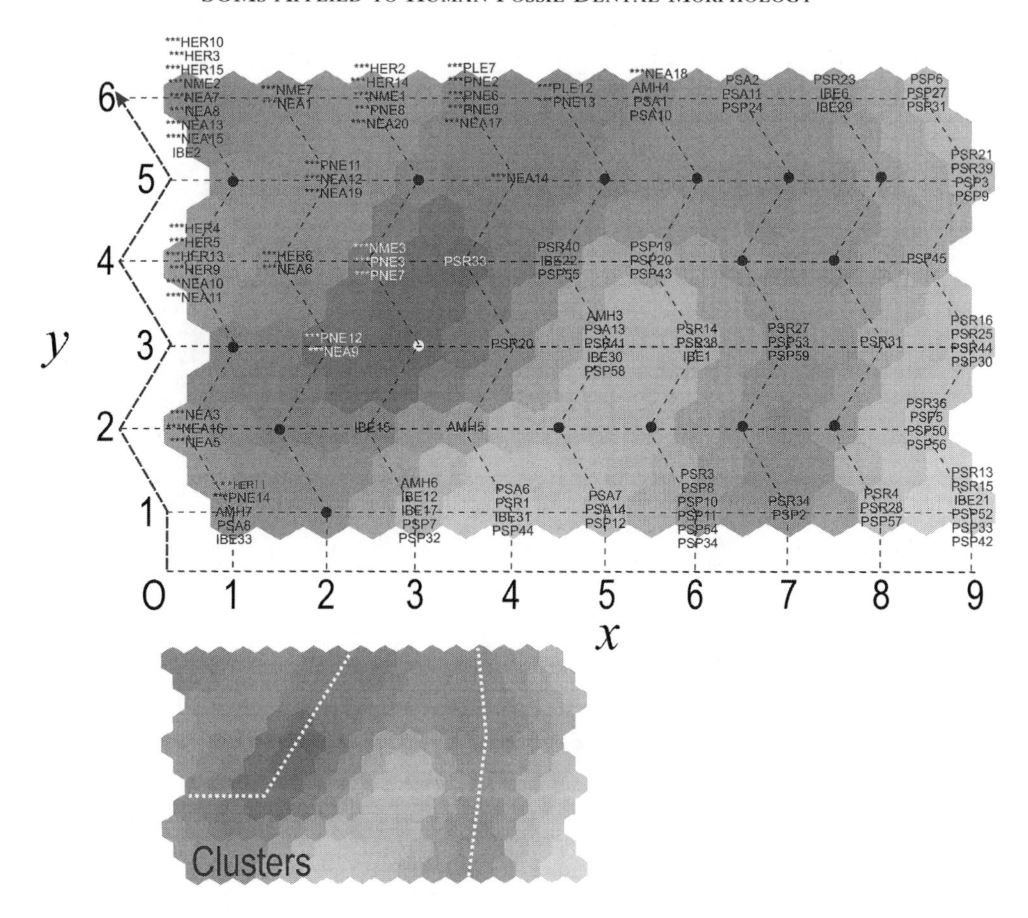

Figure 4. Self organizing map based on mandibular dentitions. This 6 × 9 neuron map visualizes 125 specimens scored for 28 traits (28.3 8% of missing data). For details concerning the graphic conventions of the figure please refer to the caption of Figure 3. To increase readability, the labels of analysed Middle and Late Pleistocene samples are preceded by three asterisks (***XXX). Middle Pleistocene samples are located on the left part of the map (x=1; y=1, x=1; y=2, x=1; y=4, x=1; y=6, x=2; y=3, x=2; y=4, x=2; y=5, x=2; y=6, x=3; y=4, x=3; y=6) as well as on the top part of the SOM (x=4; y=5, x=4; y=6, x=5; y=6 and x=6; y=6). The majority of more ancient samples lie on neurons belonging to a well defined area of the map which is surrounded by a "boundary" as visible at the bottom of the figure.

and in general lighter than those connecting the AMHS, suggests that the maxillary morphology of these samples is, in most cases, different from later ones.

Similar to the maxillary dataset, data for the mandibular dentition of Early-Archaic *Homo* and Neandertal (Figure 4) is mapped on 14 neighboring cells parting from the top left portion of the map (x=1; y=1, x=1; y=2, x=1; y=4, x=1; y=6, x=2; y=3, x=2; y=4, x=2; y=5, x=2; y=6, x=3; y=4, x=3; y=6, x=4; y=5, x=4; y=6, x=5; y=6 and x=6; y=6). The number of neurons to which specimens are mapped is quite high when compared to the maxillary data of Figure 3. Furthermore, *Homo erectus s.l.* and Middle Pleistocene specimens (labeled as "HER and "PLE") are clustered together with much later samples (see neurons x=1; y=1 and x=1; y=6). Lower- and Middle-Pleistocene morphologies are neither clustered together to a specific neuron (being always

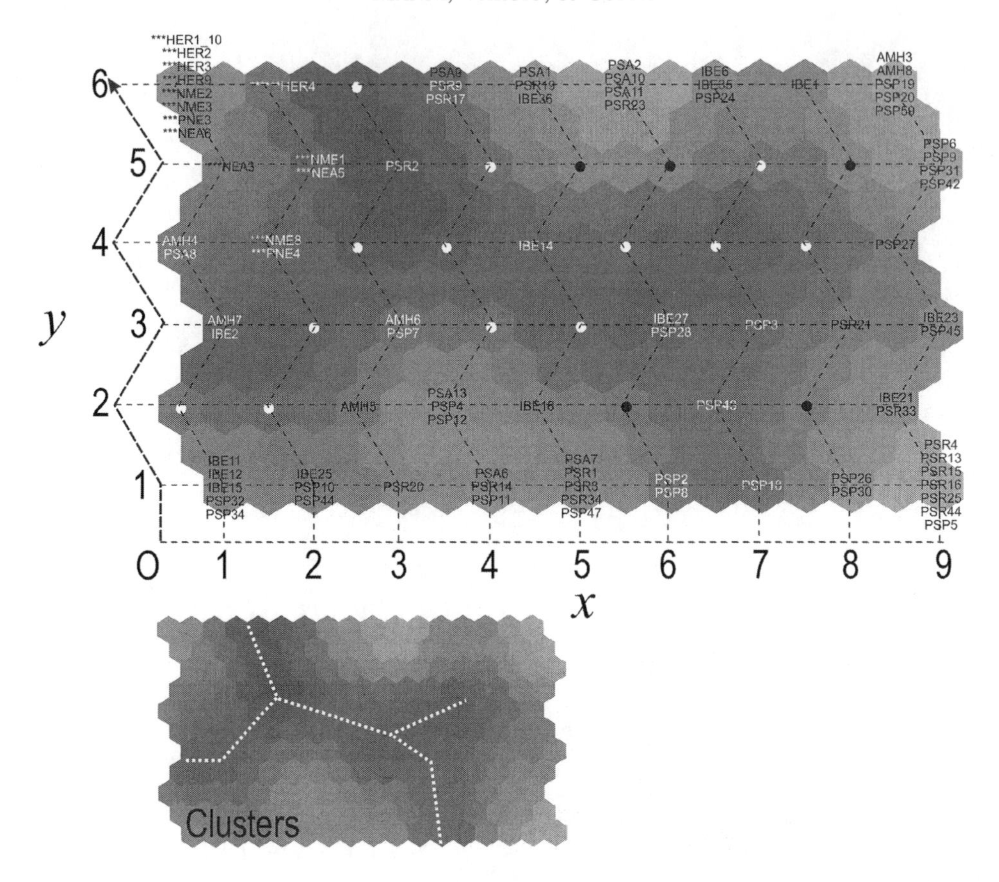

Figure 5. Self organizing map based on individuals with both maxilla and mandibles preserved. This 6×9 neuron map visualizes 89 specimens scored for 46 traits (20.8% of missing data). For details concerning the graphic conventions of this figure please refer to the caption of Figure 3. To increase readability, the labels of analyzed Middle Pleistocene samples are preceded by three asterisks (***XXX). Middle Pleistocene samples are exclusively located in the left top corner of the map (x=1; y=5, x=1; y=6, x=2; y=4, x=2; y=5, x=2; y=6) inside a well defined cluster that is surrounded by a "boundary" as visible on the small map at the bottom of the figure.

associated with Late- Pleistocene specimens), nor exclusively mapped inside the major map cluster outlined at the bottom of Figure 4. The topology of samples in such a SOM suggests that the dental morphology of mandibular traits, corresponding to Early-Archaic *Homo* and Neandertal specimens is, overall, different from later morphologies. To conclude, it must be noted that, when compared to the maxillary dataset (Figure 3), the depiction of patterns of mandibular archaic morphologies provided by SOMs is weaker.

The last Kohonen map (Figure 5) visualizes the similarities/dissimilarities of a dataset corresponding to specimens with both maxilla and mandible represented. It shows a clearer mapping of Early-Archaic *Homo* and Neandertal samples to specific neurons (x=1; y=5, x=1; y=6, x=2; y=4, x=2; y=5, x=2; y=6). Further, nearly all *Homo erectus s.l.* samples ("HER") are mapped, together with Neandertal samples, to the neuron x=1; y=6 — the only exception being the specimen "HER4", which is mapped alone on the neighboring neuron x=2; y=6, All the other Neandertal samples ("NME", "NEA" and "PNE") lie on neurons (x=1; y=5, x=2; y=4, x=2; y=5) that are specific to them since no specimens belonging to other morphologic groups are mapped there. When

Table 4. List of the Lower-Middle - (Early) Late Pleistocene specimens of archaic Homo *and anatomically modern* Homo sapiens *of Middle Palaeolithic analysed with SOMs*

Label	*Site*	*Country*	*Source*	*Individual*	*Maxilla Figure 3*	*Mandible Figure 4*	*Complete Dentition Figure 5*
HER 1–10	Dmanisi	Georgia	original	D2282+D211		X	X
HER 2	Sangiran	Indonesia	cast	A*	X	X	X
HER 3	Sangiran	Indonesia	cast	B*	X	X	X
HER 4	Sangiran	Indonesia	cast	C*	X	X	X
HER 5	Sangiran	Indonesia	cast	D*	X	X	
HER 6	Nariokotome	Kenya	cast	KNM-WT 15000	X	X	
HER 9	Zhoukoudian	China	cast	A*	X	X	X
HER 11	Koobi Fora	Kenya	cast	ER 992		X	
HER 13	Tighenif	Algeria	original	1		X	
HER 14	Tighenif	Algeria	original	2		X	
HER 15	Tighenif	Algeria	original	3		X	
PLE 7	Rabat	Morocco	original	1		X	
PLE 8	Visogliano	Italy	original	A	X		
PLE 12	Mauer	Germany	original	1		X	
PNE 1	Krapina	Croatia	original	2	X		
PNE 3	Krapina	Croatia	original	4	X	X	X
PNE 4	Krapina	Croatia	original	6	X		X
PNE 5	Krapina	Croatia	original	19	X		
PNE 6	Krapina	Croatia	original	8		X	
PNE 7	Krapina	Croatia	original	9		X	
PNE 8	Krapina	Croatia	original	10		X	
PNE 9	Krapina	Croatia	original	13		X	
PNE 11	Krapina	Croatia	original	23		X	
PNE 12	Krapina	Croatia	original	27		X	
PNE 13	Vindija	Croatia	original	206		X	
PNE 14	Vindija	Croatia	original	231		X	
NME 1	Amud	Israel	original	1	X	X	X
NME 2	Shanidar	Iraq	original	2	X	X	X
NME 3	Tabun	Israel	original	C1	X	X	X
NME 4	Tabun	Israel	original	BC7	X		
NME 5	Tabun	Israel	original	B1	X		
NME 6	Teshik Task	Uzbekistan	cast	A	X		
NME 7	Kebara	Israel	original	2		X	
NME 8	Tabun	Israel	original	B4			X
NEA 1	Hortus	France	original	VIII	X		
NEA 2	Monsempron	France	original	3	X		
NEA 3	Le Moustier	France	cast	1	X	X	X
NEA 4	La Quina	France	original	H18	X		
NEA 5	Saint Cesaire	France	original	A	X	X	X
NEA 6	Spy	Belgium	original	2	X	X	X
NEA 7	Circeo	Italy	original	3		X	
NEA 8	Gibraltar	Gibraltar	original	2		X	
NEA 9	Hortus	France	original	II–III		X	
NEA 10	Hortus	France	original	IV		X	
NEA 11	Hortus	France	original	V		X	
NEA 12	Moulin du milieu	France	original	A		X	
NEA 13	Petit-Puymoyen	France	original	1		X	
NEA 14	Petit Puymoyen	France	original	3		X	
NEA 15	La Quina	France	original	H5		X	
NEA 16	Regourdou	France	original	1		X	

(Continued)

Table 4. (Continued)

Label	*Site*	*Country*	*Source*	*Individual*	*Maxilla Figure 3*	*Mandible Figure 4*	*Complete Dentition Figure 5*
NEA 17	Subalyuk	Hungary	cast	1		X	
NEA 18	Sclayn	Belgium	original	A		X	
NEA 19	Spy	Belgium	original	1		X	
NEA 20	Zafarraya	Spain	cast	A		X	
NEA 21	Ochoz	Czeck Republic	original	1		X	
AMH 1	Qafzeh	Israel	original	5	X		
AMH 2	Qafzeh	Israel	original	6	X		
AMH 3	Qafzeh	Israel	original	7	X	X	X
AMH 4	Qafzeh	Israel	original	9	X	X	X
AMH 5	Qafzeh	Israel	original	11	X	X	X
AMH 6	Qafzeh	Israel	original	27	X	X	X
AMH 7	Qafzeh	Israel	original	8B	X	X	X
AMH 8	Skhul	Israel	original	5	X		X

(*)
Sangiran A: Sangiran 4, Sangiran 8, isolated teeth 1, 2, 18, 25, 26, 57, 59
Sangiran B: Sangiran 1, Sangiran 7, isolated teeth 48, 69, 88
Sangiran C: isolated teeth 6, 8, 43, 53, 61, 64, 75, 85
Sangiran D: isolated teeth 9, 17, 20, 42, 86, 89
Zhoukoudian A: L 2.PA.99 Skull XI, isolated teeth 1949 AN 519, A 1.57, C 2.29, C 3.45, B 4.75, L 4.303

we compare this SOM with those of Figures 3 and 4, we note that the Early-Archaic *Homo* and Neandertal specimens are no longer mapped with AMHS and later fossils. This suggests that the self organizing algorithm is now able to depict a clear distinction between archaic and AMHS morphologies as is also shown by the location of mentioned neurons inside the cluster in the upper left part of Figure 5 (clusters are visualized at the bottom of figures).

Conclusions

As the reader may note, certain specimens were represented by both maxillary and mandibular dentitions, others by only maxillary dentitions and, still others by only mandibular dentitions. Since Self Organizing Maps handle vectors with missing descriptors, that is jaws with missing teeth (and/or traits that cannot be scored), we decided to test the outcomes of the self-organizing algorithm with only mandibular dentitions, only maxillary dentitions and with both dentitions to see if the patterns of variation among specimens differing in time space, and taxonomic group were similar and strong enough to emerge from split dentition data. The results indicate that all the three SOMs (Figures 3–5) enable the discrimination between Early-Archaic *Homo* and Neandertal samples from AMHS suggesting that the aforementioned Neandertal Dental Complex can be identified even with a limited number of traits.

Nevertheless, the patterns of morphological variation are mapped in a more straightforward manner when both dentitions are preserved and can be analyzed together. This is logical because, in this latter case, more observations are available. Support for the fact that an analysis of both dentitions is more refined can be found in the topology of mapped items (Figure 5). In fact, there is a clear area of the map where archaic specimens are mapped (*H. erectus s.l.*, Middle Pleistocene and Neandertal samples).

The results listed above arise from the analysis of individual specimens, in what can be considered as an alternative approach to classical multivariate analyses where missing data need to be interpolated by comparing pre-determined groups (i.e., Neandertals, *H. erectus s.l.* specimens, etc.) from which a 'typical' value of missing descriptors is established.

A comparison of the SOMs to the MDS plot of Figure 2 shows that the depiction of variability provided by the latter method appears less ambiguous while SOMs appear more complex. The clear dichotomy of Figure 2 is related to specimens arranged into predefined groups, thus minimizing inter-individual variability. As a general rule, it is not advisable to analyze individuals, unless we admit that a single specimen can be representative of an entire group of specimens. Nevertheless, the possibility to 'explode' predefined groups, in order to independently analyze the individuals that compose them, can provide important clues into the reliability of such groups. The overall agreement between the outcome of the MDS plot of Figure 2 and the SOMs of Figures 3, 4 and 5 confirm the existence of a Neandertal Dental Complex, which may also be extended to Early-Archaic *Homo* samples when using only ASUDAS traits (Figure 5). By enabling the analysis of specimens with missing descriptors, SOMs make this additional step possible, which can be valuable in palaeoanthropological studies. However it is important to keep in mind that most of the diagnostic Neandertal traits identified by Bailey and colleagues (Bailey, 2002a, 2004; Bailey and Lynch, 2005) were not included in this analysis, because they were not available in the literature when this work started.

We have to consider that so far this is only a phenetic analysis. However, it may be possible through such this kind of study to obtain an suggestion of character polarity with an aim to carry out a cladistic phylogenetic analysis. But, in order to do that, we need to have a much larger non-modern data set composed of many more individuals. Considering the poor state of preservation of the fossils, this kind of analysis could be carried out only applying those methods which can make it possible to use a matrix with missing data, such as the SOMs can do (among others). This will be, therefore, the next step of our research.

Acknowledgments

We thank the conference organizers for the invitation and in particular Shara Bailey for her substantial editorial assistance. The authors are also grateful to the two anonymous reviewers for their helpful suggestions, which improved the manuscript. We also thank Pascal Croiseau and Bruno Toupance for their suggestions and comments. This research was supported by a MIUR COFIN2005 grant and "Progetti di Ateneo" Università di Roma "La Sapienza".

References

Bailey, S.E., 2002a. A closer look at Neandertal postcanine dental morphology. I. The mandibular dentition. Anatomical Record (New Anat.) 269, 148–156.

Bailey, S.E., 2002b. Neandertal dental morphology: implications for modern human origins. Ph.D. Dissertation, Arizona State University, Tempe.

Bailey, S.E., 2004. A morphometric analysis of maxillary molar crowns of Middle-Late Pleistocene hominins. Journal of Human Evolution 47, 183–198.

Bailey, S.E., Lynch, J., 2005. Diagnostic differences in mandibular P4 shape between Neandertals and anatomically modern humans. American Journal of Physical Anthropology 126, 268–277.

Bailey, S.E., Turner, G.C., 1999. A new look at some old teeth: an analysis of non-metric dental traits in Neandertals and Old World modern humans. American Journal of Physical Anthropology 27, 86.

Ben-Hur, A., Horn, D., Siegelmann H., Vapnik, V., 2001. Support vector clustering. Journal of Machine Learning Research 2, 125–137.

Coppa, A., Cappello, N., Cucina, A., Lucci, M., Mancinelli, D., Rendine, S., Vargiu, R., Piazza, A., 1997. Analysis of phenetic distances between living and ancient populations through the study of non-metric dental traits. Human Evolution Meeting, Cold Spring Harbor, New York (U.S.A.), New York, p. 24.

Coppa, A., Cappello, N., Cucina, A., Lucci, M., Mancinelli, D., Rendine, S., Vargiu, R., Piazza, A,. 1999a. Phenetic relationships of human populations by means of morphological dental traits. 2nd International Congress on Science and Technology for the Safeguard of Cultural Heritage in the Mediterranean Basin, C.N.R., C.N.R.S, Abaco Ed. Forlì, Paris 5–9 July 1999, pp. 268–269.

Coppa, A., Cucina, A., Lucci, M., Vargiu, R. 1999b. The Pavlov/Dolní Věstonice human remains in the context of the European Pleistocene populations. IVth International Anthropological Congress of Ales Hrdlička "World Anthropology at the Turn of the Century". Prague, Czech Republic, p. 32.

Coppa, A., Cucina, A., Vargiu, R., Mancinelli, D., Lucci, M. 1999c. The Pleistocene-Holocene transition in Italy. The contribution of the morphological dental traits. American Journal of Physical Anthropology 28, 111.

Coppa, A., Cucina, A., Vargiu, R., Mancinelli, D., Lucci, M. 1999d. The Krapina remains: the morphological dental evidence. International Conference "The Krapina Neandertals and Human Evolution in Central Europe". Zagreb, Croatia, pp. 18–19.

Coppa, A., Cucina, A., Vargiu, R., Mancinelli, D., Lucci, M., 2000a. The Pleistocene-Holocene transition in Italy. The evidence of morphological dental traits. Science and Technology for the Safeguard of Cultural Heritage in the Mediterranean Basin (A. Guarino ed.), Proceedings of 2nd International Congress, C.N.R., C.N.R.S, Paris, Vol. II, pp. 1009–1013.

Coppa, A., Cucina, A., Vargiu, R., Mancinelli, D., Lucci, M., 2000b. The Upper Paleolithic-Mesolithic dental complex in Europe. American Journal of Physical Anthropology 30, 130.

Coppa, A., Di Cintio, F., Vargiu, R., Lucci, M., Cucina, A., 2001. Morphological dental traits to reconstruct phenetic relationships between Late Pleistocene – Ancient Holocene human groups from Eurasia and North Africa. American Journal of Physical Anthropology. 32, 54.

Coppa, A., Oujaa, A., Petrone, P.P., Mehdi M., Vargiu, R., 2004. Les populations préhistoriques du Maroc: Relations phénétiques avec les populations de la fin du Pléistocène-début de l'Holocène en Eurasie et en Afrique du Nord. 1829ème Séance Réunion annuelle de la Société d'Anthropologie de Paris, Paris 22–24 janvier 2004, Bulletins et Mémoires de la Société d'Anthropologie de Paris 2004, 16, 32.

Gabriel, K.R., 1968. The biplot graphical display of matrices with application to principal component analysis. Biometrika, 58, 453–467.

Irish, J.D., 1997. Characteristic high-and-low-frequency dental traits in sub-saharan African populations. American Journal of Physical Anthropology 102, 455–467.

Irish, J.D., 1998. Ancestral dental traits in recent Sub-Saharan Africans and the origin of modern humans. Journal of Human Evolution 34, 81–98.

Irish, J.D., 2000. The Iberomauresian enigma: North African progenitor or dead end? Journal of Human Evolution 39, 393–410

Irish, J.D., Guattelli-Steinberg, D., 2002. Ancient teeth and modern human origin: an expanded comparison of African Plio-Pleistocene and recent world dental samples. Journal of Human Evolution 45, 113–144.

Kaski, S., 1997. Data exploration using Self-Organizing-Maps. Acta Polytechnica Scandinavica (Mathematics, computing and management in engineering series) 82, 1–57.

Kaski, S., Kohonen, T., 1994. Winner-take-all networks for physiological models in competitive learning. Neural Networks 7, 973–984.

Kohonen, T., 1982. Self-organized formation of topologically correct feature maps. Biological Cybernetics 43, 59–69.

Kohonen, T., 1984. *Self Organization and Associative Memory*. Springer Verlag, Berlin.

Kohonen, T., 1993. Physiological interpretation of the self-organizing map algorithm. Neural Networks 6, 895–905.

Kohonen, T., 2001. *Self-Organizing Maps. 3rd ed.* Springer, Berlin.

Kohonen., T., Hynninen, J., Kangas, J., Laaksonen, J., 1996. SOM PAK: The Self-Organizing Map program package. Report A31, Helsinki University of Technology, Laboratory of Computer and Information Science (January vol.).

Samad, T. and Harp, S.A., 1992. Self-organization with partial data. Network: Computation in Neural Systems, 3, 205–212.

Schmitt, A., Le Blanc, B., Corsini, M-M., Lafond, M. and Bruzek, J., 2001. Les réseaux de neurones artificiels: un outil de traitement des données prometteur pour l'Anthropologie. Bulletins et Mémoires de la Société d'Anthropologie de Paris 13, 143–150.

Scott, G.R., Turner II, C.G., 1997. *The Anthropology of Modern Human Teeth. Dental Anthropology and its Variation in Recent Human Populations*. Cambridge University Press, Cambridge.

Seber, G.A.F., 1984. *Multivariate Analysis*. Wiley, New York.

Stringer, G.B., Humphrey, L.T., Compton, T., 1997. Cladistic analysis of dental traits in recent humans using a fossil outgroup. Journal of Human Evolution 32, 389–402.

Torgerson, W.S., 1958. *Theory and Methods of Scaling*. Wiley, New York.

Turner II, C.G., Nichol, C.R., and Scott, G.R., 1991. Scoring procedures for key morphological traits of the permanent dentition: The Arizona State University Dental Anthropology System. In: Kelley M.A., Larsen C.S. (Eds.), Advances in Dental Anthropology. Wiley-Liss. New York, pp. 13–31.

7. Micro-computed tomography of primate molars: Methodological aspects of three-dimensional data collection

A.J. OLEJNICZAK
Human Evolution Department,
Max Planck Institute for Evolutionary
Anthropology, Deutscher platz 6, D-04103
Leipzig, Germany
olejniczak@eva.mgp.de

F.E. GRINE
Departments of Anthropology and Anatomical Sciences
Stony Brook University
Stony Brook, NY 11794-4364, USA
Frederick.Grine@stonybrook.edu

L.B. MARTIN
Departments of Anthropology and Anatomical Sciences
Stony Brook University
Stony Brook, NY 11794, USA
lmartin@notes.cc.sunysb.edu

Keywords: mCT; measurement accuracy; three-dimensional measurements; pixels; voxels; molars; 3-D rendering; molar enamel volume; slice thickness; image resolution

Abstract

Phylogenetic, paleodietary, and developmental studies of hominoid primates frequently make use of the post-canine dentition, in particular molar teeth. To study the thickness and shape of molar enamel and dentine, internal dental structures must be revealed (e.g., the location of dentine horn apices), typically necessitating the production of physical sections through teeth. The partially destructive nature of such studies limits sample sizes and access to valuable fossil specimens, which has led scholars to apply several methods of radiographic visualization to the study of teeth. Radiographic methods aimed at visualizing internal dental structures include lateral flat-plane X-rays, ultrasound, terra-hertz imaging, and computed tomography. Each of these techniques has resolution limitations rendering them inadequate for accurately reconstructing both the enamel-dentine junction and the outer enamel surface; the majority of studies are thus performed using physical sections of teeth. A comparatively new imaging technique, micro-computed tomography (mCT), accurately portrays the enamel-dentine junction of primate molars, and provides accurate measurements of enamel cap thickness and morphology. The research presented here describes methodological parameters pertinent to mCT studies of molars (slice thickness and pixel resolution), and the observable impact on measurement accuracy when these parameters

S.E. Bailey and J.-J. Hublin (Eds.), Dental Perspectives on Human Evolution, 103–115.

are altered. Measurements taken on a small, taxonomically diverse sample of primate molars indicate that slice thickness should be conservatively set at approximately 3.45 % of specimen length, and image resolution should be maximized (ideally, greater than or equal to 2048 × 2048 pixels per image) in order to ensure measurement accuracy. After discussing this base-line protocol for future mCT studies of the primate dentition, illustrative applications of this imaging technology are presented.

Introduction

Background

Internal dental structures viewed by radiographic means have been employed in taxonomic, phylogenetic, and functional studies of primate and human evolution for nearly 100 years (Miller, 1915). The scientific value of studies of internal dental morphology is evident from even a cursory review of the scientific literature pertaining to human evolution, and is only briefly recounted here. The thickness of molar enamel, for instance, is one of the most widely-cited morphological characters relevant to hominoid and hominin evolution (e.g., Gantt, 1977; Martin 1985; Grine and Martin, 1988; Andrews and Martin, 1991; Andrews, 1992; White et al., 1994; Senut et al., 2001; Brunet et al., 2002; Martin et al., 2003; Grine, 2004; Smith et al., 2005). Measurement of enamel thickness necessitates clearly portraying the enamel-dentine junction, so that internal structures of the tooth are revealed. It has further been demonstrated that controlled planes of section are necessary to accurately describe the thickness of enamel, especially in inter-species comparisons (Martin, 1983). In order to achieve uniform planes of section from which to record enamel thickness measurements, physical sections through teeth are often produced, which partially destroy the tooth.

Another fruitful line of investigation, the quantification of enamel-dentine junction (EDJ) morphology, also involves destroying dental material in order to view and measure internal surfaces. Korenhof (1960, 1961, 1978, 1982), Corruccini (1987), and Olejniczak et al. (2004) each employed destructive or semi-destructive techniques in order to study the morphology of the coronal dentine surface. Although EDJ morphology is well suited to studying taxonomic affiliation and mechanisms of dental development, partial destruction of dental tissues is necessary in order to view this surface. Sample sizes in these studies have been limited, despite the promising results produced thus far.

Techniques that involve full or partial destruction of a tooth in order to accurately visualize the EDJ and measure the thickness of enamel limit access to museum samples, and preclude the study of important fossil specimens. In order to acquire large samples of teeth (including fossil material), several non-destructive techniques have been employed, including ultrasound imaging (Yang, 1991) and tera-hertz pulse imaging (Crawley et al., 2003). Despite a long history of innovation, few radiographic methods have proven to accurately measure internal dental structures, and some methods are subject to high degrees of measurement error (e.g., lateral flat-plane X-rays (Grine et al., 2001) and medical CT (Grine, 1991; Spoor et al., 1993).

A relatively new radiographic technique, micro-computed tomography (mCT), has recently been applied to the study of the primate dentition (e.g., Chaimanee et al., 2003; Kono, 2004; Olejniczak and Grine, 2005, 2006). We have reported elsewhere that the measurement accuracy of mCT cross-sections is within 3% of equivalent measurements taken on physically prepared tooth sections, when linear and area measurements are considered (Olejniczak and Grine, 2006). Thus, mCT is capable of providing non-destructive high-resolution measurably accurate visualizations of primate teeth

(Figures 1 and 2). It is also capable of separating dental tissues (i.e., enamel, dentine, pulp) for analysis (Figure 3). Moreover, mCT facilitates the visualization of some

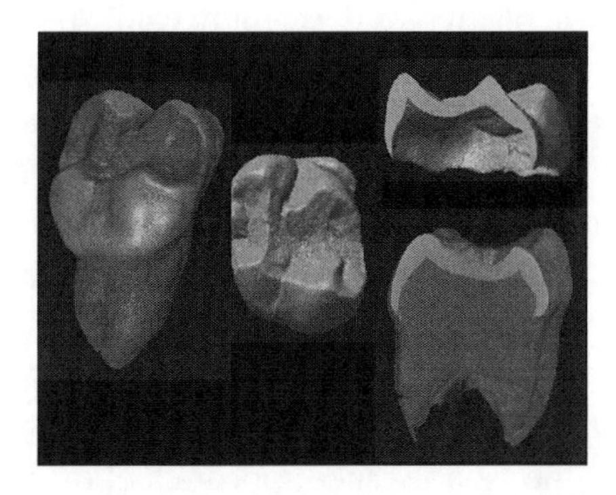

Figure 1. Three-dimensional volume rendering of the second molar of a tufted capuchin (*Cebus apella*) produced via mCT scanning. These models are reconstructed from scans that were 6 μm in thickness containing pixels that were 6 × 6 μm in dimension. The images at right show a cross-section through the tooth, both with the dentine (lower right) and with the dentine component removed via segmentation (upper right).

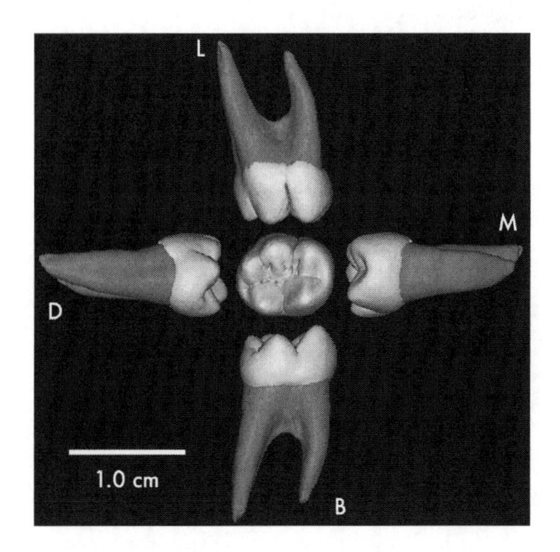

Figure 2. Surface model of a siamang (*Symphalangus syndactylus*) lower second molar in standard anatomical views. This surface model was created by threshold segmentation of an image stack containing cubic voxels of 10.5 μm per side.

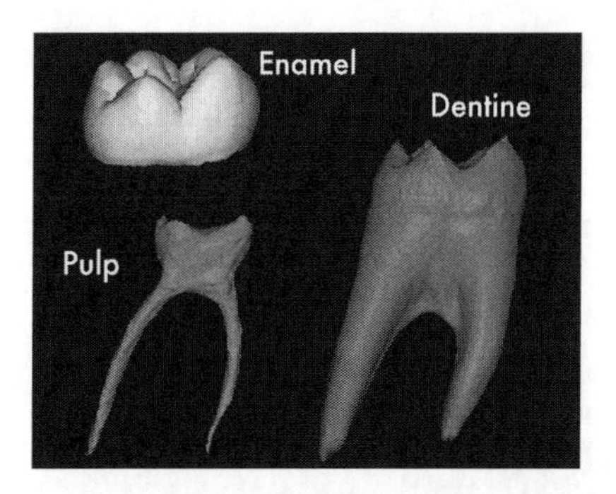

Figure 3. Surface models of the different dental tissues that are perceptible and segmentable via mCT scanning (enamel, dentine, and pulp), demonstrating the ability of this visualization technique to non-destructively expose areas of dental morphology that are inaccessible using standard odontometric methods. This is the same tooth as in Figure 2.

fossil teeth, making them available for non-destructive analysis (Figures 4 and 5). However, diagenetic re-mineralization of enamel and dentine must be considered when scanning fossil specimens, as density differences between the two tissues may be reduced through the absorption of minerals in the process of fossilization, rendering the enamel cap indistinguishable from the underlying dentine. (Olejniczak and Grine, 2006).

In a previous analysis (Olejniczak and Grine, 2006), we compared measurements from cross-sectional planes of physically

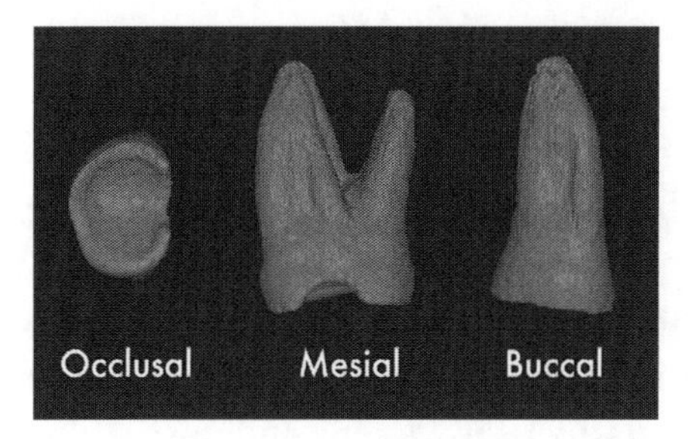

Figure 4. High-resolution surface models of the Shanidar 3 Neanderthal upper third molar in various views, demonstrating the potential application of mCT scanning to Pleistocene fossil specimens.

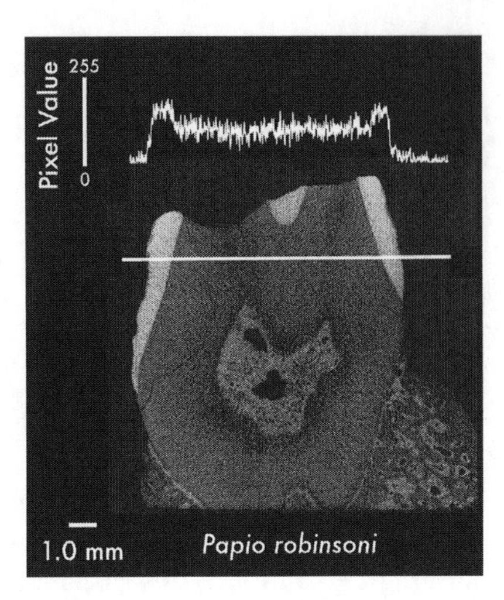

Figure 5. Cross-section through a molar of the fossil taxon *Papio robinsoni* (approximately 2.0 Myr old; E. Delson, pers. com.) derived from mCT scanning. The white line through the molar specifies a region of interest, and the pixel value plot above shows the pixel values along that line (0 is black and 255 is white). The enamel pixels show a peak compared to the pixels representing the dentine, demonstrating that mCT is able to resolve dental tissues in this Pliocene fossil tooth.

Image Parameters

Serial-imaging techniques, like mCT, involve iteratively recording cross-sectional images along the long axis of a specimen (the entire collection of images describing each specimen is referred to as the *image stack*). For example, the length of a molar may be traversed by the mCT X-ray source, stopping every 10.5 μm to record a high-resolution image of the tooth in cross-section at that location (Figure 6). The distance traversed by the X-ray source between consecutive scans is called *slice thickness*. Slice thickness is known to impact the shape and appearance of mCT-based rendering of primate teeth (Figure 7), but the specific impact of this parameter on measurements of enamel thickness is unknown. Increasing slice thickness is equivalent to reducing the resolution of the image stack, such that a greater distance must be interpolated between slices.

sectioned molars to equivalent measurements recorded on mCT planes of section from virtual models of the same teeth. We printed hard-copies of both the physical and mCT planes of section and measured the images using a digitization tablet. In the aforementioned analysis we also demonstrated that computer-based measurements (i.e., measurements taken on-screen) are as accurate as hand digitization. This suggests that, in the future, measurements may be recorded entirely on-screen, eliminating potential sources of error stemming from the printing, and then tracing of images. In order to automate the measurement of dental tissues, however, and also to ensure the accuracy of such computer measurement processes, it is necessary to understand the key image parameters associated with mCT, and then to determine the optimal settings for these parameters.

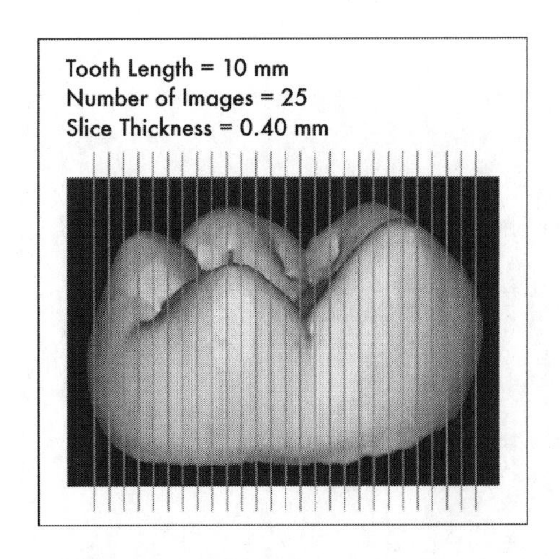

Figure 6. Schematic demonstrating slice thickness in a hypothetical scan of a *Symphalangus* molar (slices depicted here are thicker than they would be in typical scanning for illustrative purposes). Each line through the tooth represents a location where the mCT X-ray source took an image. In this case, the length of the tooth is 10 mm, and 25 images were taken, resulting in a slice thickness of 0.40 mm.

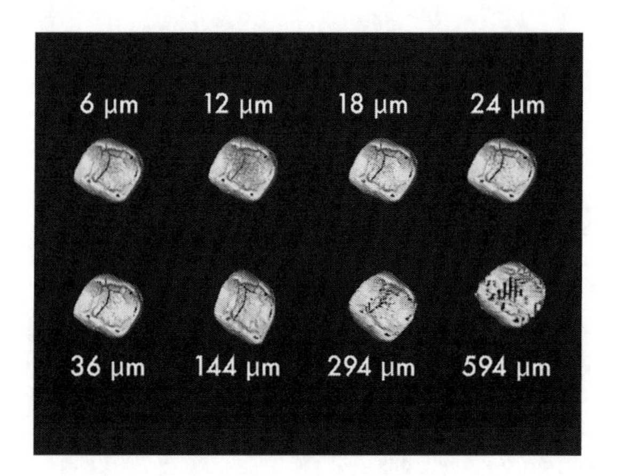

Figure 7. Surface models of an *Ateles paniscus* molar generated from image stacks with varying slice thicknesses. The tooth was worn and broken prior to scanning (note the pools of dentine at the cusp tips and post mortem cracks). Not only the image quality, but also the shape of the tooth is changed when models are based on thicker slices, which require software to interpolate increasing amounts of missing (unscanned) morphology.

The resultant image produced at each slice location is comprised of some number of square pixels of known dimension (e.g., each pixel may be $10.5 \times 10.5\,\mu m$). The number of pixels within an image is known as *image resolution*, and also serves to calibrate measurements taken on that image because the size of the pixels is known (Figure 8). For example, each of the serial images may be 15×15 pixels, so given a pixel dimension of $6 \times 6\,\mu m$, the image is known to be $90 \times 90\,\mu m$ (Figure 8).

Slice thickness also acts as a depth parameter for the serial images. For instance, if slice thickness is $6\,\mu m$, and each pixel is $6 \times 6\,\mu m$, then each pixel may be considered a cube (called a *voxel*, derived from the term *volumetric pixel*) that is $6 \times 6 \times 6\,\mu m$. In this way, each cross-sectional image is actually a cubic rectangle (Figure 8).

Slice thickness and image resolution have a substantial impact on one's ability to manipulate and measure image stacks in three dimensions because of computational limitations. For instance, the image stack of a single *Symphalangus syndactylus* molar scanned for this study is approximately 3.6 GB in size, and requires 85.3 GB of random access memory (RAM) in order to manipulate it in three dimensions at an image resolution of 2048×2048 pixels. If one reduces the resolution of each image in the stack by 75 % to 512×512 pixels, the required RAM is reduced to 5 GB, enabling standard desktop computers to manipulate and measure the image stack with only minor modification to memory handling settings (Figure 9). An alternative to reducing image resolution is to record fewer images during the scanning process, which is equivalent to increasing the slice thickness. Thus, a compromise between image quality (minimizing slice thickness and maximizing image resolution) and computational ability (increasing slice thickness and reducing image resolution) is sought in order to ensure that mCT scanning is not only accurate, but also viable for researchers without access to costly computational facilities.

One final consideration regarding the use of radiographic images such as those produced via mCT is the *pixel value*. Each pixel in every image (assuming hereafter that images are 8-bit) has an associated value, ranging from 0 to 255. A pixel value of 0 indicates that the pixel is black; a pixel value of 255 indicates that the pixel is white. Pixel values between 1 and 254 indicate that the pixel is some shade of grey, with the low end of the range indicating darker values of grey and the higher end of the range indicating brighter shades (see also Figure 5). Pixel value is a function of density as perceived by the X-ray receptor apparatus of the mCT machine, such that the hydroxyapatite crystals comprising the enamel cap (which appears nearly white in mCT images) are denser than the underlying dentine (which appears grey in mCT images). Pixel values are useful not only in visually interpreting the image stacks produced by mCT scanning, but also in the computer-based identification of tissues (i.e., image

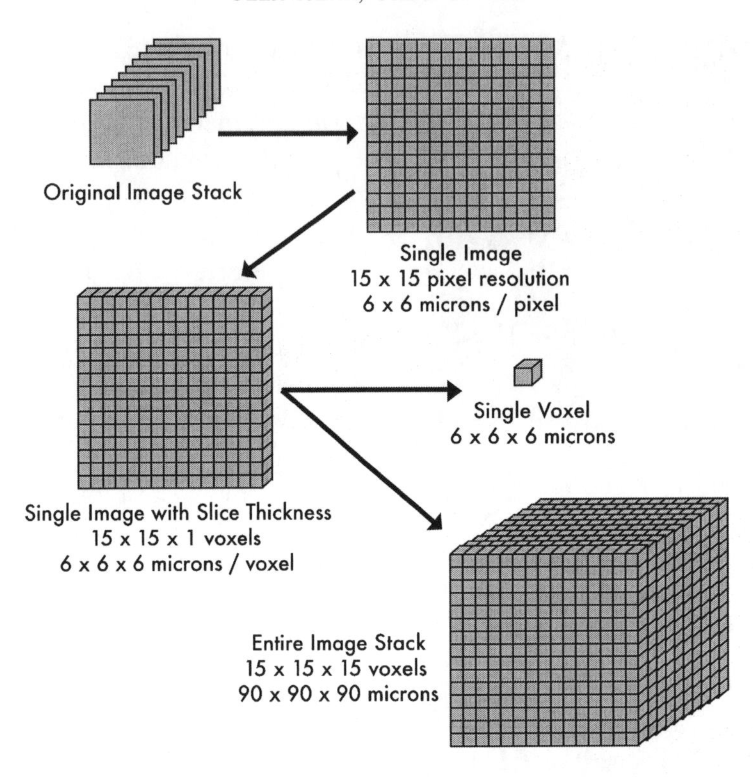

Figure 8. Schematic describing the different levels of data used in analysis of mCT image stacks. The original image stack contains the serial cross-sectional images. Each single image from within that stack is of known dimension, because the number of pixels and the size of each pixel are known. The slice thickness is used to turn each image into a cubic rectangle comprised of voxels, or three-dimensional pixels, the smallest volumetric units in the mCT image stack. The entire volume is visualized by giving each image from the original stack a thickness, and then viewing the resultant voxels in three-dimensions.

segmentation), where software may be used to automatically associate all of the nearly white pixels (e.g., pixels with values greater than 215) with one another, thus identifying the enamel cap. Because each pixel corresponds to a three-dimensional voxel of known volume (when slice thickness is considered), pixel value segmentation is a crucial step in calculating tissue volumes.

Since most mCT machines allow for customization of slice thickness and image resolution settings, and because these settings could potentially alter computer-based measurements of dental tissues regardless of segmentation techniques, the focus of the research presented here is on the ideal settings of these two parameters to ensure measurement accuracy. The experiments described here therefore address the following question: in order to accurately measure dental tissues based on pixel values in images produced by mCT scanning, what are the ideal values for slice thickness and image resolution?

Materials and Methods

Study Sample

The sample studied here contains molars from a taxonomically broad range of primates (Table 1). The sample comprises taxa that are known from previous two-dimensional analyses to exhibit a range of enamel thicknesses and enamel cap shapes (Martin, 1985; Shellis et al., 1998; Ulhaas et al.,

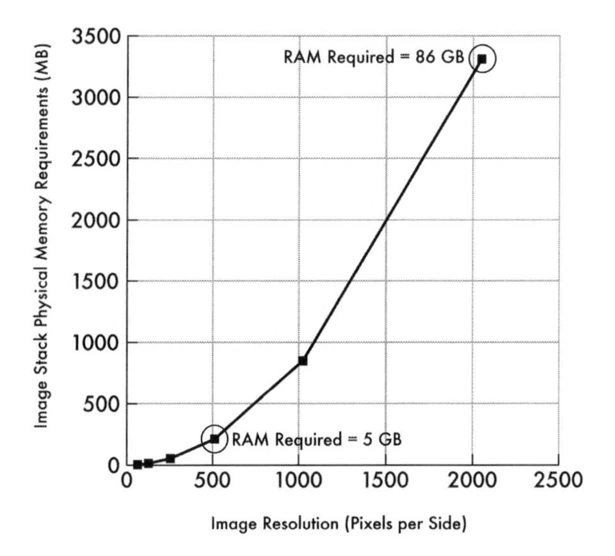

Figure 9. Plot depicting the physical memory required to store an image stack versus the resolution of the images in that stack, based on a *Symphalangus syndactylus* molar (the pattern depicted here is similar for any image stack, only the absolute values change). In the case of images stored at 2048 × 2048 pixels, which occupy 3.31 GB of hard drive space, approximately 86 GB of RAM are required to manipulate and measure the image stack in three dimensions. When the images are reduced in resolution to 512 × 512 pixels, the required RAM is reduced to approximately 5 GB. Due to the hardware limitations of modern desktop computers, smaller images are preferred because of one's ability to manipulate them with standard computational facilities. The research presented here suggests that reducing the size of the image stack (either by recording fewer images through or by reducing the resolution of those images) may have a substantial impact on measurements and should be avoided to ensure measurement accuracy.

1999; Schwartz, 2000; Martin et al., 2003; Olejniczak et al., 2004).

Each molar was scanned using a Scanco mCT 20 machine (Scanco Medical, Switzerland). The thinnest possible slices were taken, and images with the highest possible resolution were acquired for each specimen. These parameters are partially dependent upon the size of the individual specimen; hence the diameter of the mCT tube that will accommodate this size; slice thickness and image resolution for each specimen are reported in Table 1. Scanning resulted in an image stack for each specimen comprised of serial LZW compressed TIFF images, which were uncompressed prior to analysis using GraphicConvertor software (Lemke Software GmbH), and converted using NIH ImageJ software to unsigned character raw image format for volumetric analysis using Slicer3-D software.

Slice Thickness

In order to assess the impact of different slice thicknesses on computer-based measurements of mCT scans, the volume of the enamel cap in each specimen was measured at 15 different slice thicknesses. The range of slice thicknesses corresponded to the lowest possible thickness based on the mCT machine's capability on the low end, up to 1.00 mm on the high end. For example, a molar scanned at a slice thickness of 0.01 mm had volume measurements taken at the following 15 slice thicknesses for comparison: 0.01 mm, 0.02 mm, 0.03 mm, 0.04 mm, 0.05 mm, 0.06 mm, 0.07 mm, 0.08 mm, 0.09 mm, 0.10 mm, 0.20 mm, 0.30 mm, 0.40 mm, 0.50 mm, and 1.00 mm. Volume measurements were performed using Slicer3-D software, ensuring that identical pixel threshold values were assigned to the image stack during each measurement of the enamel cap volume. The volume of the enamel cap was calculated for each tooth at each slice thickness, and the length of the specimen was also noted in order to create the index [slice thickness/specimen length]. Using this percentage, slice thickness may be viewed as a function of tooth size, allowing a researcher to choose the thickest possible slices that still yield an accurate (within 1%) measurement of volumes, given the size of the tooth as the only parameter known *a priori*. Because scanning is both

Table 1. Study sample composition and scanning parameters

Taxon	*Tooth*	*Voxel Cubic Dimension (mm)*	*Image Resolution (pixels)*	*Number of Images*
Ateles geoffroyi	M_1	0.011	2048 × 2048	756
Ateles paniscus	M_1	0.006	2048 × 2048	1097
Cebus apella	M_2	0.006	2048 × 2048	663
Homo sapiens	M^3	0.010	2048 × 2048	1314
Homo sapiens	dm_1	0.008	2048 × 2048	839
Homo sapiens	M_3	0.008	2048 × 2048	1589
Papio ursinus	M^2	0.010	2048 × 2048	1449
Symphalangus syndactylus	M_2	0.010	2048 × 2048	847
Symphalangus syndactylus	M_2	0.010	2048 × 2048	834
Symphalangus syndactylus	M_1	0.010	2048 × 2048	807

time-consuming and expensive, selecting the maximum slice thickness that provides accurate measurements is important, and an effective way to reduce resultant file size as well as the cost and time devoted to scanning.

Image Resolution

A single cross-sectional image from within each specimen's image stack was examined at the highest resolution (2048 × 2048 pixels). We deliberately chose images from the center of each image stack so that both enamel and dentine were visible. NIH ImageJ software was employed to threshold the image, so that only enamel pixels were visible. The area of enamel was calculated by counting the number of pixels describing the enamel (the number of thresholded pixels) and multiplying this number by the area of a single pixel. This process was repeated on each image five more times, after halving the resolution each time. Thus the area of enamel was recorded for each tooth cross-section at the following pixel resolutions: 2048 × 2048, 1024 × 1024, 512 × 512, 256 × 256, 128 × 128, and 64 × 64 pixels. Identical thresholding was used for all six measurements of each tooth image. Finally, the 2048 × 2048 pixel image was printed and the hard-copy output was measured using a digitizing tablet interfaced with SigmaScan Pro software (v5.0, SPSS, Inc.).

Results

Slice Thickness

Table 2 shows the slice thickness at which volume measurements were 1 % different than those recorded from the thinnest slices possible. Table 2 also presents a percentage calculation of slice thickness versus the specimen length (mesio-distal length). Results of the slice thickness experiment suggest that volume is a rather robust measurement, such that a slice thickness of 8 % of the length of the specimen, on average, results in scans that are accurate to within 1 % of the best possible (thinnest) scans for that specimen (Table 2). Moreover, there is no trend for thinly enameled species (e.g., *Ateles paniscus*) to require more scans than more thickly enameled species (e.g.,*Cebus apella*) in order to achieve measurement accuracy. Nor is there a relationship between specimen length and required slice thickness.

Image Resolution

Results of the image resolution experiment indicate that area measurements taken on images to which identical thresholding algorithms had been applied tend to become less similar to the actual area value (as measured by digitization) as resolution is decreased. Table 3 depicts the value of

Table 2. Slice thickness at which volume measurement is 1 % different than ideal

Taxon	Tooth	Thickness at 1 % Different (mm)	Slice Thickness/Specimen Length
Ateles geoffroyi	M_1	0.42	9.83 %
Ateles paniscus	M_1	0.30	4.55 %
Cebus apella	M_2	0.60	15.08 %
Homo sapiens	M^3	0.80	7.60 %
Homo sapiens	dm_1	0.80	11.76 %
Homo sapiens	M_3	1.00	6.29 %
Papio ursinus	M^2	0.50	3.45 %
Symphalangus syndactylus	M_2	0.50	5.90 %
Symphalangus syndactylus	M_2	1.00	11.90 %
Symphalangus syndactylus	M_1	0.60	7.43 %
Mean			8.38 %

Table 3. Enamel area from digitization (mm^2) and thresholding (% different than digitization) at different pixel counts

Taxon	Tooth	Digitized Area (mm^2)	2048 × 2048	1024 × 1024	512 × 512	256 × 256	128 × 128	64 × 64
Ateles geoffroyi	M_1	1.56	0.64 %	4.49 %	8.33 %	9.62 %	25.00 %	44.23 %
Ateles paniscus	M_1	2.14	0.00 %	1.40 %	2.34 %	4.67 %	10.28 %	22.43 %
Cebus apella	M_2	3.54	0.28 %	0.85 %	0.56 %	0.85 %	2.54 %	7.34 %
Homo sapiens	M^3	17.68	0.79 %	1.81 %	1.92 %	5.77 %	4.07 %	1.41 %
Homo sapiens	dm_1	8.42	1.07 %	1.19 %	0.48 %	0.24 %	2.85 %	12.00 %
Homo sapiens	M_3	40.52	0.79 %	1.16 %	1.23 %	9.45 %	12.22 %	8.66 %
Papio ursinus	M^2	21.90	1.92 %	4.20 %	4.93 %	4.89 %	6.62 %	10.68 %
Symphalangus syndactylus	M_2	9.12	0.22 %	4.50 %	7.13 %	6.69 %	5.92 %	6.80 %
Symphalangus syndactylus	M_2	9.11	0.66 %	0.11 %	0.33 %	1.43 %	3.84 %	5.60 %
Symphalangus syndactylus	M_1	11.00	1.00 %	0.27 %	0.64 %	2.09 %	4.55 %	7.82 %
Mean			0.74 %	2.00 %	2.79 %	4.57 %	7.79 %	12.70 %

enamel area obtained by digitization for each tooth section, and the percent difference of computer-based measurements of the same section at different resolutions. There is a wide range of percent differences from the digitized area value, but the general trend is towards decreased measurement accuracy (i.e., increased percent differences from the value obtained via digitization) with reduced pixel counts. The average percent difference of all the specimens at a given resolution is approximately 1.5 times the percent difference at the previous (better) resolution: from 2.00 % at 1024 × 1024 pixels, to 2.79 % at 512 × 512 pixels, to 4.57 % at 256 × 256 pixels, and so on. The digitized values were almost always closest to the values obtained via thresholded pixel counts at 2048 × 2048 pixels, suggesting that 2048 pixels per side is adequate for accurate measurements compared to a hand-digitization standard (less than 1.00 % measurement difference).

Discussion

Parameter Calibration

Results of the experiments conducted here suggest that while image resolution should be maximized in order to achieve automated image measurement accuracy (e.g., counting pixels or voxels after thresholding to measure areas and volumes, respectively), slice thickness may be sacrificed while maintaining accurate measurements in some cases. That is, enamel volume may be calculated accurately in permanent human molars with slices that are between 0.80 and 1.00 mm thick, even though it is possible to achieve slice thicknesses of 0.008 mm for these specimens. However, because of its three-dimensional nature, volume is less influenced by shape irregularities caused by interpolation of thicker slices than two-dimensional measurements such as areas or distances. In a previous analysis, we found that slice thickness ranging from 0.006 mm to 0.010 mm facilitate accurate measurements of distances within mCT image stacks of teeth (Olejniczak and Grine, 2006), suggesting that volumes and distances necessitate different slice thicknesses in order to ensure measurement accuracy. The slice thicknesses reported here (Table 2) thus represent only the maximum thicknesses that will enable the accurate measurement of enamel volumes; thinner slices are required for the accurate measurement of distances.

Example Applications of mCT

Having elsewhere demonstrated the measurement accuracy of mCT visualization and its ability to visualize fossil specimens (Olejniczak and Grine, 2005, 2006), and having considered here the ideal image resolution and slice thickness parameters, we present some illustrative applications of this technology.

Martin (1983) suggested that the scaled volume of enamel on a tooth crown was the ideal measurement of overall enamel thickness, and noted that two-dimensional measurements from physical sections were proxies for overall crown volumes. Whether measurements recorded on the widely used mesial cusp ideal plane of section are reliable proxies for overall crown volume may be explored with mCT analysis, given this technology's unique ability to produce accurate, user-defined planes of section (see also Shimizu, 2002; Suwa and Kono, 2005). The relative impact of slight deviations from the ideal plane of section may also be addressed using mCT models (e.g., Olejniczak, 2005), and standard reference planes may be established in order to limit inter-observer error in planar studies. Furthermore, the volumes of dental tissues themselves may be analyzed, permitting a whole-tooth perspective on enamel thickness (e.g., Kono, 2004).

Another study facilitated by mCT technology is the examination of both mandibular architecture (e.g., Daegling and Hylander, 1997) and molar tissue patterning (e.g., enamel thickness distribution, Schwartz, 2000) within the same specimen (Figure 10). Currently, data sets produced for analyses of mandibular morphology, and those produced by other researchers to examine molar enamel thickness must be integrated in order to

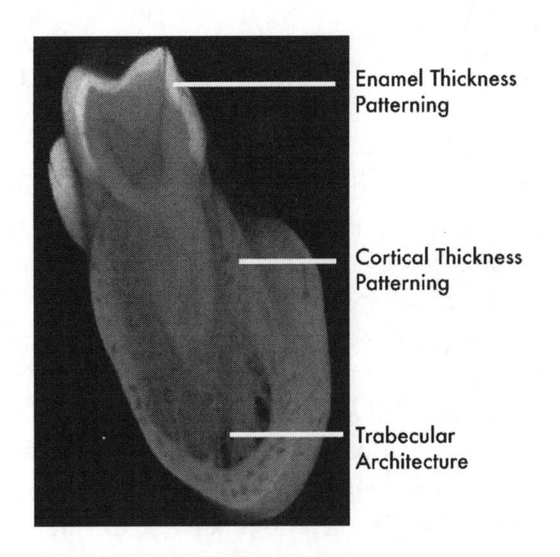

Figure 10. Section through a volume model of a *Symphalangus syndactylus* mandible produced by mCT scanning. The ability of mCT to visualize accurately the internal morphology of both tooth and jaw facilitates simultaneous study of mandibular and molar architecture.

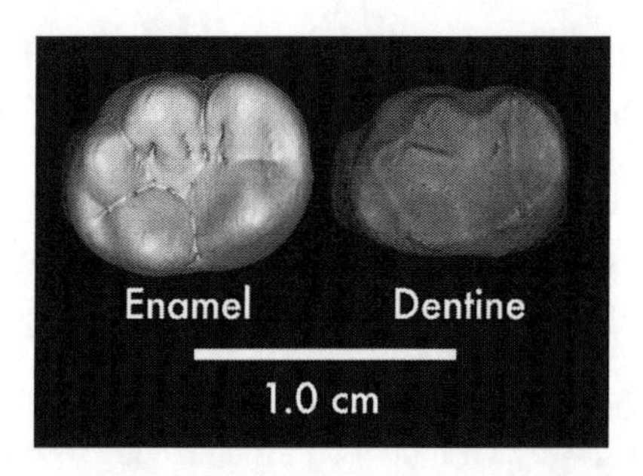

Figure 11. Occlusal views of surface models of the outer enamel surface and the enamel-dentine junction in a *Symphalangus syndactylus* lower second molar.

examine the internal morphology of both tooth and jaw within a species. mCT visualization thus affords the potential refinement of functional models of mastication by allowing the ideal plane of section through a molar to also expose the morphology of the mandible in the same plane, and in the same individual.

Finally, mCT imaging allows different dental tissues to be separated via image segmentation, so previously inaccessible areas of morphology may be measured. The morphology of the EDJ across the entire molar crown, for instance, has been implicated as an important character in studies of taxonomic affinity and dental development, although only partially destructive means (Korenhof, 1961; Corruccini 1987; Olejniczak et al., 2004) or enamel attrition patterns (Simons, 1976) have been used to access and interpret this morphology (but see Skinner and Kapadia, 2005). Measurement-accurate images of the outer enamel surface and the EDJ may be compared using mCT imaging, facilitating further analysis of this potentially important suite of characters (Figure 11).

Conclusions

1. Slice thickness in mCT studies of primate molars aimed at measuring volumes may conservatively be set at 3.45 % of the length of the specimen (the most conservative thickness reported in Table 2), although thinner slices (approximately 10 times thinner) are required for area, linear, and coordinate measurements. Slice thickness and image resolution are also functions of the size of each tooth to be scanned, such that larger teeth require fewer lower resolution scans than do smaller teeth.
2. Image resolution in mCT studies of primate molars should be maximized in order to ensure measurement accuracy, ideally at 2048×2048 pixels or greater if computer-based measurement techniques are to be performed.
3. Inaccessible areas of anatomy, such as pulp cavities, radicular canals, and the morphology of the enamel-dentine junction, may be accurately visualized using mCT methods, facilitating non-destructive and high-resolution studies of internal tooth morphology.
4. Non-destructive paleoanthropological investigations of hominid molars may be performed at high levels of accuracy using mCT, although some drawbacks exist to using this method. First, there is a negative relationship between specimen size and scanning resolution, such that isolated teeth are more easily scanned at high resolutions than are crania or mandibles. Second, mCT (and the associated analysis software) is costly and time-consuming, so the minimum resolution and number of scans necessary for accurate measurements should be taken. Nonetheless, the ability of mCT to non-destructively produce measurable models of fossil specimens promises to increase sample sizes and open new research questions about internal dental morphology, improving our understanding of hominid dental evolution.

Acknowledgments

The authors are most grateful to Dr. Shara Bailey and Prof. Jean-Jacques Hublin, as well as our fellow conference participants at the Max Planck Institute for Evolutionary Anthropology. T. Smith and A. Cleveland provided constructive criticism and logistical support, for which we are thankful. M. Skinner provided assistance measuring inter-observer error. E. Mitra scanned the Shanidar 3 Neandertal molar. Access to specimens was kindly provided by the American Museum of Natural History (New York), the Smithsonian National Museum of Natural History (Washington, DC), and the University of California Museum of Paleontology (Berkeley, CA). S. Judex and S. Xu provided access to and assistance with mCT facilities.

References

Andrews, P., 1992. Evolution and environment in the Hominoidea. Nature 360, 641–646.

Andrews P., Martin L., 1991. Hominoid dietary evolution. Philosophical Transactions of the Royal Society of London B. Biological Sciences 334, 199–209.

Brunet, M., Guy, F., Pilbeam, D., Mackaye, H.T., Likius, A., Ahounta, D., Beauvilain, A., Blondel, C., Bocherens, H., Boisserie, J.R., De Bonis, L., Coppens, Y., Dejax, J., Denys, C., Duringer, P., Eisenmann, V.R., Fanone, G., Fronty, P., Geraads, D., Lehmann, T., Lihoreau, F., Louchart, A., Mahamat, A., Merceron, G., Mouchelin, G., Otero, O., Campomanes, P.P., De Leon, M.P., Rage, J.C., Sapanet, M., Schuster, M., Sudre, J., Tassy, P., Valentin, X., Vignaud, P., Viriot, L., Zazzo, A., Zollikofer, C., 2002. A new hominid from the Upper Miocene of Chad, central Africa. Nature 418, 145–151.

Chaimanee, Y., Jolly D., Benammi, M., Tafforeau, P., Duzer, D., Moussa, I., Jaeger, J.J., 2003. A Middle Miocene hominoid from Thailand and orangutan origins. Nature 422, 61–65.

Corruccini, R.S., 1987. The Dentinoenamel Junction in Primates. International Journal of Primatology 8, 99–114.

Crawley, D., Longbottom, C., Wallace, V.P., Cole, B., Arnone, D., Pepper, M., 2003. Three-dimensional terahertz pulse imaging of dental tissue. Journal of Biomedical Optics. 8, 303–307.

Daegling, D.J., Hylander, W.L., 1997. Occlusal forces and mandibular bone strain: Is the primate jaw "overdesigned"? Journal of Human Evolution 33, 705–717.

Gantt, D.G., 1977. Enamel of primate teeth: its thickness and structure with reference to functional and phyletic implications. Ph.d. Dissertation, Washington University, St. Louis, MO.

Grine, F.E., 1991. Computed tomography and the measurement of enamel thickness in extant hominoids: implications for its paleontological application. Palaeontologia Africana 28, 61–69.

Grine F.E., 2004. Geographic variation in tooth enamel thickness does not support Neandertal involvement in the ancestry of modern Europeans. South African Journal of Science 100, 389–394.

Grine, F.E., Martin, L., 1988. Enamel thickness and development in *Australopithecus* and *Paranthropus*. In: Grine, F.E. (Ed.), Evolutionary History of the "Robust" Australopithecines. Aldine de Gruyter, New York.

Grine, F.E., Stevens, N.J., Jungers, W.L., 2001. An evaluation of dental radiograph accuracy in the measurement of enamel thickness. Archives of Oral Biology 46, 1117–1125.

Kono, R., 2004. Molar enamel thickness and distribution patterns in extant great apes and humans: new insights based on a 3-dimensional whole crown perspective. Anthropological Science 112, 121–146.

Korenhof, C.A.W., 1960. *Morphogenetical Aspects of the Human Upper Molar*. Uitgeversmaatschappij Neerlandia, Utrecht.

Korenhof, C.A.W., 1961. The enamel-dentine border: a new morphological factor in the study of the (human) molar pattern. Proceedings of Koninklijke Nederlands 64B, 639–664.

Korenhof, C.A.W., 1978. Remnants of the trigonid crests in Medieval molars of Man in Java. In: Butler, P.M., Joysey, K.A. (Eds.), Development, Function, and Evolution of Teeth. Academic Press, New York, pp.157–170.

Korenhof, C.A.W., 1982. Evolutionary trends of the inner enamel anatomy of deciduous molars from Sangiran (Java, Indonesia). In: Kurten, B. (Ed.), Teeth: Form, Function, and Evolution. Columbia University Press New York, pp. 350–365.

Martin, L.B., 1983. Relationships of the later Miocene Hominoidea. Ph.D. Dissertation, University College London, England.

Martin, L.B., 1985. Significance of enamel thickness in hominoid evolution. Nature 314, 260–263.

Martin, L.B., Olejniczak, A.J., Maas, M.C., 2003. Enamel thickness and microstructure in pitheciin primates, with comments on dietary adaptations of the middle Miocene hominoid *Kenyapithecus*. Journal of Human Evolution 45, 351–367.

Miller, G.S., 1915. The jaw of Piltdown Man. Smithsonian Miscellaneous Collections 65, 1–31.

Olejniczak, A.J., 2005. Mesiodistal and angular obliquity in studies of dental sections. American Journal of Physical Anthropology supplement, 126, 160.

Olejniczak, A.J., Grine, F.E., 2006. Assessment of the accuracy of dental enamel thickness using high-resolution micro-focal X-ray computed tomography. Anatomical Record Part A 288A, 263–275.

Olejniczak, A.J., Grine, F.E., 2005. High-resolution measurement of Neandertal tooth enamel thickness by micro-focal computed tomography (mCT). South African Journal of Science 101, 219–220.

Olejniczak, A.J., Martin, L.B., Ulhaas, L., 2004. Quantification of dentine shape in anthropoid primates. Annals of Anatomy 186, 479–485.

Schwartz, G.T., 2000. Taxonomic and functional aspects of the patterning of enamel thickness distribution in extant large-bodied hominoids. American Journal of Physical Anthropology 111, 221–244.

Senut, B., Pickford, M., Gommery, D., Mein, P., Cheboi, K., Coppens, Y., 2001. First hominid from the Miocene (Lukeino Formation, Kenya). Comptes Rendus de L'Academie Des Sciences Serie II Fascicule A- Sciences De La Terre Et Des Planetes 332, 137–144.

Shellis, R.P., Beynon, A.D., Reid, D.J., Hiiemae, K.M., 1998. Variations in molar enamel thickness among primates. Journal of Human Evolution 35, 507–522.

Shimizu, D., 2002. Functional implications of enamel thickness in the lower molars of red colobus (*Procolobus badius*) and Japanese macaque (*Macaca fuscata*). Journal of Human Evolution 43, 605–620.

Simons, E.L., 1976. The nature of the transition in the dental mechanism from pongids to hominids. Journal of Human Evolution 5, 511–528.

Skinner, M.M., Kapadia, R., 2005. An evaluation of microCT for assessing in 3-D the concordance of dental trait expression between the dentin-enamel junction and the outer enamel surface of modern human molars. American Journal of Physical Anthropology Supplement, 126, 191–192.

Smith, T.M., Olejniczak, A.J., Martin, L.B., Reid, D.J., 2005. Variation in hominoid molar enamel thickness. Journal of Human Evolution 48, 575–592.

Spoor, C.F., Zonneveld, F.W., Macho, G.A., 1993. Linear measurements of cortical bone and dental enamel by computed-tomography – applications and problems. American Journal of Physical Anthropology 91, 469–484.

Suwa, G., Kono, R., 2005. A micro-CT based study of linear enamel thickness in the mesial cusp section of human molars: reevaluation of methodology and assessment of within-tooth, serial, and individual variation. Anthropological Science 113, 273–289.

Ulhaas, L., Henke, W., Rothe, H., 1999. Variation in molar enamel thickness in genera *Cercopithecus* and *Colobus*. Anthropologie 37, 265–271.

White, T.D., Suwa, G., Asfaw, B., 1994. *Australopithecus ramidus*, a new species of early hominid from Aramis, Ethiopia. Nature 371,306–312.

Yang, Z., 1991. Ultrasound surface imaging and the measurement of tooth enamel thickness. Ph.D. Dissertation, University of Manchester, UK.

8. HRXCT analysis of hominoid molars: A quantitative volumetric analysis and 3D reconstruction of coronal enamel and dentin

D.G. GANTT
Department of Anatomy
Georgia Campus-Philadelphia College of Osteopathic Medicine
625 Old Peachtree Road NW
Suwanee, GA 30024 USA
davidga@pcom.edu

J. KAPPELMAN
Department of Anthropology
The University of Texas at Austin
TX 78712-1086 USA
jkappelman@mail.utexas.edu

R.A. KETCHAM
High-Resolution X-ray CT (Computed Tomography) Facility
Department of Geology
University of Texas, Austin
TX 78712-1100 USA
richk@mail.utexas.edu

Keywords: 3D reconstruction; HRXCT; enamel thickness; enamel and dentin volumes; hominoids

Abstract

The significance of enamel thickness in hominoid evolution has been plagued by the absence of nondestructive quantitative analysis. The aim of this investigation was to use a nondestructive method of analysis to document the volume of enamel and dentin, thereby providing a quantitative method for comparing both extant and extinct hominoid dentition. High-resolution X-ray computer tomography (HRXCT) is a nondestructive technique for visualizing and quantifying the interior of objects such as bone, teeth and minerals. HRXCT is also capable of obtaining digital information on their 3D geometries and volumetric properties. HRXCT differs from conventional medical CT-scanning in its ability to resolve details as small as a few microns in size, even when imaging objects made of high density materials like enamel and dentin. HRXCT also differs from micro-CT in its ability to examine large specimens up to 1.5 meters, with higher energy sources (typically 125–450 keV) that make the instrument capable of penetrating much denser objects including teeth and very heavily mineralized fossils. HRXCT offers several advantages over both the medical and micro-CT systems. Hominoid teeth used in this study consisted of a collection of extant hominoids as well as a number of fossil hominoids (*Proconsul and Sivapithecus*). HRXCT was used to obtain a data set of serially sectioned digitized images at a slice thickness of

S.E. Bailey and J.-J. Hublin (Eds.), Dental Perspectives on Human Evolution, 117–136.

approximately 50 micrometers per section. The digital images were analyzed, and 3D reconstructions allowed for the collection of volumetric data for coronal enamel and dentin. HRXCT provides researchers with capabilities not found in other CT systems, which allows for a wider range of specimens sizes, with customizable scanning parameters unique to each specimen type. Our results demonstrate that HRXCT is an effective means by which volumetric data and 3D reconstruction of dental hard tissues are obtained from both extant and extinct hominoid dentitions.

Introduction

The importance of enamel thickness in hominoids was first demonstrated by Gantt and colleagues (1976, 1977; Molnar and Gantt, 1977). Its significance and phyletic relationship in both extant and extinct hominoids have since been studied by a number of investigators over the past thirty years (Martin, 1983, 1985, 1986; Ward and Pilbeam, 1983; Beynon, 1984; Martin and Boyde, 1984; Beynon and Wood, 1986; Gantt, 1986, 1998; Grine and Martin, 1988; Beynon et al., 1991; Macho and Thackeray, 1992; Molnar et al., 1993; Ramirez Rozzi, 1993; Smith and Zilberman, 1994; Bromage et al., 1995; Macho, 1995, Schwartz et al., 1995, 1998, 2003; Cameron, 1997, Macho, 1994; Schwartz, 1997, 2000a; Strait et al., 1997; Gantt and Rafter, 1998; Reid et al., 1998; Risnes, 1998; Schwartz and Dean, 2000; Andrews and Alpagut, 2001; Bonis and Koufos, 2001; Grine et al., 2001; Grine, 2002; Martin et al., 2003; Smith et al., 2003, 2004, 2005). These investigators documented that humans have thick enamel, while in *Gorilla* and *Pan* the enamel is thin in comparison. Additionally, many of these investigators consider the differences in enamel thickness, especially between extant and extinct hominoids, to be a significant phyletic character and a means by which isolated teeth may be assigned to specific taxa (Gantt, 1981, 1982, 1986, 1998; Martin, 1983, 1985, 1986; Grine and Martin, 1988; Ramirez Rozzi, 1993; Schwartz, 1997, 2000a; Schwartz et al., 2003; Smith et al., 2003). Some have attempted to describe the range of enamel thickness, its pattern of distribution and its functional significance (Gantt, 1977, 1986; Macho and Berner, 1993, 1994; Spears and Macho, 1995, 1998; Schwartz, 1997, 2000a, b; Gantt and Rafter, 1998; Macho and Spears, 1999; Schwartz et al., 2003).

These endeavors, however, have led to continued controversy as to the significance of enamel thickness, the methods and procedures of analysis (see Grine, 2002, for a discussion of scaling of tooth enamel thickness and molar crown size), and criticism of using destructive techniques in order to produce sectioned teeth or thin sections, especially in extinct hominoids (Macho and Berner, 1993; Dumont, 1995). Thickness data have primarily been obtained from thin sectional and/or tooth section analysis (Gantt, 1977; Martin, 1983; Shellis, 1984, 1998; Beynon et al., 1991) and information obtained in the course of determining crown formation times (Shellis, 1984, 1998; Risnes, 1986; Beynon and Dean, 1987, 1988; Dean, 1988, 1998; Dean et al., 1993, 2001; Ramirez Rozzi, 1993; Beynon et al., 1997; Ramirez Rozzi et al., 1997; Schwartz, 1997, 2000a; Reid et al., 1998; Smith et al., 2005; Reid and Dean, 2006; Reid and Ferrell, 2006). These methods necessitate the destruction of the tooth or part of it, and thus limit the specimen's availability for use in studies of morphometry, enamel distribution, or in obtaining volumetric data.

The development of nondestructive methods for studying dental hard tissues has long been a goal of dental anthropologists and paleoanthropologists. Most researchers are hesitant to allow destructive techniques to be used on their material, especially hominoid finds. Specimens that have been sectioned have not been generally available to other investigators (Martin, 1983; Benyon and Wood, 1986; Grine and Martin, 1988; Schwartz, 1997, 2000a; Beynon et al., 1998; Dean et al., 2001; Dean and Schrenk, 2003; Martin et al., 2003; Schwartz et al., 2003; Smith et al., 2003, 2004). Attempts to use a nondestructive method to obtain enamel thickness began in the mid 1980s and early 1990s, with the use of medical computed tomography (CT) scanning systems. Some success was achieved, but not at the resolution necessary to replace the use of destructive methods and techniques (Zonneveld and Wind, 1985; Grine, 1989, 1991; Zonneveld et al., 1989; Conroy, 1991; Conroy and Vannier, 1991; Macho and Thackeray, 1992; Spoor et al., 1993, 2000a, b; Schwartz, 1997; Schwartz et al., 1998).

CT Scanning Systems

We believe that it is essential that we present a discussion of CT systems and the methods and procedures used to reconstruct the digital images into 2D and 3D images, as we believe that these technologies will significantly impact dental anthropology and paleontology. Our focus is upon the systems that are currently being used to study dental hard tissues in hominoids, both extant and extinct. We will describe their advantages and disadvantages for such investigations. At the core of all the types of CT scanning systems are: (1) an X-ray's source and detector(s) with computer technology capable of analyzing large amounts of data, and (2) digital imaging and analysis software to interpret the results and to display them in 2D and 3D. Four types of CT scanning systems are presently available: (1) medical CT systems, (2) micro-CT systems, (3) industrial CT systems, and (4) synchrotron X-ray CT systems. The latter two are custom made for specific research endeavors and are quite expensive.

Medical CT Systems

The first category, the medical CT scanning system, is the type of CT system that we are most familiar with and that is most commonly used in both medical and anthropological research, especially in the study of skeletal material. Developed in the early 1970s, medical CT systems have undergone continued improvements and enhancements, both to the equipment and to the computer, display and software platforms. At present, medical CT systems are in their sixth generation, resulting in lower patient X-ray dosage, faster acquisition and processing of the digital data. To maximize their effectiveness in differentiating tissues while minimizing patient exposure, medical CT systems need to use a limited dose of relatively low-energy X-rays (<140 keV), and to acquire data rapidly because the *patient should not move* during scanning, which would lead to distorted results. In order to obtain the best data possible given these requirements, they use relatively large mm-scale slices, and large (mm-scale) focal spots that require that the array of detectors rotate around, while the patient remains still.

At present, the best possible spatial resolution in the plane of the scan is about 0.3–0.5 mm. This resolution is determined by the X-ray beam geometry and can only be achieved when it is not limited by the pixel size of the image. CT scans of highly mineralized specimens such as teeth and/or matrix-filled fossils, made with medical scanners, show reduced image quality. Studies by Grine (1989, 1991; Grine et al., 2001) have demonstrated that the use of both lateral radiographs and medical CT scanning systems tend to overestimate enamel thickness values and should be viewed with circumspection.

Micro-CT or μCT Systems

Micro-CT scanning systems have recently become affordable and streamlined. This instrumentation is a CT scanning technology primarily designed to image small laboratory animals, such as mice, and is an ideal instrument for biomedical research laboratories due to small size and cost. Micro-CT allows for the acquisition of digital images and the processing of 3D images of both *in vitro* and *in vivo* specimens. Examples of commercial systems include SkyScan-1172 high-resolution desk-top micro-CT system (SkyScan, Aartselaar, Belgium), Scanco μCT 40 (SCANCO USA, Inc., Southeastern, PA) and the GE eXplore Locus SP (GE Healthcare Technologies, Waukesha, WI), to mention a few. However, there are also a number of custom made systems, such as the one used by Shimizu (2002) and that used by Kono (2004; Suwa and Kono, 1998, 2000, 2005).

Unlike medical CT systems, which require fast data acquisition, micro-CT systems use data acquisition times of one or more hours and incorporate micro focus X-ray sources capable of energies from 80–160 keV. Confusion abounds in the literature, due in part to the nomenclature vendors and researchers use to describe their instruments (e.g., Micro-CT, Micro-focal Computed CT, μCT, Micro Computed Tomography, Microtomography and Microfocal Tomography, etc.). There are also variations in both equipment design and software programming, from turnkey systems such as the SkyScan, Scanco and GE eXplore microCT scanning systems, to custom designed systems and software like that used by Shimizu (2002) and Kono (2004).

The primary advantages of the micro-CT systems are: (1) the availability of instruments, (2) volume digitization (volumetric data), (3) non-deteriorative scanning, and (4) high precision (>5 μm). The disadvantages however, are: (1) specimen size limitation (from single teeth to mice), (2) the use of polychromatic radiation, and (3) hardware and software system design variations that incorporate algorithms designed for soft-tissue and bone research.

Micro-CT systems have focused upon the niche of higher resolution and 3D reconstruction of small objects and animals, with the major limitation being that of specimen size. In addition, these systems often incorporate with turnkey software programs used to do the 3D reconstructions and volumetric analysis automatically. The best micro-CT images are obtained from objects in which microstructure coincides with contrast in X-ray absorption of the sample's constituent materials. The primary applications for which micro-CT systems are designed are to study bone, implants and soft tissues research on small animals. However, it should be noted that newer systems are being designed for the dental research market (see Olejniczak et al., 2007). The algorithms used by commercial systems to calculate volumetric data are based on these materials and not enamel and/or dentin. Care must be taken in data interpretation (Rowe et al., 1997; Ryan, 2000; Fajardo et al., 2002; Bernhardt et al., 2004).

Synchrotron X-ray Microtomography (XMT) or SRμCT

X-ray synchrotron microtomography (SRμCT) or synchrotron X-ray microtomography (XMT) is based on synchrotron monochromatic radiation. SRμCT based systems have recently been made available for use by dental anthropologists and paleontologists. These systems are mainly large custom designed experimental research facilities. Examples include the 390 meter, Advanced Photon Source (APS) at the U.S. Department of Energy's Argonne National Laboratory (Argonne, IL), the European Synchrotron Radiation Facility (ESRF), Grenoble, France, which is a joint facility supported and shared by 18 European countries, or custom designed systems like that described by (Elliott et al., 1998). The high brilliance of synchrotron monochromatic radiation, with a spatial resolution of 1 μm and a simple configuration, make SRμCT systems ideal for studying mineral composition of enamel (Elliott et al., 1998; Dowker et al., 2003, 2004, 2006). Tafforeau et al. (2002; Tafforeau, 2004), working at ESRF, have applied SRμCT to the study of fossilized primate materials and particularly, dental hard tissues (Tafforeau et al., 2002; Chaimanee et al., 2003; Brunet et al., 2004; Carvalho et al., 2004; Brunet et al., 2005; Feist et al., 2005; Boller et al., 2006; Marivaux et al., 2006; Tafforeau, 2006; Tafforeau et al., 2006a, 2006b). This technology is especially important in the study of fossilized dental material when attempting to discriminate between the enamel and dentin. Tafforeau et al. (2002; Tafforeau, 2004) were able to obtain differentiation between enamel and dentin, as well as to identify the dentinoenamel junction (DEJ) in all extant and in most of the fossil primate teeth studied. They were also able to obtain linear measurements of enamel thickness, but did not report on volumetric analysis even though produced exquisite 3D images (Tafforeau et al., 2002; Tafforeau, 2004).

Comparison of three micro-CT systems to SRμCT in the study of bone implants found that SRμCT eliminated specimen preparation artifacts and revealed that results from micro-CT overestimated the amount of bone formed (Bernhardt et al., 2004). Boller et al. (2006) have suggested that the best quality images in terms of signal-to-noise ratio and spatial resolution are obtained with SRμCT. The primary disadvantages of SRμCT are limited access to facilities and high cost of imaging.

Industrial CT Systems

Industrial CT systems are also custom designed for specific applications of nonliving objects due to the use of high radiation doses. Such systems are available from BIR (Bio-Imaging Research, Inc., Lincolnshire, Illinois) and from Universal Systems (Solon, Ohio) and employ the following optimizations: (1) the use of higher-energy X-rays (>450 keV), which are more effective at penetrating dense materials, (2) the use of smaller X-ray focal spots, providing increased resolution, (3) the use of finer, more densely packed X-ray detectors, with increased resolution, and (4) the use of longer exposure times, increasing the signal-to-noise ratio. Industrial CT scanners are typically custom built; therefore, no detailed description of their principles and operation will apply in all cases. Instead, we provide a description of each component of the CT-scanning process, both in general terms and as specifically applied at the University of Texas High-resolution X-ray CT (HRXCT) facility used in this investigation.

Imaging Reconstruction and 3D Visualization

In biomedical imaging, it is often necessary to acquire data from methods that essentially acquire the data one slice at a time in order to be able to view what is inside. In addition, a significant part of reconstruction is the ability to visualize all the data once put back together again. CT imaging requires the scanner to acquire a large number of projections, much like a common X-ray machine, but from many different positions. These different views through the object (or person) must be "deconvolved" (combined) to reconstruct the three-dimensional object. All CT systems produce digital images that must be processed in order to obtain quantitative data and 3D reconstructions (the abstract "rebuilding" of something that has been torn apart).

Spatial resolution describes the ability of a piece of imaging equipment to transfer spatial information from the object to the image without blurring. This is particularly important where objects with fine detail are being analyzed, such as the inner ear bones, sinuses, and the DEJ (Spoor et al., 1994, 2000a, b; Spoor and Zonneveld, 1999). The development of CT imaging over the past decade has increased in the role of digital techniques and applications. CT is an intrinsically digital modality, and has always required a way to store and analyze images electronically. CT scans produce two-dimensional images or, "slices", that display differences in X-ray attenuation (intensity loss) arising mainly from differences in density within an object. These images are spaced at intervals or made contiguous to one another. This rapid, nondestructive technology allows visualization and quantification of 3D geometry and properties. A typical digital image (2D) is divided into a square matrix of *pixels* (picture elements). In CT images, the data also have a third dimension, corresponding to the slice thickness. A CT image pixel is therefore, referred to as a *voxel* Zonneveld, 1987 (volume element).

Measurements obtained from the CT scanner are the decrease in X-ray intensity along thousands of rays traversing the slice at all angles. These measurements are then converted into images using a reconstruction algorithm(s). HRXCT differs from other CT systems in that scaling is arbitrary and customizable; it can be calibrated for obtaining accurate density and dimensional measurements of enamel and dentin, but it also retains flexibility that is useful when imaging fossils that have undergone one or more episodes of mineral replacement, potentially changing the density and composition of the constituent materials.

Terminology

In addition to the use of destructive methods, a second issue that has hindered the interpretation of enamel thickness significance in hominoids is a lack of a standard terminology used to describe differences between and among various hominoid taxa. At present, there is an array of descriptive terms used to describe enamel thickness: thin, intermediate thin, thick, intermediate thick, very thick, thick-hyperthick, and hyperthick, etc. What do these terms mean and how quantifiable are they? Do these terms truly reflect differences among both extant and extinct taxa? A standardized terminology, based on large samples of both extant and extinct hominoid molars and dentitions is needed. To achieve this goal we must use a nondestructive method such as HRXCT, which yields quantitative data that exceed the results obtained from destructive methods, and that is capable of sampling larger collections nondestructively.

Volumetric Data

The volume of the dental hard tissues is the most useful measurement for determining the amount of enamel and dentin that contributes to the overall crown size. The volume of enamel represents the total amount of enamel secreted by the ameloblasts, while the volume of coronal dentin represents the total amount of coronal dentin secreted by the odontoblasts during tooth development (in unerupted and/or unworn teeth). Tooth size is governed by three parameters: (1) the amount of enamel, (2) the amount of dentin, and (3) the size of the pulp cavity within the crown (Sasaki and Kanazawa, 1998; Avishai et al., 2004). Since teeth must withstand the forces of mastication and are necessary for health and longevity, changes in tooth size and crown morphology are the result of changes in the volume and pattern of distribution of enamel and dentin. However, we have not been able to obtain volumetric data for these components even by means of destructive techniques. This has resulted in the use of a variety of scaling techniques in an attempt to describe enamel area, thickness and its distribution (Grine, 2002). One limitation of analyzing volume is that tooth development must be complete, with little to no wear. Martin (1983, 1985; Martin et al., 2003) attempted to derive an estimate of the enamel volume. He proposed that an approximation could be obtained from a section by dividing the area of the enamel cap by the length of the DEJ. This dimension yielded an average enamel thickness, which he considered to be equivalent to the average straight-line distance traveled by the ameloblasts from the DEJ to the enamel surface. Ameloblasts however, do not travel in a straight-line course, but vary in course direction within the tooth and even within parts of the tooth crown (Dean and Shellis, 1998). Furthermore, there is a debate over whether the area of the crown's dentin core or the linear distance between the buccal and lingual cervical margins is the most appropriate scaling factor (Grine, 2002)

Purpose

The purpose of this investigation was to use a nondestructive method that was accurate, quantifiable, and repeatable to measure the volume of enamel and dentin, especially in fossil hominoids. This investigation incorporates the use of industrial high-resolution X-ray computed tomography (HRXCT) to obtain volumetric data from both extant and extinct hominoid molars. The HRXCT facility at the University of Texas at Austin is a custom designed unit for anthropological, geological and paleontological research and is capable of analyzing changes in crown morphology, enamel and dentin distribution, as well as linear measurements of enamel and dentin thickness and mineral density. This nondestructive method exceeds the accuracy currently achieved by destructive means. In fact, HRXCT expands the number of measurements that can be made because of its ability to provide complete 3D representation of the tooth or jaw and allow for comparison of these measurements to actual thin section data of the same specimen (see Olejniczak and Grine, 2006, in review; Suwa and Kono, 2005; Olejniczak et al., 2007). Therefore, our objectives were to: (1) demonstrate the utilization of the HRXCT, and (2) obtain quantitative volumetric analysis and 3D visualization of enamel and dentin in both extant and extinct hominoids (Rowe et al., 1997; Kappelman, 1998; Gantt et al., 2002, 2003; Ketcham and Ryan, 2004).

Materials and Methods

A collection of extant and extinct hominoid teeth were obtained from the personal collections of the authors, together with a HRXCT image of a specimen of *Sivapithecus parvada* (courtesy of Jay Kelley). This collection included samples of extant molars from modern humans, the chimpanzee (*Pan*) and gorilla (*Gorilla*), as well as several fossil hominoids (*Sivapithecus* and *Proconsul*) (Gantt, 1986). The extant sample consisted of mostly unerupted specimens, while the fossil samples were broken and worn specimens.

The descriptions of the technology presented in this and subsequent sections are a combination of information provided by the University of Texas system, by the manufacturer (Bio-Imaging Research, Inc., of Lincolnshire, Illinois), insights gained from experience, and general principles derived from the literature (Ryan and Ketcham, 2002a, b). The HRXCT scanner is capable of slice thicknesses from 5 μm or less to over 100 μm, depending on object size and density. Specimens may range from the size of small mice teeth to as large as a tapir's skull (current maximum size is 1.5 meters).

The X-ray source for the HRXCT is a Fein-Focus FXE-200.20 X-ray tube, which is capable of producing voltages from 50 keV to 200 keV with beam currents from 0.04 to 1.0 mA. The source utilizes a tungsten target, and the X-ray focal spot is manually adjustable from 3 to 200 microns. The detector is a Toshiba AI-5764-HVP image intensifier with software-controlled slice thickness ranging from 0.01 to 10 mm, with output via a CCD (charge-coupled device) camera. Data for this study were acquired using X-rays at 150 keV and 0.053 mA, which provided a <10 micron focal spot. Slices of 70, 50 and 30 microns were produced in serial sequence (see Table 1).

3D Reconstruction

Several software programs were used to obtain 3D reconstructions from the data sets for each taxon: *VoxBlast* (VayTek, Inc., Fairfield, IA), *Mimics* (Materialise, Inc., Ann Arbor, MI), and *Vitrea*®2 (Vital Images, Inc., of Minnetonka, MN). These programs are fully integrated, user-friendly 3D image processing and editing software systems designed to analyze medical CT data. Dr. Tom Deahl of the Department of Dental Diagnostic Sciences, The University of Texas Health Science Center at San Antonio, TX, conducted the 3D and volumetric analysis using *Mimics, Vitrea*®2 and *VoxelView* (Vital Images). These programs were use to produced 3D images and movies to allow for color enhancements for improved visualization (Kappelman, 1998, Gantt et al., 2002, 2003). Subsequent to these studies, additional analysis was provided by Mercury Computer Systems (San Diego, CA) using *Amira* imaging and by VayTech, Inc. (Fairfield, IA) with *VoxBlast 3D*. Both products are commercially available from these companies. In addition, a custom designed micro-CT imaging software program from Scanco USA, Inc. (Wayne, PA) was also used.

Volumetric Analysis

Volumetric analysis was carried out in *Mimics* and *VoxelView* imaging software programs by Dr. Deahl of the Department of Dental Diagnostic Sciences, The University of Texas Health Science Center at San Antonio, TX (Gantt et al., 2002, 2003). We also sent out our data sets for independent confirmations of our results. Two commercial imaging processing and analysis software programs, *VoxBlast 3D* and *Amira*, and one custom designed micro-CT imaging program from Scanco were used to reevaluate our original findings (see Table 2). These programs provided additional 3D visualization and quantitative volumetric analyses, which allowed us to determine if our original values, obtained using medical CT imaging software, were valid. Confirmation of these values became increasingly important as published data did not support our original findings (see Kono et al., 2002; Kono, 2004). All data were expressed in cubic millimeters.

Table 1. HRXCT scan data parameters for each specimen with slice thickness and field of reconstruction presented in mm

Taxa	*Specimen*	*Slice Thickness*	*Field of Reconstruction*	*# of Views*
Modern *Homo*	M_3	0.051	14.7	246
Pan	M_3	0.051	14.0	238
Gorilla	M_3	0.073	18.6	231
[1]*Sivapithecus sivalensis*	M_2	0.048	NA	20
[2]*Sivapithecus parvada*	I^1	0.048	NA	10
[3]*Proconsul africanus*	M_1	0.051	NA	3

Test scans of a various teeth of primates and experimental and control knockout mice were conducted from 1998 to 2004 (Gantt et al., 2003; Gantt et al., 2002). All digital images were in raw format with both 16bit and 8bit versions of the data. HXRCT scans were made with the following settings: 120 keV, 0.133 mA, no filter, air wedge, 0% offset, 2 samples per view, scanned in three-slice mode.

1. A mandibular first molar fragment of *Sivapithecus sivalensis* from Potwar Plateau, Pakistan, provided by David Pilbeam, Department of Anthropology and Peabody Museum, Harvard University, Boston, MA (see Gantt, 1986 for details).
2. A maxillary right I^1 germ of *Sivapithecus parvada*, from locality Y311 (10.0 Ma) in the Potwar Plateau, Pakistan, provided by Jay Kelley, Department of Oral Biology, College of Dentistry, University of Illinois, Chicago, IL.
3. A mandibular first molar fragment of *Proconsul africanus*, SO 901, East Africa, provided by David Pilbeam, Department of Anthropology and Peabody Museum, Harvard University, Boston, MA (see Gantt, 1986 for details).

*Table 2A. HRXCT volumetric data (expressed in mm^3) for modern humans, gorilla and chimpanzee mandibular molars, derived from manual measurements by Dr. Richard Ketcham using **IDL** imaging software. Reconstructions of the image stacks were based on the methods described and illustrated in Figure 5*

Specimens	*Enamel volume*	*Dentin volume*
Pan M_3	176.48	262.00
Gorilla M_3	399.00	725.07
Homo M_3	235.00	410.00

*Table 2B. HRXCT volumetric data (expressed in mm^3) for modern humans and gorilla, obtained by using object count function in **Amira** (Mercury Computer Systems imaging software)*

Specimens	*Enamel volume*	*Dentin volume**
Homo M_3	233.00	457.00

* Volume for dentin represents total dentin volume within the field of view and does not restrict measurements to only the coronal dentin.

*Table 2C. HRXCT volumetric data (expressed in mm^3) for gorilla provided by **Scano** (Scanco Medical, μCT proprietary image software)*

Specimens	*Enamel volume*	*Dentin volume**
Gorilla M_3	523.15	1070.19

* Volume for dentin represents total dentin volume within the field of view and does not restrict measurements to only the coronal dentin, but includes the pulp cavity as well.

Table 3. Comparison of volumetric data (expressed in mm^3) for extant hominoid mandibular molars by HRXCT and μCT studies (Evol = enamel volume; Dvol = Dentin volume)

Specimens	*Gantt et al. Evol*	*KonoEvol*	*Gantt et al. Dvol*	*Kono Dvol**
Pan	176.48	171.4 S.D. 19.1	262.00	289.3 S.D. 38.8
Gorilla	399.00	515.4 S.D. 141.6	725.07	1093 S.D. 289.8
Homo	235.00	265.7 S.D. 40.8	410.00	281.8 S.D. 55.0

* Volume for dentin represents total dentin volume within the field of view and does not restrict measurements to only the coronal dentin, but includes the pulp cavity as well. (Gantt et al., this paper; Kono, 2004)

Digital images in 8-bit and 16-bit format were converted into contiguous stacks to reconstruct the 3D geometries of the crown for each specimen. The 3D rendering allowed us to separate the enamel cap from the dentin cap for visualization of the dentinoenamel junction (DEJ), and to facilitate quantitative analysis of the volume of enamel and dentin. To insure that we were measuring only coronal dentin, a line was drawn across the cementoenamel junction (CEJ). This allows us to delineate the area of coronal enamel, dentin and pulp, within each slice to insure the accuracy for determining the correct location of the CEJ and to eliminate the area of pulp from any volume measures of the dentin, thereby obtaining the most accurate volumetric calculations of both enamel and dentin (as is illustrated in Figure 1). All specimens were examined in the same manner using automated software programs *Mimics* and *VoxelView*, and manually using *IDL* software (a visualization and analysis software program, Research Systems, Inc., Boulder, CO). Analysis by (1) VayTech, Inc. (*VoxBlast 3D*), (2) Mercury Computer Systems (*Amira*), and (3) Scanco USA, Inc. (using their own custom designed imaging software) to produce quantitative volume measurements.

Hrxct Analysis

The HRXCT results for humans were based upon the analysis of 246 HRXCT digital images with a scan slice thickness of 0.051 mm and a field of reconstruction of 14.7 mm. 3D reconstruction of the enamel cap and the occlusal

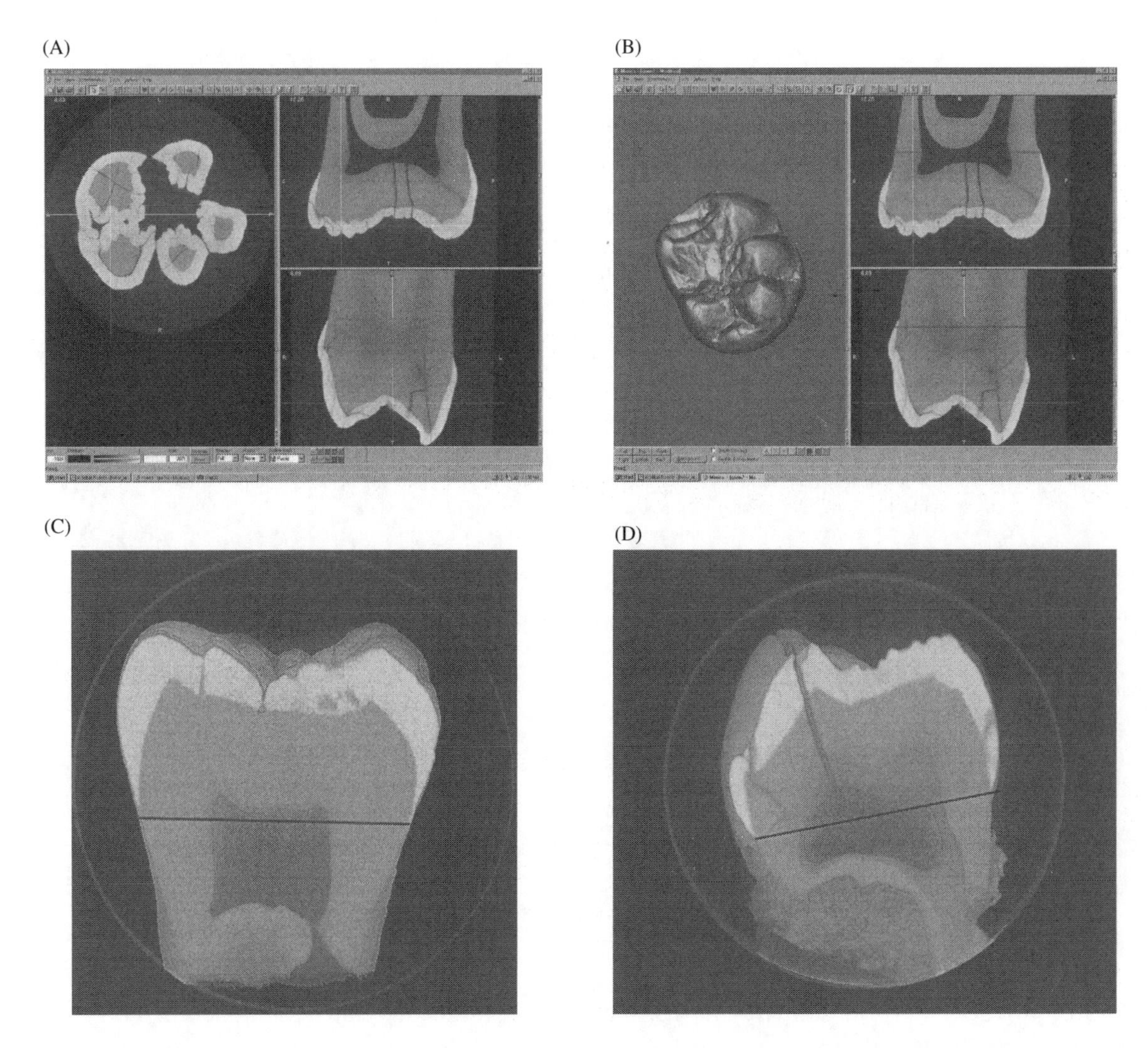

Figure 1A–D. A. Mimics 3D reconstruction software images showing three views of a slice from the lower mandibular molar of *Gorilla* and, B. The 3D reconstruction based upon the entire stack of HRXCT slices showing the enamel cap with the dentinoenamel junction and lateral surface enamel of the crown. C. A section of the 3D reconstruction of the stack of HRXCT images with a line (black) manually drawn across each section to mark the region of coronal enamel and dentin of *Homo* and, D. *Gorilla*.

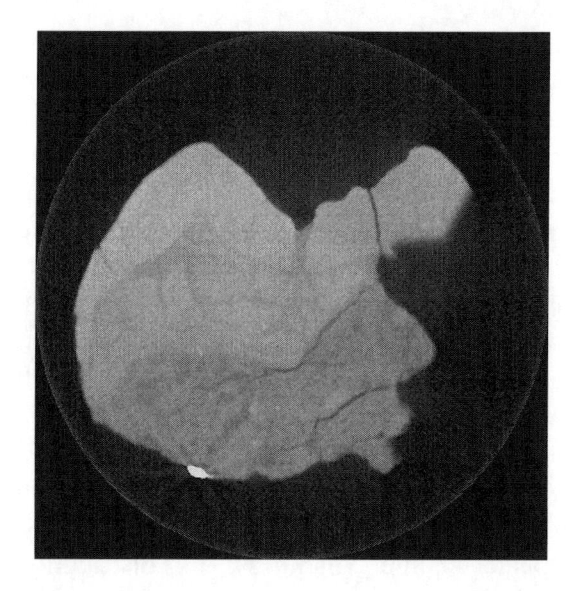

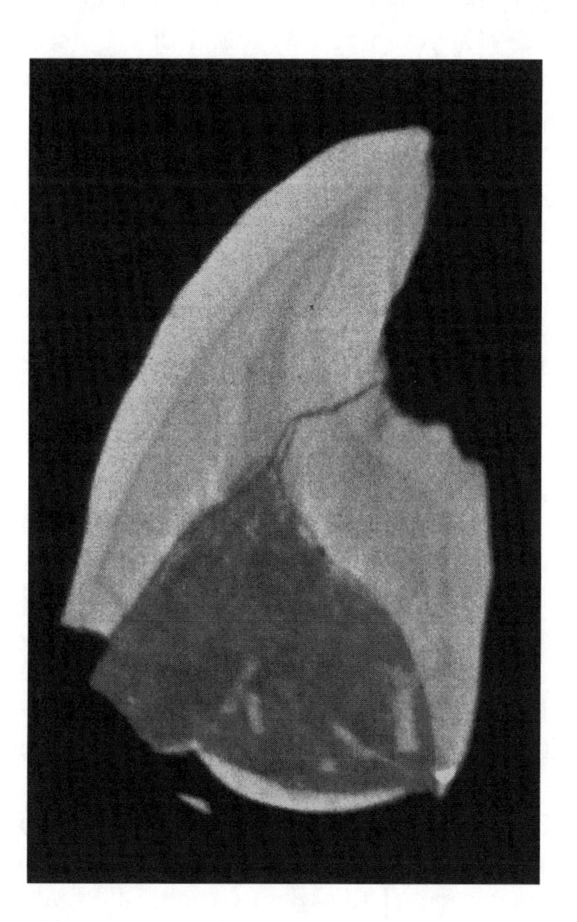

Figure 2A–B. A. HRXCT image of one slice through a maxillary right I^1 germ of *Sivapithecus parvada,* and B. A scan of the lower molar fragment of *Sivapithecus sivalensis* to demonstrate the power of the HRXCT to visualize enamel and dentin in fossil hominoids.

view of the dentin surface provides a unique perspective of the DEJ. The data set for *Gorilla* was based upon the analysis of 231 HRXCT digital images with a scan slice thickness of 0.073 mm, resulting in a field of reconstruction of 18.6 mm. Color enhancement of the 3D reconstructions of the enamel and dentin caps allows comparison of the 3D images obtained from the human molar. The results for *Pan* were based upon the analysis of 238 HRXCT digital images with a scan slice thickness of 0.051 mm and a field of reconstruction of 14.0 mm.

Results

Enamel thickness data for the scanned specimens were compared with published data (see Table 3). No variation was observed in our data set compared to published data on thin sections or sectioned teeth by Gantt (1977) and Martin (1983).

Analysis of the fossil hominoids was limited to fragmentary specimens or only a few scans; therefore, no volumetric data were obtained. However, it was possible to clearly delineate the DEJ, even with differences in mineralization due to fossilization. The enamel and dentin layers could be clearly distinguished, as could the matrix intrusions in *Sivapithecus* (Figure 2). It is evident that linear measurements could easily be made with precision. In addition, the lower contrast between enamel and dentin suggests that one or both of these materials has been, to some extent, replaced with other minerals complicating analysis, but by no means precluding reliable measurements. Increase in voltage (keV) and calibration to a hydroxyapitate standard would also facilitate data acquisition.

Discussion

The first HRXCT studies on enamel volumes were presented and published by Gantt et al. in 2002 and 2003. Previous work on trabecular bone by Fajardo et al. (2002) established the accuracy of HRXCT in determining volume ratios, trabecular bone thickness and spacing compared to histological thin sections. They suggested that, given the higher energy levels that are found in HRXCT scanners, it is also likely that this technique can be used to image fossilized trabecular bone in extinct primates (Rowe et al., 1997; Ryan, 2000; Fajardo and Muller, 2001; Fajardo et al., 2002). The combining of 3D morphometric data allowed measures to be obtained for bone volume fractions (Kappelman,1998; Fajardo et al., 2002). The advantage of the HRXCT over medical CT and micro-CT is the combination of higher energy sources with modular linear and areal detector arrays, which allows for the handling of a variety of very dense samples across a wide range of resolutions (Rowe et al., 1997). This is especially important in the study of fossilized material and large specimens.

At present, we are unaware of any existing studies that have used HRXCT imaging to study enamel and/or dentin in extant and extinct hominoids. However, there are studies that have used either micro-CT or SRμCT systems, or a combination of both, to study fossil hominoid and primate dental materials (Brunet et al., 2004; Hlusko et al., 2004; Brunet et al., 2005; Bloch and Silcox, 2006; Olejniczak and Grine, 2006). 3D laser scanning of demineralized teeth and studies of the occlusal surface morphology, enamel thickness, enamel volume, topography of the DEJ, and the pattern of enamel distribution in modern humans and extant great apes teeth have also been investigated with micro-CT (Kono et al., 1997, 2002; Suwa and Kono, 1998, 2000, 2005; Kono and Suwa, 2000, 2002; Avishai et al., 2004; Kono, 2004; Olejniczak and Grine, 2006, in review; Olejniczak et al., 2007).

A 3D scanning system that incorporated a custom 3D image data analysis software program was used to analyze enamel distribution patterns in human permanent first molars (Kono et al., 1997, 2002; Suwa and Kono, 1998). The authors used paired casts of the outer enamel surface and DEJ of each molar after removing the enamel layer by formic acid. Analysis of Kono et al.'s (2002) enamel thickness map for extant hominid mandibular first molars appears to us to represent the following pattern: *Homo>Pongo>Gorilla>Pan* concerning overall enamel thickness (Figure 15, p. 128), as well as revealing differences in enamel distribution patterns among the taxa.

A second study was performed by Kono (2004) on 74 extant hominoid molars to evaluate enamel thickness, its distribution pattern and volumes of enamel and dentin by micro-CT. Kono (2004) found discrepancies in 2D and 3D average enamel thickness, and attributed these to local differences within the tooth enamel distribution pattern and DEJ topography. Her work provides, for the first time, a detailed investigation of molar enamel thickness patterns of the entire tooth crown in extant hominoids. The use of micro-CT, together with 3D imaging software, allowed for 3D reconstructions and assessment of enamel patterning of the crown, including obtaining quantitative enamel and dentin volumes. The enamel volumes were different from those previously gathered from 2D-based studies. The "average enamel thickness" (AET) reported previously (Martin, 1983; Grine and Martin, 1988; Schwartz, 1997, 2000a; Grine et al., 2001; Martin et al., 2003; Schwartz et al., 2003; Smith et al., 2003, 2004, 2005) based on the 2D analysis of buccal-lingual sections of the mesial cusps (MCS) was found not to be an appropriate substitute or estimator of whole crown average enamel thickness (Kono, 2004; also see Suwa and Kono, 2005). The mean enamel volume values based upon analysis of first and second molars revealed the following pattern: *Gorilla>Pongo>Homo>Pan* (Appendix 1, p. 143), which is in agreement with our revised calculations.

The values from Kono's study (2004) are comparable to those obtained by HRXCT based upon the reevaluation of the volumetric data, which was performed manually (using *IDL*) and by automated analysis by *Amira* and by *Scanco* volumetric software programs.

Current investigations by Olejniczak and colleagues (Olejniczak and Grine, 2006, in review; Olejniczak et al., 2007) have also provided support for Kono's work using a Scanco micro-CT (model μCT-20). They have established an excellent protocol and methodology for the use of micro-CT in the study of teeth and have analyzed the impact on measurement accuracy produce by altering parameters. Their volumetric values are consistent with the micro-CT values obtained by Kono (2004) and with our current HRXCT values.

In an attempt to understand why our original values were significantly different from those of Kono (2004) and those presented by Olejniczak et al. (2007) we manually analyzed the data set and also sent it out for independent assessment to Scanco Medical and Mercury Computer Systems by automated volumetric analysis. Based on our reanalysis and the results provided by Scanco Medical and Mercury Computer Systems, we have rejected our previous values and agree with the results obtained by Kono and Olejniczak. These independent studies confirm that the HRXCT volumetric values are consistent and in line with reported values. Moreover, we have identified that variation exists between various automated 3D reconstruction programs, and that care must be taken to insure accurate quantitative volumetric values. Since *VoxelView* imaging software is designed for medical CT, which produce DICOM formatted digital images, data conversion of our raw tiff images was necessary for analysis, which may have resulted in the larger values.

Our work has clearly established the viability of HRXCT as a nondestructive means of analyzing dental hard tissues, especially volumetric analysis and strongly supports the use of micro-CT as well. HRXCT is capable of quantitative data acquisition of the area, density, distribution, thickness and volume of dental tissues. The utilization of HRXCT and micro-CT to study fossils and teeth is only recently beginning to be explored. (Rowe et al., 2001; Ryan and Ketcham, 2002a, b 2005; Clack et al., 2003; Alonso et al., 2004; Ketcham and Ryan, 2004; Summers et al., 2004; Olejniczak and Grine, 2006, in review; Olejniczak et al., 2007). The intent of this work is to show the potential of HRXCT, and hopefully to provide inspiration for new investigations that will benefit from further utilization of this breakthrough technology.

Conclusions

Recent CT developments (micro-CT, SRμCT and HRXCT) provide a nondestructive means by which dental hard tissues and fossilized dental specimens may be quantitatively studied to obtain linear measures and volumetric data on enamel and dentin. These technologies allow imagery of denser objects across a range of size classes, and provide increases in resolution that provide key advances, greatly enhancing the potential for applications of these technologies in biomedical, anthropological, geological and paleontological investigations. The advantages of HRXCT are that it increases power of resolution and specimen size ranges. The HRXCT allows us to study whole jaws and skulls, and to scan specimens at varying stages of development to help reveal the dynamics of craniofacial growth by obtaining data on the changes in the dental tissue volumes and mineral densities pre- and post-natally or across life histories.

In order to understand the significance of enamel thickness and its patterning, it is necessary to examine the volumes of enamel, dentin and pulp that are the components governing the overall size of the tooth. We can also analyze the entire craniofacial complex to investigate the interrelations of roots and alveolar bone. Since there are very few differences between micro-CT and HRXCT, except as noted, we are currently comparing CT digital images of the same specimens obtained from HRXCT, micro-CT and SRμCT systems, to further refine the utilization of these types of instrumentation.

It is also necessary to establish enamel thickness distribution patterns over the entire crown of each tooth type of each species being studied based upon large samples. Presently, this is the only way by which we will be able to understand the relationship of cuspal morphology and the significance of enamel, dentin, coronal pulp and root volumes, as well as their patterns of distribution in both extant and extinct hominoids. These data will significantly contribute to our understanding of dental development, functional morphology and phyletic relationships. Once these data are available, we will have a better understanding of the significance of these characters within Hominoidea, as well as in other nonhuman primates.

Acknowledgments

We thank Jean-Jacques Hublin and Shara Bailey as well as the staff of the Max Planck Institute for Evolutionary Anthropology, Leipzig, Germany, for inviting our participation and for hosting the conference. We also thank Jay Kelly for giving us permission to use the HRXCT images of *Sivapithecus parvada*, as well as David Pilbeam for the specimens of *Proconsul africanus* and *Sivapithecus sivalensis*. The hard work of the reviewers and our colleagues, David McWhorter, Brian Matayoshi and Mary Owen have greatly improved the manuscript, and we thank all of them for their time and effort. We also thank Patrick Guerin of VayTek, Inc., (*VoxBlast 3D*); Patrick Barthelemy of Mercury Computer Systems (*Amira*); and by Rasesh Kapabia of Scanco Medical, Inc., (using their custom imaging software provided with their μCT systems) for providing quantitative volumetric analysis and 3D reconstruction of our data sets.

Support for our research was provided in part by the University of Texas at Austin High-Resolution X-ray CT Facility. The HRXCT facility was created with support from the W.M. Keck Foundation, the National Science Foundation (NSF EAR-9406258; NSF EAR-0345710), and the Geology Foundation of the University of Texas.

References

Alonso, P.D., Milner, A.C., Ketcham, R.A., Cookson, M.J., Rowe, T.B., 2004. The avian nature of the brain and inner ear of *Archaeopteryx*. Nature 430, 666–669.

Andrews, P., Alpagut, B., 2001. Functional morphology of *Ankarapithecus meteai*. In: de Bonis, L., Koufos, G.D., Andrews, P. (Eds.), Hominoid Evolution and Climatic Change in Europe, Vol. 2: Phylogeny of the Neogene Hominoid Primates of Eurasia. Cambridge University Press, Cambridge, pp. 213–230.

Avishai, G., Muller, R., Gabet, Y., Bab, I., Zilberman, U., Smith, P., 2004. New approach to quantifying developmental variation in the dentition using serial microtomographic imaging. Microscopic Research Techniques 65, 263–269.

Bernhardt, R., Scharnweber, D., Muller, B., Thurner, P., Schliephake, H., Wyss, P., Beckmann, F., Goebbels, J., Worch, H., 2004. Comparison of microfocus – and synchrotron X-ray tomography for the analysis of osteointegration around TI6AL4V-implants. European Cells and Materials 7, 42–51.

Beynon, A., 1984. Preliminary observations on enamel structure and thickness in fossil hominids. Journal of Dental Research 63(505), 505.

Beynon, A.D., Dean, M.C., 1987. Crown-formation time of a fossil hominid premolar tooth. Archives of Oral Biology 32(11), 773–780.

Beynon, A.D., Dean, M.C., 1988. Distinct dental development patterns in early fossil hominids. Nature 333, 509–514.

Beynon, A.D., Dean, M.C., Leakey, M.G., Reid, D.J., Walker, A., 1997. Histological study of *Proconsul* teeth from Rusinga Island, Kenya. American Journal of Physical Anthropology Suppl. 24, 77.

Beynon, A.D., Dean, M.C., Leakey, M.G., Reid, D.J., Walker, A., 1998. Comparative dental development and microstructure of *Proconsul* teeth from Rusinga Island, Kenya. Journal of Human Evolution 35, 163–209.

Beynon, A.D., Dean, M.C., Reid, D.J., 1991. On thick and thin enamel in hominoids. American Journal of Physical Anthropology 86, 295–309.

Beynon, A.D., Wood, B.A., 1986. Variations in enamel thickness and structure in east African hominids. American Journal of Physical Anthropology 70, 177–193.

Bloch, J.I., Silcox, M.T., 2006. Cranial anatomy of the Paleocene plesiadapiform *Carpolestes simpsoni* (Mammalia, Primates) using ultra high-resolution X-ray computed tomography, and the relationships of plesiadapiforms to Euprimates. Journal of Human Evolution 50, 1–35.

Boller, E., Cloetens, P., Mokso, R., Tafforeau, P., Thibault, X., Peyrin, F., Marmottant, A., Pernot, P., Barcuchel, J., 2006. Fast acquisition high-resolution synchrotron radiation X-ray microtomography for academic and industrial purposes. In: Brebbia, C.A., Mammoli, A.A. (Eds.), Computational Methods and Experiments in Materials Characterisation II. WIT Press, Billerica, MA, pp. 197–207.

de Bonis, L., Koufos, G.D., 2001. Phylogenetic relationships of *Ouranopithecus macedoniensis* (Mammalia, Primates, Hominoidea, Hominidae) of the late Miocene deposits of Central Macedonia (Greece). In: de Bonis, L., Koufos, G.D., Andrews, P. (Eds.), Hominoid Evolution and Climatic Change in Europe, Vol. 2: Phylogeny of the Neogene Hominoid Primates of Eurasia. Cambridge University Press, Cambridge, pp. 254–268.

Bromage, T., Schrenk, F., Zonneveld, F., 1995. Paleoanthropology of the Malawi Rift: An early hominid mandible from the Chiwondo Beds, northern Malawi. Journal of Human Evolution 28, 71–108.

Brunet, M., Guy, F., Boisserie, J.R., Djimdoumalbaye, A., Lehmann, T., Lihoreau, F., Louchart, A., Schuster, M., Tafforeau, P., Likius, A., Taisso, Mackaye H., Blondel, C., Bocherens H., De Bonis, L., Coppens, Y., Denis, C., Duringer, P., Eisenmann, V., Flisch, A., Geraads, D., Lopez-Martinez, N., Otero, O., Pelaez Campomanes, P., Pilbeam, D., Ponce de Léon, M., Vignaud, P., Viriot, L., Zollikofer, C., 2004. "Toumaï", Miocène supérieur du Tchad, le nouveau doyen du rameau humain. Comptes Rendus Palevol 3, 277–285.

Brunet, M., Guy, F., Pilbeam, D., Lieberman, D. E., Likius, A., Mackaye, H.T., Ponce de Leon, M. S., Zollikofer, C., Vignaud, P., 2005. New material of the earliest hominid from the Upper Miocene of Chad. Nature 434, 752–755.

Cameron, D.W., 1997. A revised systematic scheme for the Eurasian Miocene fossil Hominidae. Journal of Human Evolution 33, 449–477.

Carvalho, M.L., Marques, J.P., Marques, A.F., Casaca, C., 2004. Synchrotron microprobe determination of the elemental distribution in human teeth of the Neolithic period. X-Ray Spectrometry 33(1), 55–60.

Chaimanee, Y., Jolly, D., Benammi, M., Tafforeau, P., Duzer, D., Moussa, I., Jaeger, J.J., 2003. A Middle Miocene hominoid from Thailand and orangutan origins. Nature 422, 61–65.

Clack, J.A., Ahlberg, P.E., Finney, S.M., Dominguez Alonso, P., Robinson, J., Ketcham, R.A., 2003. A uniquely specialized ear in a very early tetrapod. Nature 425, 65–69.

Conroy, G.C., 1991. Enamel thickness in South African australopithecines: noninvasive evaluation by computed tomography. Palaeontologica Africana 28, 53–59.

Conroy, G.C., Vannier, M., 1991. Dental development in South African australopithecines. Part 1: Problems of pattern and chronology. American Journal of Physical Anthropology 86, 121–136.

Dean, M.C., 1988. Growth of teeth and development of the dentition in *Paranthropus*. In: Grine, F.E. (Ed.), Evolutionary History of the "Robust" Australopithecines. Aldine de Gruyter, New York, pp. 43–53.

Dean, M.C., 1998. A comparative study of cross striation spacings in cuspal enamel and of four methods of estimating the time taken to grow molar cuspal enamel in *Pan, Pongo* and *Homo*. Journal of Human Evolution 35, 449–462.

Dean, M.C, Beynon, A., Thackeray, J., Macho, G., 1993. Histological reconstruction of dental development and age at death of a juvenile *Paranthropus robustus* specimen,

SK 63, from Swartkrans, South Africa. American Journal of Physical Anthropology 91, 401–419.

Dean, C., Leakey, M.G., Reid, D., Schrenk, F., Schwartz, G.T., Stringer, C., Walker, A., 2001. Growth processes in teeth distinguish modern humans from *Homo erectus* and earlier hominins. Nature 414, 628–631.

Dean, M.C., Schrenk, F., 2003. Enamel thickness and development in a third permanent molar of *Gigantopithecus blacki*. Journal of Human Evolution 45, 381–387.

Dean, M.C., Shellis, R.P., 1998. Observations on stria morphology in the lateral enamel of *Pongo, Hylobates* and *Proconsul* teeth. Journal of Human Evolution 35(4–5), 401–410.

Dowker, S.E., Elliott, J.C., Davis, G.R., Wassif, H.S., 2003. Longitudinal study of the three-dimensional development of subsurface enamel lesions during *in vitro* demineralization. Caries Research 37, 237–245.

Dowker, S.E., Elliott, J.C., Davis, G.R., Wilson, R.M., Cloetens, P., 2004. Synchrotron x-ray microtomographic investigation of mineral concentrations at micrometer scale in sound and carious enamel. Caries Research 38, 514–522.

Dowker, S.E.P., Elliott, J.C., Davis, G.R., Wilson, R.M., Cloetens, P., 2006. 3D study of human dental fissure enamel by synchrotron X-ray microtomography. Eurupean Journal of Oral Science 114, 353–359.

Dumont, E.R., 1995. Enamel thickness and dietary adaptation among extant primates and chiropterans. Journal of Mammalogy 76, 1127–1136.

Elliott, J.C., Wong, F.S., Anderson, P., Davis, G.R., Dowker, S.E., 1998. Determination of mineral concentration in dental enamel from X-ray attenuation measurements. Connective Tissue Research 38, 61–72; discussion 73–69.

Fajardo, R.J., Muller, R., 2001. Three-dimensional analysis of nonhuman primate trabecular architecture using micro-computed tomography. American Journal of Physical Anthropology 115, 327–336.

Fajardo, R.J., Ryan, T.M., Kappelman, J., 2002. Assessing the accuracy of high-resolution X-ray computed tomography of primate trabecular bone by comparisons with histological sections. American Journal of Physical Anthropology 118, 1–10.

Feist, M., Liu, J., Tafforeau, P., 2005. New insights into Paleozoic charophyte morphology and phylogeny. American Journal of Botany 92, 1152–1160.

Gantt, D.G., 1976. Enamel thickness: its significance and possible phyletic implications. American Journal of Physical Anthropology 44, 179–180.

Gantt, D.G., 1977. Enamel of primate teeth: Its thickness and structure with reference to functional and phyletic implications. Ph.D. Dissertation, Washington University.

Gantt, D.G., 1981. Enamel thickness and neogene hominoid evolution. American Journal of Physical Anthropology 54, 222.

Gantt, D.G., 1982. Hominid evolution – a tooth's inside view. In: Kurten, B. (Ed.), Teeth: Form, Function, and Evolution. Columbia University Press, New York, pp. 107–206.

Gantt, D.G., 1986. Enamel thickness and ultrastructure in hominoids: with reference to form, function and phylogeny. In: Swindler, D.R. (Ed.), Systematics, Evolution and Anatomy. Alan R. Liss, Inc., New York, pp. 453–475.

Gantt, D.G., 1998. *Proconsul* – thick or thin? A study of enamel thickness and its significance. American Journal of Physical Anthropology Supplement 26, 103.

Gantt, D.G., Kappelman, J., Ketcham, R.A., Alder, M., Deahl, T., 2003. 3D approach to interpret enamel thickness and volume. American Journal of Physical Anthropology 70, 124.

Gantt, D.G., Kappelman, J., Ketcham, R.C., Colbert, M., 2002. 3D reconstruction of enamel volume in human and gorilla molars. American Journal of Physical Anthropology 55, 135.

Gantt, D.G., Rafter, J.A., 1998. Evolutionary and functional significance of hominoid tooth enamel. In: Veis, A. (Ed.), Sixth International Symposium on the Composition, Properties and Fundamental Structure of Tooth Enamel. Connective Tissue Research, Lake Arrowhead, CA, pp. 195–206 [499–510].

Grine, F.E., 1989. Use of computed tomography to measure tooth enamel thickness. American Journal of Physical Anthropology 12, 84.

Grine, F.E., 1991. Computed tomography and the measurement of enamel thickness in extant hominoids: implications for it paleontological application. Palaeontolgia Africana 28, 61–69.

Grine, F.E., 2002. Scaling of tooth enamel thickness, and molar crown size reduction in modern humans. South African Journal of Science. 98, 503.

Grine, F.E., Martin, L.B., 1988. Enamel thickness and development in *Australopithecus* and

Paranthropus. In: Grine, F.E. (Ed.), Evolutionary History of the "Robust" Australopithecines. Aldine de Gruyter, New York, pp. 3–42.

Grine, F.E., Stevens, N.J., Jungers, W.L., 2001. An evaluation of dental radiograph accuracy in the measurement of enamel thickness. Archives of Oral Biology 46, 1117–1125.

Hlusko, L.J., Suwa, G., Kono, R.T., Mahaney, M.C., 2004. Genetics and the evolution of primate enamel thickness: a baboon model. American Journal of Physical Anthropology 124, 223–233.

Kappelman, J., 1998. Advances in three-dimensional data acquisition and analysis. In: Rosenberger, A., Fleagle, J.G., McHenry, M., Strasser, E. (Eds.), Primate Locomotion. Plenum, New York, pp. 205–222.

Ketcham, R.A., Ryan, T.M., 2004. Quantification and visualization of anisotropy in trabecular bone. J. Microsc. 213, 158–171.

Kono, R.T., 2004. Molar enamel thickness and distribution patterns in extant great apes and humans: new insights based on a 3-dimensional whole crown perspective. Anthropological Science 112, 121–146.

Kono, R.T., Suwa, G., 2000. Appearance patterns of molar enamel thickness in tooth sections: Analysis based on 3-dimensional data. Anthropological Science 108, B10.

Kono, R.T., Suwa, G., 2002. Analysis of enamel distribution patterns in human first molars. Anthropological Science 110, C19.

Kono, R.T., Suwa, G., Kanazawa, E., Tanijiri, T., 1997. A new method of evaluating enamel thickness based on a three-dimensional measuring system. Anthropological Science 105, 217–229.

Kono, R.T., Suwa, G., Tanijiri, T., 2002. A three-dimensional analysis of enamel distribution patterns in human permanent first molars. Archives of Oral Biology 47, 867–875.

Macho, G.A., 1994. Variation in enamel thickness and cusp area within human maxillary molars and its bearing on scaling techniques used for studies of enamel thickness between species. Archives of Oral Biology 39, 783–792.

Macho, G.A., 1995. The significance of hominid enamel thickness for phylogenetic and life-history reconstruction. In: Moggi-Cecchi, J. (Ed.), Aspects of Dental Biology: Palaeontology, Anthropology and Evolution. International Institute for the study of Man, Florence, pp. 51–68.

Macho, G.A., Berner, M.E., 1993. Enamel thickness of human maxillary molars reconsidered. American Journal of Physical Anthropology 92, 189–200.

Macho, G.A., Berner, M.E., 1994. Enamel thickness and the helicoidal occlusal plane. American Journal of Physical Anthropology 94, 327–337.

Macho, G.A., Spears, I.R., 1999. Effects of loading on the biomechanical [correction of biochemical] behavior of molars of Homo, Pan, and Pongo [published erratum appears in American Journal of Physical Anthropology1999 Sep; 110(1): 117]. AJPA 109(2), 211–227.

Macho, G.A., Thackeray, J.F., 1992. Computed tomography and enamel thickness of maxillary molars of Plio-Pleistocene hominids from Sterkfontein, Swartkrans, and Kromdraai (South Africa): an exploratory study. American Journal of Physical Anthropology 89, 133–143.

Marivaux, L., Chaimanee, Y., Tafforeau, P., Jaeger, J.-J., 2006. New strepsirrhine primate from the late Eocene of Peninsular Thailand (Krabi Basin). American Journal of Physical Anthropology 130, 425–434.

Martin, L.B., 1983. The relationship of Late Miocene Hominoidae. Ph.D., Dissertation, University College London.

Martin, L.B., 1985. Significance of enamel thickness in hominoid evolution. Nature 314, 260–263.

Martin, L.B., 1986. Relationships among extant and extinct great apes and humans. In: Wood, B., Martin, L., Andrews, P. (Eds.), Major Topics in Primate and Human Evolution. Cambridge University Press, New York, pp. 161–187.

Martin, L.B., Boyde, A., 1984. Rates of enamel formation in relation to enamel thickness in hominoid primates. In: Fearnhead, R.W., Suga, S. (Eds.), Tooth Enamel IV. Elsevier Science Publishers, New York, pp. 447–451.

Martin, L.B., Olejniczak, A.J., Maas, M.C., 2003. Enamel thickness and microstructure in pithecine primates, with comments on dietary adaptations of the middle Miocene hominoid *Kenyapithecus*. Journal of Human Evolution 45, 351–367.

Molnar, S., Gantt, D.G., 1977. Functional implications of primate enamel thickness. American Journal of Physical Anthropology 46, 447–454.

Molnar, S., Hildebolt, C., Molnar, I.M., Radovčić, J., Gravier, M., 1993. Hominid enamel thickness: I. The Krapina Neandertals. American Journal of Physical Anthropology 92, 131–138.

Olejniczak, A.J., Grine, F.E., 2006. High-resolution measurement of Neandertal tooth enamel thickness by micro-focal computed tomography (μ-CT). South African Journal of Science 101, 219–2205.

Olejniczak, A.J., Grine, F.E., In review. Assessment of the accuracy of dental enamel thickness using high-resolution micro-focal X-ray computed tomography. Anatomical Record 288A, 263–275.

Olejniczak, A.J., Grine, F.E., Martin, L., 2007. Micro-computed tomography of primate molars: Methodological aspects of three-dimensional data collection. In: Bailey, S.E., Hublin, J.-J. (Eds.), Dental Perspectives on Human Evolution: State of the Art Research in Dental Paleoanthropology. Springer, Dordrecht. pp. 117–136.

Reid, D.J., Dean, M.C., 2006. Variation in modern human enamel formation times. Journal of Human Evolution 50, 329–346.

Reid, D.J., Ferrell, R.J., 2006. The relationship between number of striae of Retzius and their periodicity in imbricational enamel formation. Journal of Human Evolution 50, 195–202.

Reid, D.J., Schwartz, G.T., Dean, C., Chandrasekera, M.S., 1998. A histological reconstruction of dental development in the common chimpanzee, *Pan troglodytes*. Journal of Human Evolution 35, 427–448.

Risnes, S., 1986. Enamel apposition rate and the prism periodicity in human teeth. Scandinavian Journal of Dental Research 94, 394–404.

Risnes, S., 1998. Growth tracks in dental enamel. Journal of Human Evolution 35, 331–350.

Rowe, T., Kappelman, J., Carlson, W., Ketcham, R.A., Denison, C., 1997. High resolution computed tomography: a breakthrough technology for earth scientists. Geotimes 42, 23–27.

Rowe, T., Ketcham, R.A., Denison, C., Colbert, M., Xu, X., Currie, P.J., 2001. Forensic palaeontology: The Archaeoraptor forgery. Nature 410, 539–540.

Ramirez Rozzi, F.V., 1993. Tooth development in East African Paranthropus. Journal of Human Evolution 24, 429–454.

Ramirez Rozzi, F.V., Bromage, T., Schrenk, F., 1997. UR 501, the Plio-Pleistocene hominid from Malawi. Analysis of the microanatomy of the enamel. Comptes Rendus de l'Academie des Sciences, Paris', Ser II-A 325, 231–234.

Ryan, T.M., 2000. Quantitative analysis of trabecular bone structure in the femur of lorisoid primates using high resolution X-ray compute tomography. American Journal of Physical Anthropology Suppl. 30, 266–267.

Ryan, T.M., Ketcham, R.A., 2002a. Femoral head trabecular bone structure in two omomyid primates. Journal of Human Evolution 43, 241–263.

Ryan, T.M., Ketcham, R.A., 2002b. The three-dimensional structure of trabecular bone in the femoral head of strepsirrhine primates. Journal of Human Evolution 43, 1–26.

Ryan, T.M., Ketcham, R.A., 2005. Angular orientation of trabecular bone in the femoral head and its relationship to hip joint loads in leaping primates. Journal of Morphology 265, 249–263.

Sasaki, K., Kanazawa, E., 1998. Morphological traits on the dentino-enamel junction of lower deciduous molar series. In: Mayhall, J. (Ed.), Dental Morphology 1998: Proceedings of the 11th International Symposium on Dental Morphology. Oulu University Press, Oulu, Finland, pp. 167–178.

Schwartz, G.T., 1997. Taxonomic and functional aspects of enamel cap structure in South African Plio-Pleistocene hominids: a high-resolution computer topographic study, Ph.D. dissertation, Washington University.

Schwartz, G.T., 2000a. Taxonomic and functional aspects of the patterning of enamel thickness distribution in extant large-bodied hominoids. American Journal of Physical Anthropology 111, 221–244.

Schwartz, G.T., 2000b. Enamel thickness and the helicoidal wear plane in modern human mandibular molars. Archives of Oral Biology 45(5), 401–409.

Schwartz, G.T., Dean, C., 2000. Interpreting the hominid dentition: ontogenetic and phylogenetic aspects. In: O'Higgins, P., Cohn, M.J. (Eds.), Development, Growth, and Evolution. Academic Press, New York, pp. 207–233.

Schwartz, G.T., Liu, W., Zheng, L., 2003. Preliminary investigation of dental microstructure in the Yuanmou hominoid (Lufengpithecus hudienensis), Yunnan Province, China. Journal of Human Evolution 44, 189–202.

Schwartz, G.T., Thackeray, J.F., Martin, L.B., 1995. Taxonomic relevance of enamel cap shape in extant hominoids and South African Plio-Pleistocene hominids. American Journal of Physical Anthropology Suppl. 20, 193.

Schwartz, G.T., Thackeray, J.F., Reid, C., van Reenan, J.F., 1998. Enamel thickness and the topography of the enamel-dentine junction in South African Plio-Pleistocene hominids with special reference to the Carabelli trait. Journal of Human Evolution 35, 523–542.

Shellis, R.P., 1984. Variations in growth of the enamel crown in human teeth and a possible relationship between growth and enamel structure. Archives of Oral Biology 29, 697–705.

Shellis, R.P., 1998. Utilization of periodic markings in enamel to obtain information on tooth growth. Journal of Human Evolution 35, 387–400.

Shellis, R.P., Beynon, A.D., Reid, D.J., Hiiemae, K.M., 1998. Variations in molar enamel thickness among primates. Journal of Human Evolution 35, 507–522.

Shimizu, D., 2002. Functional implications of enamel thickness in the lower molars of red colobus (Procolobus badius) and Japanese macaque (Macaca fuscata). Journal of Human Evolution 43, 605–620.

Spears, I.R., Macho, G.A., 1995. The helicoidal occlusal plane- a functional and biomechanical appraisal of molars. In: Radlanski, R.J., Renz, R. (Eds.), Proceedings of the 10th International Symposium on Dental Morphology. Marketing Services, Berlin, pp. 391–397.

Spears, I.R., Macho, G.A., 1998. Biomechanical behaviour of modern human molars: implications for interpreting the fossil record. American Journal of Physical Anthropology 106, 467–482.

Smith, P., Zilberman, U., 1994. Thin enamel and other tooth components in Neanderthals and other hominids. American Journal of Physical Anthropology 95, 85–87.

Smith, T.M., Martin, L.B., Leakey, M.G., 2003. Enamel thickness, microstructure and development in *Afropithecus turkanensis*. Journal of Human Evolution 44, 283–306.

Smith, T.M., Martin, L.B., Reid, D.J., Bonis, L.de, Koufos, G.D., 2004. An examination of dental development in *Graecopithecus freybergi* (=*Ouranopithecus macedoniensis*). Journal of Human Evolution 46, 551–577.

Smith, T.M., Olejniczak, A.J., Martin, L.B., Reid, D.J., 2005. Variation in hominoid molar enamel thickness. Journal of Human Evolution 48, 575–592.

Spoor, C.F., Zonneveld, F.W., Macho, G.A., 1993. Linear Measurements of Cortical Bone and Dental Enamel by Computed Tomography: Applications and Problems. American Journal of Physical Anthropology 91, 469–484.

Spoor, F., Jeffery, N., Zonneveld, F., 2000a. Imaging Skeletal Growth and Evolution. In: O'Higgins, P., Cohn, M.J. (Eds.), Development, Growth, and Evolution. Academic Press, New York, pp. 123–185.

Spoor, F., Jeffery, N., Zonneveld, F., 2000b. Using diagnostic radiology in human evolutionary studies. Journal of Anatomy 197, 61–76.

Spoor, F., Wood, B., Zonneveld, F., 1994. Implications of early hominid labyrinthine morphology for evolution of human bipedal locomotion. Nature 369, 645–648.

Spoor, F., Zonneveld, F., 1999. Computed Tomography – Based three-dimensional imaging of hominid fossils: Features of the Broken Hill 1, Wadjak 1, and SK 47 Crania (Chapter 12). In: Koppe, T., Nagia, H., Alt, K.W. (Eds.), The Paranasal Sinuses of Higher Primates (Development, Function, and Evolution). Quintessence Publishing Co, Inc, Chicago, pp. 151–175.

Strait, D.S., Grine, F.E., Moniz, M.A., 1997. A reappraisal of early hominid phylogeny. Journal of Human Evolution 32, 17–82.

Summers, A.P., Ketcham, R.A., Rowe, T., 2004. Structure and function of the horn shark (*Heterodontus francisci*) cranium through ontogeny: development of a hard prey specialist. Journal of Morphology 260, 1–12.

Suwa, G., Kono, R.T., 1998. A refined method of measuring basal crown and cusp areas by use of a three-dimensional digitizing system. Anthropological Science 106, 95–105.

Suwa, G., Kono, R.T., 2000. Some methodological aspects of measuring molar enamel thickness. Anthropological Science 108(1), B9.

Suwa, G., Kono, R.T., 2005. A micro-CT based study of linear enamel thickness in mesial cusp section of human molars: reevaluation of methodology and assessment of within-tooth, serial, and individual variation. Anthropological Science, 113, 273–289.

Tafforeau, P., 2004. Aspects Phylogenetiques et Fonctionnels de la Microstructure de l'Email Dentaire et de la Structure Tridimensionnelle des Molaires Chez les Primates Fossiles et Actuels: Apports de la Microtomographie a Rayonnement X Synchrotron. Ph.D. dissertation, Universite de Montpellier II.

Tafforeau, P., 2006. X-ray Imaging (Fossil Organ-Utan). European Synchrotron Radiatin Facility (ESRF), Grenoble, p. 1.

Tafforeau, P., Boistel, R., Boller, E., Bravin, A., Brunet, M., Chaimanee, Y., Cloetens, P., Feist, M., Hoszowska, J., Jaege, r.J.-J., Kay, R.F., Lazzari, V., Marivaux, L., Nel, A., Nemoz, C., Thibault, X., Vignaud, P., Zabler, S., 2006a. Applications of X-Ray synchotron microtomography for non-destructive studies of paleontological specimens. Applied Physics A 83, 195–202.

Tafforeau, P., Boistel, R., Boller, E., Bravin, A., Brunet, M., Chaimanee, Y., Cloetens, P., Feist, M., Hoszowska, J., Jaeger, J.J., Kay, R.F., Lazzari, V., Marivaux, L., Nel, A., Nemoz, C., Thibault, X., Vignaud, P., Zabler, S., 2006b. Some applications of X-ray Synchrotron microtomography for non-destructive 3D studies of paleontological specimens. ESRF Publications.

Tafforeau, P., Ducrocq, S., Marivaux, L., 2002. Phylogeny of contemporary and fossil primates, a non destructive study of teeth structurefor the knowledge of the Anthropoids origin. European Synchrotron Radiation Facility (ESRF), Grenoble, p. 1.

Ward, S.C., Pilbeam, D.R., 1983. The enamel of neogene hominids: structural and phyletic implications. In: Ciochon, R.L., Corruccini, R.S. (Eds.), New Interpretations of Ape and Human Ancestry. Plenum Press, New York, pp. 211–238.

Zonneveld, F.W., 1987. *Computed Tomography of the Temporal Bone and Orbit*. Lippincott Williams and Wilkins, Munich.

Zonneveld, F.W., Spoor, C.F., Wind, J., 1989. The use of CT in the study of the internal morphology of hominid Fossils. Medicamundi. 34, 117–128.

Zonneveld, F.W., Wind, J., 1985. High-resolution computed tomography of fossil hominid skulls: A new method and some results. In: Tobias, P.V. (Ed.), Hominid Evolution: Past, Present and Future. Alan Liss, New York, pp. 427–436.

PART II

DENTAL MICROSTRUCTURE AND LIFE HISTORY

1. Introduction

R. MACCHIARELLI
Laboratoire de Géobiologie, Biochronologie
et Paléontologie humaine,
Université de Poitiers
86000 Poitiers, France
roberto.macchiarelli@univ-poitiers.fr

S.E. BAILEY
Department of Human Evolution
Max Planck Institute for Evolutionary Anthropology
Deutscher Platz 6
D-04103 Leipzig, Germany
and
Center for the Study of Human Origins,
Department of Anthropology, New York University,
25 Waverly Place
New York, NY 10003, USA
sbailey@nyu.edu

Because of their structural nature, teeth undoubtedly constitute the most abundant fossil evidence for mammal evolution, and are the most investigated elements in paleoanthropology. Recent and ongoing advances in developmental biology, quantitative genetics, and structural microanatomy illustrate the extraordinary amount of information preserved in their tissues (e.g., Dean, 2000; Jernvall and Jung, 2000; Jung et al., 2003; Hlusko, 2004; Mitsiadis and Smith, 2006; Pereira et al., 2006). However, a critical portion of this invaluable record, which is crucial to model/reconstruct the phylogenetic relationships, dispersal routes and evolutionary pathways, paleoecological contexts, adaptive strategies, health conditions, age- and sex-related growth and variation patterns of extinct taxa, and even to outline at least fragments of individual life-histories – is hidden at microstructural level within the crown and the root(s). Additionally, because of taphonomic dynamics and diagenetic changes, fossil signals from this record are almost invariably intermittent and noisy.

During the last two decades, routine research in dental (paleo) anthropology has extended from the outer tooth surface to the inner microstructural morphology. Accordingly, researchers agree that the possibility to properly exploit the hidden dental archive for reliable comparative investigation and unambiguous paleobiological interpretation is reliant on at least two issues: a detailed knowledge of the extant and recent reference evidence and of its pattern of (normal and pathological) variation based on large,

S.E. Bailey and J.-J. Hublin (Eds.), Dental Perspectives on Human Evolution, 139–146.

controlled series, and the development of appropriate research strategies and sharper analytical tools.

In this perspective, the five papers forming the section on dental microstructure and life history of this volume (Bromage et al., 2007; Guatelli-Steinberg et al., 2007; Ramirez Rozzi and Lacruz, 2007; Schwartz et al., 2007; Smith et al., 2007) are of paradigmatic value.

Schwartz et al. provide an excellent example of the kinds of information teeth can provide about the life history of extinct primates. Dental development has been shown to be a good proxy for the pace of life history and recent studies have shown there is a high correlation between tooth development and age of weaning, age of first reproduction, and lifespan, among other variables. Life history has long played an important part in the study of evolutionary patterns and processes. Of particular interest is the question of when, during the course of human evolution, did hominins achieve a *Homo sapiens*-like life history pattern. *Homo sapiens* is remarkable among primates because of its long lifespan, early weaning of infants and late age of first reproduction, which is later than would be expected for a great ape of the same body mass. Schwartz et al. use long-period markings within the tooth enamel to reconstruct the life history of a gorilla-sized extinct lemur, *Megalapadis edwardsi*. Schwartz et al. clearly show that, while often used as a predictor for life history variables, body size may not be a good proxy for dental development. They find that *M. edwardsi* differs from other larger bodies hominoids in crown initiation, crown formation and eruption sequence. And while gestation length in this mega-lemur may be as long as that in the similarly-sized gorilla, other life history variables differ considerably (e.g., weaning time). Schwartz et al. suggest, as others have before (e.g. Hammer and Foley, 1996), that it may be more appropriate to use brain size rather than body size as a predictor of life history.

The recent landmark publication of the draft DNA sequence of the common chimpanzee (*Pan troglodytes*) genome has opened new perspectives in the quantitative study of hominid evolution (see Li and Saunders, 2005). In fact, 365 years after the first formal description of an ape by the Dutch anatomist Nicolaas Tulp, key biological aspects of our closest living relative remain poorly known. This is notably true for the comparative growth patterns in *P. troglodytes* and *P. paniscus* (e.g., Heltne and Marquardt, 1989), the latter having been the least investigated taxon to date. Ramirez Rozzi and Lacruz (2007) add a quality building block to the understanding of dental growth in the bonobo.

The authors (Ramirez Rozzi and Lacruz, 2007) finely detail the enamel microstructural features on two modestly worn permanent teeth from a young female individual. Interestingly, while quantitatively limited, their histological results show a relatively high appositional rate for both crowns compared to that reported for the common chimpanzee (cf. Reid et al. 1998). Of course, additional research on larger *P. paniscus* dental samples is needed to confirm this evidence, but this preliminary signal points to the potential value of subtle investigations on extant interspecific primate variation for appropriately interpreting the hominid fossil record.

Both inter- and intraspecific variations in tooth development are investigated by Smith et al in the paper that follows. Although chimpanzees and humans have substantially different life history patterns, previous studies have found that they have similar crown formation times (e.g., Reid et al., 1998). This is surprising given that tooth development is said to correlate highly with life history variables. Focusing on the internal structures of the crown - including Retzius line periodicity, daily secretion rate and Retzius line number - Smith et al employ larger samples to examine developmental differences in chimpanzees and humans. Their results suggest that average formation times within specific cusp types differ between chimpanzees and humans; however, they do find a great deal of variation in formation time and overlap in the ranges of the two genera. Humans appear to be more variable than chimpanzees in Retzius line periodicity. Importantly, Smith et al. further suggest that myriad factors can influence differences in tooth developmental variables. While some studies have shown that sex differences are not significant (Schwartz et al., 2001), Smith et al. found that this is not true of all human populations. Humans also appear to differ in developmental parameters according to their geographic origin. Finally, other factors including tooth size, functional differences, developmental environments and life history may impact developmental variation in chimpanzees and humans. Given the number of variables that can affect tooth crown development and the range of variation found in their human sample, Smith et al caution against drawing conclusions based on small samples. Based on a highly significant negative correlation found between Retzius line number and periodicity in the human sample, Smith et al also caution that surface manifestations of enamel growth (perikymata) may not provide information about real differences in the rate or formation period of tooth crown. They suggest that using a taxon-specific periodicity range will provide the most accurate inferences about developmental rate or time from incremental lines on the external tooth surface.

The paper by Guatelli-Steinberg and co-authors (2007) deals with the "Neandertals *vs.* modern humans" historical debate seen from the imbricational enamel formation perspective. This study brings additional data to a previous analysis from the same research team suggesting that Neandertal tooth growth is encompassed within the extant human range of interpopulation variation (Guatelli-Steinberg et al., 2005). Because of the correlation between dental and somatic growth (Smith, 1991), the evidence has critical paleobiological implications concerning the Neandertal life history profile and, potentially, the yet unresolved (for some scholars, at least) taxonomic status of Neandertals.

The investigative tools used by Guatelli-Steinberg and colleagues are the perikymata, a series of successive, regular, horizontal incremental ridges left on the outer tooth surface (imbricational enamel) by periodic brief disruptions in enamel deposition, which represent the external manifestation of the inner striae of Retzius (microfeatures obliquely running from the enamel-dentine junction to the outer enamel surface; FitzGerald, 1998). The enamel depositional front proceeds downwards from the cusp tip, where perikymata are rather spaced apart, towards the crown base, region where they are packed together (Schwartz et al., 2001). As the period of perikymata formation (periodicity) is known (Reid and Dean, 2000), they can be used to assess duration of tooth crown formation and crown extension rate (e.g., the numerical expression – in μm/day – of the newly differentiated ameloblasts developed in a given period of time at the margin of a growing tooth crown; Schwartz and Dean, 2000). In other words, they are a good proxy for dental growth chronology.

Following the early stages of crown formation, tooth growth in our own taxon, *Homo sapiens*, is typically characterized with respect to the other hominids by a marked slowing in the rate of enamel extension towards the cervix (Shellis, 1998). What about Neandertals?

Because of a number of methodological considerations and unavoidable technical constraints, Guatelli-Steinberg et al. (2007) have limited their investigation to the upper and lower permanent incisors and canines (premolars and molars are not considered). As a whole, the Neandertal sample consists of 55 teeth from 30 individuals, while modern human variation is sampled from three geographically, ecologically, and socio-economically different populations (314 teeth from 246 individuals).

According to their results, the length of time the Neandertal imbricational enamel takes to form falls within the extant human variation. Even in the case of the lower incisors, Neandertal enamel growth is slower than, for example, in Southern Africans. Nonetheless, some interesting differences between the fossil and the modern samples concern, in particular, the shape of their respective growth curves towards the crown base and the pattern of mean perikymata number per tooth type. Following Guatelli-Steinberg and collaborators (2006), this latter evidence deserves additional investigation. Another installment is expected on this thrilling paleo-biohistory.

The major interest of this contribution is that, while it confirms some previously reported results from the same authors about enamel formation times of the anterior Neandertal dentition (Guatelli-Steinberg et al., 2005), it also detects subtle qualitative and quantitative peculiarities with respect to the extant human condition, which may reflect both external morpho-dimensional and/or structural developmental tooth differences. As a whole, this important study critically revives the discussion about dental growth in Neandertals (e.g., Dean et al., 1986, see also 2001; Mann et al., 1990; Ramirez Rozzi, 1993a; Tillier et al., 1995; Ramirez Rozzi and Bermúdez de Castro, 2004).

Interestingly, previous similar analyses on the anterior permanent dentition developed by Ramirez Rozzi and Bermúdez de Castro (2004; see also Ramirez Rozzi, 1993a) have suggested a "surprisingly" rapid dental growth in Neandertals (evidence from 146 teeth from 55 individuals), an even shorter pattern than shown, for example, by Middle Pleistocene *Homo*. According to these results, the duration of imbricational enamel formation of the anterior permanent crowns is about 15 % shorter in Neandertals than in the Western European Upper Paleolithic-Mesolithic humans, the reference sample astutely used in their analysis for comparison. That is, the cervically-oriented enamel extension and crown formation slowing typical of modern humans is much less evident in Neandertals, thus entailing a shorter crown formation period (fewer perikymata) and, consequently, a faster dental development as a whole (Ramirez Rozzi and Bermúdez de Castro, 2004).

Despite the most recent significant advances in the debate on Neandertal dental growth, it seems wise to foresee that any "conclusive" statement on this subject should integrate in the future comparative direct microstructural evidence from the inner enamel (cuspal and imbricational) and root development (Kelley, 2004; Reid and Dean, 2006) and should focus specifically on molar teeth, which correlate better with life history variables. As an additional disturbance element potentially complicating the general investigative scenario of the "Neandertals vs. modern humans" dental debate, I would also like to note the rather composite chronological, geographical and ecological nature of the so-called "Neandertal sample" which, in my view, still deserves a sharper site-specific and time-related characterization (notably in terms of evolutionary trends) for the elements of both the primary and secondary dentition.

Even a superficial overview of the literature available on this subject (see the references cited above among others) suggests that, besides the obvious, predictable lack of consensus – or even explicit disagreement – among the researchers about the interpretation of specific results which are sometimes intimately related to the nature, quality, and amount of the investigated sample(s), a non-negligible methodological role is likely played by a number of variables involved in the assessment of the perikymata counting procedures (e.g., direct observation on original specimens vs. analysis of replicas, inclusive selection criteria, region of interest choice, tooth/crown axial orientation, threshold definition between microfeatures, etc.).

Since it is not possible to section every tooth we deal with, notably in the case of fossil specimens, may we hope that technological advances will significantly reduce the proportional weight of at least some of the methodological limits and disturbance factors structurally affecting dental paleobiology and paleoanthropology? The contribution on this subject by Bromage and co-authors (2007) is clearly affirmative.

Bromage et al. (2007) have successfully applied, to a set of seven hominin permanent teeth from the South African Plio-Pleistocene sites of Sterkfontein and Swartkrans, the high-performance capacities and versatility of a non-invasive investigative tool: the Portable Confocal Scanning Optical Microscope (PCSOM). Specifically designed for hard tissue microanatomy (Boyde, 1995), this original instrument assures high resolution imaging for reliable qualitative and quantitative assessment of any enamel microfeature imprinted on the external crown morphology and/or detectable on naturally fractured enamel.

Even if given as "preliminary", the results provided by the authors are already of special value in the field because they subtly precise, notably for the cross striation periodicity, the poorly known pattern of enamel development of *Australopithecus africanus* (represented here by six specimens) and corroborate its taxonomic distinction with respect to *Paranthropus* (cf. Beynon and Dean, 1987; Dean, 1987; Dean et al., 1993, 2001; Ramirez Rozzi, 1993b) on the basis of the rates of ameloblast differentiation (measured by means of the angles between striae of Retzius and the enamel-dentine junction; Grine and Martin, 1988). Clearly, this application validates the method and opens new research perspectives on the still enigmatic *P. robustus* and, as soon as will be technically possible, I hope also on other still poorly understood fossil hominin taxa.

The exceptional images of the *A. africanus* enamel microstructures and the related original results provided by Bromage et al. (2007) originate from the specific development and sharp application to the fossil record of an advanced investigative tool capable to conciliate the potentially conflicting requirements between physical safeguard (integrity) of the specimens, on one side, and the needs for their scientific exploitation (which usually foresees handling and, occasionally, even transport), on the other one. Together with the versatility of the analytical system – which is suitable to be easily transported and settled in museums and, potentially, even on field – this is a fundamental characteristic for the future development of noninvasive advanced research on odontoskeletal specimens of paleoanthropological interest. In this perspective, a new generation of studies is successfully extensively experiencing the most recent analytical potentialities offered by x-ray computed microtomography (μCT) for high resolution linear, surface, and volumetric estimates (e.g., Kono, 2004; Suwa and Kono, 2005; Mazurier et al., 2006; Olejniczak and Grine, 2006).

As noted at the beginning of this overview, reliable research on fossil dental microfeatures and inner structural tooth morphology necessarily requires substantial evidence derived from large extant and recent reference collections characterized for their subtle normal and pathological variation (e.g., FitzGerald et al., 2006), as well as pointed methodological enhancements. Of course, the tasks are not easy to attain, particularly when the nature of the questions we pose to the fossil record are precise and deal, directly or indirectly, with our own paleo-biohistory. Nonetheless, with respect to a number of other scientific domains, paleobiologists and paleoanthropologists certainly do not appear the most disadvantaged researchers. For example, none has directly measured within the Earth – as we can do rather accurately with the enamel or even the cross striations or the prisms – the topographic thickness variation occurring at the so-called D" boundary between the lower mantle and the outer core (which supplies, among the others, the Hawaiian volcanic chain), tentatively established at about 2.891 m of depth by means of subtle speed changes in the P-waves (Garnero, 2004).

In sum, even if a number of concrete structural problems exist in the field and still affect/limit the quality and quantity of our observations, dental paleoanthropology at the start of the new millennium appears in quite healthy conditions. This research domain is among the most productive in evolutionary biology and, as already occurred in the past, will likely continue to play the key role of outpost in generating advanced scientific models concerning primate evolution. In the next decade, we will certainly achieve technological progress capable of affecting, or even deeply changing, over traditional habits and research routines in this field. In my personal view, deciduous teeth will be major protagonists.

References

Beynon, A.D., Dean, M.C., 1987. Crown formation time of a fossil hominid premolar tooth. Archives of Oral Biology 32, 773–780.

Boyde, A., 1995. Confocal optical microscopy. In: Wootton, R., Springall, D.R., Polak, J.M. (Eds.), Image Analysis in Histology: Conventional and Confocal Microscopy. Cambridge University Press, Cambridge, pp. 151–196.

Bromage, T.G., Lacruz, R.S., Perez-Ochoa, A., Boyde, A., 2007. Portable confocal scanning optical microscopy of *Australopithecus africanus* enamel structure. In: Bailey, S.E., Hublin, J.-J. (Eds.), Dental Perspectives on Human Evolution: State of the Art Research in Dental Paleoanthropology. Springer, Dordrecht. pp. 139–146

Dean, M.C., 1987. Growth layers and incremental markings in hard tissues, a review of the literature and some preliminary observations about enamel structure of *Paranthropus boisei*. Journal of Human Evolution 16, 157–172.

Dean, M.C., 2000. Progress in understanding hominoid dental development. Journal of Anatomy 197, 77–101.

Dean, M.C., Beynon, A.D., Thackeray, J.F., Macho, G.A., 1993. Histological reconstruction of dental development and age at death of a juvenile *Paranthropus robustus* specimen, SK 63, from Swartkrans, South Africa. American Journal of Physical Anthropology 91, 401–419.

Dean, M.C., Leakey, M.G., Reid, D., Schrenk, F., Schwartz, G.T., Stringer, C., Walker, A., 2001. Growth processes in teeth distinguish modern humans from *Homo erectus* and earlier hominins. Nature 414, 628–631.

Dean, M.C., Stringer, C.B., Bromage, T.G., 1986. Age at death of the Neandertal child from Devil's Tower, Gibraltar and the implications for studies of general growth and development in Neandertals. American Journal of Physical Anthropology 70, 301–309.

FitzGerald, C., 1998. Do enamel microstructures have regular time dependency? Conclusions from the literature and a large-scale study. Journal of Human Evolution 35, 371–386.

FitzGerald, C., Saunders, S., Bondioli, L., Macchiarelli, R., 2006. Health of infants in an Imperial Roman skeletal sample: perspective from dental microstructure. American Journal of Physical Anthropology 130, 179–189.

Garnero, E.J., 2004. A new paradigm for Earth's core-mantle boundary. Science 304, 834–836.

Grine, F.E., Martin, L.B., 1988. Enamel thickness and development in *Australopithecus* and *Paranthropus*. In: Grine, F.E. (Ed.), The Evolutionary History of the "Robust" Australopithecines. Aldine de Gruyter, New York, pp. 3–42.

Guatelli-Steinberg, D., Reid, D.J., Bishop, T.A., Larsen, C.S., 2005. Anterior tooth growth periods in Neandertals were comparable to those of modern humans. Proceedings of the National Academy of Sciences of the USA 102, 14197–14202.

Guatelli-Steinberg, D., Reid, D.J., Bishop, T.A., Larsen, C.S., 2007. Imbricational enamel formation in Neandertals and recent modern humans. In: Bailey, S.E., Hublin, J.-J. (Eds.), Dental Perspectives on Human Evolution: State of the Art Research in Dental Paleoanthropology.

Springer, Dordrecht. pp. 139–146

Hammer M.L.A. and Foley R.A. (1996) Longevity and life history in hominid evolution. Journal of Human Evolution. 11. 61–66.

Heltne, P.G., Marquardt, L.A. (Eds.), 1989. *Understanding Chimpanzees*. Harvard University Press, Cambridge.

Hlusko, L.J., 2004. Integrating the genotype and phenotype in hominid paleontology. Proceedings of the National Academy of Sciences of the USA 101, 2653–2657.

Jernvall, J., Jung, H.-S., 2000. Genotype, phenotype, and developmental biology of molar tooth characters. Yearbook of Physical Anthropology 43, 171–190.

Jung, H.-S., Hitoshi, Y., Kim, H.-J., 2003. Study on tooth development, past, present, and future. Microscopy Research and Technique 60, 480–482.

Kelley, J., 2004. Neandertal teeth lined up. Nature 428, 904–905.

Kono, R., 2004. Molar enamel thickness and distribution patterns in extant great apes and humans: new insights based on a 3-dimensional whole crown perspective. Anthropological Science 112, 121–146.

Li, W.-H., Saunders, M.A., 2005. The chimpanzee and us. Nature 437, 50–51.

Mann, A.E., Lampl, M., Monge, J., 1990. Décompte de périkymaties chez les enfants néandertaliens de Krapina. Bulletins et mémoires de la Société d'anthropologie de Paris 2, 213–224.

Mazurier, A., Volpato, V., Macchiarelli, R., 2006. Improved noninvasive microstructural analysis of fossil tissues by means of SR-microtomography. Applied Physics A Material Science Proceedings 83, 229–233.

Mitsiadis, T.A., Smith, M.M., 2006. How do genes make teeth to order through development? Journal of Experimental Zoology (Mol. Dev. Evol.) 306B, 1–6.

Olejniczak, A.J., Grine, F.E., 2006. Assessment of the accuracy of dental enamel thickness measurements using microfocal x-ray computed tomography. The Anatomical Record 288A, 263–275.

Pereira, T.V., Salzano, F.M., Mostowska, A., Trzeciak, W.H., Ruiz-Linares, A., Chies, J.A.B., Saavedra, C., Nagamachi, C., Hurtado, A.M., Hill, K., Castro-de-Guerra, D., Silva, W.A., Bortolini, M.C., 2006. Natural selection and molecular evolution in primate PAX9 gene, a major determinant of tooth development. Proceedings of the National Academy of Sciences of the USA 103, 5676–5681.

Ramirez Rozzi, F., 1993a. Microstructure et développement de l'émail dentaire du néandertalien de Zafarraya, Espagne. Temps de formation et hypocalcification de l'émail dentaire. Comptes Rendus de l'Academie Sciences Paris 316, 1635–1642.

Ramirez Rozzi, F., 1993b. Tooth development in East African *Paranthropus*. Journal of Human Evolution 24, 429–454.

Ramirez Rozzi, F., Bermúdez de Castro, J.M., 2004. Surprisingly rapid growth in Neandertals. Nature 428, 936–939.

Ramirez Rozzi, F., Lacruz, R.S., 2007. Histological study of an upper incisor and molar of a bonobo (Pan paniscus) individual. In: Bailey, S.E., Hublin, J.-J. (Eds.), Dental Perspectives on Human Evolution: State of the Art Research in Dental Paleoanthropology. Springer, Dordrecht. pp. 139–146

Reid, D.J., Beynon, A.D., Ramirez Rozzi, F.V., 1998. Histological reconstruction of dental development in four individuals from a Medieval site in Picardie, France. Journal of Human Evolution 35, 463–477.

Reid, D.J., Dean, M.C., 2000. Brief communication: the timing of linear hypoplasias on human anterior teeth. American Journal of Physical Anthropology 113, 135–139.

Reid, D.J., Dean, M.C., 2006. Variation in modern human enamel formation times. Journal of Human Evolution 50, 329–346.

Reid, D.J., Schwartz, G.T., Dean, M.C., Chandrasekera, M.S., 1998. A histological reconstruction of dental development in the common chimpanzee, *Pan troglodytes*. Journal of Human Evolution 35, 427–448.

Schwartz, G.T., Dean, M.C., 2000. Interpreting the hominid dentition: ontogenetic and phylogenetic aspects. In: O'Higgins, P., Cohon, M.J. (Eds.), Development, Growth and Evolution. Implications for the study of the Hominid Skeleton. Academic Press, London, pp. 207–233.

Schwartz, G.T., Reid, D.J., Dean, M.C., 2001. Developmental aspects of sexual dimorphism in hominoid canines. International Journal of Primatology 22, 837–860.

Schwartz, G.T., Godfrey, L.R., Mahoney, P., 2007. Inferring primate growth, development and life history from dental microstructure: the case of the extinct Malagasy lemur, Megaladapis. In: Bailey, S.E., Hublin, J.-J. (Eds.), Dental

Perspectives on Human Evolution: State of the Art Research in Dental Paleoanthropology. Springer, Dordrecht. pp. 139–146

Shellis, R.P., 1998. Utilization of periodic markings in enamel to obtain information on tooth growth. Journal of Human Evolution 35, 387–400.

Smith, B.H., 1991. Dental development and the evolution of life history. American Journal of Physical Anthropology 86, 157–174.

Smith, T.M., Reid, D.J., Dean, M.C., Olejniczak, A.J., Ferrell, R.J., Martin, L.B. 2007. New perspectives on chimpanzee and human molar crown development. In: Bailey, S.E., Hublin, J.-J (Eds.),Dental Perspectives on Human Evolution: State of the Art Research in Dental Paleoanthropology. Springer, Dordrecht. pp. 139–146

Suwa, G., Kono, R.T., 2005. A micro-CT based study of linear enamel thickness in the mesial cusp section of human molars: reevaluation of methodology and assessment of within-tooth, serial, and individual variation. Anthropological Science 113, 273–290.

Tillier, A.-M., Mann, A.E., Monge, J., Lampl, M., 1995. L'ontogenèse, la croissance de l'émail dentaire et l'origine de l'homme moderne: l'exemple des Néandertaliens. Anthropologie et Préhistoire 106, 97–104.

2. Inferring primate growth, development and life history from dental microstructure: The case of the extinct Malagasy lemur, *Megaladapis*

G.T. SCHWARTZ
School of Human Evolution and Social Change and the Institute of Human Origins
Arizona State University
P.O. Box 872402
Tempe, AZ 85287 USA
garys.iho@asu.edu

L.R. GODFREY
Department of Anthropology
University of Massachusetts
Amherst, Massachusetts 01003
lgodfrey@anthro.umass.edu

P. MAHONEY
Department of Archaeology
University of Sheffield
Sheffield S1 4ET UK
P.Mahoney@sheffield.ac.uk

Keywords: incremental lines, dental development, eruption schedules, *Megaladapis*, gestation

Abstract

Teeth grow incrementally and preserve within them a record of that incremental growth in the form of microscopic growth lines. Studying dental development in extinct and extant primates and its relationship to life history and ecological parameters (e.g., diet, somatic growth rates, gestation length, age at weaning) holds the potential to yield unparalleled insights into the life history profiles of fossil primates. In this paper, we use the incremental growth record preserved in teeth to reconstruct dental development, and thereby infer the life history of *Megaladapis edwardsi*, a giant, gorilla-sized, extinct lemur of Madagascar. By examining the microstructure of the first and developing second molars of a juvenile individual, we establish its chronology of molar crown development (M_1 CFT = 1.04 years; M_2 CFT = 1.42 years) and determine its age at death (1.39 years). Crown initiation, formation, and completion times are short compared to *Gorilla*. Microstructural data on prenatal M_1 crown formation time allow us to calculate a minimum gestation length of 0.54 years for *M. edwardsi*, compared to 0.70 years in *Gorilla*. Postnatal crown and root formation in *M. edwardsi* data allow us to estimate the age at M_1 emergence (~0.9 years), and to establish a minimum age for M_2 emergence (>1.39 years).If *Megaladapis* were developmentally similar to large-bodied anthropoids (such as gorillas), we might expect it

S.E. Bailey and J.-J. Hublin (Eds.), Dental Perspectives on Human Evolution, 147–162.

to exhibit slow dental development coupled with relatively early replacement of its deciduous molars. This is not the case. Total molar development is comparatively rapid and poorly explained as a function of adult body mass.

Introduction

Research on the ecomorphology of subfossil lemurs has helped us paint rather detailed pictures of their adaptive profiles (positional behavior, activity rhythms, and even grooming behavior; see Jungers et al., 2002). What has been less forthcoming is information about the life histories of these organisms. Only recently has it been possible to reconstruct important aspects of life history variation in fossil prosimians. Because life histories are "manifestations of ontogenies played out within population contexts" (Godfrey et al., 2002: 114), it is possible to use ontogenetic data to recover aspects of the life history profiles of extinct species. An excellent place to begin is by deciphering the chronology of developing teeth.

Teeth are a unique biological system in that two of their component hard tissues (enamel and dentine) preserve a permanent record of their growth in the form of incremental markings. As a result, direct evidence for the timing of important developmental events during evolution is available from even fragmentary dental remains. Among the developmental hallmarks that can be inferred from the dental growth record are the overall sequence and pace of dental eruption, the timing of first molar emergence, gestation length, and weaning age to name just a few. The timing of each of these events is tightly linked to a primate's overall life history schedule (e.g., Smith, 1991, 2000). Thus, reconstructing the pattern and pace of dental development using incremental markings preserved within dental hard tissues makes it possible to reconstruct the life history of extinct primates with unparalleled accuracy.

As an example of the kinds of life history inferences that can be garnered from the internal growth record of teeth, we provide microstructural data on the incremental lines in extinct lemur teeth to reconstruct the chronology of dental development, tooth emergence, gestation length, and overall life history schedule in one of the largest Malagasy lemurs, *Megaladapis edwardsi*. We also test the correlation between body mass and the overall pace of life history by comparing the pace of dental development and emergence to a range of extant primates, including the similarly-sized *Gorilla gorilla*.

Why Teeth?

A vast literature exists on the basics of tooth growth and histology (see Aiello and Dean, 1990 and Hillson, 1996 for reviews). Essentially, teeth grow incrementally, like trees or shells, and the cells that produce the two main tissue components of tooth crowns (ameloblasts in enamel and odontoblasts in dentine) do so in accordance with the body's circadian rhythm. As these cells secrete enamel and dentine, they leave in their wake a trail of incremental markings, of which there are two types: short-period, or daily, lines (cross striations in enamel, von Ebner lines in dentine), and longer-period lines (striae of Retzius in enamel and Andresen lines in dentine) (Figure 1); see Schwartz and Dean (2000) for a review of tooth growth. These incremental features provide a "road map" for charting the total amount of time, to the day, it takes to form individual tooth crowns (termed crown formation time, CFT) and roots, the timing of tooth initiation for individual teeth, a chronology of dental development for an entire associated dentition, the

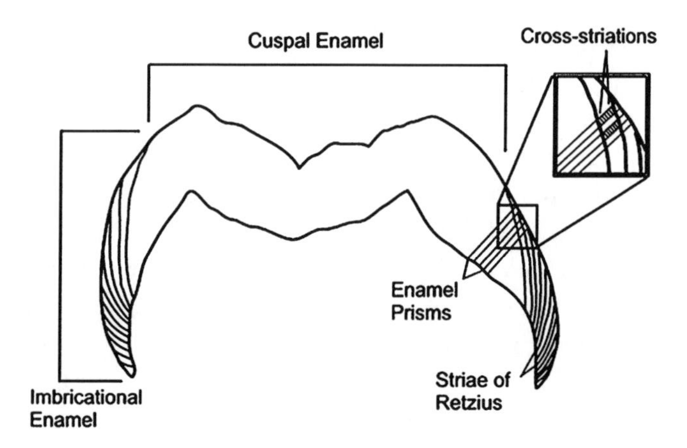

Figure 1. Schematic representation of a cross section of a molar illustrating the cuspal and imbricational (lateral) components of the crown and the major incremental features associated with enamel growth. Enamel prisms run from the enamel-dentine junction out towards the crown surface and contain short-period lines, or daily cross striations. Each stria of Retzius (long-period line) runs obliquely to the direction of prisms and is separated from adjacent striae by a certain number of days (the periodicity), which remains constant in all the teeth of one individual. Imbricational enamel formation time can be determined by multiplying the number of striae by the periodicity. In posterior teeth, such as the molar shown here, each internal stria in the imbricational enamel is continuous with an external perikyma (see text). Taken from Smith et al., 2003.

timing of birth, and possibly, for immature individuals, the age at death. Because enamel does not remodel and is only lost through wear, breakage, etc., these incremental lines are also well-preserved in fossilized teeth and ultimately allow paleontologists to reconstruct the trajectory of dental development in extinct species (e.g., Dean et al., 1993, 2001; Macho et al., 1996; Beynon et al., 1998; Macho, 2001; Schwartz et al., 2002, 2005).

Why Malagasy Lemurs?

The lemurs of Madagascar evolved in isolation on an island with a wide array of habitats, and as such exhibit a wide diversity of ecological adaptations and certain unique traits that are poorly understood (see review by Wright, 1999). Particularly enigmatic is their variation in life history characteristics; generalizations that have emerged from the study of life history variation in anthropoids hold weakly for the Malagasy lemurs, if at all. Among anthropoids, growth rates (including the pace of dental development) generally display a negative correlation with body mass (Harvey and Clutton-Brock, 1985; Harvey et al., 1987; Smith, 1991). Additionally, relatively early eruption of the permanent anterior dentition (or replacement teeth) generally signals slow growth and dental development. Thus, the sequence of dental eruption and tooth replacement correlates well with body mass in anthropoids. Larger species have slower dental development, and are characterized by relatively earlier eruption of their permanent anterior teeth compared to the permanent molar series (Smith, 2000).

In striking contrast to anthropoids, lemurs show a poor correlation between the pace of dental development and adult body mass, as well as between rates of somatic growth and dental development. Godfrey et al. (2004) describe this life-history paradox in some detail. Among the larger bodied of extant lemurs, those species with the *fastest* schedules, or pace, of dental eruption (indriids – *Avahi, Propithecus*, and *Indri*) exhibit some of the *slowest* somatic growth rates, whereas those species with the slowest schedules of dental eruption (lemurids) have the most rapid somatic growth rates (Figure 2). Furthermore, there is little relationship in lemurs between the *sequence* and *pace* of dental eruption. A rapid *pace* of dental eruption characterizes the indriids, despite their having eruption *sequences* that look much like those of large-bodied hominoids (Godfrey et al., 2005a) (Figure 3). Lemurids exhibit relatively slow dental eruption but *sequences* supposedly indicative of faster eruption.

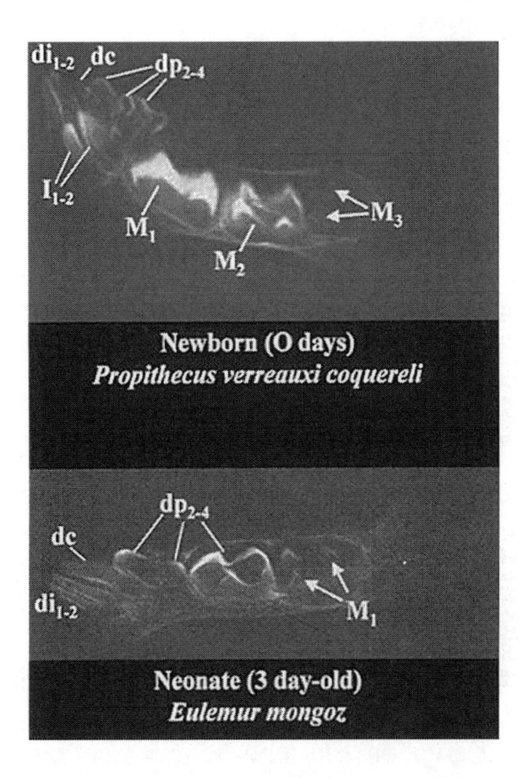

Figure 2. Comparative dental development in a neonatal indriid (*Propithecus*) above and lemurid (*Eulemur*) below. Somatic growth is more rapid in lemurids than in like-sized indriids. Nevertheless, dental development is highly accelerated in indriids, with the entire deciduous dentition (in the mandible, di_{1-2}, dc, dp_{2-4}) fully erupted (or nearly so) at birth. The crown of the permanent M_1 is virtually complete at birth, the M_2 is nearly crown complete and the M_3 is already initiated, and the crowns of the replacement incisors are well developed. This pattern is very different from lemurids, where the deciduous premolars (dp_{2-4}) are only just crown complete, the permanent incisor crowns are not yet forming, and the first permanent molar (M_1) has only just begun cusp initiation.

A growing dataset on the ontogenies of several families of extinct giant lemurs is now available (King et al., 2001; Godfrey et al., 2002; Schwartz et al., 2002; Schwartz and Godfrey, 2003; Godfrey et al., 2005b) (Figure 4). This research has served to underscore their developmental diversity. Recent work on the chimpanzee-sized *Palaeopropithecus ingens* demonstrated a pattern of extreme dental acceleration (i.e., crown formation times are rapid and all three molars initiate formation prior to birth) such as that seen in its extant sister taxon, *Propithecus* (Schwartz et al., 2002) (see Figure. 2). Yet, the pace of dental development is much slower in other large-bodied subfossil lemurs, such as members of the family Archaeolemuridae (e.g., *Hadropithecus stenognathus* and *Archaeolemur* spp.) (Godfrey et al., 2005b). *Hadropithecus* was smaller in body mass than either *Palaeopropithecus ingens* or *Megaladapis edwardsi*.

We have mentioned that, among extant lemurs, there is little relationship between the sequence and the pace of dental eruption. The same lack of relationship holds for the extinct lemurs. It is the "slow-paced" Archaeolemuridae, and not members of the "fast-paced" Palaeopropithecidae, that show "fast" eruption sequences. The *Colobus*- and *Papio*-sized species of *Archaeolemur*, as well as the large male baboon-sized *Hadropithecus stenognathus*, have relatively late emergence of the anterior teeth and early eruption of the molars (Figure 5).

Here we report new data on dental development in one of the largest-bodied extinct lemurs, *Megaladapis edwardsi*, thus adding taxonomic and size variation to our roster of lemurs whose dental developmental schedules and life histories are at least partly

Taxon							
Tupaia	M1		M2		M3	PIPIPI	"FAST"
Aotus	M1		M2	I	M3	IPPP	
Lepilemur	M1		M2	tc	M3	PPP	
Varecia	M1	tc	M2	P	M3	PP	
Eulemur	M1	tc	M2	P	M3	PP	
Saimiri	M1		M2	IIPPP	M3		
Pongo	M1		M2	IPIP	M3		
Semnopithecus	M1	II	M2	P	M3	P	
Papio	M1	II	M2	PP	M3		
Chlorocebus	M1	II	M2	PP	M3		
Macaca	M1	II	M2	PP	M3		
Gorilla	M1	II	M2	PP	M3		"SLOW"
Pan	M1	II	M2	PP	M3		
Propithecus	M1	tc P	M2	P	M3		
Avahi	M1	tc P	M2	P	M3		
Hylobates	M1	IIP	M2	P	M3		
Homo	M1	IIP	M2	P	M3		

Figure 3. Dental eruption/emergence sequences for a range of extant prosimian and anthropoid primates. The sequence can be arranged along a continuum from very "fast" to very "slow". An eruption sequence is considered "fast" (i.e., suggestive of a fast pace of life) if the permanent anterior dentition (the so-called replacement teeth, incisors and premolars) erupt later than the permanent molar series, whereas relatively early eruption of the replacement teeth is taken as signaling a "slow" pace of life. "Fast" sequences characterizes lemurids (*Eulemur, Lepilemur, Varecia*), while indriids (*Avahi* and *Propithecus*) have sequences more similar to those of hominoids. "I" = incisor, "P" = premolar, and "tc" = toothcomb.

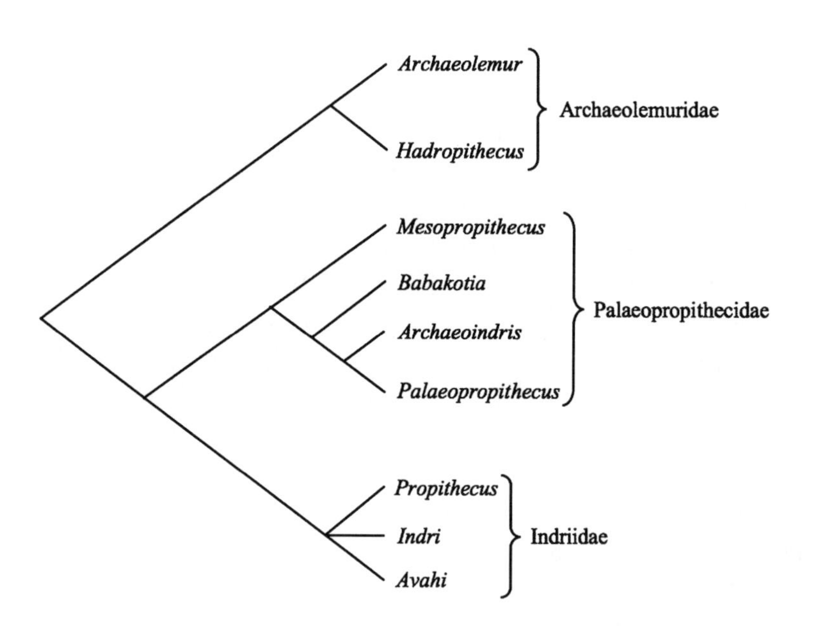

Figure 4. Cladogram depicting the relationship of two extinct families of subfossil lemurs (Archaeolemuridae and Palaeopropithecidae) to the extant Indriidae.

Taxon						
Tupaia	M1		M2		M3	PIPIPI
Aotus	M1		M2	I	M3	IPPP
Hadropithecus	**M1**		**M2**	**II**	**M3**	**PPP**
Archaeolemur majori	**M1**		**M2**	**II**	**M3**	**PPP**
A.* sp. cf. *edwardsi	**M1**		**M2**	**PP**	**M3**	**IIP**
Lepilemur	M1		M2	tc	M3	PPP
Megaladapis	**M1**	**tc**	**M2**	**P**	**M3**	**PP**
Varecia	M1	tc	M2	P	M3	PP
Eulemur	M1	tc	M2	P	M3	PP
Saimiri	M1		M2	IIPPP	M3	
Pongo	M1		M2	IPIP	M3	
Semnopithecus	M1	II	M2	P	M3	P
Papio	M1	II	M2	PP	M3	
Chlorocebus	M1	II	M2	PP	M3	
Macaca	M1	II	M2	PP	M3	
Gorilla	M1	II	M2	PP	M3	
Pan	M1	II	M2	PP	M3	
Propithecus	M1	tc P	M2	P	M3	
Avahi	M1	tc P	M2	P	M3	
Hylobates	M1	IIP	M2	P	M3	
Homo	M1	IIP	M2	P	M3	

"FAST" (shaded group: *Tupaia* through *Eulemur*)

Figure 5. Dental eruption/emergence sequences (same as Figure 3) indicating the position of *Megaladapis, Hadropithecus* and *Archaeolemur*, all of which possess a signature presumably indicative of a "fast" overall pace of life.

known. We compare the chronology of dental development in *M. edwardsi* (one of the largest-bodied extinct lemurs, ca. 80–90 kg) to *Gorilla gorilla* (whose females weigh roughly 75 kg) to further test for dissociations between body mass and the pace of development in the subfossil lemur radiation. We also evaluate the relationship between the sequence and pace of dental emergence, and estimate the age at M1 emergence and gestation length in *M. edwardsi* within the comparative context of extant primates and extinct subfossil lemurs.

Materials and Methods

The chronology of molar crown development in *Megaladapis edwardsi* was constructed using a left hemi-mandible of a juvenile specimen (UA 4620) from Beloha Anavoha, a subfossil site in southwestern Madagascar. This specimen preserves dp_4, an erupted M_1, and a partially formed M_2 still encased within an exposed crypt (Figure 6). For comparative purposes, we examined the permanent molar series (M_{1-2}) of a specimen of *Gorilla gorilla gorilla* (HT 47/90, Newcastle University). Further details about the fossil specimen, and preparation techniques can be found in Schwartz et al. (2005).

Each molar was molded, carefully removed from the alveolus, and embedded in polyester resin to reduce the risk of splintering while sectioning. Using a diamond-wafering blade saw (Buehler® Isomet 5000), longitudinal sections between 180 and 200 μm were taken through the mesial and distal cusp tips and dentine horns of both the M_1 and M_2, passing through the protoconid/metaconid and the hypoconid/entoconid, respectively (Figure 7). Each section was mounted on a microscope slide, lapped to 100–120 μm using a graded series of grinding pads (Buehler® EcoMet 4000), polished with a 3 μm aluminum-oxide powder, placed in an ultrasonic bath to remove surface debris, dehydrated through a series of

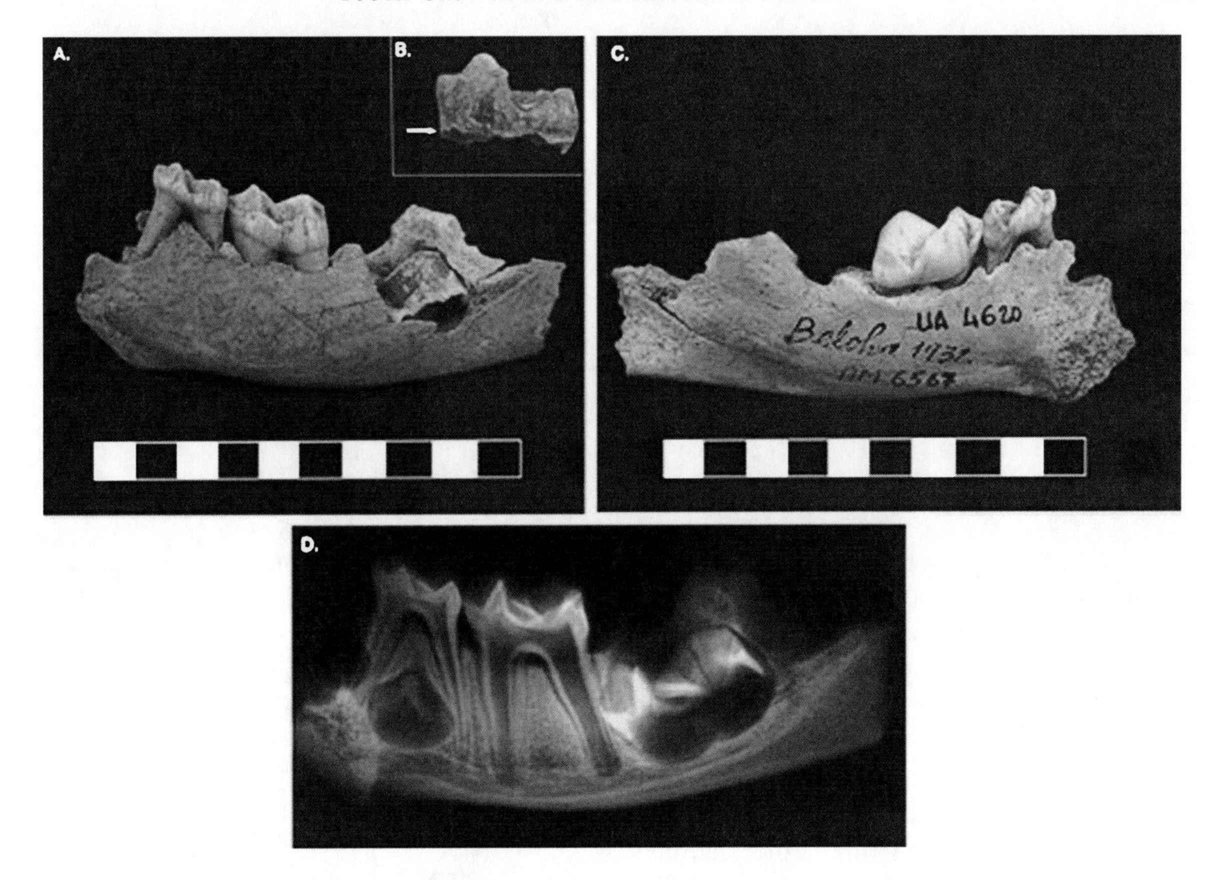

Figure 6. (a) Buccal view of the left hemi-mandible from a juvenile specimen of *Megaladapis edwardsi* (UA 4620) illustrating the erupted dp_4, and M_1, and the M_2 in its crypt. (b) (Inset) Buccal view of the unerupted M_2 after removal from the crypt. The arrow points to the few millimeters of initial root formation (see Results). Figure 6c is a lingual view of the hemi-mandible. (d) A radiograph of UA 4620 illustrating the P_4 developing within the crypt deep to the dp_4, and also the M_2 nearly crown complete in the crypt posterior to M_1. Scale bar = 10cm.

alcohol baths, cleared (using Histoclear®) and mounted with cover slips using a xylene-based mounting medium (DPX®) (Reid et al., 1998; Schwartz et al., 2001).

Dental Chronology

All sections were examined using polarized light microscopy (Olympus BX-52). Short-period (i.e., daily cross striation) and long-period (i.e., stria of Retzius) lines in enamel were used to: (1) measure daily enamel secretion rates (DSRs); (2) calculate molar cuspal and total crown formation times; (3) provide a chronology of molar crown and root development; (4) provide an age at death estimate; and (5) provide a minimum gestation length. DSRs were measured as the linear distance corresponding to five days of enamel secretions (i.e, across six daily cross striations), measured along the long axis of an enamel prism, and then divided by five to yield a daily rate. CFT was calculated by summing the time taken to form the cuspal (or appositional) and lateral (or imbricational) enamel components. Because daily lines can rarely be counted along the entire length of one cuspal prism, the cuspal enamel thickness was divided into three regions of equal thickness (inner, middle, and outer). DSRs were measured as often as possible within each region and then multiplied by the thickness of each region which, when

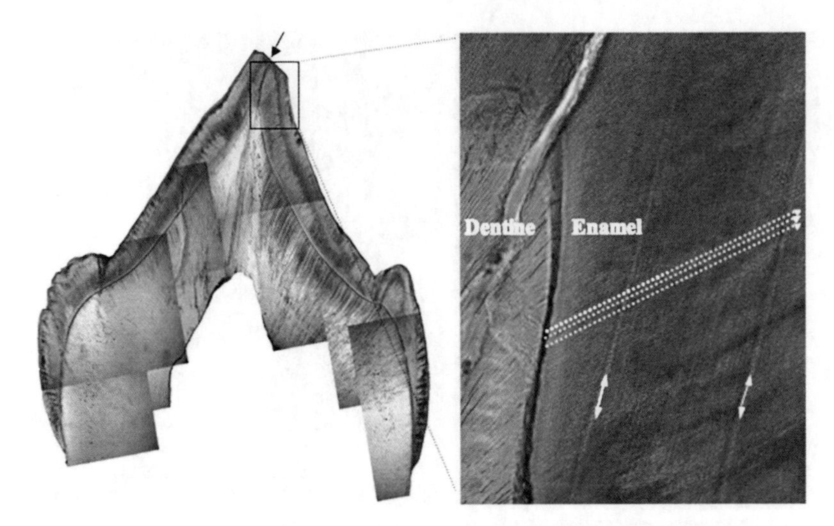

Figure 7. Left, a thin section of M_1 protoconid of UA 4620, produced using polarized microscopy at a magnification of ×10. Right, a close-up of the boxed inset, taken at a magnification of ×40. Enamel and dentine are indicated. Small white arrows point to two accentuated striae of Retzius and the dotted white lines indicate the position of enamel prisms running from the enamel-dentine junction towards the outer enamel surface. Black arrow indicates the position of a distinct cuspal wear facet (see Discussion).

summed for the three regions, yields the total cuspal enamel formation time. Imbricational enamel formation time was calculated by multiplying the total number of striae of Retzius by the striae periodicity (the number of daily cross striations between adjacent striae) (e.g., Reid et al., 1998).

To create a chronology of molar development so that pre- and postnatal molar development could be calculated, the position of the neonatal line must be determined for each tooth, and each molar crown registered in time to every other. The former was determined by charting the position of the first markedly accentuated line within each cusp. As accentuated lines are recorded in all teeth at any given point in time, they can be used to relate development in the different molars to the same point in time (Figure 8).

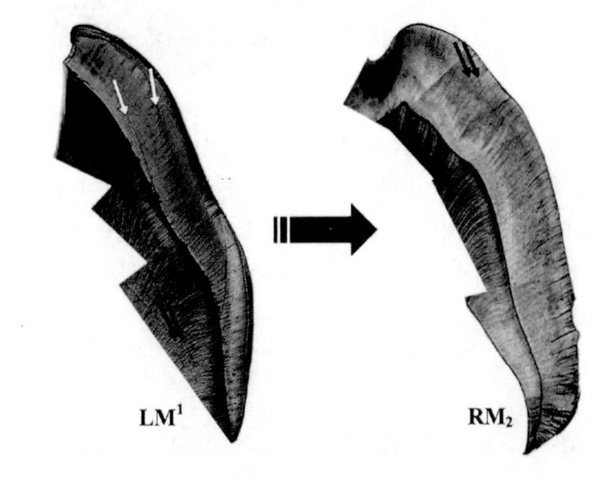

Figure 8. LM^1 and RM_2 of a *G. g. gorilla* illustrating the technique of registering one tooth to another during development (taken from Schwartz et al., 2006). Several accentuated lines appear throughout crown development (white and black arrows). A "doublet" (two black arrows), representing the two stress event close in time, is visible in both teeth, enabling the researcher to register the proportion of crown developed in the RM_2 to that in LM^1. Nearly two-thirds of the RM_2 crown is completed at the same time that the entire LM^1 crown formed.

Minimum Gestation Length and M1 Emergence Age

Minimum gestation length was calculated by assuming that the M_1 crown could not have initiated formation prior to the end of the first trimester. This is approximately when

M_1 is known to initiate in extant indriids, i.e., the group of extant lemurs with the most accelerated dental eruption schedules (Schwartz et al., 2002). The minimum gestation length is thus given by: Prenatal crown formation time +0.5 (Prenatal crown formation time). If the total amount of prenatal M1 crown formation equals two thirds of the total gestational period, then the total time of gestation is equivalent to that total amount of prenatal M1 CFT plus an additional half of that prenatal M1 CFT (i.e., an additional third of total gestation).

The age at M_1 emergence was estimated using the amount of postnatal time devoted to forming the remainder of the M_1 crown, and allowing one to several months for the initial root development prior to emergence. The age at death, which we are able to derive as this individual died at a time during M_2 crown development, could be used to derive a minimum age for gingival emergence of the M_2, as it had not yet erupted (see Figure 6).

Dental Eruption Sequences

Dental eruption sequences in the immature *Megaladapis edwardsi* were compared to other lemurs in museum collections in Europe, the USA, and Madagascar, and to a range of other primate species taken from Smith (2000), Smith et al. (1994), and Schwartz et al. (2005).

Results

Data on crown formation times in *M. edwardsi* are presented in Table 1 and a bar chart illustrating the dental chronology (compared to that for *G. gorilla*) is presented in Figure 9. Molar CFTs for *M. edwardsi, G. gorilla* and a selection of primate taxa are shown in Table 2. The periodicity for our *M. edwardsi* individual is 3, which yields an imbricational formation time between 192 days (hypoconid) and 207 days (metaconid) for the M_1, and between 225 days (protoconid) and 300 days (metaconid) for the M_2. Summing the imbricational and cuspal formation times gave CFTs of >379 days for the first molar (because the cervix was damaged slightly and missing the very last portion of enamel, our M_1 CFT is not exact), and 517 days for the second. Combining root formation times with CFTs yielded a total molar formation time of $\geq$639 days for the first molar (a virtually complete M_1 root is present [see Figure 6d] yielding a minimum root formation time of 260 days), and 583 days for crown and initial root of the second molar. By relating the neonatal and other accentuated lines among the cusps, it was determined that the M_1 initiated development 132 days before birth, and continued at least 248 days afterwards (Figure 9). The M_2 initiated development 75 days before birth and continued for 508 days (1.39 years), when the individual died.

Age at M1 Emergence

Given postnatal crown formation time along with age at death, we can estimate the age at M_1 emergence. Its age at death was estimated above at 508 days (= 1.39 years or 17 months), thereby providing a maximum age estimate for M_1 emergence. The M_1 was in full occlusion and slightly worn when this individual died. Allowing several months for the process of eruption into full occlusion and then to develop distinct wear facets, we can reconstruct gingival emergence at no later than 13 months. Age at M_1 crown completion provides a minimum age estimate for M_1 eruption. M_1 crown formation was complete around 248 days (or 8.3 months) after birth. Allowing at least one or two months for root development prior to eruption, gingival emergence can be estimated at greater than 9 months. (It is noteworthy, in this regard, that the unerupted M_2 displays two months of root extension.) Working from both directions (i.e., with estimates of M_1 eruption no earlier

Table 1. Details of molar crown formation in Megaladapis edwardsi

		M_1		M_2	
		Hypoconid	*Metaconid*	*Protoconid*	*Metaconid*
ENAMEL:	**Cuspal**				
	Daily Secretion Rates (μm per day):				
	Inner	6.8	7.1	6.6	6.3
	Mid	7.4	7.4	7.1	7.0
	Outer	7.7	7.6	7.6	7.2
	Mean	7.3	7.4	7.1	6.8
	Cuspal Thickness (μm)	1124	1263	1238	1481
	Cuspal Formation time (days)	154.4	171.9	174.5	217.0
	Imbricational				
	No. Striae of Retzius	64	69	75	100
	Striae Periodicity (days)	3	3	3	3
	Formation time from striae (days)	192	207	225	300
	Imbricational formation time (days)	192	207	225	300
	Total crown formation time (Cuspal+Imbricational)	346 days (0.95 yrs)	379 days (1.04 yrs)	399 days (1.08 yrs)	517 days (1.42 yrs)
ROOT:	Total root formation	≥260 days	(0.71 yrs)	66 days (0.18 yrs)[a]	
	Total molar formation time (Crown + Root)	>639 days (1.75 yrs)		583 days (1.60 yrs)[a]	
	Prenatal enamel formation time	132 days (0.36 yrs)		75 days (0.21 yrs)	
	Minimum Gestation length	198 days (0.54 yrs)			
	Age at death	508 days (1.39 yrs)			

[a] Indicates that root was incomplete at the time of death.

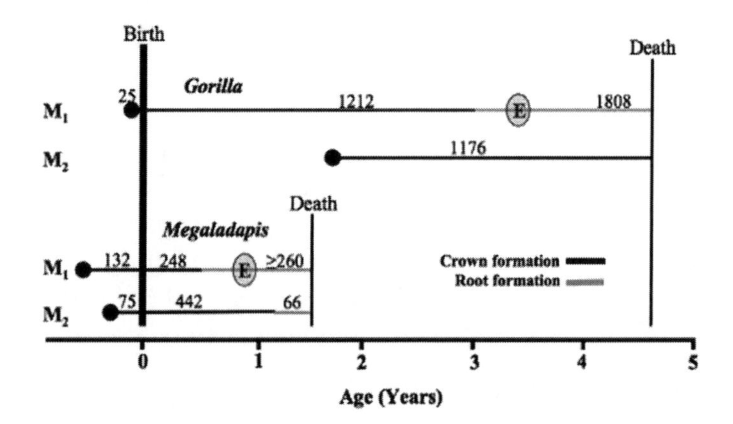

Figure 9. Dental chronology of permanent M_1 and M_2 in individuals of *Megaladapis edwardsi* compared to *Gorilla gorilla*. The black horizontal lines indicate the portion of time devoted to crown development while the grey horizontal line is the time devoted to root development. The vertical lines associated with birth and death are also indicated. The numbers above each horizontal line indicate the number of days associated with that particular portion of either crown or root development. The "E" indicates the time of M_1 emergence in each species.

Table 2. Mean maxillary and mandibular molar crown formation times for a selection of primates including several giant subfossil lemurs

	Crown Formation Times (yrs)			
	M_1	M_2	M^1	M^2
Homo sapiens (n > 10)	3.0	3.2	2.6	3.2
Pongo pygmaeus	3.1		2.8	3.5
Pan troglodytes (n = 4)	2.8	3.7	2.7	3.6
Gorilla gorilla[1] (n = 3)	3.3	3.4	3.6	3.4
Hylobates lar (n = 3)	1.3	1.2		
Hylobates syndactylus (n = 2)	1.3	1.8		
Macaca mulatta	1.0	2.0		
Papio cynocephalus	1.8	1.5		
Papio anubis (n = 2)	1.4	1.8		
Theropithecus gelada (n = 4)	1.5	2.0		
Papio hamadryas (n = 2)	1.5	1.6		
Cercocebus torquatus (n = 2)	1.4	1.8		
Semnopithecus entellus	0.9	1.3		
Macaca fascicularis	1.1			
Propithecus verreauxi	0.6	0.5		
Varecia variegata	0.6			
Palaeopropithecus ingens			0.6	0.7
Archaeolemur majori	1.4	1.3		
Hadropithecus stenognathus				2.6
Megaladapis edwardsi	**1.0**	**1.4**		

Sample size of one unless stated otherwise. Data compiled from Macho, 2001; Dirks, 2003; Dirks et al., 2002; Schwartz et al., 2002; Godfrey et al., 2005b.
[1]Schwartz, unpublished data.

than 9 months and no later than 13 months), we estimate gingival emergence of M_1 at ca. 11 months (0.9 years). M_2 emerged at 17 months (1.4 years) or later. The very open crypt suggests that eruption was imminent at the time of death.

Prenatal Dental Development

Prenatal M_1 CFT is relatively long in *Megaladapis* (132 days, out of a total crown formation of 380 days, or 35 %) (Figure 10). It is much longer in *Palaeopropithecus* both in relative and absolute terms (187 days or 85 % of a total of 221), and much shorter in *Archaeolemur* (ca. 85 days or 16 % of a total of 522). However, even in *Archaeolemur*, the percentage of total crown formation that occurs prior to birth is greater than that of any measured anthropoid. Lemurs and anthropoids are clearly different in the extent to which M_1 forms prenatally: for our sample of nine anthropoid species (all belonging to the Cercopithecoidea or the Hominoidea) and six lemur species (extinct and extant), the mean prenatal percentages of total M1 CFT are 4.6 % (SE = 1.5 %) for the former and 40.7 % (SE = 9.9 %) for the latter (Mann-Whitney test of rank difference, $p<0.001$).

Minimum Gestation Length

Observations regarding prenatal dental development can be used to estimate minimum gestation length. We estimated minimum gestation length in *M. edwardsi* to be 198 days (= 0.54 years, or 6.5 months). This value was calculated by assuming that M_1 crown initiation could not have occurred during the first trimester. The minimum gestation

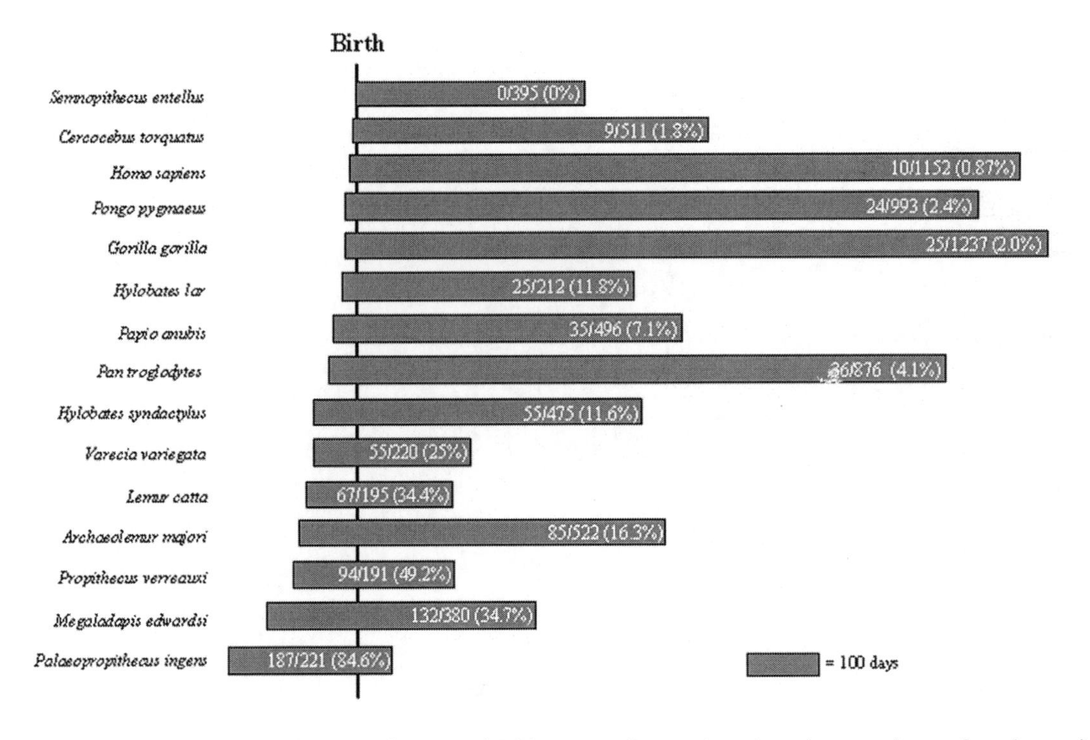

Figure 10. Variation in the degree of prenatal M1 crown formation time in a variety of anthropoids and strepsirrhines, including the subfossil lemurs *Archaeolemur, Palaeopropithecus*, and *Megaladapis*. Each scaled grey bar represents total M1 crown formation time as well as its initiation time relative to birth. Within each bar, the fraction indicates the number of days of prenatal M1 formation (numerator) relative to the total M1 CFT (denominator). The resulting percentage of total M1 CFT occurring prenatally is also provided.

is thus given by: prenatal crown formation time + (0.5 * prenatal crown formation time). If it occurred later (as is highly likely), then gestation length would have been greater than 6.5 months. Gestation length was longer than in papionins (e.g., *Papio cynocephalus* at 6 months, *Macaca mulatta* at 5.4 months), and possibly as long as in chimpanzees and gorillas (8–8.5 months). Minimum gestation length for another giant lemur, *Palaeopropithecus ingens*, was calculated in a similar manner to be greater than nine months (Schwartz et al., 2002). The emerging picture is that gestation lengths were not short in giant lemurs.

Dental Eruption Sequences

Dental eruption sequences of primates are highly variable. Some (e.g., those of *Varecia* and *Lepilemur*, wherein the permanent anterior dentition erupts after most of the permanent molars) suggest a "fast" overall pace of growth and development, while others (hominoids, *Avahi*, and *Propithecus*, wherein the permanent anterior dentition erupts relatively early) suggest a "slow" pace of growth and development (see Figure 3). Eruption sequences for *Megaladapis edwardsi* resemble those of the Archaeolemuridae and extant lemurids; they exhibit a "fast" pattern very unlike the "slow" pattern characteristic of apes rivaling *Megaladapis* in body mass (see Figure 5).

Discussion

The record of growth preserved within dental hard tissues (daily cross striations, Retzius lines, etc.) allows us to reconstruct the dental developmental schedule of fossil species with great accuracy. In certain cases, like our specimen here, we are even able to reconstruct such fundamental aspects of an organism's biology as gestation length and age at M1 eruption.

If the anthropoid relationship between body mass and dental development holds for subfossil lemurs, then larger species should exhibit slow dental development (i.e., long CFTs), coupled with relatively early replacement of their deciduous molars. Despite similarities in body mass, *G. gorilla* and *M. edwardsi* differ in M_1 crown formation times by a factor of around three (Beynon et al., 1991; Schwartz et al., 2005). Age at M_1 crown completion is made even earlier in *Megaladapis* by its early initiation of crown formation. Prenatal M_1 enamel formation is 132 days in *Megaladapis*, leaving only 248 days for postnatal development to crown completion. In contrast, in *Gorilla,* birth occurs approximately 25 days after M_1 enamel formation commences, and formation is complete only after 1212 days. Dental eruption sequences in *Megaladapis* are unlike those of larger-bodied hominoids, such as *Gorilla*, apparently corroborating a shortened period of total molar development.

Among living lemurs, variation in crown formation time does not correlate well with variation in molar development schedules. The observed acceleration of molar eruption of indriids (compared to lemurids) is not due to longer or shorter crown formation times, but to a heterochronic predisplacement – an earlier initiation of prenatal crown formation – in the former (see Figure 10). Early initiation of molar crowns, particularly when coupled with rapid root extension rates (also characteristic of lemurs; Catlett and Samonds, 2006), results in accelerated M1 emergence.

An accelerated dental developmental schedule does not necessarily imply early weaning; it can instead result in greater dental precocity at weaning (Godfrey et al., 2001). However, we have reason to believe that weaning was indeed relatively early in *Megaladapis*. The presence of a distinct wear facet on M_1 (see Figure 7), coupled with advanced root formation in M_1 and crown formation in M_2, all suggest that this 17-month-old

Megaladapis had been weaned. Wild gorillas are not weaned until after age three years (i.e., >36 months). Thus *Megaladapis* was likely very different from like-sized anthropoids in its life history characteristics. Only gestation length in *Megaladapis* may have rivaled that of gorillas.

Conclusion

Using the incremental lines preserved in enamel, we reconstructed the dental developmental schedule and life history profile of one of the largest lemurs, *Megaladapis edwardsi*, and compared it to a selected group of extant lemurs and anthropoids, including the similarly-sized *Gorilla gorilla*. We show that dental development is accelerated in *M. edwardsi* compared to the similarly-sized *G. gorilla*, and that this acceleration is achieved by initiating M_1 well prior to birth, as well as by having shorter overall molar CFTs. On the basis of prenatal crown formation time, we infer a gestation length in *Megaladapis* perhaps as long as in *Gorilla*. The combination of accelerated dental development and the wear present on the fully erupted M_1 implies earlier weaning in *M. edwardsi* than in *Gorilla*. It appears that rapid dental development in *Megaladapis* occurred in conjunction with relatively earlier age of ecological independence from the mother, despite comparable body mass and possibly comparable gestation length. Thus, *Megaladapis* likely differed from *Gorilla* in fundamental aspects of its life history. Body mass is a poor predictor of life history parameters across the Order Primates. Other variables, such as brain size, may be more appropriate (Schwartz et al., 2005)

Acknowledgments

The first author would like to extend his thanks to Professor Jean-Jacques Hublin and Dr. Shara Bailey for organizing the "Dental Perspectives on Human Evolution: State of the Art Research in Dental Anthropology" conference, to the Max Planck Institute for funding and hosting the event, and to the many colleagues for their interesting and thought-provoking presentations and stimulating discussions. Prof. Holly Smith and one anonymous reviewer provided valuable comments on our manuscript. We are also indebted to the caretakers of the collection of Malagasy subfossil lemurs at the University of Antananarivo, from which we obtained the juvenile specimen of *Megaladapis edwardsi* (UA 4620) studied here, and who generously gave us permission to study its dental microstructure. Funding for this work was provided by NSF BCS-0503988 (to GTS) and NSF BCS-0237338 (to LRG).

References

Aiello, L.C., Dean, M.C., 1990. *An Introduction to Human Evolutionary Anatomy*. Academic Press, London. 608 pp.

Beynon, A.D., Dean, M.C., Leakey, M.G., Reid, D.J., Walker, A., 1998. Comparative dental development and microstructure of *Proconsul* teeth from Rusinga Island, Kenya. Journal of Human Evolution 35, 163–209.

Beynon, A.D., Dean, M.C., Reid, D.J., 1991. Histological study on the chronology of the developing dentition in gorilla and oran-utan. American Journal of Physical Anthropology 86, 189–203.

Catlett, K.K., Samonds, K.E. 2006. Getting to the root of the matter. Evolutionary Anthropology 15, 149.

Dean, M.C., Beynon, A.D., Thackeray, J.F., Macho, G.A., 1993. Histological reconstruction of dental development and age at death of a juvenile *Paranthropus robustus* specimen, SK 63, from Swartkrans, South Africa. American Journal of Physical Anthropology 91, 401–419.

Dean, M.C., Leakey, M.G., Reid, D.J., Schrenk, F., Schwartz, G.T., Stringer, C., Walker, A.C., 2001. Growth processes in teeth distinguish modern humans from *Homo erectus* and earlier hominins. Nature 414, 628–631.

Dirks, W., 2003. Effect of diet on dental development in four species of catarrhine primates. American Journal of Primatology 61, 29–40.

Dirks, W., Reid, D.J., Jolly, C.J., Phillips-Conroy, J.E., Brett, F.L., 2002. Out of the mouths of baboons: stress, life history, and dental development in the Awash National Park hybrid zone, Ethiopia. American Journal of Physical Anthropology 118, 239–252.

Godfrey, L.R., Petto, A.J., Sutherland, M.R., 2002. Dental ontogeny and life history strategies: the case of the giant extinct indroids of Madagascar. In: Plavcan, J.M., Kay, R.F., Jungers, W.L., van Schaik, C.P. (Eds.), Reconstructing Behavior in the Primate Fossil Record. Kluwer Academic/Plenum Publishers, New York, pp. 113–157.

Godfrey, L.R., Samonds, K.E., Jungers, W.L., Sutherland, M.R., 2001. Teeth, brains, and primate life histories. American Journal of Physical Anthropology 114, 192–214.

Godfrey, L.R., Samonds, K.E., Jungers, W.L., Sutherland, M.R., Irwin, M.T., 2004. Ontogenetic correlates of diet in Malagasy lemurs. American Journal of Physical Anthropology 123, 250–276.

Godfrey, L.R., Samonds, K.E., Wright, P.C., King, S.J., 2005a. Schultz's unruly rule: dental developmental sequences and schedules in small-bodied, folivorous lemurs. Folia Primatologica 6, 77–99.

Godfrey, L.R., Semprebon, G.M., Schwartz, G.T., Burney, D.A., Jungers, W.L., Flanagan, E.K., Cuozzo, F.P., King, S.J., 2005b. New insights into old lemurs: the trophic adaptations of the Archaeolemuridae. International Journal of Primatology 26, 825–854.

Harvey, P.H., Clutton-Brock, T.H., 1985. Life history variation in primates. Evolution 39, 559–581.

Harvey, P.H., Martin, R.D., Clutton-Brock, T.H., 1987. Life histories in comparative perspective. In: Smuts, B.B., Cheney, D.L., Seyfarth, R.M., Wrangham, R.W., Struhasker, T.T. (Eds.), Primate Societies. University of Chicago Press, Chicago, pp. 181–196.

Hillson, S. 1996. *Dental Anthropology*. Cambridge University Press, Cambridge. 389 pp.

Jungers, W.L., Godfrey, L.R., Simons, E.L., Wunderlich, R.E., Richmond, B.G., Chatrath, P.S., 2002. Ecomorphology and behavior of giant extinct lemurs from Madagascar. In: Plavcan, J.M., Kay, R.F., Jungers, W.L., van Schaik, C.P. (Eds.), Reconstructing Behavior in the Primate Fossil Record. Kluwer Academic/Plenum Publishers, New York, pp. 371–411.

King, S.J., Godfrey, L.R., Simons, E.L., 2001. Adaptive and phylogenetic significance of ontogenetic sequences in *Archaeolemur*, subfossil lemur from Madagascar. Journal of Human Evolution 41, 545–576.

Macho, G.A., 2001. Primate molar crown formation times and life history evolution revisited. American Journal of Primatology 55, 189–201.

Macho, G.A., Reid, D.J., Leakey, M.G., Jablonski, N., Beynon, A.D., 1996. Climatic effects on dental development of *Theropithecus oswaldi* from Koobi Fora and Olorgesailie. Journal of Human Evolution 30, 57–70.

Reid, D.J., Schwartz, G.T., Chandrasekera, M.S., Dean, M.C., 1998. A histological reconstruction of dental development in the common chimpanzee, *Pan troglodytes*. Journal of Human Evolution 35, 427–448.

Schwartz, G.T., Dean, M.C., 2000. Interpreting the hominid dentition: ontogenetic and phylogenetic aspects. In: O'Higgins, P., Cohen, M. (Eds.), Development, Growth and Evolution: Implications for the Study of Hominid Skeleton. Academic Press, London, pp. 207–233.

Schwartz, G.T., Godfrey, L.R., 2003. Big bodies, fast teeth. Evolutionary Anthropology 12, 259.

Schwartz, G.T., Mahoney, P., Godfrey, L.R., Cuozzo, F.P., Jungers, W.L., Randria, G.F.N., 2005. Dental development in *Megaladapis edwardsi* (Primates, Lemuriformes): implications for understanding life history variation in subfossil lemurs. Journal of Human Evolution 49, 702–721.

Schwartz, G.T., Reid, D.J., Dean, M.C., 2001. Developmental aspects of sexual dimorphism in hominoid canines. International Journal of Primatology 22, 837-860.

Schwartz, G.T., Reid, D.J., Dean, M.C., Zihlman, A.L., 2006. A faithful record of stressful life events recorded in the dental developmental record of a juvenile gorilla. International Journal of Primatology 27, 1201–1219.

Schwartz, G.T., Samonds, K.E., Godfrey, L.R., Jungers, W.L., Simons, E.L., 2002. Dental microstructure and life history in subfossil Malagasy lemurs. Proceedings of the National Academy of Sciences of the USA 99, 6124–6129.

Smith, B.H., 1991. Dental development and the evolution of life history in Hominidae. American Journal of Physical Anthropology 86, 157–174.

Smith, B.H., 2000. "Schultz's Rule" and the evolution of tooth emergence and replacement in primates and ungulates. In: Teaford, M.F., Smith, M.M., Ferguson, M.W.J. (Eds.), Development, Function, and Evolution of Teeth. Cambridge University Press, Cambridge, pp. 212–227.

Smith, B.H., Crummett, T.L., Brandt, K.L., 1994. Ages of eruption of primate teeth: a compendium for aging individuals and comparing life histories. Yearbook of Physical Anthropology 37, 177–231.

Smith, R.J., Gannon, P.J., Smith, B.H., 1994. Ontogeny of australopithecines and early *Homo* – evidence from cranial capacity and dental development. Journal of Human Evolution 29, 15–168.

Smith, T.M., Martin, L.B., Leakey, M.G., 2003. Enamel thickness, microstructure and development in *Afropithecus turkanensis*. Journal of Human Evolution 44, 283–306.

Wright, P.C., 1999. Lemur traits and Madagascar ecology: coping with an island environment. Yearbook of Physical Anthropology 42, 31–72.

3. Histological study of an upper incisor and molar of a bonobo (*Pan paniscus*) individual

F. RAMIREZ ROZZI
UPR 2147 CNRS, 44 rue de l'Amiral Mouchez
75014 Paris, France
ramrozzi@ivry.cnrs.fr
and
Dept. of Human Evolution
Max-Planck-Institute for Evolutionary Anthropology
Leipzig, Germany

R.S. LACRUZ
Institute for Human Evolution
University of the Witwatersrand
P. Bag 3 WITS 2050
Johannesburg, South Africa
lacruzr@science.pg.wits.ac.za

Keywords: Striae of Retzius, appositional rate, perikymata, Bonobo, *P. paniscus, P. troglodytes*

Abstract

Work based on ground sections of teeth has provided accurate information on dental development in extant and extinct hominoid species. In contrast to radiographic studies, histological work is usually carried out using relatively small sample sizes. However, incremental lines in enamel and dentine enable us to interpret stages of crown formation and to establish patterns of dental development. Although these types of studies have been carried out in modern humans, common chimpanzees, gorillas, orangutans, and gibbons as well as in some extinct hominoids, almost nothing is known about the bonobo (*Pan paniscus*). In this paper we present some aspects of dental development for a young female with the I^1 crown just completed. Ground sections were obtained for the right I^1 and M^1. The spacing between successive cross striations was measured in the outer, middle and inner portions of occlusal, lateral and cervical thirds of the enamel. The periodicities of the striae of Retzius were obtained, and the number of striae/perikymata were used to calculate the lateral formation time. Prism length and the average distance between cross striations were used to determine the cuspal formation time. Spacing between cross striations shows a gradual increase from the inner to the outer portions, and a decrease from the occlusal to the cervical region, as observed in modern humans and great apes. It is noteworthy that average values in this *P. paniscus* individual appear to be high. Crown formation time of this *P. paniscus* I^1 was short. In addition, the perikymata packing pattern in *P. paniscus* was also different from that of *G. gorilla* and *P. troglodytes*, in that the number of perikymata increased towards the cervix.

S.E. Bailey and J.-J. Hublin (Eds.), Dental Perspectives on Human Evolution, 163–176.

Introduction

Recent studies on the tooth histology of the common chimpanzee (*Pan troglodytes*) have provided useful data for comparing the variation of microanatomical features expressed among the three great ape genera (Beynon et al., 1991a; Reid et al., 1998, Smith 2004). Additionally, these studies have increased the reliability of data on dental development in *P. troglodytes*, elucidating differences between histological and radiological methods (Reid et al., 1998). As a result, a clearer picture has emerged that allows correlations between developmental time and cuspal function in molars (Reid et al., 1998).

Most histological studies of great apes have focused on comparisons between genera (e.g. Beynon et al., 1991a,b; Dean, 1998). Little attention has been given to understanding the variation among closely related species from a histological perspective. Studies of this nature are necessary for assessments of species differences in the fossil record.

The pygmy chimpanzee or bonobo (*Pan paniscus*) is distinguished from its closest relative the common chimpanzee (*P. troglodytes*) on the bases of social behavior, morphological, and genetic differences (Johanson 1974; Shea 1984; Uchida 1992; Ruvolo 1994; Uchida 1996; White 1996; Braga 1998). Body size differences have been noted between these taxa, with *P. troglodytes* being slightly larger-bodied and showing more marked sexual dimorphism, even when the smallest of the *P. troglodytes* subspecies is considered (Jungers and Susman, 1984, Shea 1984). Some metrical and morphological differences exist in the dentitions of *P. paniscus* and *P. troglodytes* (Kinzey, 1984; Uchida, 1992). However, the greatest morphological difference appears to be the paedomorphic skull of *P. paniscus* (Shea, 1983).

The aim of this research was to provide preliminary histological data on crown formation time, age at death, variation of appositional rates and perikymata packing pattern for two tooth types of a young female *P. paniscus* individual. This information was subsequently compared with data for *P. troglodytes*. Although only one individual was available for this study, it marks the beginning of an investigation into interspecific hominoid variation.

Materials and Methods

The specimen was a young female brought from the Democratic Republic of Congo to a zoo in South Africa, that died before reaching maturity. The individual was buried at the zoo and was recently exhumed, some four years after death, due to construction plans on the zoo premises. The remains were donated to the Palaeontology Department at the University of the Witwatersrand.

Most of the post-cranial skeleton and part of the skull and face were preserved. The facial region consists of right and left maxillary fragments with deciduous dc, dm^1, dm^2 and an erupting permanent M^1. The permanent I^1, I^2, C, P^3, and P^4 crowns and one incomplete M^2 crown were encrypted in the maxilla. The roots of the first permanent molars were nearly complete but lacked apical closure (stage 6 of Demirjian et al., 1973). The M^1 protocone is the only cusp showing wear. Crown formation had just been completed on the right I^1, which shows 207 microns of root. The crowns of the remaining teeth (I^2, C, P^3, and P^4) had not yet completed their formation.

The central right incisor and the first permanent right molar were removed from the specimen and embedded in cyanoacrylate. The molar was then sectioned (150 μm in thickness) across the mesial and distal cusps and the incisor was sectioned labio-lingually. All sections were polished from both sides to a final thickness of about 100 μm. The sections were studied using polarized and transmitted light (Zeiss Universal Photomicroscope).

Daily appositional rate of ameloblasts corresponds to the cross striation repeat interval. The buccal face of the incisor and the lingual face of the M^1 protocone were divided in three equal parts (thirds): occlusal, lateral, and cervical (Figure 1). In each third, three regions of enamel were identified: outer, middle, and inner. In each region, the distance between cross striations was measured several times and an average value for the cross striation repeat interval was obtained.

The arrangement of the striae of Retzius allows the division of the enamel crown into two parts (Figure 2): (1) cuspal (appositional) enamel, where the striae do not reach the enamel surface, and which involves successive layers of enamel; and (2) lateral (imbricational) enamel, where the striae emerge at the tooth surface and form perikymata (Risnes 1985; Beynon and Wood, 1986, 1987; Dean 1987). Crown formation time (CFT) is thus the sum of both cuspal and lateral formation time. Cuspal formation time was calculated as follows: the tip of the protocone cusp of the right M^1 was first reconstructed and the first lateral stria of Retzius estimated (Figure 3). The length of a prism running from the dentine horn to the estimated first lateral stria of Retzius was measured. Close to the dentine horn, prism courses undulate, whereas in the outer part of the enamel, prisms run straight; prism course can be followed accurately on the reconstructed region of the protocone. Prism length was then divided by the average of cross striation repeat interval in outer, middle and inner regions of the occlusal third of the enamel, giving an approximate value of cuspal formation time. The same method was used to obtain the cuspal formation time on the right I^1. Lateral formation time was obtained in both teeth by direct counts of striae of Retzius, which were then multiplied by the cross striation periodicity (the number of cross striations between striae).

The first lateral stria of Retzius and the neonatal line, which forms at birth, were identified on the paracone of the M^1 (Figure 4). The length of a prism running from the neonatal line to the first lateral stria

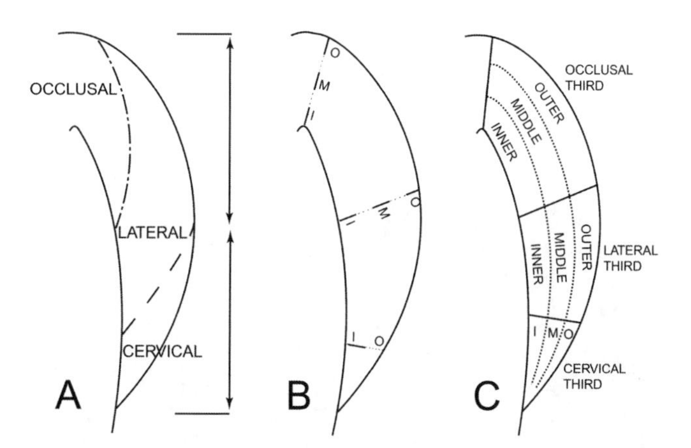

Figure 1. Divisions of the lateral crown faces. A: Beynon et al. (1991, Figure 2) proposed the division of the crown into occlusal, lateral and cervical areas. In the occlusal area, striae of Retzius do not reach the enamel surface. The lateral part is formed by the lateral striae of Retzius in the upper half of the crown whereas the lateral striae of Retzius in the lower half form the cervical region. B: Following these divisions, the appositional rate was measured in previous works in three limited areas for inner (I), middle (M), and outer (O) enamel. C: In the present work, the lateral crown face was divided in three equal thirds (occlusal, lateral, and cervical) where three regions are distinguished, outer, middle, and inner. Average appositional rate is given for each of these regions in each third.

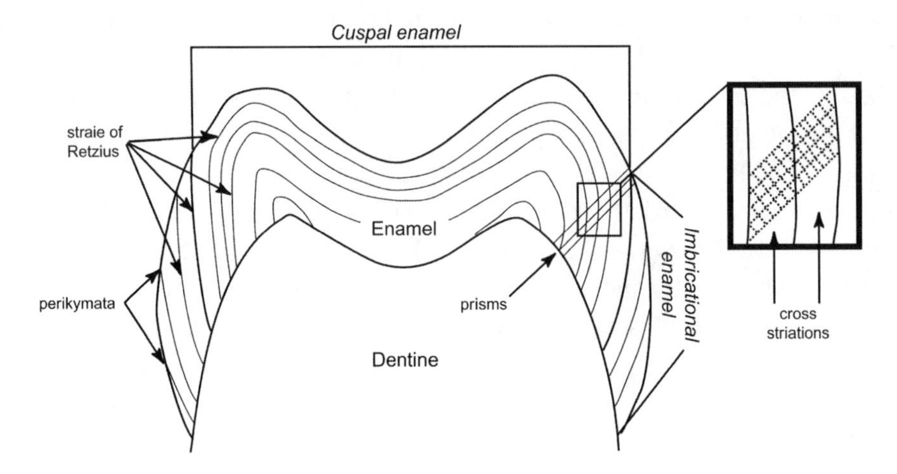

Figure 2. Division of the crown. The disposition of the striae of Retzius enables us to distinguish two areas of the enamel crown. The cuspal (appositional) area comprises the enamel where striae of Retzius do not reach the surface forming hidden successive layers of enamel. In the lateral (imbricational) part of the crown, striae of Retzius reach the enamel surface and form shallow depressions known as perikymata.

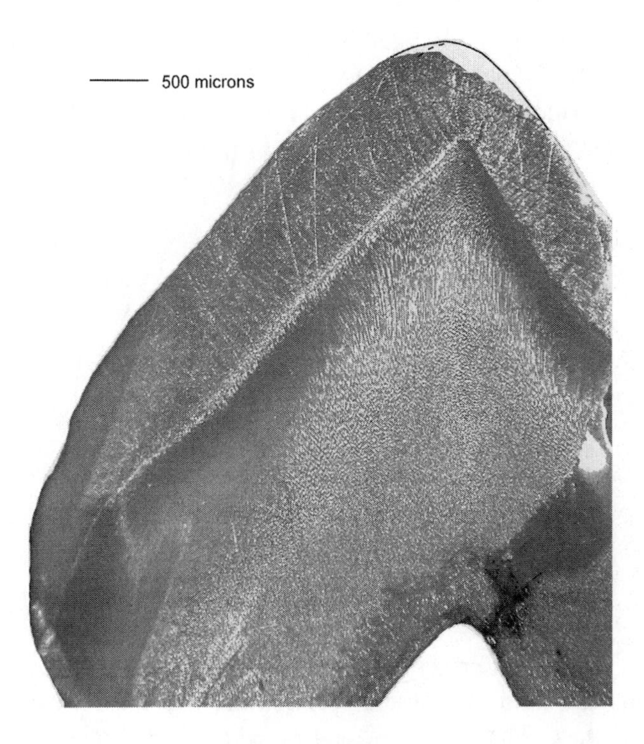

Figure 3. M^1, protocone. The protocone shows very minimal wear on the cusp tip. It is the only cusp showing any sign of wear. The number of lateral striae of Retzius around the cusp tip is very low and thus it is easy to estimate the number of lateral striae of Retzius missing due to wear. The reconstruction of the original cusp can be done by following the buccal and the occlusal enamel surface. The first striae of Retzius to reach the enamel surface (first lateral stria) can be thus identified.

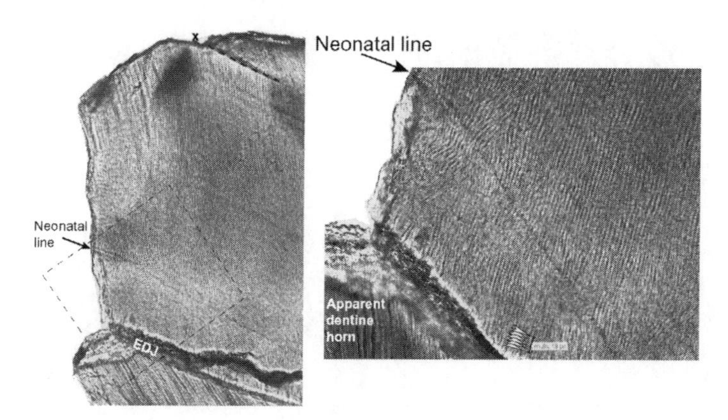

Figure 4. M^1, paracone cusp. Close to the apparent dentine horn and near the enamel dentine junction (EDJ), the distance between 5 cross striations is 19 microns, the cross striation repeat interval measures thus 3.8 microns. The neonatal line is easily observable. X indicates the possible first lateral stria of Retzius. The prism's course from the neonatal line to the first lateral stria of Retzius was followed and measured to calculate number of days and deduce the cuspal formation time after birth in order to obtain the age at death. It may be possible that our identification of the first lateral stria of Retzius in the paracone may not be completely accurate. However, this has very little effect on estimations of the age at death because the crown formation time was followed from neonatal line to the accentuated lines A and B, and from there to the end of formation of the upper first incisor, the time at which the animal died.

of Retzius was measured and divided by the average appositional rate in order to obtain cuspal formation time after birth.

Accentuated lines were identified on the four cusps of the M^1, and these lines were also observed on the I^1 (Figure 5). These marked striae enabled us to reconstruct dental formation from the M^1 to the I^1. It is worth noting that the death of the individual occurred when the crown formation of I^1 had just completed and the root measures 207 microns. Therefore, the age at death of the individual was obtained by adding: (1) the cuspal formation time after birth in the paracone of M^1, (2) the lateral enamel development between the first lateral stria of Retzius and the marked stria (accentuated line) in the paracone, (3) the crown formation time obtained in the incisor between the marked stria (accentuated line) and the end of crown formation on this tooth, and (4) time of root formation assuming a 11.7 micron/day (Beynon et al., 1991a). For each cusp, the crown was divided into ten equal zones or deciles, where counts of lateral striae of Retzius were made to assess the perikymata packing pattern along the outer surface of the enamel (Figure 6).

Results

Tables 1 and 2 provide the appositional rates in the incisor and molar protocone of this *P. paniscus* individual and in other hominoid taxa (Figures 7, 8, and 9). The length of prisms in the cuspal enamel of the I^1 was 848 microns, and was estimated to be 830 microns in the protocone of the M^1. When these values were divided by appositional rates corresponding to the three regions of the occlusal third of the enamel (Tables 1 and 2), cuspal formation times of 0.41 and 0.45 years for the I^1 and M^1, respectively, were obtained (Table 3).

The incisor and the first molar were used to obtain striae of Retzius periodicity. Periodicity could not be precisely determined for this individual. Although the most likely periodicity is 8 days, given the uncertainty in determining the exact value, crown formation time (CFT) was calculated assuming 7 and 9 day

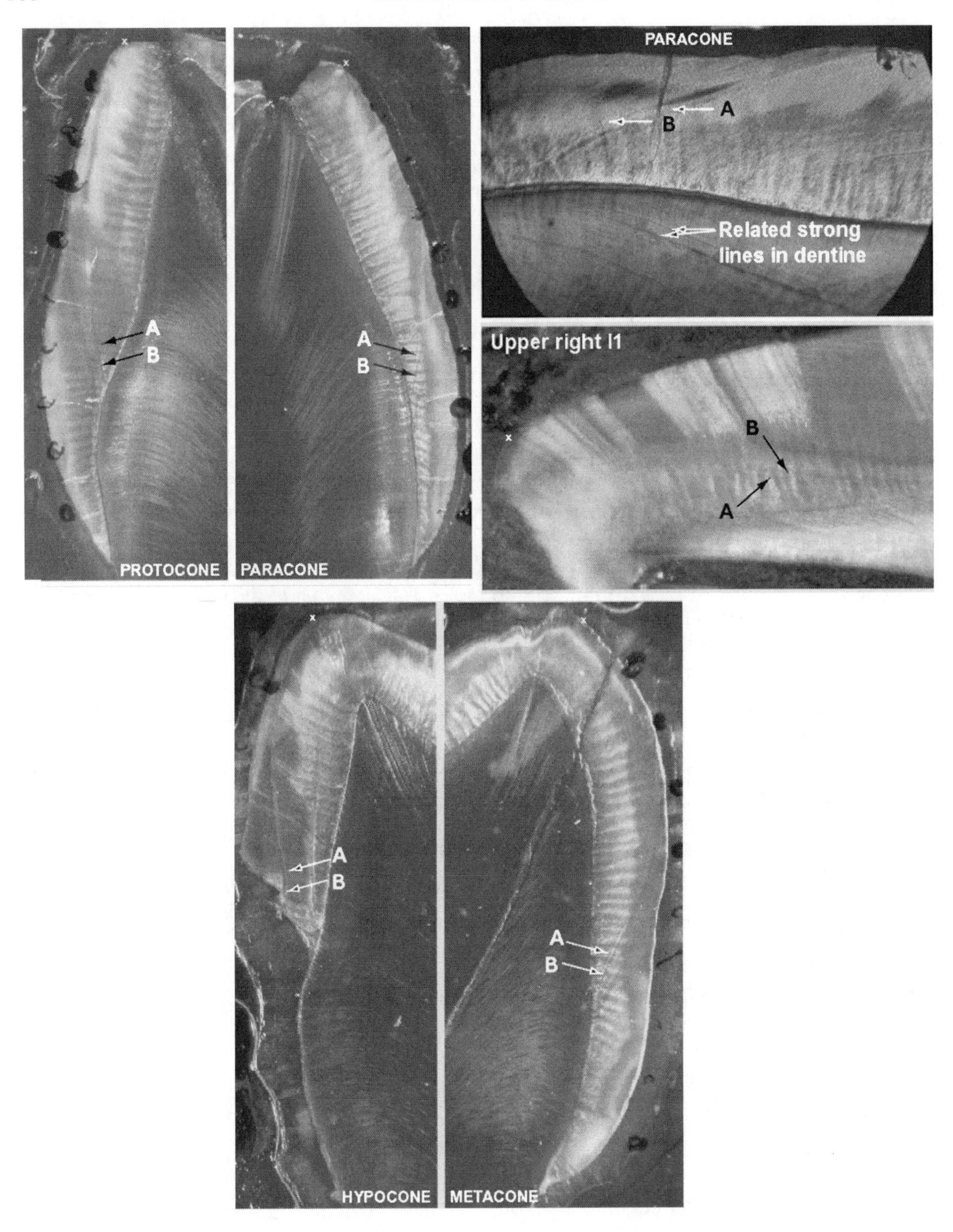

Figure 5. Section of four cusps of the M^1 and of the I^1 showing marked striae labeled as A and B. These marked striae correspond to accentuated lines in dentine. The matching of lines A and B in M^1 and in I^1 enables us to reconstruct dental formation from one tooth to the other and since the individual died at the time when I^1 crown formation had finished, the age at death can been obtained. X indicates the contact between the first lateral stria of Retzius and the enamel surface in each cusp.

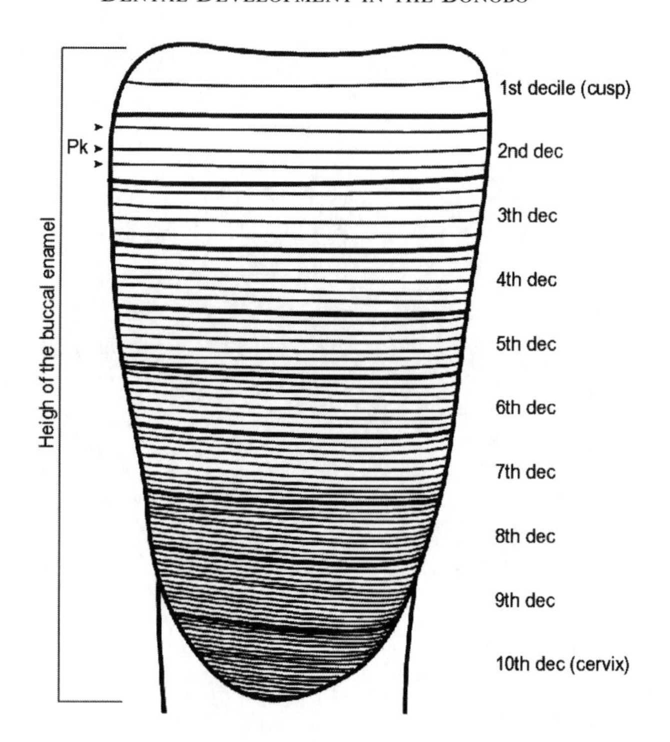

Figure 6. Perikymata packing pattern. A measure of the buccal enamel height was taken using a vernier micrometer eye-piece connected to a digital ocular measure linked, in turn, to a calculator-meter-printer RZD-DO (Leica). Buccal enamel height was divided into 10 equal divisions (deciles) from the first formed enamel at the cusp to the last formed at the cervix (Dean and Reid, 2001). Perikymata (Pk) counts were made in each of the divisions of the crown height.

Table 1. Appositional rates in I^1 in microns

	Occlusal			*Lateral*			*Cervical*		
	outer	*middle*	*inner*	*outer*	*middle*	*inner*	*outer*	*middle*	*inner*
*P. troglodytes**	4.9	4.5	3.5	4.6	4.1	3.5	3.5		3.0
P. paniscus	5.8	5.9	5.1	5.1	5.6		4.9	4.6	4.0

* From Beynon et al. (1991b). Differing from previous works, appositional rate in *P. paniscus* corresponds to an area of the lateral enamel (see Figure 1).

Table 2. Appositional rates in molars in microns

	Occlusal			*Lateral*			*Cervical*		
	outer	*middle*	*inner*	*outer*	*middle*	*inner*	*outer*	*middle*	*inner*
H. sapiens†	5.1	4.3	2.7	5.0	4.0	2.6	2.8		2.3
P. troglodytes†	5.0	4.4	3.1	4.1	4.1	3.1	3.8		2.8
*P. troglodytes**	4.8	4.0	3.5	4.8	4.4	3.3	3.2		2.9
P. troglodytes°	4.6	4.3	3.6						
Gorilla†	6.1	5.2	3.2	6.1	4.9	3.3	4.4		3.2
Pongo†	5.3	4.7	3.3	5.1	4.0	3.4	3.7		2.9
P. paniscus	6.2	5.7	4.4	5.2	5.3	4.8	4.6	4.4	3.9

† From Beynon et al. (1991b). * From Reid et al. (1998). ° From Smith (2004). Differing from previous works, appositional rate in *P. paniscus* corresponds to an area of the lateral enamel (see Figure 1).

Figure 7. I^1, buccal face, 6th decile close to the enamel surface (ES). Only a few cross striations (white arrows) are indicated but they can be seen throughout this image as well as striae of Retzius (black arrows). Seven cross striations were marked, which measured 38 microns, thus average distance between cross striations is equal to 5.43 microns. The distance between two adjacent striae of Retzius along the prism axis (white dots) is equal to 96 microns. If we assume a periodicity of 8 cross striations between striae, the average distance between cross striations is 6 microns; if a periodicity of 9 is assumed, the average distance between cross striations would be 5.33 microns. The lower average value for the outer lateral third presented in Table 3 results in a reduction of the distance between cross striations towards the EDJ and the cervix (see Dean, 1998). Cervix is upper right. Although average values presented in this work are difficult to compare with appositional rates measured in a very limited area of the enamel from other studies (Beynon et al., 1991, Reid et al., 1998), values for the 6th decile closely correspond to the lateral enamel values from previous studies. The appositional rate in this *P. paniscus* individual is higher than those reported for *P. troglodytes*, even when a periodicity of 9 is assumed.

periodicities. Given our observations, these appeared to be the minimum and maximum possible values. Counts of lateral striae of Retzius and crown formation times are shown in Table 3.

Accentuated lines in dentine and enamel allow the reconstruction of the timing of developmental events by comparing the chronology of these events in different cusps of the same tooth, and across tooth types of the same individual (Boyde, 1964, 1990). Two accentuated lines, labeled A and B, were identified in the four cusps of the M^1 and in the I^1 (Figure 5). These can be easily

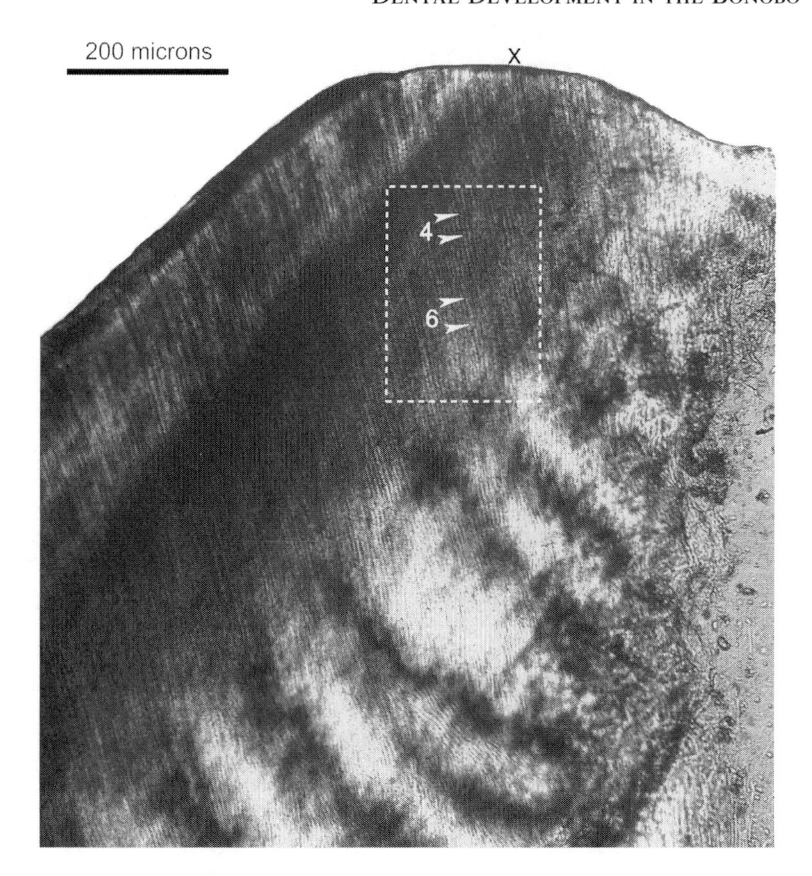

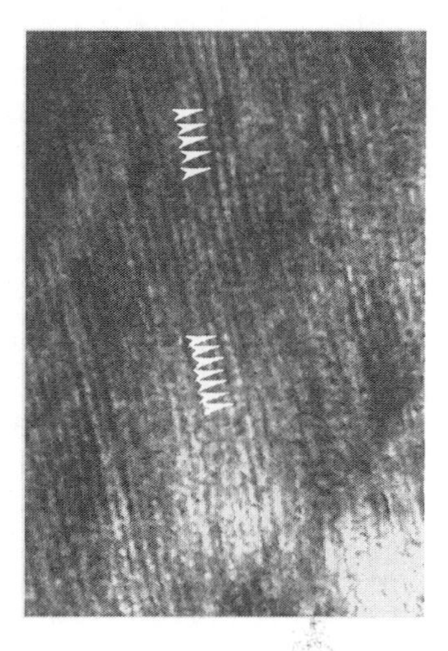

Figure 8. M[1], protocone. The first lateral stria of Retzius is indicated with an X (see figure 5). Cross striations are seen in the cuspal enamel. One group of 6 cross striations and another of 4 cross striations are shown in the picture. In the main picture only the limits of each group are identified. The distance measured for 6 cross striations is 32 microns, indicating a cross striation repeat interval of 5.33 microns. The distance measured for 4 cross striations closer to the outer enamel surface is 25 microns, equivalent to 6.25 microns for each cross striation. Prism width measured from 10 prisms corresponds to 5.3 and 5.5 microns respectively in the two groups.

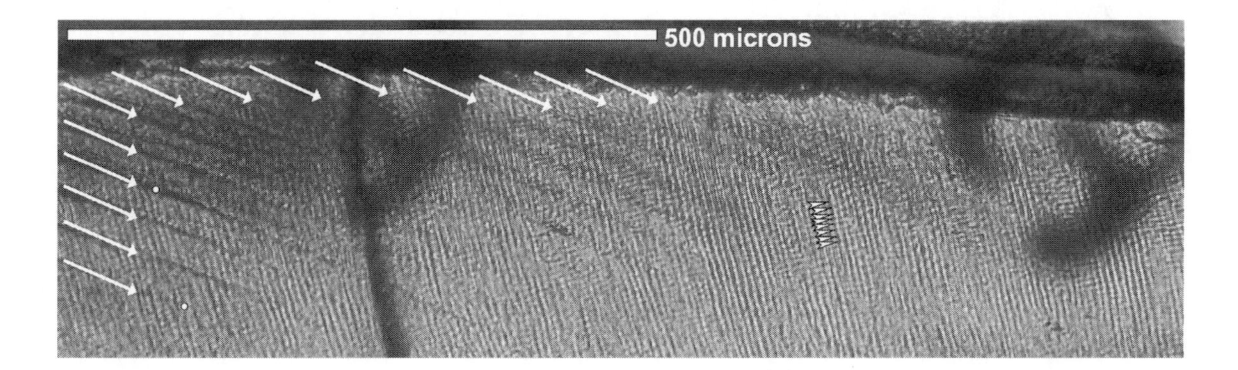

Figure 9. M[1], protocone, buccal face, 8th decile. Striae of Retzius (white arrows) are visible next to the enamel surface. Seven cross striations are marked (black arrows), the total distance is 34 microns, indicating that the average distance between them is 4.86 microns. The distance between 3 striae of Retzius along prism's axis (white dots on the left) measures 100 microns. If we assume a periodicity of 8, the average distance between cross striations corresponds to 4.17 microns.

Table 3. Crown formation time

Pan paniscus	*LatSr*	*CH*	*LFT (7)*	*LFT (9)*	*CuspFT*	*CFT (7) (9)*
I^1	153	1221	2.90	3.78	0.41	3.31/4.19
M^1 protocone	85	624	1.63	2.09	0.45	2.08/2.54
M^1 paracone	83	549	1.59	2.0		
Pan troglodytes						CFT
Mean I1*	218	1410				5.24
Mean M^1 proto ($N=2$) †	99	650				2.22–2.39
Mean M^1 para ($N=2$) †	113	625				2.06–2.92

LatSr: number of lateral striae, CH: crown height in microns, LFT: lateral formation time calculated with 7-day and 9-day periodicity in years, CuspFT: cuspal formation time in years, CFT: crown formation time in years. * From Reid et al. (1998) and †Smith (2004).

matched from one cusp to another, and they are associated with marked lines in the dentine as well. Lines A and B correspond closely to the positions of the 8th and 10th lateral striae of Retzius (counting from cusp to cervix) in the protocone and to the 18th and 20th lateral striae of Retzius in the paracone. Line A is about the 12th–13th lateral stria of Retzius in the metacone and is cuspal in the hypocone, while line B corresponds to the first lateral stria of Retzius in this last cusp. Comparisons between the positions of lines A and B on the lateral enamel suggest that this part of the crown was more developmentally advanced in the paracone than in the protocone. The lateral part of the hypocone is the last to be formed.

Lines A and B were also observed in the cuspal enamel of the central incisor. Line B was formed at 0.25 years in the incisor, whereas it appeared at 0.62–0.67 years in the M^1 protocone. In the M^1 paracone, the prism length between the neonatal line and the first lateral stria of Retzius was 697 microns. Cross striation repeat interval measured in the occlusal third of the paracone showed similar values to those obtained in the protocone. Thus, average cross striation repeat intervals of the protocone were used to calculate the period of prism formation. The first lateral regular stria in the paracone was formed 129 days (0.35 years) after the neonatal line. Line B corresponds to the 20th lateral stria of Retzius in the paracone. Assuming an 8 day periodicity, line B was formed at 0.79 years of age (0.35 yrs $+ 20 \times 8$). In the I^1, the cuspal enamel formed after 0.16 years of line B. The lateral enamel of I^1 was formed in 3.77 years. The root formed corresponds to 0.05 yrs. Therefore, this individual died at an age

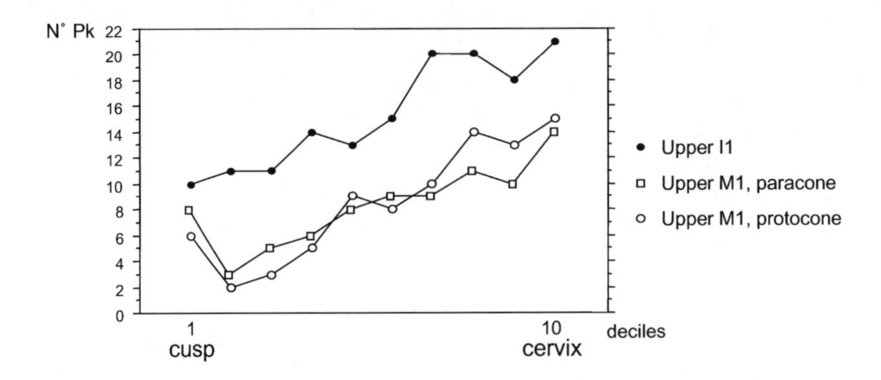

Figure 10. Perikymata packing pattern in the molar and incisor of *P. paniscus* studied here. The number of perikymata (N° Pk) increases towards the cervix.

Table 4. Perikymata packing pattern in P. paniscus

	deciles									
	1 (cuspal)	*2*	*3*	*4*	*5*	*6*	*7*	*8*	*9*	*10 (cervix)*
I^1	10	11	11	14	13	15	20	20	18	21
M^1 paracone	8	3	5	6	8	9	9	11	10	14
M^1 protocone	6	2	3	5	9	8	10	14	13	15

of approximately 4.77 years (0.79 yrs + 0.16 yrs + 3.77 yrs + 0.05 yrs).

The perikymata packing patterns follow the same outline in the I^1 and M^1 studied from this individual: the number of perikymata increases while the space between perikymata decreases towards the cervix (Figure 10, Table 4).

Discussion

Results presented in this study are preliminary in nature because they are based on only one individual. The cross striation appositional rates measured in the incisor and molar of this *P. paniscus* individual are higher than values observed in *P. troglodytes* (Beynon et al., 1991b; Reid et al., 1998; Smith, 2004). Reid et al. (1998) and Smith (2004) showed values for each tooth type (1st, 2nd or 3rd molar) while Beynon et al. (1991a) averaged values for molars. In general, the values for this *P. paniscus* individual are among the highest in great apes for both tooth types (Tables 1, 2, Figures 7, 8, 9). Differences between our values and those presented by Reid et al. (1998) and Smith (2004) could be the result of using different measurement schemes. Indeed, appositional rates in this *P. paniscus* individual for inner, middle, and outer areas shown here represent average values of measurements taken across the whole area of each third (Figure 1C). In other words, average values do not result from measurements along an imaginary line crossing a particular area of the enamel as in previous studies (Figure 1B) (Beynon et al., 1991a; Reid et al., 1998). However, it is worth noting that high values, which surpass even those reported here for *P. paniscus*, have been observed in some *P. troglodytes* teeth. For example, Dean (1998) measured daily increments following prisms from the EDJ to the outer enamel to obtain cuspal crown formation. In the outer cuspal enamel of this *P. troglodytes* individual, appositional rates were 6.5 microns and higher (Dean, 1998, Figure 2). Therefore, the use of different methods could partially explain the differences in appositional rates among the great apes shown in Tables 1 and 2, but they may also be due to interspecific variation.

Reid et al. (1998) reported that crown initiation in *P. troglodytes* begins at 0.21–0.26 years in upper first incisors, and at 0.15 years before birth (−0.15) in the M^1. The same relationship between I^1 and M^1 crown formation is found in this *P. paniscus* individual, where crown formation in M^1 is advanced by about 0.4 years relative to the I^1.

The two most recent studies on the histology of *P. troglodytes* in which the same specimens were analyzed reported different periodicities (Reid et al., 1998; Smith, 2004). The first study reported values of 7 and 8 for four individuals and included all tooth types, while the second reported 6 and 7 as the most common values in molars only ($n = 75$). If the values of Reid and co-workers are correct, an intriguing pattern emerges when molar and incisor crown formation times in the *P. paniscus* specimen are compared to data on *P. troglodytes* (Reid et al., 1998). Molar crown formation is similar to or slightly less than that observed here in *P. paniscus* (depending on the periodicity used), whereas incisor crown formation is not, being 20%–37% shorter in the former depending on striae periodicity. However, Smith (2004) reported lower periodicity, which in the case of one specimen (#88/89) was 7 instead of 8. Therefore, if this value is correct, given the number of lateral striae of Retzius on the I^1 reported by Reid et al. (1998) on the same specimen (average of left and right I^1 is 218 striae), then the differences between both specimens are less marked (only about 10%). The lowest value of lateral striae of Retzius counts reported by Smith (2004) on anterior cusps of M^1 was 91 ($n = 2$) and similar values were obtained by Reid et al. (1998) which are only slightly higher than the values reported here for *P. paniscus*.

The different positions of lines A and B in the molar cusps reflect differences in the onset of enamel formation. Kronfeld (1954) suggested that in modern humans the sequence of cusp formation in upper molars is: paracone, protocone, metacone, and hypocone. Thus, we would expect to find lines A and B situated in a more occlusal position in the metacone than in the protocone. However, the position of these lines in the metacone is more cervical than in the protocone. The lack of knowledge about cuspal formation in each cusp does not allow accurate inferences to be made regarding the complete sequence of cusp formation in the right M^1, thus it is not possible to confirm at present the same sequence of cusp formation in *P. paniscus* as in modern humans.

The perikymata packing pattern in this *P. paniscus* individual is different from that reported in *G. gorilla* and *P. troglodytes* shown in Dean and Reid (2001). In the anterior teeth of *G. gorilla* and *P. troglodytes*, the number of perikymata increases towards the second third of the crown and from about the 7th division, perikymata markedly decrease in number towards the cervix. In contrast, the number of perikymata in this *P. paniscus* individual increases towards the cervix, where the highest numbers of perikymata are found (Figure 11). It is noteworthy that the numbers of perikymata decrease slightly from the 8th to the 9th decile in the *P. paniscus* teeth studied here. The perikymata packing pattern observed in this *P. paniscus* individual is similar to extant and extinct *Homo* and australopiths (e.g., Dean and Reid, 2001): the number of perikymata increases towards the cervix and the space between them decreases.

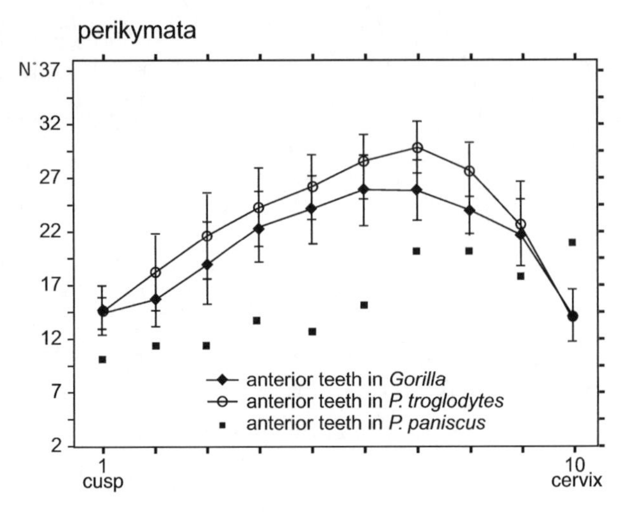

Figure 11. Plot showing the perikymata packing pattern in the I^1 of the *P. paniscus* individual compared with that in anterior teeth in *P. troglodytes* and *G. gorilla* modified from Dean et al. 2001. The pattern is different than in other great apes, with the highest number of perikymata found in the last decile.

Conclusion

We have presented preliminary results on dental growth of *P. paniscus* based on a single individual. These results require confirmation using larger sample sizes. Although there may be some differences in methodology between our study and others, it appears that the *P. paniscus* individual studied here shows high appositional rates. Molar crown formation is similar to or slightly shorter than values reported for *P. troglodytes* depending on the periodicity used. Incisor formation of the *P. paniscus* individual studied here appears to be shorter than *P. troglodytes*, regardless of the periodicity obtained in the studies of Reid et al. (1998) and Smith (2004). In contrast to other African great apes, the perikymata packing pattern in this *P. paniscus* individual is characterized by an increase in the number of perikymata towards the cervix.

Acknowledgments

We thank Jean-Jacques Hublin and Shara Bailey for their invitation to participate in the Dental Anthropology Conference that was held at the Max Planck Institute of Evolutionary Anthropology. We also thank Don Reid, Tanya Smith and Jose Braga for useful discussions on this presentation and Monique Tersis for help in the preparation of this manuscript. Additional funding to RL was provided by the Palaeo-Anthropology Scientific Trust (PAST).

References

Beynon, A.D., Dean, M.C., Reid, D.J., 1991a. Histological study on the chronology of the developing dentition in gorilla and orangutan. American Journal of Physical Anthropology 86, 189–203.

Beynon, A.D., Dean, M.C., Reid, D.J., 1991b. On thick and thin enamel in hominoids. American Journal of Physical Anthropology 86,295–310.

Beynon, A.D., Wood, B.A., 1986. Variations in enamel thickness and structure in East African hominids. American Journal of Physical Anthropology 70, 177–193.

Beynon, A.D., Wood, B.A., 1987. Patterns and rates of enamel growth in the molar teeth of early hominids. Nature 326, 493–496.

Boyde, A., 1964. The structure and development of mammalian enamel. Ph.D. Dissertation, University of London, England.

Boyde, A., 1990. Developmental interpretations of dental microstructure. In: De Rousseau, C.J. (Ed.), Primate Life History and Evolution. Wiley-Liss, New York, pp. 229–267.

Braga, J., 1998. Chimpanzee variation facilitates the interpretation of incisive suture closure in South African Plio-Pleistocene hominids. American Journal of Physical Anthropology 105, 121–135.

Dean, M.C., 1987. Growth layers and incremental markings in hard tissues; a review of the literature and some preliminary observations about enamel structure in *Paranthropus boisei*. Journal of Human Evolution 16, 157–172.

Dean, M.C., 1998. A comparative study of cross striation spacing in cuspal enamel and of four methods of estimating the time taken to grow molar cuspal enamel in *Pan, Pongo* and *Homo*. Journal of Human Evolution 35, 449–462.

Dean, M.C., Reid, D.J., 2001. Perikymata spacing and distribution on hominid anterior teeth. American Journal of Physical Anthropology 116, 209–215.

Demirjian, A., Goldstein, H., Tanner, J.M., 1973. A new system of dental age assessment. Human Biology 45, 211–227.

Johanson, D.C., 1974. Some metric aspects of the permanent and deciduous dentition of the pygmy chimpanzee (*Pan paniscus*). American Journal of Physical Anthropology 41, 39–48.

Jungers, W.L., Susman, R.L., 1984. Body size and skeletal allometry in African apes. In: Susman, R.L. (Ed.), The Pygmy Chimpanzee, Plenum Press, New York, pp. 131–177.

Kinzey, W.G., 1984. The dentition of the pygmy chimpanzee, *Pan paniscus*. In: Susman, R.L. (Ed), The Pygmy Chimpanzee. New York, Plenum Press, pp. 65–88.

Kronfeld, R., 1954. Development and calcification of the human deciduous and permanent dentition. In: Steward, T.D., Trotter M. (Eds.), Basic Readings on the Identification of Human Skeletons. Wenner-Gren Foundation for Anthropological Research, New York, pp. 3–10.

Reid, D.J, Schwartz, G.T., Dean, M.C., Chandrasekera, M.S., 1998. A histological reconstruction of dental development in the common chimpanzee, *Pan troglodytes*. Journal of Human Evolution 35, 427–448.

Risnes, S., 1985. A scanning electron microscopy study of the three dimensional extent of Retzius lines in human dental enamel. Scandinavian Journal of Dental Research 93, 145–152.

Ruvolo, M., 1994. Molecular evolutionary processes and conflicting gene trees: the hominoid case. American Journal of Physical Anthropology 94, 89–113.

Shea, B.T., 1983. Paedomorphosis and neotony in the pygmy chimpanzee. Science 222, 521–522.

Shea, B.T., 1984. Allometry in pygmy and common chimpanzees. In: Susman, R.L. (Ed.), The Pygmy Chimpanzee. Plenum Press, pp 89–130.

Smith, T.M., 2004. Incremental development of primate dental enamel. Ph.D. Dissertation, SUNY Stony Brook, New York.

Uchida, A., 1992. Intra-species variation in dental morphology of the great apes: Implications for taxonomy of fossil hominoids. Ph.D. Dissertation, Harvard University, USA.

Uchida, A., 1996. What we don't know about great ape variation. Tree 11, 163–168.

White, F.J., 1996. *Pan paniscus* 1973–1996: twenty-three years of field research. Evolutionary Anthropology 5, 11–17.

4. New perspectives on chimpanzee and human molar crown development

T.M. SMITH
Human Evolution Department, Max Planck Institute for Evolutionary Anthropology, Deutscher Platz 6, D-04103 Leipzig Germany
tsmith@eva.mpg.de

D.J. REID
Department of Oral Biology, School of Dental Sciences Newcastle University, Framlington Place Newcastle upon Tyne NE2 4BW, U.K.
D.J.Reid@newcastle.ac.uk

M.C. DEAN
Department of Anatomy and Developmental Biology University College London Gower Street, London WC1E 6BT, U.K.
ucgacrd@ucl.ac.uk

A.J. OLEJNICZAK
Human Evolution Department, Max Planck Institute for Evolutionary Anthropology, Deutscher Platz 6, D-04103 Leipzig Germany
olejniczak@eva.mpg.de

R.J. FERRELL
Assistant Professor
Department of Sociology & Anthropology
Howard University
2441 6th St. NW Washington, DC 20059
rferrell@howard.edu

L.B. MARTIN
Departments of Anthropology and of Anatomical Sciences
Stony Brook University
Stony Brook, NY 11794-4364
U.S.A.
lawrence.martin@stonybrook.edu

Keywords: molar development, incremental feature, crown formation time, cross-striation, Retzius line, periodicity, enamel thickness, dental development, hominid evolution, hominoid evolution

S.E. Bailey and J.-J. Hublin (Eds.), Dental Perspectives on Human Evolution, 177–192.

Abstract

Previous histological studies of small samples of chimpanzee and human molars suggested similarities in crown formation time, which is surprising given substantial life history differences. As part of an on-going study of hominoid molar development, we report on the largest-known sample of chimpanzee and human molars, including re-evaluation of previously examined histological sections. Variation of incremental features within and between genera is examined, including Retzius line periodicity, daily secretion rate, and Retzius line number. Differences due to population-level variation and sexual dimorphism are also considered. Significant increasing trends in daily secretion rates were found from inner to outer cuspal enamel, ranging from approximately 3–5 microns/day in chimpanzees. Humans demonstrate slightly lower and higher mean values at the beginning and end of cuspal formation, respectively, but both genera show an overall average of approximately 4 microns/day. Retzius line periodicity ranges from 6–7 days within chimpanzees and 6–12 days within humans. Within upper molars, mesiopalatal cusps (protocones) show thicker cuspal enamel and longer crown formation time than mesiobuccal cusps (paracones). Within lower molars, mesiobuccal cusps (protoconids) show greater Retzius line numbers, longer imbricational formation time, and thicker cuspal enamel than mesiolingual cusps (metaconids), resulting in longer formation times. A negative correlation was found between Retzius line number and periodicity in the human sample, resulting in similar crown formation times within cusp types, despite the range of individual periodicities. Few sex differences were found, but a number of developmental differences were apparent among human populations. Cusp-specific formation time in chimpanzees ranges from 2–3 years on average. Within specific cusp types, humans show greater average formation times than chimpanzees, due to higher mean periodicity values and/or thicker cuspal enamel. However, formation time within specific cusp types varies considerably, and the two genera show overlapping ranges, which has implications for the interpretation of small samples.

Introduction

Recent histological studies of chimpanzee and human enamel crown formation have suggested a number of similarities in molar formation time (Reid et al., 1998a,b). This is unexpected given their life history differences, which include an almost two-fold difference in the age at M1 eruption (reviewed in Kelley and Smith, 2003; but see Zihlman et al., 2004). However, these studies were based on small samples, and included several individuals with moderate to heavy dental wear. Recent studies of hominoid anterior tooth formation have yielded evidence of developmental differences among taxa (e.g., Schwartz and Dean, 2001; Schwartz et al., 2001). Yet little is known about the variation in molar development within and among taxa; postcanine development had been described for only four individuals of *Pan*, two individuals of *Gorilla*, and a lone representative of *Pongo* (Beynon et al., 1991; Reid et al., 1998a; Schwartz et al., 2006). Recently, several researchers have analyzed large collections of diverse groups of modern humans and chimpanzees (Thomas, 2003; Smith, 2004; Reid and Dean, 2006; Reid and Ferrell, 2006; Smith et al., 2007). The current study aims to use these data to highlight and contrast some of the developmental variables in molar enamel within and between modern humans and our closest-living primate relatives. We also attempt to generate a broad comparative framework for the interpretation of developmental data on more limited fossil collections.

The specific objectives are (1) to determine the appropriate units of analysis for developmental variables (e.g., individual, molar, specific molar, specific cusp), and (2) to investigate variation in enamel development at hierarchical levels. Within cusp types and tooth types, we examine several developmental variables: cuspal daily secretion rate, Retzius line periodicity, Retzius line number, imbricational enamel formation time, cuspal enamel thickness, and cusp-specific enamel formation time. We also explore the potential influence of sex and population differences on these variables. Our final comparison is between chimpanzees and humans. This is of particular interest given previous studies that have suggested overall developmental similarities in enamel formation among hominoids, as well as known life history differences between these two genera.

Incremental Development

Tooth mineralization begins over the location of the future dentine horns, where cuspal enamel is secreted in an appositional manner by ameloblasts (enamel-forming cells) as they migrate away from the enamel-dentine junction (EDJ). Epithelial cells continue to differentiate into secretory ameloblasts along the future EDJ, permitting the growth of enamel through extension from the horn tip to the enamel cervix. Enamel development is characterized by the formation of short- and long-period incremental lines, representing rhythmic changes or disturbances in enamel secretion (Figure 1). Short-period lines, known as cross-striations, show a circadian repeat interval, and may be used to determine the daily secretion rate. Long-period lines are known as Retzius lines (or striae of Retzius), which run from the EDJ to the surface of the tooth and form perikymata. This region, referred to as imbricational enamel, includes both the lateral and cervical enamel. Within an individual dentition, a consistent number of cross-striations can be counted along enamel

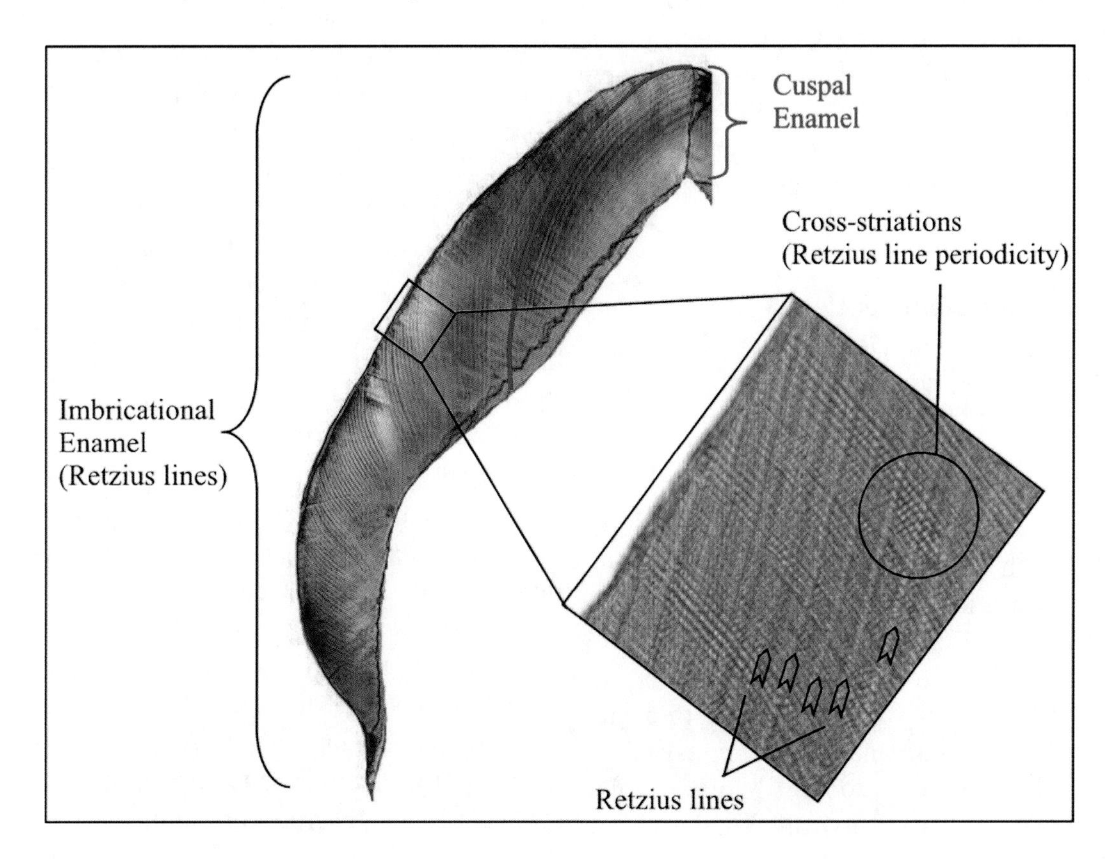

Figure 1. Incremental enamel development illustrated in a chimpanzee third molar. A portion of the enamel crown is shown divided into cuspal and imbricational areas. Cuspal enamel is defined (in blue) as the first formed enamel over the dentine horn that does not show Retzius lines running to the tooth surface. Imbricational enamel is characterized by the presence of Retzius lines running from the enamel-dentine junction to the surface of the tooth. At a higher magnification (lower right), daily cross-striations can be seen between Retzius lines; this consistent relationship is termed the Retzius line periodicity.

prisms between Retzius lines, (FitzGerald, 1998; Smith, 2004) and this number is known as the Retzius line periodicity.

From these developmental variables, the duration of crown formation may be estimated as the sum of cuspal and imbricational enamel formation times. Cuspal formation time is frequently estimated by dividing the linear cuspal enamel thickness by the average cuspal daily secretion rate (or cross-striation spacing). Imbricational formation time is often assessed by counting Retzius lines from the cusp tip to the cervix, and multiplying this number by the Retzius line periodicity. This yields a cusp-specific crown formation time. Because crown mineralization generally begins and ends at different times in different cusps, crown formation times derived from different cusps should not be directly compared (Reid et al., 1998a, b; Smith et al., 2007). Examination and registration of the first- and last-formed cusps are required to assess total crown formation time accurately.

Material and Methods

The chimpanzee sample consists of wild-shot, captive, and unknown provenience individuals from four collections, and includes *Pan troglodytes verus* and *Pan troglodytes* of unknown subspecies attribution (see Smith, 2004; Smith et al., 2007). Sex was unknown for the majority of individuals. A total of 269 histological sections of 134 molars from 75 individuals were prepared according to standard histological procedures (Reid et al., 1998a,b). The human sample consists of four collections of modern human molar teeth, including material from South Africa, Northern England, medieval Denmark, and North America (described further in Thomas, 2003; Reid and Dean, 2006; Reid and Ferrell, 2006). Sex was known for two of the four populations. Approximately 500 histological sections were generated from 420 molars of 365 individuals. Unfortunately, relatively few unworn or lightly worn cusps were available for the assessment of formation time. The approach taken in this study was intentionally conservative, as worn or missing enamel was not estimated for cusps showing more than very slight wear. Sections that were clearly oblique to an ideal plane or moderately to heavily worn were used only to determine the periodicity of Retzius lines (described below). Oblique sections were defined as those that did not show a relatively complete dentine horn tip, which reduces the accuracy of estimates (Smith et al., 2006).

Data are reported here for mesial cusps of maxillary and mandibular first through third molars, and includes the following variables: daily secretion rate, Retzius line periodicity, Retzius line number, and linear cuspal enamel thickness. In addition, cuspal and imbricational formation times were determined and summed to yield the cusp-specific crown formation time. Cuspal enamel formation was determined for both taxa with the use of taxon-specific regression equations requiring linear cuspal enamel thickness derived from a non-oblique plane of section (Dean et al., 2001). Imbricational (lateral and cervical) enamel formation time was determined as the number of Retzius lines multiplied by the individual's periodicity.

Data analysis included both parametric and rank-based (non-parametric) tests where appropriate. Rank-based statistical methods are appropriate given relatively small sample sizes or data that are not distributed normally, which often fail to meet the assumptions of parametric analyses. Rank-based tests are also appropriate because many of the data collected are discrete or categorical (non-continuous) (Conover, 1999). Comparisons of the developmental variables detailed below were made between mesial cusps, among molar types, between sexes, among populations, and between genera. Data on certain developmental variables are reported here for the larger human sample, where comparisons are more robust, but the trends are consistent with the chimpanzee sample unless indicated.

(1) For daily secretion rate (DSR), cross-striations were measured where possible in the cuspal enamel from the enamel near the dentine horn to the surface of the cusp (near the point where the first imbricational Retzius line meets the tooth surface), and were grouped into equivalent inner, middle, and outer zones (with a minimum of three sequential cross-striation measurements in a minimum of four areas per zone). Data were combined for all cusp and molar types, as Smith (2004) demonstrated in the chimpanzee sample that mean rates did not differ within zones among cusp types or molar types. Means, ranges, and standard deviations were determined for inner, middle, and outer cuspal enamel zones, and *t*-tests were performed to test for differences within zones between genera. Conover's (1999) adaptation of the Jonckheere- Terpstra test for trends was used to test for a gradient in rate from inner to outer cuspal enamel within each genus. Spearman's Rho is the statistic of choice for assessing the level of significance of the Jonckheere-Terpstra test, and is a more appropriate test for trends than the parametric ANOVA model (Jonckheere, 1954; Smith et al., 2005).

(2) Retzius line periodicity (the number of cross-striations between Retzius lines) was determined from each section when possible, including multiple sections and/or teeth per individual, to confirm the uniformity of this feature and the accuracy of the count. The mean, mode, and range were determined at the genus level. For the larger (and more variable) human sample, periodicity was plotted as a frequency histogram, and normality was tested using the one-sample Kolmogorov-Smirnov test for normality (Conover, 1999). Differences between the periodicities of the two known-sex human samples were assessed with the Mann-Whitney *U*-test (Sokal and Rohlf, 1995). Differences in periodicity among (combined-sex) human populations were assessed with the Kruskal-Wallis test, with population as the factor; if overall significance was achieved, then the multiple comparisons technique described by Conover (1999) was performed to determine which populations differed from one another. Differences between the periodicities of the two genera were assessed with the Mann-Whitney *U*-test.

(3) Differences between mesial cusps were tested in the four remaining developmental variables: Retzius line number, imbricational formation time, cuspal enamel thickness, and (cusp-specific) crown formation time. Means and ranges were determined for each variable in each cusp and tooth type, and two-sample *t*-tests were employed to assess differences between cusps, resulting in six comparisons for each of the four variables.

(4) Differences in the four aforementioned developmental variables were also examined among molar types. Kruskal-Wallis tests for molar differences were performed using molar type as the factor, testing each upper and lower mesial cusp type separately, resulting in four comparisons for each of the four variables. When significance was achieved, the multiple comparisons technique described above was performed in order to determine which molar types differed from one another significantly. The Jonckheere-Terpstra test for trends was also used to test for increasing or decreasing trends in mean variables along the molar row in both mesial cusp types.

(5) Differences in the four developmental variables were also examined between human populations. Kruskal-Wallis tests for population differences were performed using population as the factor, testing both mesial cusps in each upper and lower molar type separately, resulting in 12 comparisons for each of the four variables. When significance was achieved, Conover's (1999) multiple comparisons technique was performed to determine which populations differed from one another.
(6) The relationships between Retzius line periodicity and number was also examined within each mesial cusp and molar type in the human sample using Kendall's Coefficient of Concordance (Conover, 1999). The intention was to assess whether the negative correlation between these two variables reported by Reid et al. (2002) and Reid and Ferrell (2006) in canines from the Danish sample could be detected in a larger sample of diverse human molars.
(7) Finally, comparisons of the four developmental variables were made between chimpanzees and humans. However, because of the inequality of sample sizes (particularly the lack of maxillary *Pan* molars), any statistical results must be treated with due caution. Therefore, we only report the trends in mean differences, the relationships of the ranges of variables, and the patterning of developmental variables throughout the molar row.

Results

Daily Secretion Rate

Data on cuspal daily secretion rate (cross-striation spacing) are reported for inner, middle, and outer enamel zones in Table 1, along with the results of comparisons between genera.

Table 1. Mean molar cuspal daily secretion rate in Homo *and* Pan *(in microns/day), with t-test results for inter-generic differences*

	N	*Inner*	*Middle*	*Outer*	*Overall*
Homo	21	2.55	4.34	5.45	4.11
Pan	69	3.62	4.28	4.61	4.17
t	88	−10.886	0.524	7.112	−0.652
p	88	**<0.001**	0.602	**<0.001**	0.516

N is the total number of molar cusps sampled (above the line), or the degrees of freedom for the *t*-test (below the line). Inner, middle, and outer zones represent thirds in the enamel from the dentine horn to the cusp tip. Raw data are shown above the dotted line, and the results of a *t*-test for each specific zone are shown below the line. Significant differences are indicated in bold.

Humans showed a significantly lower mean value for inner enamel secretion rates, and a greater mean value for outer enamel rates, although the overall mean values were not significantly different. A statistically significant increasing trend in daily secretion rate from inner to outer enamel was found in both genera ($p < 0.001$ in both cases).

Retzius Line Periodicity

The Retzius line periodicity, or number of cross-striations between Retzius lines, is shown here for humans and chimpanzees (Table 2).

Table 2. Retzius line periodicity (in days) for Homo *and* Pan *individuals*

Taxon	*Group*	*N*	*Mean*	*Mode*	*Range*
Homo	South Afri.	121	8.6	8	6–12
Homo	Danish	61	8.4	9	7–11[1]
Homo	North. Eng.	83	8.1	8	6–11
Homo	North Am.	100	7.9	8	7–9
Homo	all	365	8.3	8	6–12
Pan	all	61[2]	6.4	6	6–7

For Group, South Afri. – South African, Danish – medieval Danish, North. Eng. – Northern England, North Am. – North American. N is the total number of individuals sampled. [1] The complete sample of anterior and posterior teeth from 84 Danish individuals reported in Reid and Ferrell (2006) ranged from 6–12 with a mean of 8.5. [2] It was not possible to determine conclusively the periodicity in 14 of the 75 individuals (see Smith, 2004).

No evidence was found to suggest that periodicity is variable within an individual's dentition when multiple teeth were compared. The human sample was plotted as a histogram (Figure 2); moment statistics indicate that this is a leptokurtic distribution skewed to the right. Further, the Kolmogorov-Smirnov statistic was significantly different ($p < 0.001$) from four hypothetical distributions (normal, uniform, exponential, and Poisson).

Of the two known-sex human samples (North American and South African), females were found to have a significantly higher periodicity than males in the South African sample only ($p < 0.010$). Nonetheless, sexes were combined for the following analyses, as sex was unknown for the remaining two human populations and the majority of the chimpanzee sample. Statistical differences were found among human groups ($p < 0.001$). Post hoc comparisons indicated that the South African sample showed a greater mean periodicity than all other groups, and the Danish sample was additionally greater than the Northern English and North American samples. When humans and chimpanzees were compared, humans showed a significantly greater mean periodicity ($p < 0.001$), despite the fact that the range of the chimpanzee sample fell entirely within the human range.

Variation Between Mesial Cusps

Results of comparisons of developmental variables between mesial cusps in the human sample are shown in Table 3. Within upper molars, mesiopalatal cusps (protocones)

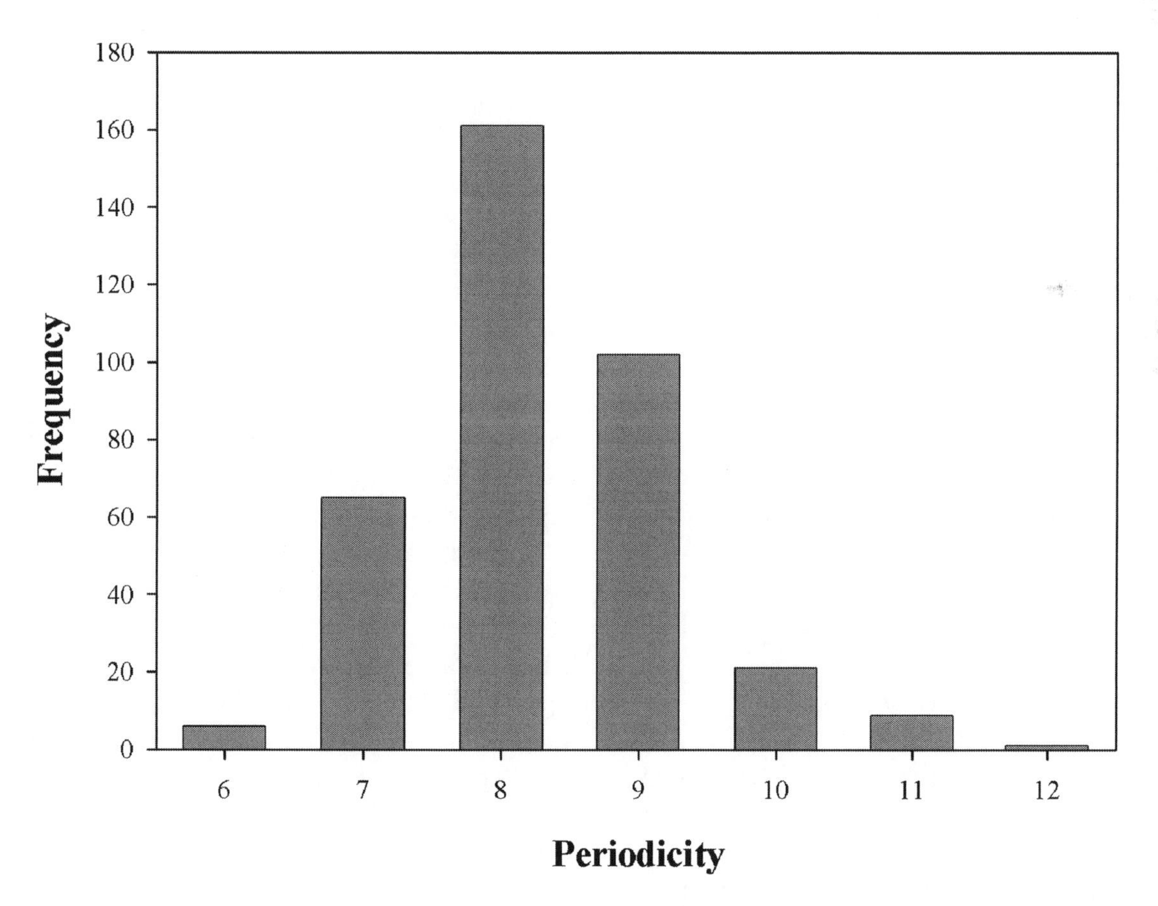

Figure 2. Histogram of human Retzius line periodicity data. Each value represents one of 365 individuals from diverse human populations. See text for details about the non-normality of this distribution.

Table 3. Mann-Whitney U-test comparisons of developmental variables between mesial cusps in Homo

Position	*Tooth*	*Statistic*	*Retz*	*Imb*	*Thick*	*CFT*
Upper	M1	*Z*	−0.092	−0.563	−4.272	−3.080
		p	0.927	0.573	**<0.001**	**0.002**
	M2	*Z*	−0.557	−0.832	−4.747	−3.337
		p	0.577	0.405	**<0.001**	**0.001**
	M3	*Z*	−1.166	−1.696	−4.782	−4.298
		p	0.244	0.090	**<0.001**	**<0.001**
lower	M1	*Z*	−5.855	−7.436	−2.293	−6.151
		p	**<0.001**	**<0.001**	**0.022**	**<0.001**
	M2	*Z*	−5.971	−6.822	−5.280	−6.580
		p	**<0.001**	**<0.001**	**<0.001**	**<0.001**
	M3	*Z*	−3.920	−6.560	−4.389	−6.031
		p	**<0.001**	**<0.001**	**<0.001**	**<0.001**

Z – the Mann-Whitney test statistic, *p* - *p*–value of the given comparison. The results of the comparison between mesial analogues are shown for the following variables: Retz – Retzius line number, Imb – imbricational formation time, Thick – cuspal enamel thickness, and CFT – crown formation time. Significant differences are indicated in bold.

had significantly thicker cuspal enamel and greater crown formation time than the respective mesiobuccal cusps (paracones). Mean values for Retzius line number and imbricational formation time were also greater in mesiopalatal cusps than in mesiobuccal cusps, although the differences were not significant. The opposite pattern was found in lower molars; mesiobuccal cusps (protoconids) had significantly greater numbers of Retzius lines, longer imbricational time, thicker cuspal enamel, and greater crown formation time than mesiolingual cusps (metaconids).

Variation Within the Molar Row

Results of comparisons of the four developmental variables among molars are shown in Tables 4 and 5, including the results of tests for trends in the molar row. Within upper molars, cuspal enamel thickness and crown formation time showed a significant increasing trend from first to third molars in both mesiopalatal and mesiobuccal cusps. Mean Retzius line number also increased posteriorly, although this was not significant. Within lower molars, both Retzius line number and imbricational formation time decreased from first to third molars in mesiobuccal cusps. Post-hoc tests revealed that these trends were generally the result of significant differences between first and second molars and between first and third molars. Significant increasing trends were also observed in cuspal enamel thickness and crown formation time in both mesial cusps in lower molars, largely due to significant increases between first and second and/or first and third molars.

In light of these trend results, a non-parametric correlation analysis was conducted between imbricational formation time and cuspal enamel thickness in the human sample, and a significant negative correlation was found in upper first molar mesiobuccal cusps ($p < 0.050$), and in all lower mesiobuccal cusps (M1: $p < 0.010$; M2: $p < 0.001$, M3: $p < 0.050$). Cusps with thicker cuspal enamel show shorter imbricational formation times. The result of this correlation appears to be that cusp-specific formation time is constrained to a degree within mesiobuccal cusps (although

Table 4. Kruskal-Wallis tests for differences among molar types in the Homo *sample*

Position	*Cusp*	*Statistic*	*Retz*	*Imb*	*Thick*	*CFT*
Upper	mb	*H*	1.793	3.370	49.702	23.579
		p	0.408	0.185	**<0.001**	**<0.001**
	mp	*H*	2.188	2.723	17.491	6.799
		p	0.335	0.256	**<0.001**	**0.033**
Lower	mb	*H*	12.154	37.191	49.267	6.128
		p	**0.002**	**<0.001**	**<0.001**	**0.047**
	ml	*H*	4.665	1.747	31.140	14.329
		p	0.097	0.418	**<0.001**	**0.001**

For upper molars, mb – mesiobuccal cusp, and mp – mesiopalatal cusp. For lower molars, mb – mesiobuccal cusp, and ml – mesiolingual cusp. H – the Kruskal-Wallis test statistic, *p* - *p*-value of the given comparison. Variables are defined in Table 3. Significant differences are indicated in bold.

Table 5. Kruskal-Wallis post-hoc comparisons for differences in developmental variables between molar types, with the directionality of significant trends indicated

Position	*Cusp*	*Var*	*M1 v. M2*	*M1 v. M3*	*M2 v. M3*	*JT*
Upper	mb	Thick	<	<	<	+
		CFT	<	<	–	+
Upper	mp	Thick	<	<	–	+
		CFT	<	<	–	+
Lower	mb	Retz	>	>	–	–
		Imb	>	>	–	–
		Thick	<	<	–	+
		CFT	–	<	<	+
	ml	Thick	<	<	–	+
		CFT	–	<	–	+

Cusp and Var- cusp type and developmental variables are defined in Tables 3 and 4. The direction of the sign below each bivariate comparison indicates which molar position is significantly greater. No significant difference is indicated by '–'. JT indicates the significant results of the Jonkheere-Terpstra test for trends ($p < 0.05$), with increasing trends illustrated by '+' and decreasing trends illustrated by '–'.

an overall increasing trend in cusp-specific crown formation time was significant from first to third molars in each case). It was not possible to test for similar trends within chimpanzee molars as sample sizes were too small.

Variation Within Homo

Sex differences in the four main developmental variables were examined within mesial cusps and molar types between the two known-sex human populations, and significant differences were found in only one of 64 comparisons. In this instance, the upper third molar mesiobuccal cusp in South African males showed a significantly greater imbricational formation time than females ($p < 0.050$), due to a greater number of Retzius lines. However, given the lack of consistent developmental differences between males and females, the sexes were combined for the subsequent analyses.

When human populations were compared, a complicated pattern of similarities and differences became apparent. Of the 48 comparisons of the four developmental variables, 18 comparisons showed significant differences among populations (Table 6). In general, no single cusp or tooth type showed more frequent inter-population differences, except for mesiobuccal cusps in upper third molars. The most frequent developmental variable to differ among populations was cuspal enamel thickness, in six of 12 comparisons, followed by imbricational formation time and cusp-specific crown formation time in five of 12 comparisons. Retzius line number appeared to be the least variable among populations, with differences in only two of 12 comparisons. Post-hoc tests showed that variation was found between several different population combinations (Table 7). Excluding any single population still resulted in a proportionate number of differences among the remaining populations. A few general trends were evident, as the North American sample consistently showed the thickest cuspal enamel in third molars (data were not available for first or second molars from this sample), and the Danish sample tended to show the thinnest cuspal enamel for most tooth positions. Despite these differences, the populations were combined, as the remaining issues concerned specific trends and generic differences.

Correlations Between Variables

Table 8 shows a highly significant negative correlation between Retzius line number and periodicity for each cusp and molar

Table 6. Kruskal-Wallis test for differences in developmental variables among populations of Homo

Pos	*Tooth*	*Cusp*	*Statistic*	*Retz*	*Imb*	*Thick*	*CFT*
Upper	M1	mb	*H*	2.349	4.793	1.705	1.792
			p	0.309	0.091	0.426	0.408
		mp	*H*	2.299	5.369	5.647	6.099
			p	0.317	0.068	0.059	**0.047**
	M2	mb	*H*	2.203	11.652	6.000	9.548
			p	0.332	**0.003**	0.050[1]	**0.008**
		mp	*H*	0.280	3.923	2.329	3.485
			p	0.870	0.141	0.312	0.175
	M3	mb	*H*	9.216	17.261	21.305	20.033
			p	**0.027**	**0.001**	**<0.001**	**<0.001**
		mp	*H*	0.992	0.978	22.005	8.407
			p	0.803	0.807	**<0.001**	**0.038**
Lower	M1	mb	*H*	1.232	8.893	7.703	1.170
			p	0.540	**0.012**	**0.021**	0.557
		ml	*H*	4.076	5.634	4.591	6.117
			p	0.130	0.060	0.101	**0.047**
	M2	mb	*H*	4.653	1.236	2.764	0.491
			p	0.098	0.539	0.251	0.782
		ml	*H*	9.147	8.980	9.776	2.668
			p	**0.010**	**0.011**	**0.008**	0.263
	M3	mb	*H*	5.535	10.787	13.228	3.147
			P	0.137	**0.013**	**0.004**	0.370
		ml	*H*	7.184	2.653	8.588	3.862
			p	0.066	0.448	**0.014**	0.145

Cusp, Statistic, and variables are defined in Table 3 and 4. Significant results are in bold. The degrees of freedom for each cusp and molar type comparison are 2 for first and second molars, and 3 for third molar comparisons (as the North American sample consisted of third molars only). [1]The post-hoc results for this comparison are also given in the following table.

Table 7. Kruskal-Wallis post-hoc comparisons for significant differences in developmental variables between populations

Pos	*Tooth*	*Cusp*	*Var*	*1 v. 2*	*1 v. 3*	*1 v. 4*	*2 v. 3*	*2 v. 4*	*3 v. 4*
Upper	M1	mp	CFT	n/a	<	–	n/a	n/a	–
	M2	mb	Imb	n/a	>	>	n/a	n/a	–
			Thick	n/a	–	>	n/a	n/a	–
			CFT	n/a	>	>	n/a	n/a	–
	M3	mb	Retz	>	>	–	>	–	–
			Imb	>	>	–	>	–	<
			Thick	<	–	–	>	>	>
			CFT	–	–	–	>	>	–
		mp	Thick	<	–	–	>	>	–
			CFT	<	–	–	–	–	–
Lower	M1	mb	Imb	n/a	–	–	n/a	n/a	<
			Thick	n/a	–	–	n/a	n/a	>
		ml	CFT	n/a	<	–	n/a	n/a	–
	M2	ml	Retz	n/a	<	<	n/a	n/a	–
			Imb	n/a	–	<	n/a	n/a	<
			Thick	n/a	–	>	n/a	n/a	–
	M3	mb	Imb	–	–	–	<	–	–
			Thick	–	–	–	–	>	–
		ml	Thick	–	–	–	–	>	–

Cusp and Var are indicated for Tables 3 and 4. Population codes are: 1- South African, 2- North American, 3- Northern England, 4- medieval Danish. The direction of the sign below the bivariate comparisons indicates which population is significantly greater. No significant difference is indicated by '–,' n/a indicates that comparisons were not possible due to the lack of first and second molar data for population 2 (North American).

Table 8. Correlations between Retzius line number and periodicity in Homo

Position	*Tooth*	*Cusp*	*N*	*Correlation*	*p-value*
Upper	M1	mb	26	−0.628	< **0.001**
		mp	19	−0.771	< **0.001**
	M2	mb	29	−0.611	< **0.001**
		mp	22	−0.776	< **0.001**
	M3	mb	81	−0.648	< **0.001**
		mp	68	−0.762	< **0.001**
Lower	M1	mb	34	−0.793	< **0.001**
		ml	50	−0.527	< **0.001**
	M2	mb	43	−0.679	< **0.001**
		ml	42	−0.560	< **0.001**
	M3	mb	50	−0.694	< **0.001**
		ml	41	−0.758	< **0.001**

Cusps are defined in Table 4. N equals the number of cusps (individuals). Correlation is Kendall's Tau correlation coefficient. Significant results are indicated in bold.

type in the combined human sample. The result of this correlation is that imbricational formation time is constrained in a given cusp and molar type. Individuals with a smaller number of Retzius lines have a greater periodicity, on average, than individuals with a greater number of Retzius lines.

Inter-Generic Comparisons

The final comparison is between human and chimpanzee developmental variables. Several trends are evident, and are summarized in Table 9. One of the notable differences between the two genera is in cuspal enamel thickness, which was generally two to three times greater in humans than in chimpanzees. Additionally, humans consistently showed longer mean crown formation times than chimpanzees, ranging from several months to more than a year of difference for certain cusp types. In general, there was little overlap in formation time ranges, although exceptions were found for certain cusp types and tooth types.

It also appeared that humans and chimpanzees showed differences in trends of developmental variables along the molar row (Figures 3 and 4). Humans commonly showed the greatest mean values in third molars (or less commonly in first molars), while chimpanzees appeared to show the greatest values in second molars (or less commonly in third molars). A particularly clear difference was apparent in imbricational formation time from lower first to third molars, which was greatest in human first molars, and least in chimpanzee first molars relative to their respective posterior molars.

Table 9. Summary of developmental variable comparisons between Pan *and* Homo

Variable	*Trend*	*Range Overlap*
Periodicity	*Homo* > *Pan*	*Pan* within *Homo*
Daily secretion rate		
Inner	*Homo* < *Pan*	little
Middle	*Homo* = *Pan*	*Homo* within *Pan*
Outer	*Homo* > *Pan*	partial
Overall	*Homo* = *Pan*	*Pan* within *Homo*
Retzius line number	*Homo* < *Pan*	partial
Imbricational time	*Homo* < *Pan*	partial
Cuspal thickness	*Homo* > *Pan*	none to little
Crown formation time	*Homo* > *Pan*	little

See text for description of variables.

Discussion

This study is a methodological analysis of enamel formation that has demonstrated that specific developmental variables should be assessed at differing hierarchical levels. Daily secretion rate does not vary within cusps or molar types, Retzius line periodicity varies at the individual (dentition) level, and the four additional developmental variables: Retzius line number, imbricational formation time, cuspal enamel thickness, and crown formation time all vary within cusps and among molar types (also see Smith et al., 2007 for additional data on distal cusps and total crown formation time in the chimpanzee sample). These factors each need to be considered when discussing variation in enamel developmental, or when making comparisons between molars. In light if this, some of the early work characterizing enamel development in naturally fractured hominid teeth should be viewed with caution (e.g., Beynon and Wood, 1987; Ramirez Rozzi, 1993).

The results of comparisons between mesial cusp pairs and among molar types may also have functional implications. As reported above, mesiopalatal cusps show significantly thicker cuspal enamel and longer crown formation times than mesiobuccal cusps in upper molars. Within lower molars, mesiobuccal cusps show significantly greater Retzius line numbers, longer imbricational formation times, thicker cuspal enamel, and longer formation times than mesiolingual cusps. This implies differential patterning in mesial cusp development between upper and lower molars, trends that are consistent with functional models of thicker enamel, and thus prolonged development, in principle or functional cusps (also see Reid et al., in 1998a; Suwa and Kono, 2005). When differences were examined among molar types, a similar pattern was revealed; increasing trends were found in cuspal enamel thickness and crown formation time from first to third molars

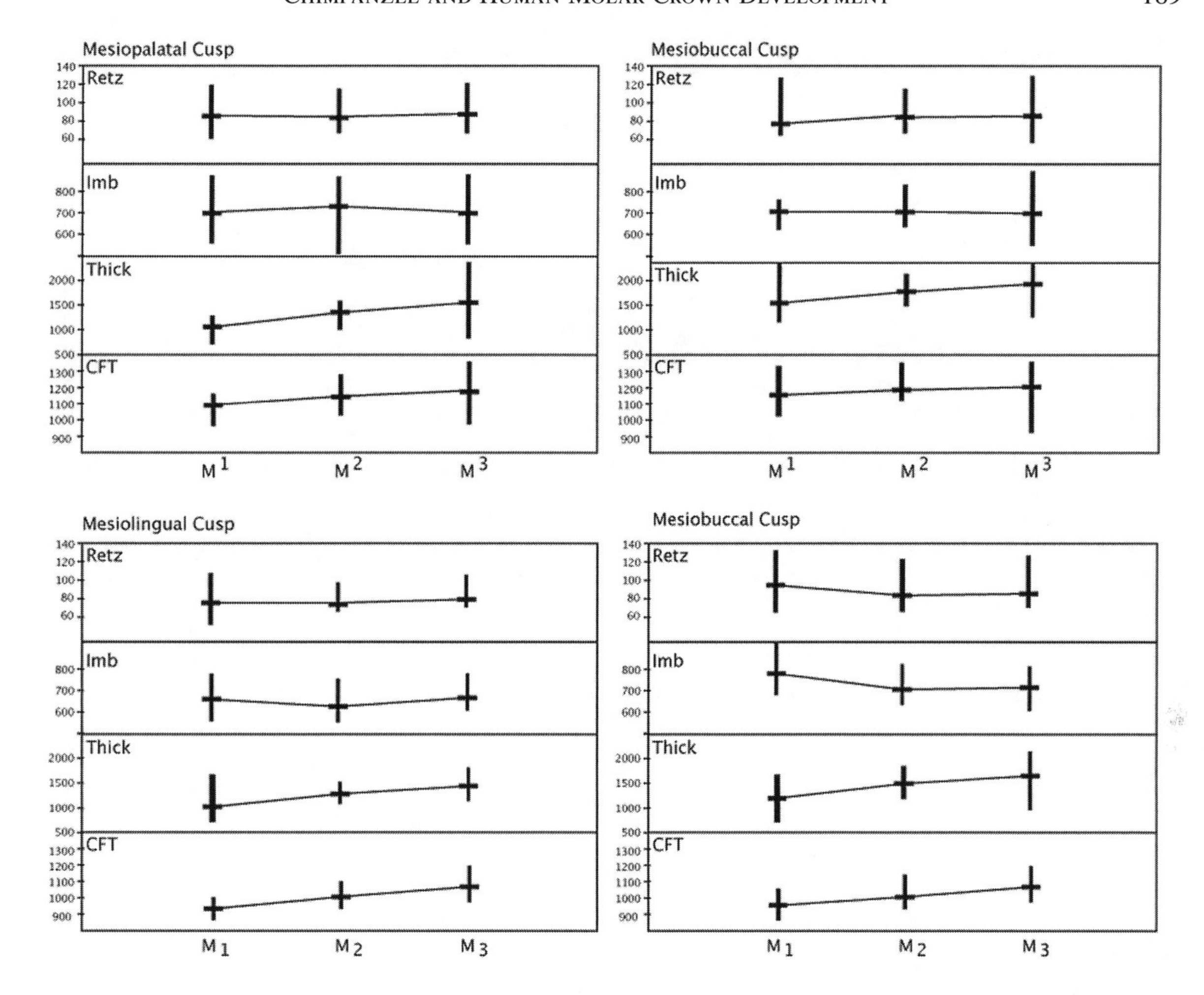

Figure 3. Anterior-to-posterior trends in developmental variables in human molars. Retz- Retzius line number, Imb- imbricational formation time (in days), Thick- cuspal enamel thickness (in microns), and CFT- crown formation time (in days). The range of values was plotted from left (M1) to right (M3) to show directional trends (upper molars above lower molars). Ends of vertical boxes represent 25% and 75% ranges of the data; the horizontal line is the mean.

in both mesial cusps for upper and lower molars. Smith et al. (2005) also showed an increasing trend for enamel cap area and/or average enamel thickness in mesial sections of the same sample of chimpanzees, which may relate to trends in overall tooth size throughout the molar row. Work in progress may shed more light on the relationship between tooth size and formation time variation.

Comparisons of developmental variables among human populations showed a complex pattern of developmental differences and similarities (also see Reid and Dean, 2006). Cuspal enamel thickness and cusp-specific crown formation time were most commonly different among populations. The North American third molar sample showed the most notable differences when compared to other populations, which may be due in part to their extremely thick enamel. If is unclear why this population showed a restricted range of periodicity values (including the lowest mean value) and very thick cuspal enamel. The most geographically similar European populations (Northern England and medieval Danish) showed relatively few differences, despite being separated by approximately 800 years. Future work is required to

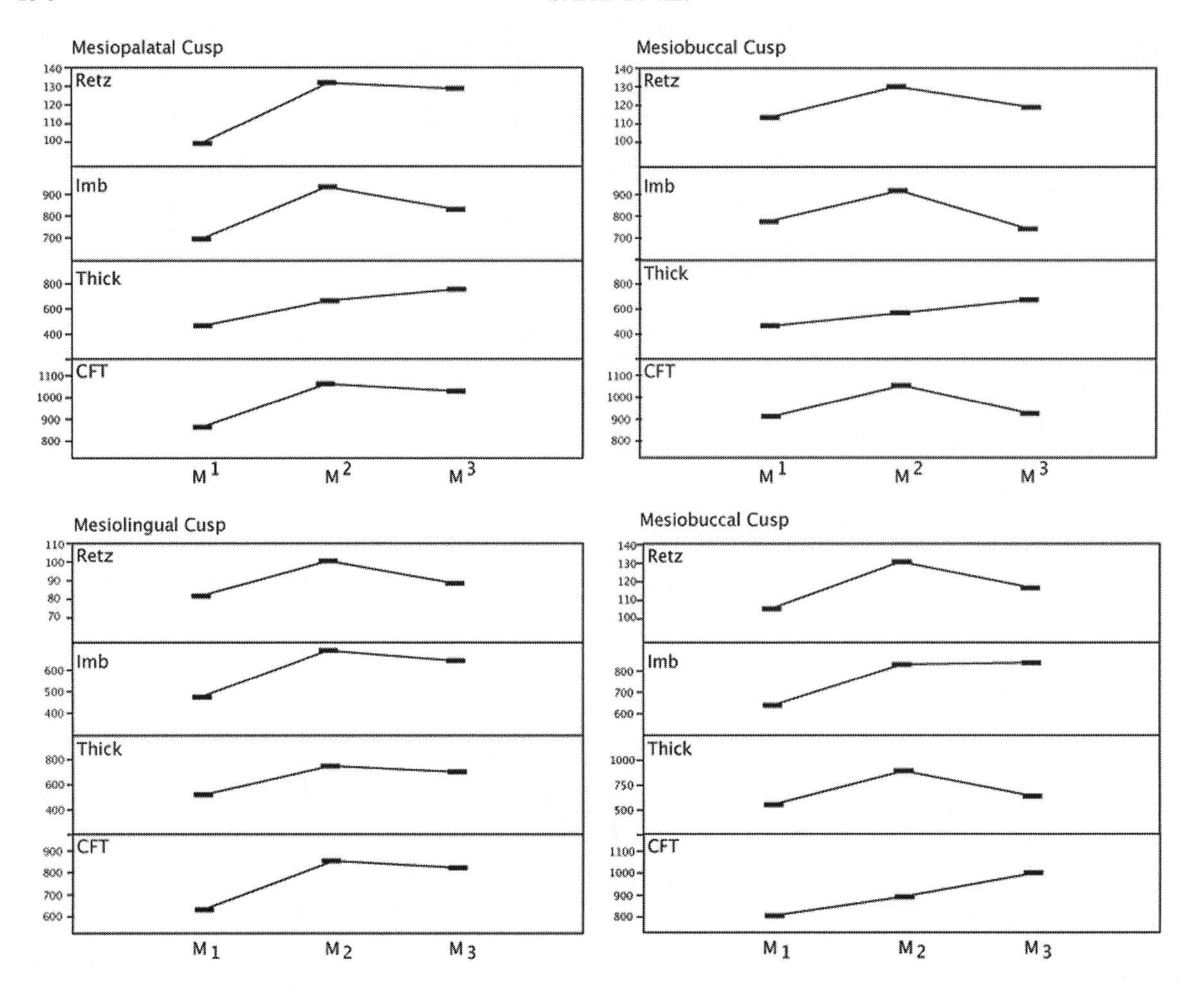

Figure 4. Anterior-to-posterior trends in developmental variables in chimpanzee molars. Retz- Retzius line number, Imb- imbricational formation time (in days), Thick- cuspal enamel thickness (in microns), and CFT- crown formation time (in days). The range of values was plotted from left (M1) to right (M3) to show directional trends (upper molars and lower molars). Due to the limited sample sizes, only mean values are given for each variable.

clarify the degree of variation among diverse human populations as well as among non-human primate sub-species/species groups. The influence of different developmental environments may also be a source of variation, as captive animals may show earlier ages of dental eruption than wild animals (Phillips-Conroy and Jolly, 1988; Zihlman et al., 2004). Additional studies controlling for these factors are required to facilitate a better understanding of how developmental variables may be influenced by phylogenetic, functional, or environmental factors.

An additional finding that warrants comment was the highly significant negative correlation between Retzius line number and periodicity in the human sample (also see Reid and Dean, 2006; Reid and Ferrell, 2006). This finding has serious implications for studies of perikymata, external manifestations of Retzius lines, and suggests that comparisons of perikymata numbers without knowledge of the individual periodicities may *not* provide information about actual differences in the rate or period of formation. For example, comparing counts of Retzius lines holding cusp and molar types constant could yield values that differ by as much as a factor of two, such as 70 lines and 140 lines, but the periodicities are likely to also differ by a factor of two, such as 12 and 6 days, respectively. In this example, what appears different on the surface of a tooth may actually represent the same developmental time. We advocate consideration of the taxon-specific periodicity range when making inferences about developmental rate or time from incremental lines on the surface of teeth.

Previous work on a large collection of hominoid canines by Schwartz and colleagues addressed issues of inter-generic differences and sex-differences (Schwartz and Dean, 2001; Schwartz et al., 2001), and the present study confirms a number of their conclusions. In terms of daily secretion rate, Schwartz et al. (2001) also found that humans have lower and higher values in the respective inner and outer cuspal enamel than these regions in chimpanzees. In regards to Retzius line periodicities, they also found that humans had a significantly greater mean periodicity than chimpanzees. However, Schwartz et al. (2001) reported periodicity ranges for humans and chimpanzees of 7–11 days and 6–9 days, respectively, which are different than the ranges found in this study. It is likely that differences between studies are due to the inclusion of outliers, which are present in very low frequencies. Cuspal enamel thickness in the former study was found to be greater in humans, also consistent with the present study. However, canine crown formation times were greater in chimpanzees, which appeared to be related to differences in canine crown height (Schwartz and Dean, 2001). Work by Reid and Dean (2000), Reid and Ferrell (2006), and Reid and Dean (2006) has shown some degree of overlap in canine formation time between humans and female chimpanzees, but not for male chimpanzees.

It is unclear why females in one of the two known-sex human populations in this study showed significantly greater Retzius line periodicity than males. Schwartz et al. (2001) did not find sex differences in their sample of great apes and humans, and in most instances males showed slightly higher mean values (although their female mean was higher in the human sample). It would be interesting to investigate periodicity in other samples of known sex (and known mass) humans and non-human primates. In conclusion, it appears that tooth size, sexual dimorphism, functional differences, and life history may impact developmental variation between chimpanzees and humans, particularly when anterior and posterior teeth are considered together.

Acknowledgments

Pam Walton is thanked profusely for her expert technical assistance. In addition, the following individuals provided important material support: Peter Andrews, Jesper Boldsen, Stefan Bracha, Vivian Fisher, Daphne Hill, Paula Jenkins, Kevin Kuykendall, Michelle Morgan, David Pilbeam, and Cynthia Reid. We would like to thank Jean-Jacques Hublin and Shara Bailey for inviting us to take part in this conference, and two anonymous reviewers for comments on this paper. In addition, we would like to acknowledge assistance from Annette Weiske, Silke Streiber, Diana Carstens, and last but never least, Allison Cleveland. This research has been supported by the IDPAS Research Awards, NSF Dissertation Improvement Grant, the Leverhulme Trust, and the Max Planck Society.

References

Beynon, A.D., Dean, M.C., Reid, D.J., 1991. Histological study on the chronology of the developing dentition in gorilla and orangutan. American Journal of Physical Anthropology 86, 189–203.

Beynon, A.D., Wood, B.A., 1987. Patterns and rates of enamel growth in the molar teeth of early hominids. Nature 326, 493–496.

Conover, W.J., 1999. Practical Nonparametric Statistics. New York: John Wiley and Sons, Inc.

Dean, M.C., Leakey, M.G., Reid, D., Schrenk, F., Schwartz, G.T., Stringer, C., Walker, A., 2001. Growth processes in teeth distinguish modern humans from *Homo erectus* and earlier hominins. Nature 414, 628–631.

FitzGerald, C., 1998. Do enamel microstructures have regular time dependency? Conclusions from the literature and a large-scale study. Journal of Human Evolution 35, 371–386.

Jonckheere, A.R., 1954. A distribution-free k-sample test against ordered alternatives. Biometrika 41, 133–145.

Kelley, J., Smith, T.M., 2003. Age at first molar emergence in early Miocene *Afropithecus turkanensis* and life-history evolution in the Hominoidea. Journal of Human Evolution 44, 307–329.

Phillips-Conroy, J.E., Jolly, C.J., 1988. Dental eruption schedules of wild and captive baboons. American Journal of Primatology 15, 17–29.

Ramirez Rozzi, F.V., 1993. Tooth development in east African *Paranthropus*. Journal of Human Evolution 24, 429–454.

Reid, D.J., Beynon, A.D., Ramirez Rozzi, F.V., 1998b. Histological reconstruction of dental development in four individuals from a Medieval site in Picardie, France. Journal of Human Evolution 35, 463–477.

Reid, D.J., Dean, M.C., 2000. Brief communication: the timing of linear hypoplasias on human anterior teeth. American Journal of Physical Anthropology 113, 135–139.

Reid, D.J., Dean, M.C., 2006. Variation in modern human enamel formation times. Journal of Human Evolution 50, 329–346.

Reid, D.J., Ferrell, R.J., 2006. The relationship between number of striae of Retzius and their periodicity in imbricational enamel formation. Journal of Human Evolution 50, 195–202.

Reid, D.J., Ferrell, R.J., Walton, P., 2002. Histologically derived canine crown formation times from a medieval Danish sample. American Journal of Physical Anthropology 34 (Suppl.), 129.

Reid, D.J., Schwartz, G.T., Dean, C., Chandrasekera, M.S., 1998a. A histological reconstruction of dental development in the common chimpanzee, *Pan troglodytes*. Journal of Human Evolution 35, 427–448.

Schwartz, G.T., Dean, C., 2001. Ontogeny of canine dimorphism in extant hominoids. American Journal of Physical Anthropology 115, 269–283.

Schwartz, G.T., Reid, D.J., Dean, M.C., 2001. Developmental aspects of sexual dimorphism in hominoid canines. International Journal of Primatology 22, 837–860.

Schwartz, G.T., Reid, D.J., Dean, M.C., Zihlman, A.L., 2006. A faithful record of stressful life events preserved in the dental development of a juvenile gorilla. International Journal of Primatology 22, 837–860.

Smith, T.M., 2004. Incremental development of primate dental enamel. Ph.D. Dissertation, Stony Brook University.

Smith, T.M., Olejniczak, A.J., Martin, L.B., Reid, D.J., 2005. Variation in hominoid molar enamel thickness. Journal of Human Evolution 48, 575–592.

Smith, T.M., Reid, D.J., Dean, M.C., Olejniczak, A.J., Martin, L.B., 2006. Molar crown development in common chimpanzees (*Pan troglodytes*). Journal of Human Evolution 52, 201–206.

Smith, T.M., Reid, D.J., Sirianni, J.E., 2006. The accuracy of histological assessments of dental development and age at death. Journal of Anatomy 208, 125–138.

Sokal, R.R., Rohlf, F.J., 1995. *Biometry*. W.H. Freeman and Company, New York.

Suwa, G., Kono, R.T., 2005. A micro-CT based study of linear enamel thickness in the mesial cusp section of human molars: reevaluation of methodology and assessment of within-tooth, serial, and individual variation. Anthropological Science 113, 273–290.

Thomas, R.J., 2003. Enamel defects, well-being and mortality in a medieval Danish village. Ph.D. Dissertation, Pennsylvania State University.

Zihlman, A., Bolter, D., Boesch, C., 2004. Wild chimpanzee dentition and its implications for assessing life history in immature hominin fossils. Proceedings of the National Academy of Sciences of the USA 101, 10541–10543.

5. Portable confocal scanning optical microscopy of *Australopithecus africanus* enamel structure

T.G. BROMAGE
Hard Tissue Research Unit
Dep'ts of Biomaterials and Basic Sciences
New York University College of Dentistry
345 East 24th Street, New York
NY 10010-4086, USA
tim.bromage@nyu.edu

R.S. LACRUZ
Institute for Human Evolution
B.P.I. for Palaeontological Research
University of the Witwatersrand
P. Bag 3 WITS 2050
Johannesburg, South Africa
LacruzRS@science.pg.wits.ac.za

A. PEREZ-OCHOA
Instituto de Postgrado y Extension Universitaria
Centro Superior de Estudios Universitarios
LA SALLE Universidad Autonoma de Madrid
Av. Lasalle, 10 Madrid 28003, Spain
perezochoa@msn.com

A. BOYDE
Hard Tissue Research Unit
Dental Biophysics
Queen Mary University of London
New Road, London E1 1BB, England
a.boyde@qmul.ac.uk

Keywords: portable confocal microscope, hominid skeletal microstructure

Abstract

The study of hominid enamel microanatomical features is usually restricted to the examination of fortuitous enamel fractures by low magnification stereo-zoom microscopy or, rarely, because of its intrusive nature, by high magnification compound microscopy of ground thin sections. To contend with limitations of magnification and specimen preparation, a Portable Confocal Scanning Optical Microscope (PCSOM) has been specifically developed

S.E. Bailey and J.-J. Hublin (Eds.), Dental Perspectives on Human Evolution, 193–209.

for the non-contact and non-destructive imaging of early hominid hard tissue microanatomy. This unique instrument can be used for high resolution imaging of both the external features of enamel, such as perikymata and microwear, as well as internal structures, such as cross striations and striae of Retzius, from naturally fractured or worn enamel surfaces. Because there is veritably no specimen size or shape that cannot be imaged (e.g. fractured enamel surfaces on intact cranial remains), study samples may also be increased over what would have been possible before. We have applied this innovative technology to the study of enamel microanatomical features from naturally occurring occluso-cervical fractures of the South African hominid, *Australopithecus africanus* representing different tooth types. We present for the first time detailed information regarding cross striation periodicity for this species and, in addition, we present data on striae-EDJ angles in a large sample of teeth and crown formation time for a molar of *A. africanus*. Our results characterize a pattern of enamel development for *A. africanus*, which is different to that reported for the genus *Paranthropus*, as previously observed.

Introduction

Most fossils are either translucent or, if they are surface reflective, are not flat. In both cases, light interacts with the sample over a considerable vertical range and is reflected (or the fluorescent light emanates) from a thick layer. The challenge we face for the non-destructive examination of enamel is how to obtain research-grade images of microantomical features in the field setting from such surfaces and sub-surface volumes. We have found a solution in development of portable confocal microscopy for the evaluation of rare and unique early hominid fossils. Our ultimate objective is to image features, such as cross-striations and striae of Retzius, for the purpose of describing aspects of the hard tissue biology and the organismal life and evolutionary histories of our extinct ancestors.

The principle of the Portable Confocal Scanning Optical Microscope (PCSOM) is to eliminate the scattered, reflected, or fluorescent light from out of focus planes, allowing only light originating from the plane of focus of the objective lens to contribute to image formation. It does this at the several conjugate focal planes (each plane representing the image of the other, that is intermediate, eye point, and image recording device), and thus eliminates light coming from all out of focus planes. In practice, an illuminated spot in the plane of focus is scanned across the field of view and an image is compiled. Confocal scanning optical microscopy thus differs from conventional light microscopy, where light from the focus plane of the objective lens, as well as from all out of focus planes across the entire field of view, is observed. The history and various technical achievements in confocal microscopy are summarized in Boyde (1995).

There is much interest in obtaining details of hominid enamel microanatomy from fractured surfaces, but such surfaces are rarely giving of all the desired detail; amongst existing instruments, the resolving power, such as that of stereo-zoom microscopy, and detail from below the surface, limited as in scanning electron microscopy, has been wanting. However, the PCSOM provides Z-axis through focus imaging of topographically complex surfaces at relatively high magnifications revealing a plane view of enamel microstructure (e.g. striae of Retzius and cross striations). Further, with the employ of circularly polarized light, the PCSOM provides some sub-surface enamel crystallite orientation contrast as well (Bromage et al., 2005).

Beyond a simple description of the PCSOM, we report here initial studies using this technology to assess *Australopithecus africanus* crown formation time, cross striation periodicity, and variation on the enamel extension rate for selected teeth. The phylogenetic relationships of *A. africanus* with other broadly contemporaneous hominids remain moot (Tobias, 1980; White et al., 1981; Berger et al., 2002), thus it is our aim to employ enamel developmental parameters potentially useful in Plio-Pleistocene hominid species comparisons (Grine and Martin, 1988).

Materials and Methods

We employ a PCSOM based on the Nipkow disk technique (Nipkow, 1884) described in detail by Petran and Hadravsky (e.g., 1966) and first commercialized in the early 1980's. The Petran and Hadravsky design uses a so-called "two-sided" disk; the specimen is illuminated through an array of pinholes on one side of the disk whilst detected through a conjugate array of pinholes on the other (via a number of delicately aligned mirrors). Applications of this technology to bone and tooth microanatomy were described by Boyde et al. (1983). Another Nipkow disk design (the one used here) employs a "single-sided" disk in which the illumination *and* detection pinhole is one in the same (Kino, 1995); that is, illuminating light and its reflections from the object pass through the same pinhole, which is imaged by the eyepiece objective or camera. This latter design is robust and able to tolerate our relatively extreme portable applications.

To date we have developed two versions of the PCSOM; the 1K2 (Figure 1) and the 2K2 (Figure 2). Both employ a one-sided Nipkow disk Technical Instrument Co. K2S-BIO confocal module (Zygo Corp., Sunnyvale, CA), specifically configured for paleoanthropological research problems (Bromage et al., 2003). Like other confocal scanning optical microscopes, the final image derives from the plane of focus, thus it eliminates the fog due to the halo of reflected, scattered or fluorescent light above and below the plane of focus, which otherwise confounds image content in conventional light microscopy.

An interesting feature of the single-sided disk design by Kino (1995) is the approach taken to suppress internal, non-image-related reflections that are a significant problem in this type of system; light reflecting from internal components of the microscope, having nothing to do with forming an image, degrades the

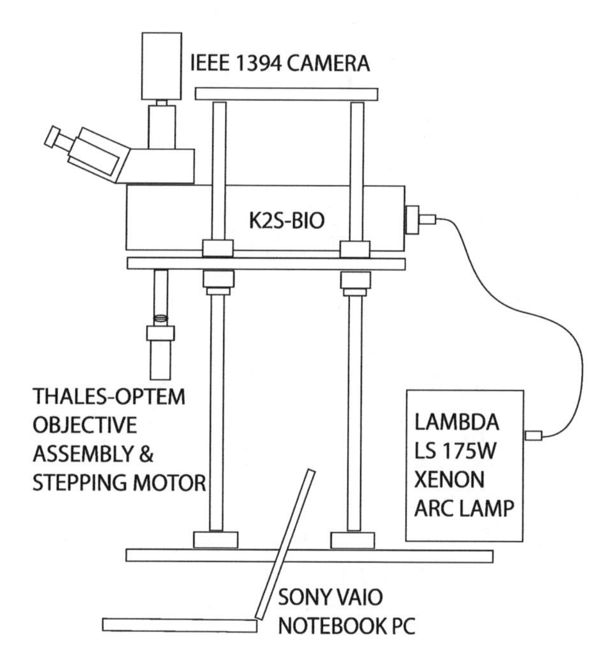

Figure 1. Diagram of the 1K2 PCSOM (see text for details).

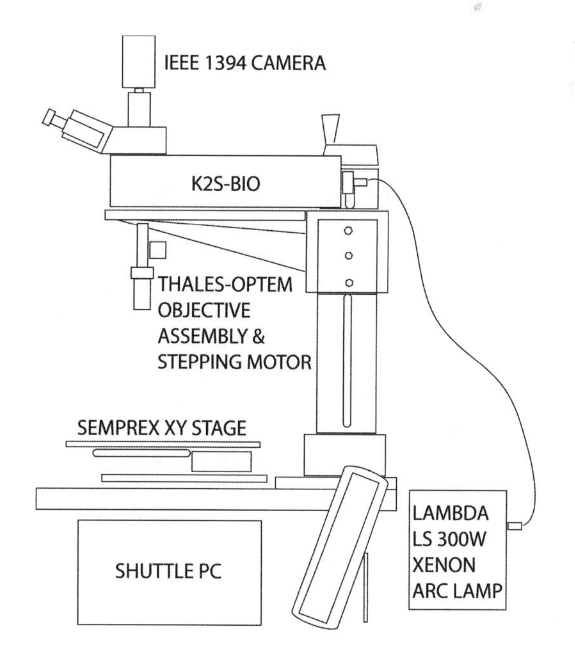

Figure 2. Diagram of the 2K2 PCSOM (see text for details).

image. This method is the classical method of illuminating with polarized light to stop light reflections from within the optical system (e.g. from optical hardware within the body of the microscope), but not the useful light reflecting from the specimen and returning through the objective lens. Linear polarizing light filters and a single quarter-wave plate filter, described further below, provide the means for eliminating the unwanted reflected light. A consequence of the single-sided disk design is that it significantly reduces the number of mirrors in the light path making the alignment of the optics less critical. The result is a very robustly constructed instrument able to withstand transport and relatively rough handling (e.g., as checked-in baggage for air travel).

The microscope configurations include several other features critical to our research. Consideration was given to obtaining objective lenses with relatively long working distances (i.e., ca. 20 mm) because often we have little control over the geometry of broken fossil bone surfaces examined under remote field or museum conditions, and so we must be prepared to image through long Z-height positions to avoid interference between the fossil surface and the objective nosepiece. Objectives chosen include 5x and 10x lenses (34 mm and 19 mm working distances respectively; Thales-Optem Inc., Fairport, NY, USA) and Mitutoyo 20x and 50x lenses (20 mm and 13 mm working distances respectively; Mitutoyo Asia Pacific Pte Ltd, Singapore). Flexibility in magnification is achieved by both the introduction of a Thales-Optem 0.5x or 1.9x CCD adapter or by converting the fixed magnification optical assembly described above into a zoom system, which involves the introduction of a Thales-Optem 70XL zoom module (1–7x) between the K2S-BIO module coupler and the manual coarse/fine focus module. For fully automated image acquisition, we motorized the Z focus (below).

Automation in X, Y, and Z axes has been variously implemented onto the PCSOM. The 1K2 includes a motorized RS232 Z-stepping motor control setup (Thales-Optem Inc., Fairport, NY, USA) in place of the manual coarse/fine focus module when automation is desired. This setup includes an independently powered OEM (original equipment manufacturer) computer controller board connected to a stepping motor, which moves in small discrete steps, fitted to the Z-focus module and the serial port of the computer. Included software permits one to drive the focus to stored set positions between the desired ends of travel, or to incrementally drive the focus by any stipulated distance until all optical planes within the field of view have been imaged. Movement in X and Y-axes are carried out on a manual microscope stage. The 2K2 includes a KP53 motorized precision micro-stepping X-Y stage from the Semprex Corporation (Campbell, CA, USA), and a Vexta 2-phase Z-axis stepping motor (Oriental Motor USA Corp., Torrance, CA, USA). Integrated XYZ movement is performed by an Oasis 4i PCI stepper motor controller board for XY stage and Z focus. A three-axis trackball/mouse control of XYZ axes allows manual stage and focus movement to aid real-time viewing.

Portable image acquisitions are transmitted through the FireWire™ IEEE 1394 digital interface now common on notebook and desktop computers, thus eliminating the need for a framegrabber. The 1K2 uses a 4-pin IEEE 1394 high resolution 12 bit monochrome QIMAGING Retiga 1300 camera (Burnaby, BC, Canada), which has a 2/3″ monochrome progressive scan interline CCD containing 1280 × 1024 pixels. Real-time image previewing capability facilitates camera setup conditions, which are adjusted by software interface. Adjustments include integration time, gain, and offset. The 2K2 uses a JVC KY-F1030U 6-pin IEEE 1394 digital camera containing a 1/2″

color progressive scan interline CCD and 1360 × 1024 output pixels, operating at 7.5 frames per second live.

The 175W (1K2) and 300W (2K2) Lambda LS Xenon Arc Lamps (Sutter Instrument Company, Novato, CA, USA) transmit a flat and intense beam of light via a liquid light guide. It operates at wavelengths suitable for both fluorescence and white light illumination (320nm to 700nm output in an ozone-free bulb), is robustly constructed and prealigned, and is economically packaged and lightweight, housing its own power supply.

The 1K2 employs A Sony VAIO Mobile Pentium notebook PC computer for image capture. We currently use a VAIO SRX27 (800MHz; 256k RAM; Windows XP). It weighs less than 3 pounds, thus satisfying our need for maximum portability, and it contains a 4-pin IEEE 1394 interface. A Shuttle XPC SB52G2 computer with a Pentium4 Intel processor and Windows XP Professional (Shuttle Computer Group Inc., Los Angeles, CA, USA) supports fully automated XYZ stage movement and image acquisition. A reasonably lightweight and thin standard 1024 × 768 15″ monitor (Dell Inc., Round Rock, TX, USA) was chosen for our real-time viewing.

The microscope returns image detail from a very thin optical plane at and immediately below the object surface (1-50 micrometers, depending upon specimen characteristics). To obtain two- or three-dimensional projections from a surface which is anything but perfectly flat, potential fields of view must be compiled from a through-series of captured images at all optical planes represented in the Z-axis. Computerized control over image acquisition for both the 1K2 and 2K2 using Syncroscopy Auto-Montage software (Syncroscopy Inc., Frederick, MD, USA) permits an even and fully representative image of either a pseudo-planar field of view or a three-dimensional reconstruction of surface or sub-surface details. Figure 3 is a completely in-focus surface reflection image of fractured and topographically complex *Paranthropus robustus* molar enamel. Application of a coverslip and clearing medium (see below) permits this field of view to be collapsed into a 2D image of its contained enamel microanatomy (Figure 4). For extensive automated XY image montaging

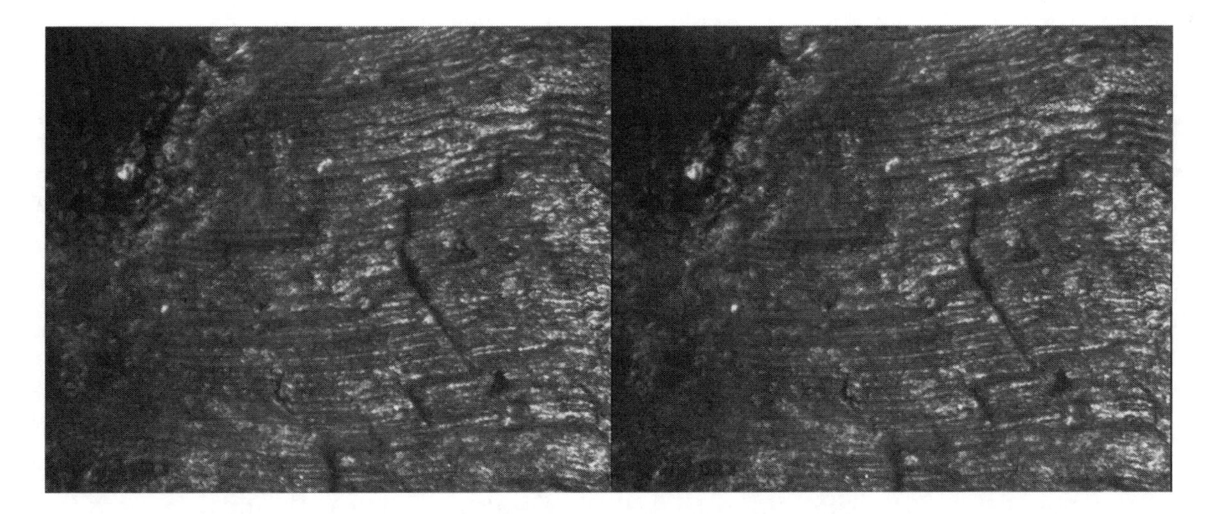

Figure 3. Fractured enamel surface of a *Paranthropus robustus* molar (SKW 4769; Transvaal Museum). A three-dimensional view of topographic relief in this surface reflection image may be obtained by mental reconstruction of left and right images into one stereoscopic image. FW = 450 μm each frame.

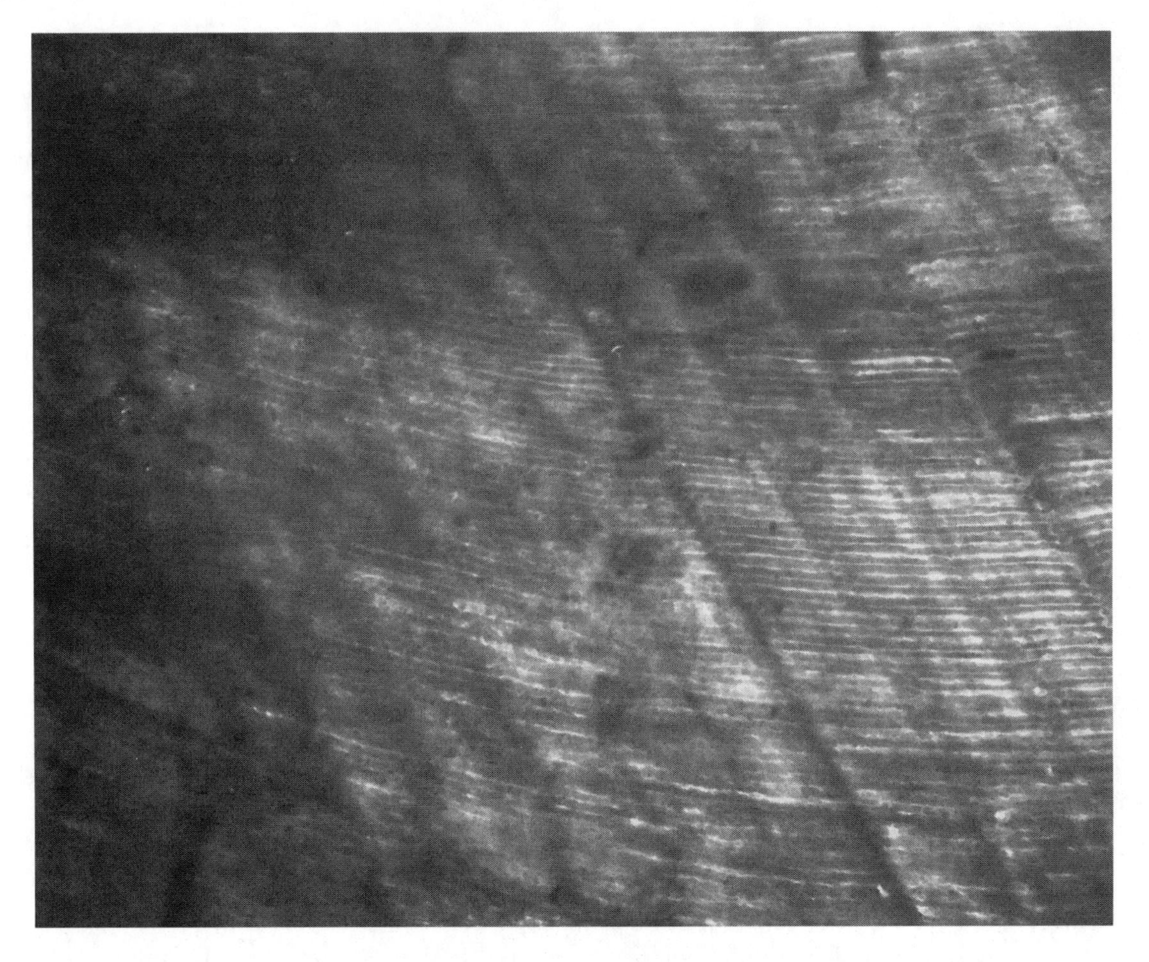

Figure 4. Same field of view as Figure 3, imaging deep to the surface and revealing incremental enamel microanatomy. FW = 450 μm.

with the 2K2, Syncroscopy Montage.Explorer (Syncroscopy Inc., Frederick, MD, USA) software is employed, which can operate in "3D mode" to acquire useful Z focal planes over fields as large as 40,000 × 40,000 pixels.

The custom stands for both the 1K2 and 2K2 are simple and lightweight. The 1K2 stand consists of three 1/2″ thick Garolite sheet grade platforms supported by four 1″ diameter, 24″ length, ceramic coated hardened precision aluminum shafts. The top platform bolts to the ends of the shafts in order to stabilize the stand. The central platform slides along the four shafts through 1″ bore Frelon-lined fixed alignment anodized aluminum linear bearings to facilitate the vertical repositioning of the the confocal module. This sliding platform is secured at any desired vertical position by a 1″ bore aluminum clamp on each shaft below the platform bearings. This platform has a forward aperture through which the K2S-BIO objective assembly passes. The bottom platform has an identical aperture through which the objective assembly can be lowered to image objects of any size permitted below the table top.

The 2K2 stand is composed of aluminum and includes an upright cylinder, containing within a lead screw operable from above, which drives the Nipkow disk module platform up or down; the drive is sensitive enough to be used as a coarse focus adjustment. The cylinder inserts into a sleeve at the base from which two hollow rectangular feet slide forward and rotate out at any angle appropriate for the balance of weight and required workspace. The platform for holding the K2S-BIO attaches to a sleeve around the cylinder, which rides on a bearing that conveys the module in any rotational position within the workspace.

Each microscope automatically switches between 110V and 220V electrical supplies (only the Nipow disk motor requires an optional 110V/220V adaptor), fits into two suitcases (Pelican Products, Inc., Torrance, CA, USA), and may be set up and tested within one hour of arrival at museum locations.

Enamel Microstructure

Naturally fractured *Australopithecus africanus* teeth from Member 4 of the Sterkfontein Formation, dated to approximately 2.5 my (Vrba, 1995) were examined. The work has only begun, and to date four naturally fractured molars, one previously sectioned molar and one canine have been imaged for this preliminary study. They include: STW 11 (RM^3), STW 90 (RM_3), STW 190 (Left maxillary molar fragment), STW 284 (LM^2), STW 37 (LM^3), and STW 267 (canine). The fractured surface of the tooth, exposing enamel in cross section, was placed approximately perpendicular to the optical axis, over which was placed a drop of immersion oil, and over this, a glass cover slip according to standard microscopal investigation. Because the fractured surfaces were not perfectly flat, images were Z-montaged in Syncroscopy Auto-Montage.

It is generally accepted that the angles formed between striae of Retzius and the enamel dentine junction (EDJ) provide useful information on the variation of differentiation rates of enamel forming cells (ameloblasts) (Boyde, 1964), which is of value to understand mechanisms of enamel development. To study striae/EDJ angles, the EDJ was divided into three equal sections along its length: cuspal, middle and cervical, following Beynon and Wood (1986) and Ramirez Rozzi (2002) (Figure 5). The angles were measured as illustrated in Schwartz et al. (2003: Figure 2A). In addition, we provide crown formation time for the molar STW 284. This specimen was selected because it had been sectioned previously (Grine and Martin, 1988) and thus there was good control over the plane of section, and because striae of Retzius were visible through the entire length of the protocone cusp.

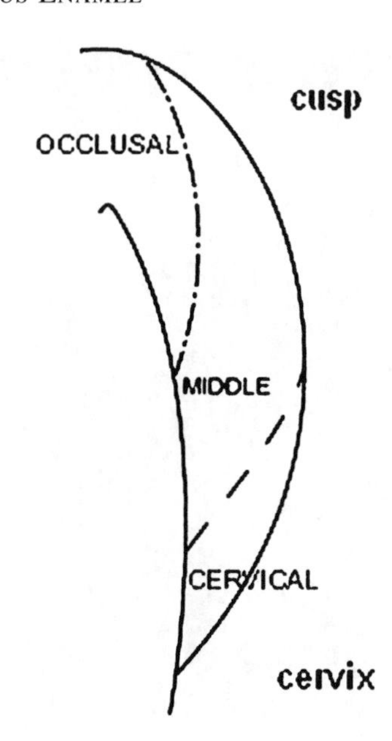

Figure 5. Diagram of the divisions of the EDJ within which striae/EDJ angles are measured.

Results

Cross striations were identified as varicosities and constrictions along a prism. Our study recorded 6 cross striations between adjacent striae of Retzius for the *A. africanus* M^3 STW 11 (Figure 6) and 6 or 7 for the M^2 STW 284 (Figure 7). It was difficult to ascertain better which number of cross striations is correct for Stw 284 because fields of view showed either cross striations or striae of Retzius, but not both, in one field. The anterior dentition represented by the single canine STW 267 was very difficult to image as most of the outer enamel surface is damaged and

Figure 6. Cross striations between striae of Retzius on outer enamel of the *Australopithecus africanus* molar (STW 11; University of the Witwatersrand). Six cross striations were counted between adjacent striae of Retzius in this specimen. FW = 130 μm.

overlaid with matrix. However, measurements of cross striation repeat intervals and distances between adjacent striae of Retzius permitted us to calculate a value of 9. Dean et al. (1993b) observed the same number of cross striations in their original study of the *Paranthropus* canine SK 63.

We also measured the angles formed between striae of Retzius and the EDJ. Table 1 shows the values of striae/EDJ angles obtained for each of the specimens considered in this study, which included the canine SK 63 (Dean et al., 1993b). As shown by Beynon and Wood (1986) and Ramirez Rozzi (2002),

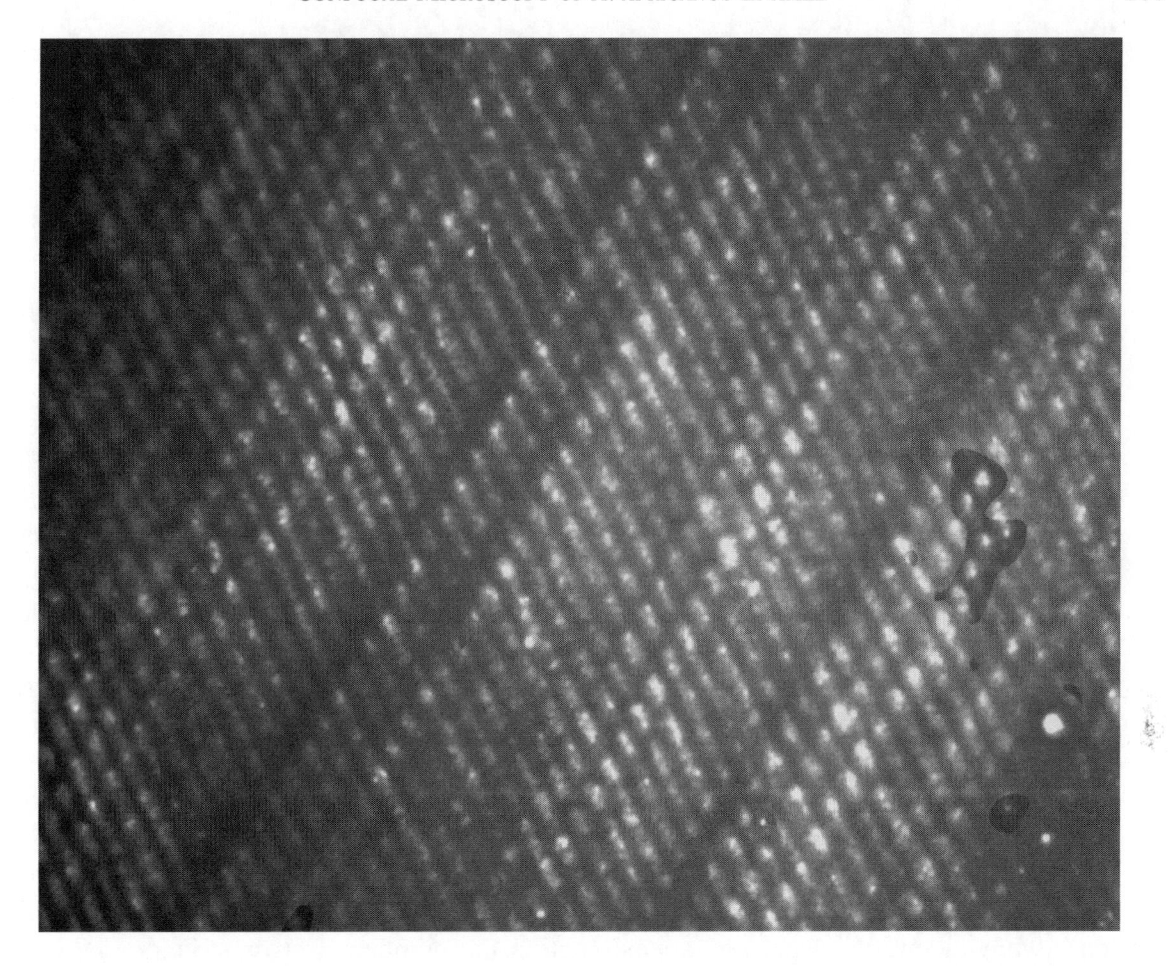

Figure 7. Cross striations on cuspal enamel of the *Australopithecus africanus molar* (STW 284; University of the Witwatersrand). Prisms course from upper left to lower right and several marked Retzius lines can be seen coursing from lower left to upper right. FW = 190 μm.

molar angles in *Australopithecus* are more obtuse compared to *Paranthropus*. For the canine the values increase in both genera from the cusp to the cervix, as previously noted for other taxa (Beynon et al., 1991; Macho and Wood, 1995). However, the change observed in the *Australopithecus* specimen is of a lesser magnitude than reported for *Paranthropus*.

For molars, *A. africanus* striae/EDJ angles may be compared with those of east African *Paranthropus* specimens investigated by Ramirez Rozzi (2002). Table 2 indicates the results of the non-parametric Mann-Whitney test between each section along the EDJ in molars of each genus. The differences are statistically significant in the cervical third, but not significant in the middle or cuspal thirds. As in the canine (Table 1), the striae/EDJ angles in *A. africanus* molars ($n = 5$) increase from the cusp to the cervix more than it does in *Paranthropus*, which shows almost no change in the mean values from the middle section of the crown to the cervix ($n = 12$). The differences between sections are statistically significant in *Australopithecus* ($p < 0.05$).

Crown Formation Time of STW 284

The first stria of Retzius reaching the enamel surface divides the enamel crown into two

Table 1. Striae-EDJ angle values for each specimen studied at each division along the EDJ. The numbers in brackets indicate the number of angles measured for each section

Specimen number	*Tooth*	*Area*	*Mean (n)*
SK 63	**UC**	Cuspal	14.5 (4)
		Middle	31.8 (9)
		Cervical	34.6 (8)
STW 279	**UC**	Cuspal	?
		Middle	34.2 (4)
		Cervical	41.4 (5)
STW 284	**UM2**	Cuspal	20.0 (2)
		Middle	31.3 (6)
		Cervical	40.6 (5)
STW 190	**frag**	Cuspal	14.6 (3)
		Middle	28.0 (5)
		Cervical	39.3 (4)
STW 90	**Lm3**	Cuspal	21.0 (2)
		Middle	30.7 (7)
		Cervical	34.7 (7)
STW 11	**UM3**	Cuspal	16.5 (2)
		Middle	28.5(7)
		Cervical	51.4 (7)
STW 37	**UM3**	Cuspal	18.0 (2)
		Middle	35.0 (5)
		Cervical	38.0 (4)

portions, which identify cuspal (or appositional) and cervical (or imbricational) developmental periods (Beynon and Wood, 1987). To calculate crown formation time in STW 284, we multiplied the number of striae in the protocone (e.g. Figure 8) by the cross striation periodicity. In addition, cuspal enamel thickness was measured following the path of enamel rods from the point where the first lateral stria appears, to the EDJ (e.g. Figure 9). Because prisms decussate near the EDJ in Stw 284, this measurement was multiplied by the Risnes (1986) correction factor, which takes into account the fact that the prism orientation is not straight from the EDJ outward, which was then divided by the observed average cross striation repeat interval of cuspal enamel. This value was obtained by measuring many groups of three to five adjacent cross striations identified through the thickness of enamel in various fields of inner (near the EDJ), mid (central portions of enamel) and outer (near the external enamel surface) cuspal enamel. Six or 7 cross striations were identified between striae of Retzius in the upper second molar STW 284. Counts of striae on the protocone gave a total of 82. Taking into consideration the number of cross striations between striae (6 or 7), this gives a range of 492 or 594 days, or 1.34 to 1.62 years, respectively, for the formation of lateral enamel. Cuspal enamel thickness was estimated to be 2670 microns, which was then multiplied by the Risnes (1986) correction factor. The average value of daily secretion rates of cuspal enamel, which included inner, mid and outer values, was 5.6 microns. The duration of cuspal enamel was thus estimated to be 1.5 years. As cusp formation time is the sum of cuspal and lateral enamel, this gives a total of 2.8 (6 cross striations) or 3.1 (7 cross striations) years for the development of the protocone.

Table 2. Results of Mann-Whitney test of Paranthropus and A. africanus for the striae-EDJ angle values at each division along the EDJ

	Sample	*Mean*	*SD*	*p value*
Cuspal				
EA *Paranthropus*	12	13.2	5.2	
A. africanus	5	18.2	2.6	**N.S.**
Middle				
EA *Paranthropus*	12	26.7	6.9	
A. africanus	5	33.9	4.5	**N.S.**
Cervical				
EA *Paranthropus*	12	26.0	6.5	
A. africanus	5	42.0	6.0	$p < \mathbf{0.05}$

The values shown for E.A. *Paranthropus* were taken from Ramirez Rozzi (2002) and include, on the lower dentition, six M3, a possible M2 or M3, and two M1 or M2. The upper dentition consists of one M2, one M3, and a possible M2 or M3 (Ramirez Rozzi 2002: Table 15.2)

As noted before (Ramirez Rozzi, 1993), using counts of Striae or perikymata on anterior cusps alone to determine crown formation time can underestimate the total period of formation as posterior cusps complete their formation with some delay

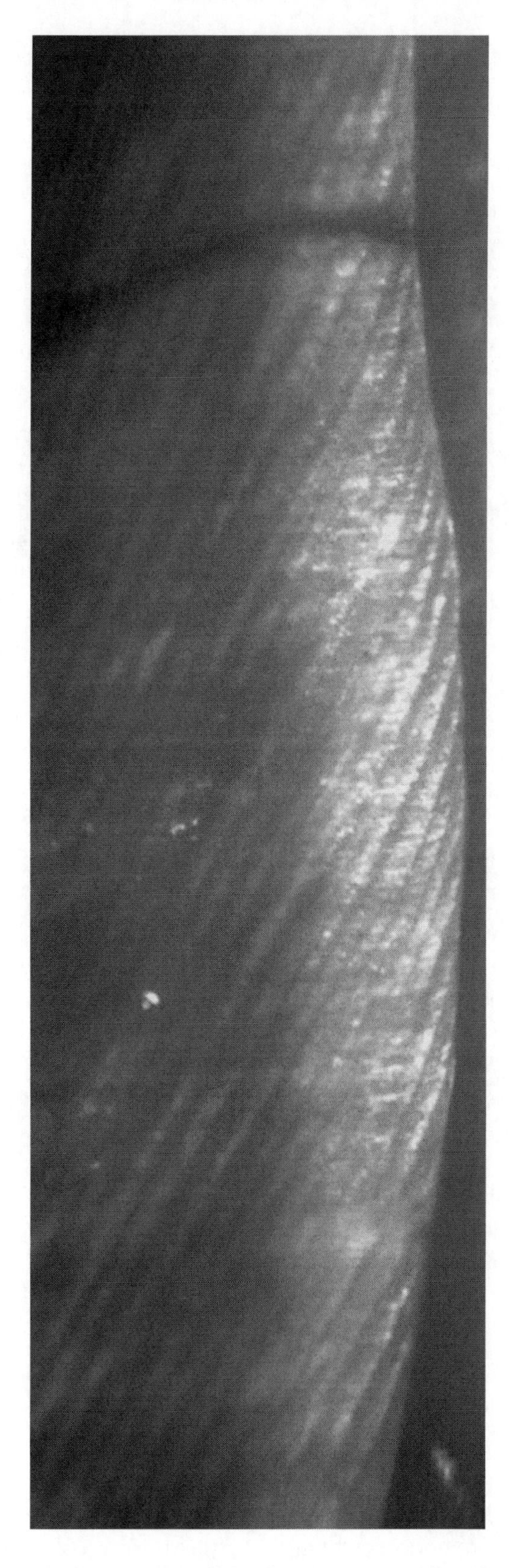

Figure 8. Striae of Retzius can be seen reaching the outer enamel surface of the *Australopithecus africanus* molar (STW 284; University of the Witwatersrand). FW = 504 μm.

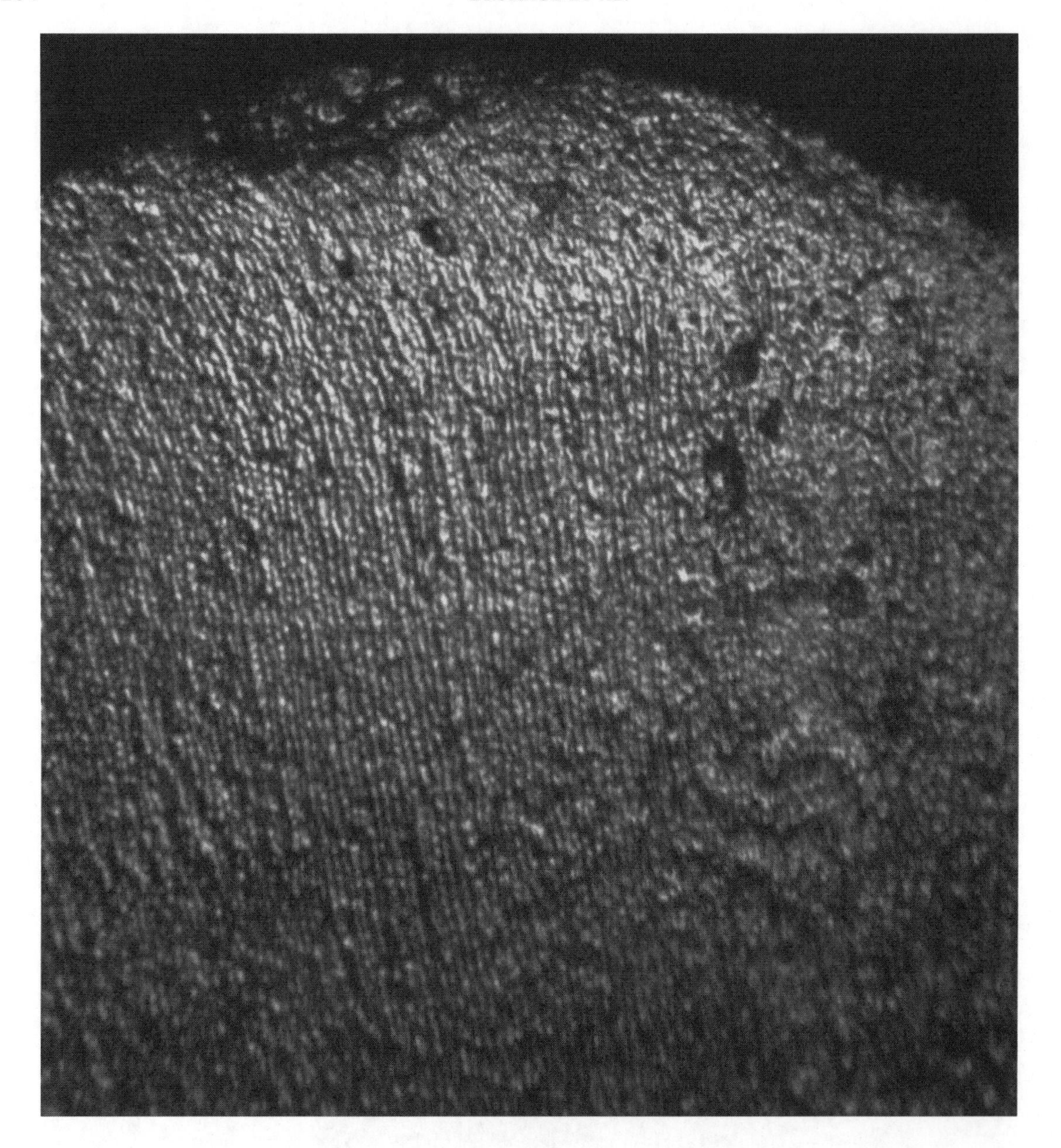

Figure 9. Cuspal enamel of *Australopithecus africanus* molar (STW 284; University of the Witwatersrand). Prisms can be identified running almost vertically towards the outer enamel surface (top). Cross striations are seen along each prism as dark horizontal lines. FW = 1.3 mm. The boundary between lateral and cuspal enamel is located slightly more cervically and could not be imaged here.

relative to the anterior cusps (e.g., Kraus and Jordan, 1965). The last visible stria on the protocone was followed to its corresponding perikyma and this was followed to the hypocone (e.g., Figure 10). The perikymata cervical to it on this cusp were counted giving a total of 12 perikymata, or an additional 0.2 years of growth. This gives a total of 3.0 (6 cross striations) or 3.2 (7 cross striations) years for the crown development of STW 284.

Discussion

While the improvement over conventional light microscopy in imaging thin sections may not be substantial, the improvement made by the

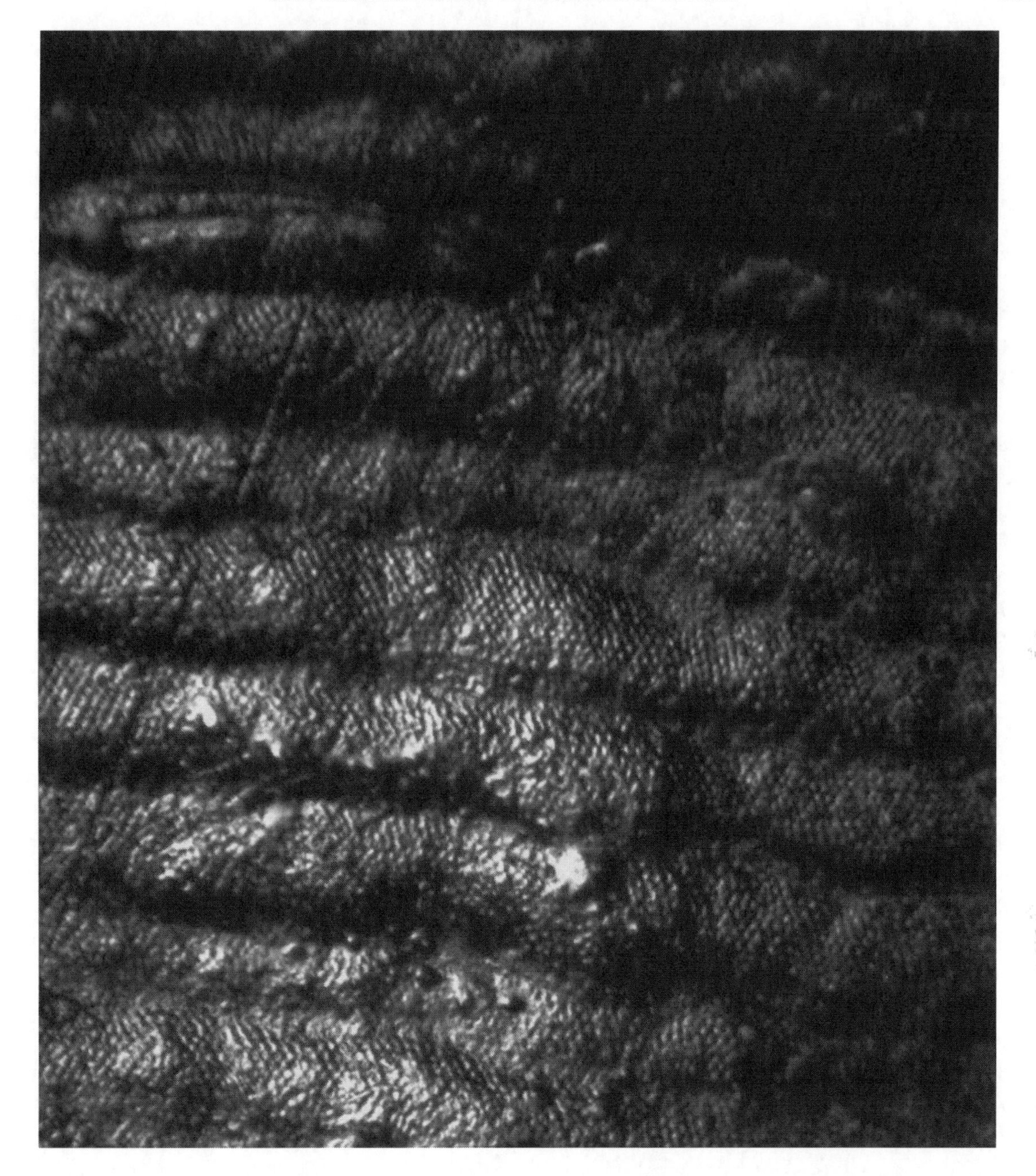

Figure 10. Perikymata on the cervical enamel of *Australopithecus africanus* (Stw 284). Tomes process pits may be observed as specs between perikymata. FW = 325 μm.

Portable Confocal Microscope for the examination of the surface layers of bulk samples non-destructively is nothing short of revolutionary. Even if images cannot be obtained through a great depth, the convenience factor of not having to produce a thin section as a prerequisite for excellent optical microscopy is a very great advantage in our research.

Two PCSOM microscopes are in service to date. The first (1K2) was described by Bromage et al. (2003); it is automated in Z and operates a notebook-based PC monochrome image acquisition system. The work reported here was performed with this system. This microscope is dedicated to specific long-term projects (e.g. dissertations). The other microscope (2K2) is fully automated in X, Y, and Z. With development of the PCSOM the potential for non-destructive mineralized tissue research on rare and unique early hominid remains is great.

Enamel development (amelogenesis) is a function of the number of cells involved in matrix secretion, secretion rates, and the rate at which these cells become differentiated along the enamel-dentine junction (EDJ). All active cells during amelogenesis periodically stop normal secretory activity, creating features known as striae of Retzius (cf. Boyde, 1990). Their periodicity can be calculated by recording the number of daily cell secretions or cross striations between each stria (cf. Bromage 1991).

Originally, Boyde (1964) proposed a method to estimate the rate of cell differentiation based on the angles formed between the striae of Retzius and the EDJ. More acute angles indicate a higher ameloblast differentiation rate. A later study (Beynon and Wood, 1986) made use of this method in an analysis of isolated teeth attributed to *Paranthropus* and *Homo* to assess differences between these two genera. They found that the angles formed between the EDJ and the striae were more acute in *Paranthropus* than in *Homo*, with means of 23 and 31 degrees respectively. Their measurements were taken on the occlusal third of the crown. Ramirez Rozzi (1993, 1998, 2002) used a larger sample derived from the Omo Shungura Formation, Ethiopia, to assess possible temporal changes in the rates of ameloblast differentiation in isolated teeth from a well stratified and dated chronological sequence. Measurements were taken on three sections along the EDJ (cuspal, central – equivalent to our "middle" region – and cervical areas). In agreement with Beynon and Wood (1986), Ramirez Rozzi found fast rates of enamel differentiation in the genus *Paranthropus*, though he noted differences between two sets identified within the Omo sample, of *P. aethiopicus* and *P. boisei*. The only published record of angles formed by striae of Retzius and the EDJ in *A. africanus* is that of Grine and Martin (1988) and, although no measurements were given, they observed that *Paranthropus* showed more acute angles than *A. africanus*. Thus, at present, there is almost no data on the microanatomical features of this species.

An important aspect in studies of dental development using microanatomy is the cross striation periodicity. Most commonly, this value is assessed from histological ground sections or by scanning electron microscopy, but both methods are intrusive and laborious. Thus only four studies to date have included information regarding cross striation periodicity on hominid fossils; a *P. boisei* premolar (Beynon and Dean, 1987), a molar of *P. boisei* (Dean 1987), a *P. robustus* canine (Dean et al., 1993b), and a Neandertal molar (Dean et al., 2001). The periodicities recorded in these samples range from 7 to 9 cross striations. Here we report a relatively significant sample of cross striation periodicities for a single hominid species.

Values of cross striation periodicity observed in the small sample of teeth attributed to *A. africanus* are highly variable. For the two molars observed the numbers ranged from 6 to 7. The anterior dentition, represented here by a single canine (STW 267), presented a calculated value of nine cross striations which is the same number observed by Dean et al. (1993b) in the *Paranthropus* canine SK 63. This variation falls within the cross striation periodicity values recorded for modern humans (6–12) (Dean and Reid, 2001) and is similar to chimpanzees (6–8) (Reid et al., 1998b; Smith, 2004) and other hominids.

Results from this preliminary study indicate that there may be some differences in growth mechanisms of enamel tissue between *A. africanus* and *Paranthropus*. In general, rates of ameloblast differentiation in *A. africanus*, measured as the angles formed between striae of Retzius and the EDJ, decrease as the development of the crown approaches the more cervical aspects of the tooth; that is, the angles have higher values in *A. africanus* than in *Paranthropus* (Table 1). It could be argued that some of these differences are the result of studying naturally fractured teeth where there is no control of the plane of section. However, the fact that all *A. africanus* molars studied show the same pattern of difference from East African *Paranthropus* molars, suggests that the plane of fracture does not significantly affect the results.

The crown formation time of a single molar of *A. africanus* was estimated to be 3.0 to 3.2. years. The former value is similar to the mean of crown formation time of molars attributed to *P. boisei* and greater than values of *P. aethiopicus* (Ramirez Rozzi, 1993). However, the crown formation time of STW 284 is slightly less than reported for modern human second molars (Beynon and Wood, 1987; Dean et al., 1993a; Reid et al., 1998a) in spite of the fact that *A. africanus* molars have thicker enamel and greater occlusal area. All of this taken together emphasizes differences already noted between extant and extinct taxa on the one hand, and between different hominid species on the other (Beynon and Wood, 1987; Beynon and Dean, 1988; Bromage and Dean, 1985; Dean et al., 2001).

Conclusions

The Portable Confocal Scanning Optical Microscope was specifically developed to offer superb analytical light microscopy of early hominid skeletal material. Limitations over the handling and transport of rare fossils have motivated its development so that specimens may be examined by whatever the circumstances dictate.

This study has added new information on the growth processes of enamel identified in the southern African hominid taxa *A. africanus*. Given the results obtained here, it would be important to assess growth processes for the South African taxon *P. robustus*, for which almost no information on molar development is available, to possibly help better establish relationships among early African hominids.

Acknowledgments

Support for this work was generously provided by the L.S.B. Leakey Foundation, the Blanquer and March Foundations (Spain), the Palaeoanthropology Scientific Trust (PAST, South Africa) and Dr. D. McSherry. For the availability of hominid specimens and assistance, the Department of Palaeontology, Transvaal Museum, Pretoria, South Africa, and the Palaeoanthropology Research Unit, Department of Anatomy, University of the Witwatersrand, Johannesburg, South Africa, are gratefully acknowledged. Much appreciation to Shara Bailey and Jean-Jacques Hublin for organizing the symposium, "Dental perspectives on human evolution: State of the art research in dental paleoanthropology" and to Shara Bailey and anonymous reviewers for critical comments on the manuscript.

References

Berger, L., Lacruz, R.S., de Ruiter, D.J., 2002. Revised age estimates of *Australopithecus* bearing deposits at Sterkfontein, South Africa. American Journal of Physical Anthropology 119, 192–197.

Beynon, A.D., Dean, M.C., 1987. Crown formation time of a fossil hominid premolar tooth. Archives of Oral Biology 32, 773–780.

Beynon, A.D., Dean, M.C., 1988. Distinct dental development patterns in early fossil hominids. Nature 335, 509–514.

Beynon, A.D., Dean, M.C., Reid, D.J., 1991. A histological study on the chronology of the developing dentition of gorilla and orangutan. American Journal of Physical Anthropology 86, 295–309.

Beynon, A.D., Wood, B., 1986. Variations in enamel thickness and structure in East African hominids. American Journal of Physical Anthropology 70, 177–193.

Beynon, A.D., Wood, B., 1987. Patterns and rates of enamel growth on the molar teeth of early hominids. Nature 326, 493–496.

Boyde, A., 1964. The structure and development of mammalian enamel. Ph.D. Dissertation, University of London.

Boyde, A., 1990. Developmental interpretations of dental microstructure. In: Jean de Rousseau, C. (Ed.), Primate Life History and Evolution. Wiley-Liss Publ., New York, pp. 229–267.

Boyde, A., 1995. Confocal optical microscopy. In: Wootton, R., Springall, D.R., Polak, J.M. (Eds.), Image Analysis in Histology: Conventional and Confocal Microscopy. Cambridge University Press, Cambridge, UK, pp. 151–196.

Boyde, A., Petran, M., Hadravsky, M., 1983. Tandem scanning reflected light microscopy of internal features in whole bone and tooth samples. Journal of Microscopy 132, 1–7.

Bromage, T.G., 1991. Enamel incremental periodicity in the pig-tailed macaque: a polychrome fluorescent labelling study of dental hard tissues. American Journal of Physical Anthropology 86, 205–214.

Bromage, T.G., Dean, M.C., 1985. Re-evaluation of the age at death of immature fossil hominids. Nature 317, 525–527.

Bromage, T.G., Perez-Ochoa, A., Boyde, A., 2003. The portable confocal microscope: scanning optical microscopy anywhere. In: Méndez Vilas, A. (Ed.), Science, Technology and Education of Microscopy: An Overview. Formatex, Badajoz, Spain, pp. 742–752.

Bromage, T.G., Perez-Ochoa, A., Boyde, A., 2005. Portable confocal microscope reveals fossil hominid microstructure. Microscopic Analysis 19, 5–7.

Dean, M.C., 1987. Growth layers and incremental markings in hard tissues, a review of the literature and some preliminary observations about enamel structure of *Paranthropus boisei*. Journal of Human Evolution 16, 157–172.

Dean, M.C., Beynon, A.D., Reid, D.J., Whittaker, D.K., 1993a. A longitudinal study of tooth growth in a single individual based on long and short period markings in dentine and enamel. International Journal of Osteoarchaeology 3, 249–264.

Dean, M.C., Beynon, A.D., Thackeray, J.F., Macho, G.A., 1993b. Histological reconstruction of dental development and age at death of a juvenile *Paranthropus robustus* specimen, SK 63, from Swartkrans, South Africa. American Journal of Physical Anthropology 91, 401–419.

Dean, M.C., Leakey, M., Reid, D., Schrenk, F., Schwartz, G., Stringer, C., Walker, A., 2001. Growth processes in teeth distinguish modern humans from *Homo erectus* and earlier hominins. Nature 44, 628–631.

Dean, M.C., Reid, D.J., 2001. Perikymata and distribution on Hominid anterior teeth. American Journal of Physical Anthropology 116, 209–215.

Grine, F.E., Martin, L.B., 1988. Enamel thickness and development in *Australopithecus* and *Paranthropus*. In: Grine, F.E. (Ed.), The Evolutionary History of the "Robust" Australopithecines. Aldine de Gruyter, New York, pp. 3–42.

Kino, G.S., 1995. Intermediate optics in Nipkow disk microscopes. In: Pawley, J.B. (Ed.), Handbook of Biological Confocal Microscopy. Plenum Press, New York, pp. 155–165.

Kraus, B.S., Jordan, R.E., 1965. The human dentition before birth. Lea & Febiger Publ., Philadelphia.

Macho, G.A., Wood, B.A., 1995. The role of time and timing in hominid dental evolution. Evolutionary Anthropology 4, 17–31.

Nipkow, P., 1884. Elektrisches teleskop. Patentschrift 30105 (Kaiserliches Patentamt, Berlin), patented 06.01.1884.

Petran, M., Hadravsky, M., 1966. Method and arrangement for improving the resolving power and contrast. United States Patent No. 3,517,980, priority 05.12.1966, patented 30.06.1970 US.

Ramirez Rozzi, F., 1993. Tooth development in East African *Paranthropus*. Journal of Human Evolution 24, 429–454.

Ramirez Rozzi, F., 1998. Can enamel microstructure be used to establish the presence of different species of Plio-Pleistocene hominids from Omo, Ethiopia? Journal of Human Evolution 35, 543–576.

Ramirez Rozzi, F., 2002. Enamel microstructure in hominids: New characteristics for a new paradigm. In: Minugh-Purvis, N., McNamara, K.J. (Eds.), Human Evolution Through Developmental Change. Johns Hopkins University Press, Baltimore, pp. 319–348.

Reid, D.J., Beynon, A.D., Ramirez Rozzi, F.V., 1998a. Histological reconstruction of dental development in four individuals from a Medieval site in Picardie, France. Journal of Human Evolution 35, 463–478.

Reid, D.J., Schwartz, G.T., Dean, M.C., Chandrasekera, M.S., 1998b. A histological reconstruction of dental development in the common chimpanzee. Journal of Human Evolution 35, 427–448.

Risnes, S., 1986. Enamel apposition rate and the prism periodicity in human teeth. Scandinavian Journal of Dental Research 94, 394–404.

Schwartz, G.T., Liu, W., Zheng, L. (2003). Preliminary investigation of dental microstructure in the Yuanmou hominoid (*Lufengpithecus hudienensis*), Yumn Province, China. Journal of Human Evolution 44, 189–202.

Smith, T.M., 2004. Incremental development of primate dental enamel. Ph.D. Dissertation, State University of New York, Stony Brook.

Tobias, P.V., 1980. *Australopithecus afarensis* and *A. africanus*: critique and an alternative hypothesis. Palaeontologica Africa 23, 1–17.

Vrba, E.S., 1995. The fossil record of African antelopes (Mammalia, Bovidae) in relation to human evolution and paleoclimate. In: Vrba, E.S. (Ed.), Paleoclimate and Evolution, With Emphasis on Human Origins. Yale University Press, New Haven, pp. 385–424.

White, T.D., Johanson, D.C., Kimbel, W.H., 1981. *Australopithecus africanus*: its phyletic position reconsidered. South African Journal of Science 77, 445–471.

6. Imbricational enamel formation in Neandertals and recent modern humans

D. GUATELLI-STEINBERG
Department of Anthropology
Department of Evolution, Ecology, and Organismal Biology
The Ohio State University
Columbus, OH 43210, USA
guatelli-steinbe.1@osu.edu

D.J. REID
Department of Oral Biology
School of Dental Sciences
Newcastle University, Framlington Place
Newcastle-upon-Tyne, UK
D.J.Reid@newcastle.ac.uk

T.A. BISHOP
Department of Statistics
The Ohio State University
Columbus, OH 43210, USA
tab@stat.ohio-state.edu

C. SPENCER LARSEN
Department of Anthropology
Department of Evolution, Ecology, and Organismal Biology
The Ohio State University
Columbus, OH 43210, USA
larsen.53@osu.edu

Keywords: perikymata, enamel, Neandertals, growth

Abstract

Aspects of imbricational enamel growth are important for two reasons. First, they may be species-typical, providing insight into taxonomic questions. Second, because dental and somatic growth are linked, aspects of imbricational enamel growth may also provide insights into species-typical rates of growth and development. The present study investigates aspects of imbricational enamel formation in Neandertal anterior teeth relative to three modern human population samples from diverse regions (Point Hope, Alaska; Newcastle-upon-Tyne, England; Southern Africa). A recent study by the same authors (Guatelli-Steinberg et al., 2005) focuses on evaluating how different Neandertals were from these modern human populations in the number of

S.E. Bailey and J.-J. Hublin (Eds.), Dental Perspectives on Human Evolution, 211–227.

perikymata on their anterior teeth and in their imbricational enamel formation times. The present study integrates the results and conclusions of that study with research on imbricational enamel growth curves across deciles within anterior teeth and the pattern of imbricational enamel growth across anterior tooth types. The central findings of the present study are: (1) Neandertal anterior teeth overlap with those of the modern human samples in mean perikymata numbers and estimates of imbricational enamel formation times; (2) the modern human population samples show greater similarity to each other than any of them do to Neandertals in their enamel growth curves across deciles within tooth types; and (3) Neandertals exhibit a pattern in the mean number of perikymata across anterior tooth types that appears to diverge from that of the modern human samples.

Introduction

The long childhood growth period of modern humans has been interpreted as an adaptation that provides time for brain growth (Sacher, 1975; Martin, 1983; Leigh, 2001; Crews and Gerber, 2003) and/or extensive learning (Mann, 1972; Gould, 1977; Bogin, 1997; Leigh and Park, 1998). The complex behaviors made possible by large brains and protracted growth provide selective advantages that accrue over the extended human lifespan (Smith and Tompkins, 1995). Thus, a reduction in adult mortality rates may have been a pre- or co-requisite for the evolution of long childhoods in humans (Trinkaus and Tompkins, 1990; Kelley, 2002; Crews and Gerber, 2003). While the evolutionary conditions and causes of human life history characteristics continue to be debated (Trinkaus and Tompkins, 1990; Bogin, 1997; Leigh and Park, 1998; Leigh, 2001; Crews and Gerber, 2003), studies of dental development in fossil hominins are providing insight into when prolonged growth periods emerged during human evolutionary history.

Across the primate order, dental and somatic development are closely linked because teeth develop as part of growing organisms. Weaning, for example, cannot take place until teeth erupt, and molars can only erupt when the jaw has grown large enough to accommodate them (Smith, 1989; Smith, 1991). Aspects of dental development (such as M1 eruption) are highly correlated with the length of growth periods and brain size (Smith, 1989; Smith, 1991). Relative to other primates, modern humans erupt their first molars later, have larger brains, and experience extended periods of somatic growth (Smith, 1989; Smith, 1991; Smith and Tompkins, 1995). Given the relationships between somatic development, dental development and brain size across the primate order, it is not surprising that small-brained Plio-Pleistocene australopiths may have erupted their first molars two-and-a-half to three years earlier than do modern humans (Bromage and Dean, 1985) and formed their anterior tooth crowns in significantly shorter periods of time (Dean et al., 2001).

Investigations into Neandertal growth and development, however, have not yet clearly answered the question of whether Neandertals experienced periods of growth comparable to the long childhoods of modern humans. With their large cranial capacities, Neandertals might be expected to have taken a long time to grow; on the other hand, if Neandertals had short life spans, they would be expected to have grown up quickly relative to modern humans (Trinkaus and Tompkins, 1990). To date, the dental evidence bearing on this question remains equivocal. It has been suggested that Neandertal molar eruption times might have been accelerated with respect to those of modern humans (Wolpoff, 1979; Dean et al., 1986), although Neandertals and modern humans appear to share similar dental eruption patterns (Tompkins, 1996). Anterior crown formation times in the Krapina Neandertals were found to be within the modern human range (Mann et al., 1991), but crown formation times in the teeth of a Zafarraya Neandertal were found to be abbreviated (Ramirez-Rozzi, 1993).

Most recently, Ramirez-Rozzi and Bermúdez de Castro (2004) provided evidence that, relative to the anterior teeth of Upper Paleolithic-Mesolithic *Homo sapiens*, those of Neandertals grew in 15% less time and exhibited enamel growth curves indicative of more rapid rates of enamel extension. The authors argued that these two characteristics were Neandertal autapomorphies indicating that Neandertals had abbreviated periods of somatic growth and were a separate species from *Homo sapiens*. Clearly however, to determine if Neandertals did indeed differ from *Homo sapiens* in these aspects of enamel growth, it is first necessary to determine the range of variation in enamel growth that exists among modern human populations. We recently published a paper (Guatelli-Steinberg et al., 2005) focusing on the duration of anterior tooth growth in Neandertals relative to three modern human population samples. In the present paper, we incorporate these results for the purpose of integrating them with additional data addressing imbricational enamel growth curves across deciles within anterior teeth and the pattern of imbricational enamel growth across anterior tooth types. Thus, this paper brings together data bearing on two related questions. Do Neandertals exhibit anterior tooth imbricational enamel formation spans that are significantly shorter than those of modern humans? Do Neandertals exhibit unique patterns of anterior tooth imbricational enamel growth with respect to modern humans?

Background: Enamel Growth

Assessment of enamel formation times in fossil teeth is usually limited to counting incremental structures, called perikymata, on the enamel surface (e.g., Bromage and Dean, 1985; Mann et al., 1991; Ramirez Rozzi, 1993; Dean and Reid, 2001 a,b, Dean et al., 2001; Ramirez Rozzi and Bermúdez de Castro, 2004). As shown in Figure 1, perikymata are external outcroppings of striae of Retzius, which are internal enamel growth layers that appear as dark lines in transmitted light microscopy of thin sections (Hillson, 1996; Aiello and Dean, 2002). The exact growth period of each stria, its periodicity, can be determined by counting the daily growth increments, or cross-striations, that lie between them (Bromage, 1991; Hillson, 1996; Fitzgerald, 1998; Aiello and Dean, 2002). Striae outcrop as perikymata only in the imbricational enamel, enamel covering the sides rather than cusps of teeth (Hillson, 1996; Aiello and Dean, 2002). This fact limits

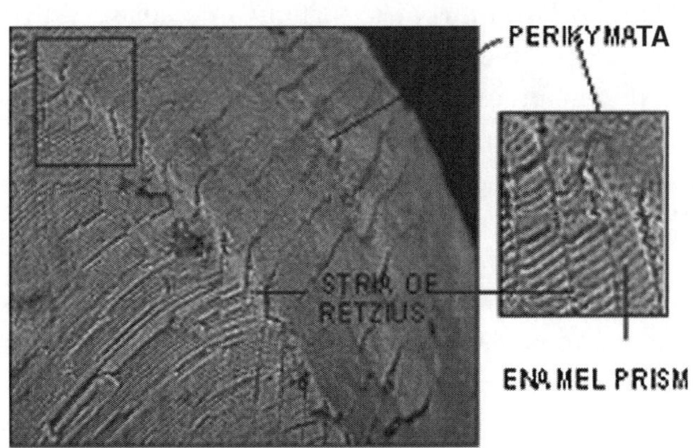

Figure 1. Relationship between striae of Retzius and perikymata. Cross striations (not shown) would appear at higher magnifications as varicosities and constrictions along the enamel prisms. Images courtesy of Tanya Smith and Jay Kelley (Kelley and Smith, 2003).

the assessment of crown formation times from fossil tooth surfaces to the imbricational enamel only, which fortunately constitutes the majority of the enamel surface in the anterior teeth (Hillson, 1996; Aiello and Dean, 2002). The total length of time for imbricational enamel to form has recently been estimated by counting the total number of perikymata and multiplying by a periodicity of nine (Dean and Reid, 2001a; Dean et al., 2001; Ramirez Rozzi and Bermúdez de Castro, 2004), the mean and modal periodicity found in a sample of 184 African apes and humans (Dean and Reid, 2001a).

There are three problems with the assumption of a nine day average periodicity in all hominins. First, there is a relationship between body size and periodicity across catarrhine species, most clearly evident within the Hominoidea, such that small bodied species tend to have lower average periodicities than those with larger bodies (Smith et al., 2003). This relationship is not as clear across the primate order as a whole, as large-bodied sub-fossil lemurs had low periodicities (Schwartz et al., 2002). The relationship between body size and mean periodicity in hominoids suggests that small bodied extinct hominins may have had lower average periodicities than modern humans, but periodicities of seven and nine were noted for *Paranthropus boisei* (Dean, 1987) and *Paranthropus robustus* (Dean et al., 1993) respectively, both falling within the modern human range. In the case of late archaic Neandertals, whose average body mass exceeded that of modern humans (Ruff et al., 1997), average periodicity might have been higher than it is in modern humans. A second problem with the assumption of a nine-day periodicity is that the average and modal periodicity for a sample of 365 modern human teeth was recently found to be eight days rather than nine (Smith et al., 2007). Third and most importantly, applying any average periodicity to a sample of teeth ignores the internal variability in periodicities within a sample. It is known that modern humans have highly variable periodicities, ranging between six to twelve days (see Smith et al., 2007). Because fossil sample sizes are often small, and periodicities, at least in humans, are so variable it is not reasonable to assume that the average periodicity for a small sample is the same as that of the species as a whole. Variation of periodicities within a sample can now be incorporated into analyses of perikymata numbers by taking advantage of a recently established relationship between total striae numbers and periodicity (Reid and Ferrell, 2006) as we describe in the Materials and Methods section below.

In consideration of these problems with using a mean periodicity of nine days, we first simply compare the total number of perikymata between Neandertals and each modern human group for each tooth type. Next, we make use of the significantly negative correlation between total striae numbers and periodicity to analyze Neandertal imbricational enamel formation times with respect to modern humans.

Imbricational enamel formation times are in part related to the rates of enamel formation. In extant great apes, some fossil hominins, and humans, perikymata/striae become increasingly compressed in the cervical regions of teeth, representing a slowing of enamel extension and secretion rates as the tooth grows (Beynon and Wood, 1987; Dean et al., 2001; Dean and Reid, 2001a, b; Ramirez Rozzi and Bermúdez de Castro, 2004). The degree to which perikymata compress in the cervical regions appears to vary across hominin species (Beynon and Wood, 1987; Dean et al., 2001; Dean and Reid 2001a, b) and indeed "perikymata packing," the number of perikymata within each decile of crown height (see Ramirez Rozzi, 2007), can be plotted graphically from the first to last decile to form enamel growth curves that appear to differentiate extinct hominin species (Dean and Reid, 2001b). In part, the shorter estimated crown formation times of *Paranthropus* relative to *Australopithecus*, and of *Australopithecus* relative to modern *Homo sapiens,* appear to result from faster rates of enamel extension, particularly in the cervical regions of their teeth (Beynon and Wood, 1987; Dean and Reid, 2001b). Ramirez Rozzi and Bermúdez de Castro (2004) find perikymata to be less densely packed in the cervical regions of Neandertal teeth relative to Upper Paleolithic-Mesolithic modern humans, suggesting more rapid rates of enamel extension in the former. To determine if this pattern of enamel growth does indeed differentiate Neandertals from modern humans, we compare enamel growth curves of Neandertals to those of three modern human groups. We believe that in addition to species and/or population differences in enamel growth curves, there might also be species and/or population differences in the average number of perikymata across teeth. We investigate this possibility here as well, noting that enamel growth patterns across teeth have not previously received much attention in the literature.

Materials and Methods

Samples

Table 1 lists the Neandertal teeth and sample sizes for all samples. The Neandertal sample spans approximately 150 ka to 40 ka. Note that Tabun II is included in the Neandertal sample. Tabun II has long been viewed as Neandertal (McCown and Keith, 1939; Trinkaus, 1983, 1987, 1993; Stefan and Trinkaus, 1998; Schwartz and Tattersall, 2000) but only recently has been considered to be anatomically modern, specifically by Quam and Smith (1998), Rak (1998), and Rak et al. (2002). Stefan and Trinkaus (1998) in particular argue that dentally this specimen is "closest to the Near Eastern late archaic human lineage and the Krapina sample" (1998: 443). It is because of the dental affinities of this specimen to the Krapina sample that we include Tabun II in our Neandertal dental sample.

The sample from England derives from a single living population from Newcastle-upon-Tyne, which includes individuals of different ethnicities. The Southern African sample comprises several indigenous populations, with different ethnic backgrounds. The Alaskan sample is an archaeological one of Point Hope Inuit, spanning six culture periods: the Near Ipiutak (500–100 BC), Ipiutak (100BC–500AD), Birnirk (500–900 AD), Western Thule (900–1300 AD), Tigara (1300–1700 AD) and Recent (1700–present) (Schwartz et al., 1995).

Methods of Counting Striae/Perikymata

Only one tooth (right or left) was chosen for each individual for each tooth type. Choice of right or left teeth was made on the basis of which antimere was most complete. If both antimeres were equally complete, we alternated between choosing right or left teeth. Only teeth estimated to have 80% or more of their crown heights intact (i.e., minimally worn teeth) were included. Crown heights were measured using a reticule calibrated to a magnification of 50x. Original crown heights were reconstructed by following the contour of each side of the tooth cusp and projecting it until the sides meet. Both measured and reconstructed crown heights were recorded.

For the Neandertal and Inuit samples, high-resolution polyvinyl siloxanes (Coltene's President Jet and Struer's RepliSet) were used to make dental impressions from the buccal surfaces of anterior teeth. These were cast in high-resolution epoxy (Struer's Epofix), and were coated with a gold-palladium alloy for microscopic observation. Perikymata were counted under a light microscope, while a scanning electron microscope was used to create a micrographic record of tooth surfaces. Each replica was oriented

Table 1. Sample composition

Tooth Type	*Neandertal specimens*	*Neandertal (N)*[1]	*Inuit (N)*	*S. African (N)*	*Newcastle (N)*
UI1	Krapina 91, 93, 94, 123, 126, 155, 194, 195; Devil's Tower 1, Le Moustier 1	10	10	20	19
UI2	Krapina 122, 128, 130, 131, 148, 156, 160, 196; Le Moustier 1	9	10	21	16
UC	Krapina Maxilla E, 37, 56, 76, 102, 103, 141, 142, 144, 146, 191; Kůlna, Le Moustier 1, La Quina 5	14	9	26	39
LI1	Krapina Mandible E, 73; Ochoz, Tabun II, Le Moustier 1	5	12	20	15
LI2	Krapina Mandible C, Mandible D, Mandible E, 71, 90; Ochoz, Tabun II, Le Moustier 1	8	14	23	13
LC	Krapina Mandible D, Mandible E, 75, 119, 120, 121, 145; Ochoz, Tabun II, Le Moustier 1	10	10	24	13
Total Teeth		55	65	134	115
Total Individuals		30	17	114	115

[1] Total number of Krapina Neandertals based on designation of associated teeth as "Krapina Dental People" (Radovčić et al., 1988)

so it was perpendicular to the microscope's optical axis. The samples from Newcastle and Southern Africa are thin sections on which striae of Retzius were counted under transmitted light microscopy. Only striae clearly outcropping onto the surface as perikymata were counted, thus assuring comparability between perikymata and striae data. For the Newcastle and Southern African samples, it was possible to count cross-striations to determine the periodicity for each tooth. Hence the number of days it took to form the imbricational enamel in these teeth could be calculated directly by multiplying the periodicity by the number of striae.

Starting at the cusp tip, each tooth was divided into 10% increments (deciles) of the reconstructed crown height, and perikymata/striae were counted within the increments. This procedure allows data to be gathered on enamel growth curves from cusp to cervix. For teeth missing up to 20% of their crowns owing to wear, estimates of perikymata/striae were made for the first two deciles. The number of perikymata within the first two deciles of complete crowns used in this study have very low standard deviations, ranging from one to two perikymata for each tooth type within each population. Such low variation within the first two deciles of growth makes it possible to accurately estimate growth in slightly worn teeth based on the number of perikymata/striae in the first and second deciles of complete crowns for each tooth type and each population sample. Teeth were excluded from the study if more than one decile beyond the first two deciles contained indistinct or missing perikymata. For teeth in which a single decile contained indistinct perikymata, the number of perikymata was estimated from adjacent deciles (following Dean and Reid, 2001b). While both first and second authors counted perikymata, the second author counted both perikymata and striae. In order to eliminate interobserver error, only counts by the second author were used in the statistical analysis. Intraobserver error for the second author is less than 5% for counting both perikymata and striae (Dean et al., 2001).

Counting Perikymata and Assessing Imbricational Enamel Formation Times in Neandertals

We test the hypothesis that Neandertals were growing their teeth in shorter time periods than the modern human comparative samples in two ways. First, using the method of Ramirez Rozzi and Bermúdez de Castro (2004), we compare the means of the total perikymata numbers per tooth type of the Neandertal sample with each of the comparative samples. Under the assumption that the mean periodicity of the Neandertal sample is equivalent to that of the modern human comparative samples, Neandertals can be inferred to be forming their imbricational enamel in shorter periods of time only if they have a significantly lower mean number of perikymata than the comparative samples. For each tooth type, we conducted one-way ANOVAs of the total number of perikymata and Dunnett's simultaneous *t*-tests to determine if there were significant differences in the mean number of perikymata between Neandertals and each comparative sample. The random error terms in the statistical models describing the total perikymata numbers for a specific population and tooth were normally distributed, justifying our use of parametric tests (Rupert, 1981).

The second way we test the hypothesis of abbreviated growth in Neandertal teeth takes into account the unknown periodicities in our Neandertal sample. For each tooth type from a given modern human population, there is a significant negative correlation (r from -0.90 to -1.00) between the total number of perikymata on teeth and their periodicities (Reid and Ferrell, 2006). Recall that the time it takes for imbricational enamel to form is equal to the total number of perikymata on a tooth multiplied by its periodicity. The result of the negative correlation between periodicity and total perikymata number is that crown formation times are fairly constant for each tooth type in a population, with only small standard deviations from the mean (Reid and Ferrell, 2006). Regression equations of the following form can be used to describe this relationship for each tooth type in the Newcastle and Southern African samples: Periodicity $= \alpha + \beta$ (perikymata number), where $\alpha =$ the regression constant, and $\beta =$ the coefficient of the total perikymata number. For each population (Newcastle and Southern African) every point on the regression line for a particular tooth type represents a value for periodicity and total perikymata number which, when multiplied together, result (approximately) in that tooth type's average imbricational enamel formation time. The R^2 values for these twelve regression equations (two populations, each with six tooth types) range from 0.828 to 0.949, all with p values <0.001. They are therefore highly predictive of the relationship between the total number of perikymata and periodicity for a given tooth type, a relationship that is determined by the time it takes for a particular tooth type to form (Reid and Ferrell, 2006). We note that the residuals of these regression equations are distributed normally. Linear regression is justified under this condition (Kutner et al., 2004).

As shown in the results section, the mean numbers of perikymata for each tooth type in our Neandertal sample are higher than those of the Southern African sample; nevertheless, it could be argued that if the Neandertals in our sample had lower periodicities than Southern Africans, they might be growing their teeth in equivalent periods of time. It could further be argued that the Neandertals in our sample might be forming their teeth in shorter periods than Southern Africans, even 15% shorter, if the Neandertals in our sample had still lower periodicities.

Because we are chiefly concerned with the question of whether Neandertal imbricational enamel formation times are encompassed within the modern human range of variation, we have limited our analysis of Neandertal periodicities to the comparison of Neandertals with Southern Africans. Therefore, we use the Southern African regression equations for the six anterior teeth relating periodicity to perikymata number to determine what the periodicities in our Neandertal samples would have to be if their imbricational enamel formation times were equivalent to or 15% shorter than those of Southern Africans. We refer to these conditional periodicities as "hypothetical periodicities."

We obtain hypothetical periodicities for Neandertals under the assumption of equivalence to Southern African imbricational enamel formation times by inserting Neandertal perikymata numbers into the Southern African regression equations. We also determine what the periodicities for the Neandertals would be if their teeth grew in 15% less time than those of the Southern African sample in the following way. We first multiply the hypothetical Neandertal periodicities based on South African regression equations by the number of perikymata. This gives an estimated time in days for Neandertal imbricational enamel formation that is comparable to that of a Southern African tooth with the same number of perikymata. We then multiply this estimated enamel formation time by 0.85 and divide by the number of perikymata for each Neandertal tooth to obtain a second set of hypothetical periodicities for Neandertals.

The key to our analysis is that the known lower limit of periodicities in modern humans and African apes is six. In fact, there are only two individuals out of 184 in Dean and Reid (2001a) combined African ape and human sample exhibiting periodicities of six. In the combined Newcastle and Southern African samples used in our study, there is only one case out of 229 individuals in which the periodicity was six. Smith et al. (2007) have found no chimpanzee with periodicities of less than six. Thus we reject the hypothesis that Neandertals grew their teeth in the same period of time, or in 15% less time, than Southern Africans if the only way for these hypotheses to be true is to assume that any of our Neandertal specimens had periodicities less than six.

Imbricational Enamel Growth Curves Across Deciles Within Tooth Types

To compare growth patterns within teeth, we plotted mean perikymata/striae numbers in each decile for each tooth type for each population sample. We visually inspect these plots for differences between population samples in the shape of their growth curves.

The Pattern of Imbricational Enamel Growth Across Teeth as Evidenced by the Mean Number of Perikymata Across Tooth Types

To assess growth patterns across teeth, we used a general linear model and performed an ANOVA of total perikymata/striae per tooth in which the covariate was reconstructed crown height, and the factors were tooth type, population, and the interaction of tooth type and population. If the pattern of mean perikymata/striae numbers across anterior teeth differed across samples, we would expect the interaction of tooth type and population to be a statistically significant source of variance. We used reconstructed crown height as a covariate because in our preliminary analysis we found that within populations, crown heights are generally positively correlated with total perikymata/striae numbers. We wished to know if the pattern of mean perikymata/striae numbers across tooth types differs between Neandertals and the three modern human population samples, when potential population differences in crown height variation are statistically controlled.

Results

Assessing Imbricational Enamel Formation Times in Neandertals

All one-way ANOVAs for difference in mean perikymata/straie numbers among populations for each tooth type were statistically significant, with p values <0.001. Table 2 shows the Dunnett's simultaneous t-tests by tooth type for mean differences in perikymata/striae numbers between Neandertals and the modern human samples. The Inuit have statistically significantly higher numbers than Neandertals on UI1, UC, and LC, while the two populations do not show a statistically significant difference on UI2, LI1, and LI2. The Newcastle sample has statistically significantly higher perikymata/striae numbers than Neandertals on UI1 and LC, but significantly lower numbers on LI2, and these two populations do not show significantly different numbers on UI2, UC, and LI1. Lastly, the Southern African sample shows statistically significantly lower numbers than Neandertals on all incisors, with statistically insignificant differences on both canines.

Table 2. Dunnett's t-tests for differences in perikymata/striae number means across populations: by tooth type

UI1: *Neandertal subtracted from:*	*Difference of means*	*T-value*	*P-value*
Inuit	32.30	4.259	0.0002
Newcastle	27.18	4.103	0.0004
South African	−20.60	−3.137	0.0073
UI2: Neandertal subtracted from:			
Inuit	7.96	1.148	0.4936
Newcastle	−9.69	−1.542	0.2729
South African	−28.63	−4.765	0.0001
UC: Neandertal subtracted from:			
Inuit	25.738	3.0512	0.0082
Newcastle	10.584	1.7206	0.2007
South African	−4.121	−0.6296	0.8481
LI1: Neandertal subtracted from:			
Inuit	−0.67	−0.105	0.9986
Newcastle	2.10	0.344	0.9591
South African	−25.22	−4.286	0.0002
LI2: Neandertal subtracted from:			
Inuit	−12.36	−1.763	0.1743
Newcastle	−22.60	−3.185	0.0061
South African	−42.23	−6.461	0.0000
LC: Neandertal subtracted from:			
Inuit	38.500	4.2728	0.0002
Newcastle	40.085	4.7299	0.0001
South African	3.825	0.5044	0.9128

In some cases, the comparisons at the level of tooth type are made with small sample sizes, potentially contributing to non-significant results. However, when the differences are significant they reveal that Neandertal mean perikymata numbers are lower than the mean perikymata/striae numbers of the Inuit, higher than those of Southern Africans, and variable with respect to those of the sample from Newcastle. Thus, assuming equivalent mean and modal periodicities across the samples, as did Ramirez Rozzi and Bermúdez de Castro (2004), Neandertals would have formed their imbricational enamel more quickly than the Inuit, more slowly than Southern Africans, and either more quickly or slowly than the sample from Newcastle depending on tooth type.

Hypothetical periodicities calculated for Neandertal teeth using the regression equations relating periodicity to the total number of striae in the Southern African sample reveal distributions for two tooth types (LI1 and LI2) that include periodicities below six, the lower limit of the known range of periodicities in African apes and humans (Figure 2). Neandertal LI1 and LI2 are therefore likely to be growing over longer periods of time than the LI1 and LI2 of Southern Africans. It is possible that Neandertal UI1, UI2, UC, and LC are growing in the same time periods as the UI1, UI2, UC, and LC of Southern Africans, because their hypothetical periodicities fall in the range of six to ten days. Hypothetical periodicities under the assumption of 15% faster growth in Neandertal teeth relative to Southern Africans are also shown in Figure 2: in this scenario, all teeth (but especially incisors) include periodicities below six. It is therefore extremely unlikely that Neandertal imbricational enamel formation times were 15% shorter than those of the Southern Africans.

Imbricational Enamel Growth Curves Across Deciles Within Tooth Types

Figure 3 presents the mean number of perikymata per decile for each population sample by tooth type. It is notable that the number of perikymata in the first four deciles of growth in Neandertals is slightly higher than it is in any of the modern human populations, a pattern similar to that found by Ramirez Rozzi and Bermúdez de Castro (2004) in Neandertals vs. Upper Paleolithic-Mesolithic *Homo sapiens*. On the other hand, while these authors find Neandertals to have mean numbers of perikymata in the last three deciles of growth that are consistently lower than those of Upper Paleolithic-Mesolithic *Homo sapiens*, our data show Neandertals to have lower means in these deciles only for the canine teeth. Finally we note that the shapes of the Neandertal growth curves generally appear to be straighter (less curvilinear) than those of the modern human groups, especially for the canines and upper incisors.

Table 3. ANOVA results for total perikymata/striae numbers

	With crown height as covariate: $R^2 = 70.82\%$		
Factors	*df*	*F*	*P*
Population	3	90.59	0.000
Tooth Type	5	31.20	0.000
Population * Tooth Type	15	5.46	0.000
Crown Height	1	52.08	0.000

This difference seems to result primarily from the fact that the mean number of perikymata in the first four deciles of growth is higher for Neandertals than it is for all three of the modern human samples.

The Pattern of Imbricational Enamel Growth Across Teeth as Evidenced by the Mean Number of Perikymata Across Tooth Types

Table 3 presents the ANOVA results for total perikymata numbers (controlling for variation in crown heights) and Figure 4 shows the interaction plot. Statistically significant differences in total perikymata numbers were found among populations and among tooth types. The interaction of tooth type and population is also a significant source of variance in perikymata numbers ($F = 5.46$, $p < 0.001$). Examination of the interaction plot shows that the modern human samples are much more similar to each other in terms of the pattern of mean perikymata numbers across different tooth types than any of them are to the Neandertals. Anterior teeth in Neandertals show a more restricted range of mean perikymata numbers across tooth types than do the modern human samples.

This divergent pattern of Neandertals helps to explain why the mean number of perikymata on Neandertal teeth is higher or lower than that of Newcastle depending

Assuming 85% South African growth
Periods for Neandertals

UI1:

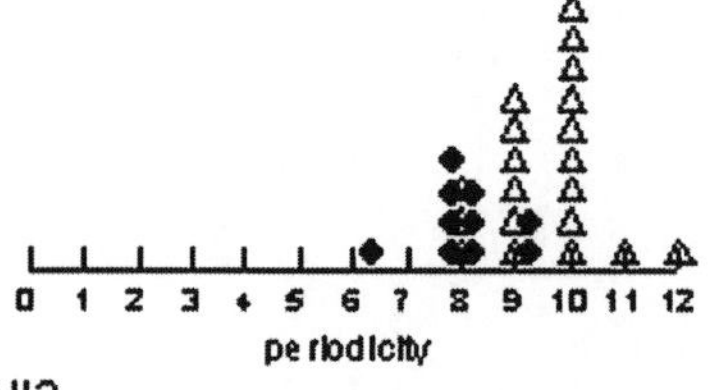

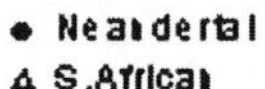

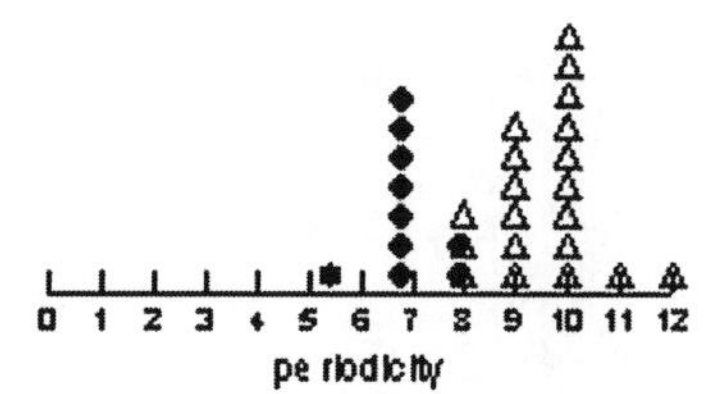

UI2

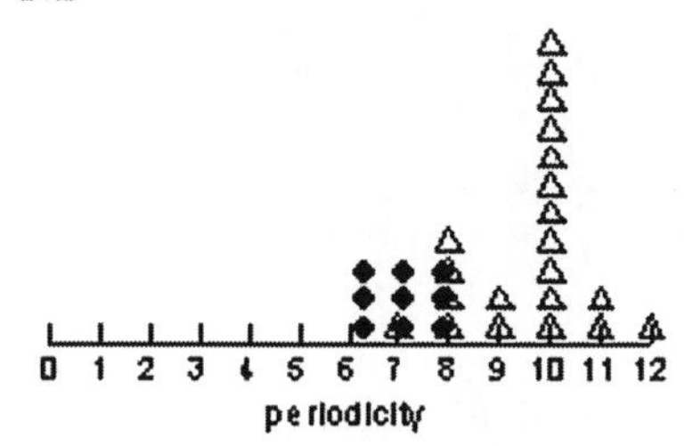

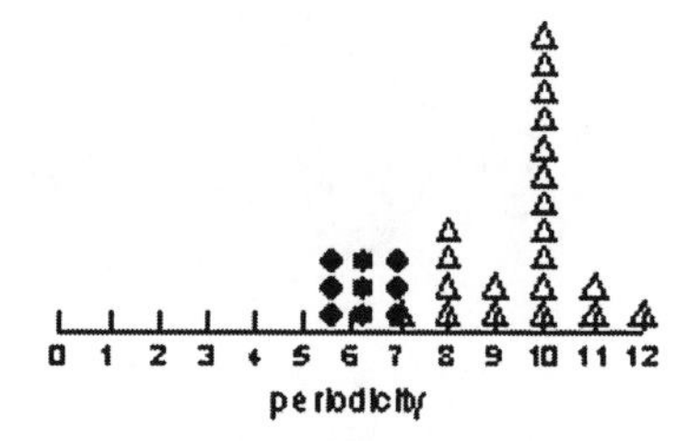

UC

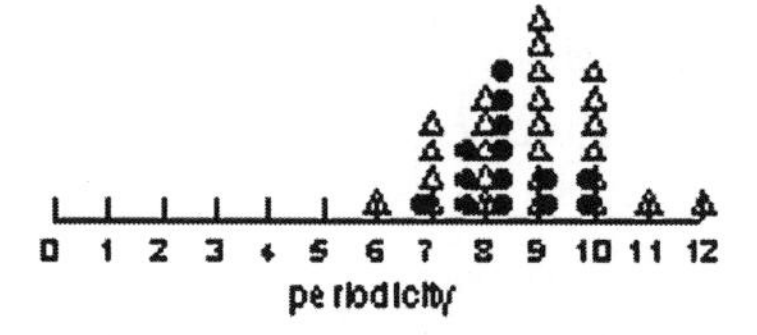

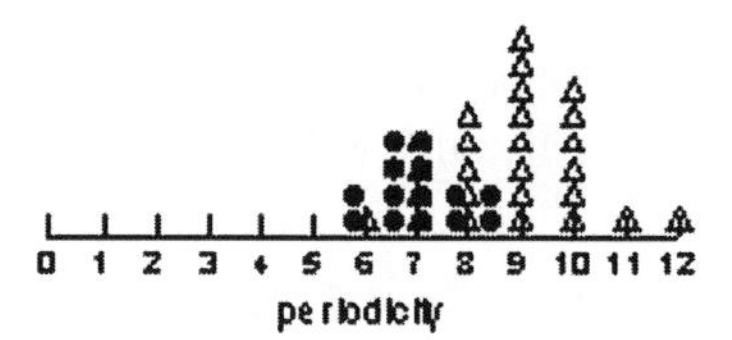

LI1

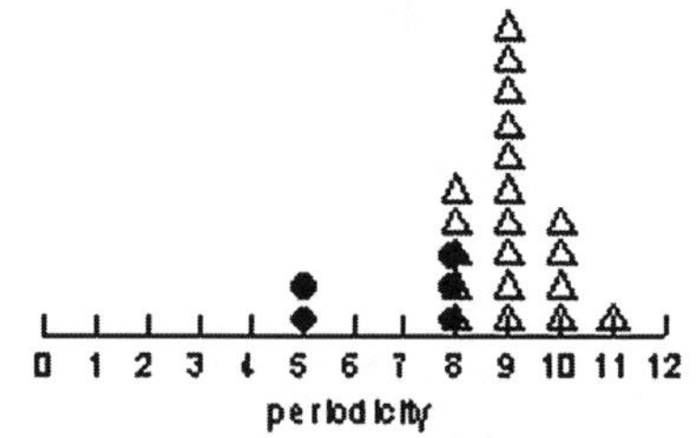

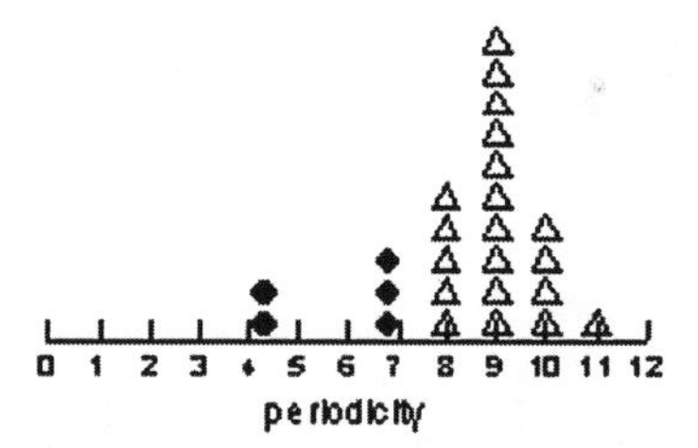

LI2

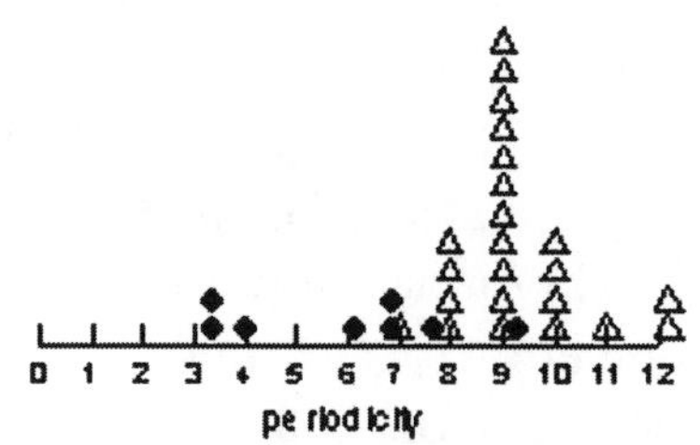

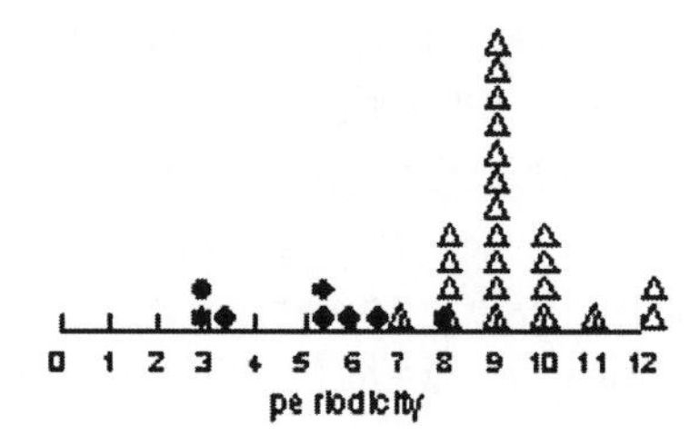

LC

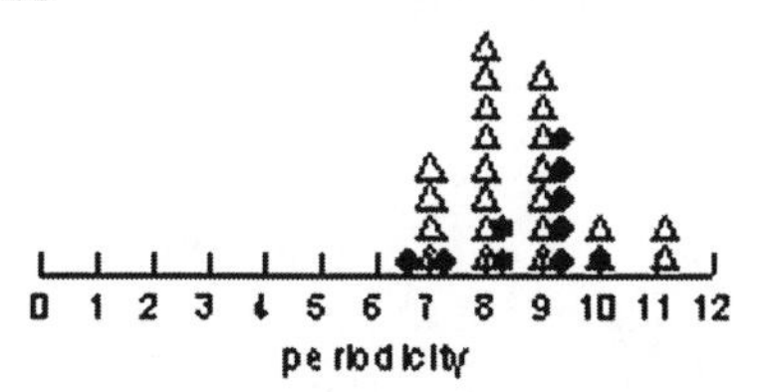

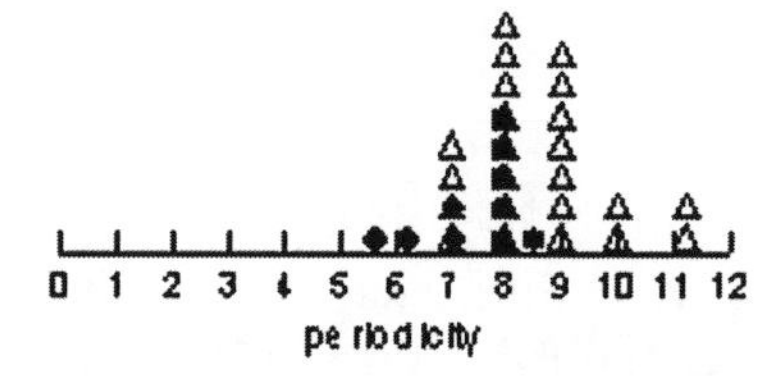

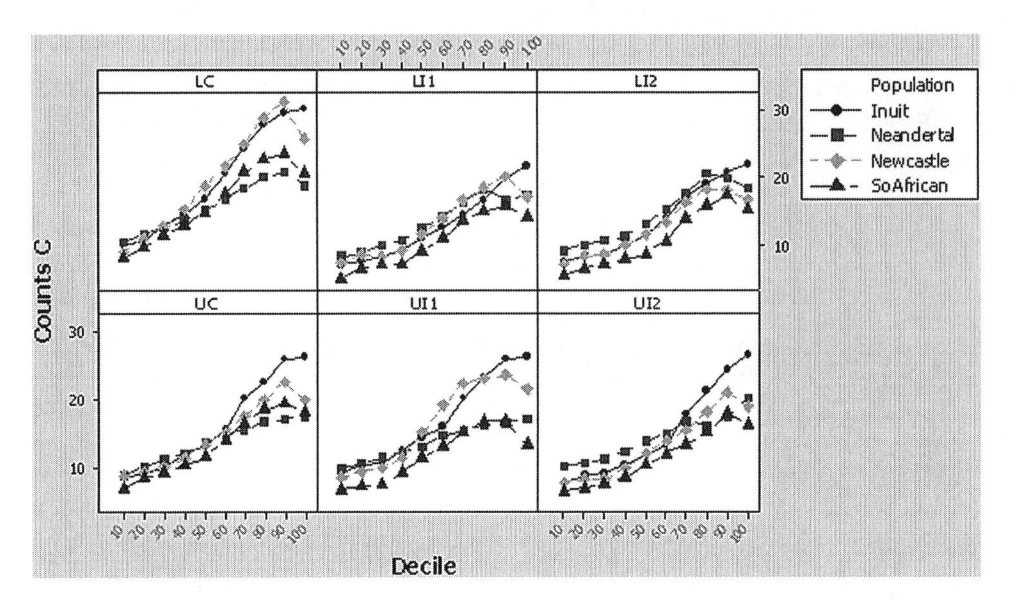

Figure 3. Growth curves for Neandertals and comparative modern human populations samples by tooth type. The mean number of perikymata/striae is shown from the first decile of growth (cusp tip) to the last (cervix).

on tooth type. In other words, the statistically significant interaction of population and tooth type indicates that the numbers of perikymata on anterior teeth do not increase or decrease uniformly for each tooth type from one population to the next. This is particularly evident when Neandertals are compared to our modern human population samples.

Discussion

This study demonstrates that Neandertal anterior tooth imbricational enamel formation times are within the range of variation that three modern human populations exhibit. Based on applying a nine day periodicity to perikymata numbers, Ramirez Rozzi and Bermúdez de Castro (2004) found Neandertals to have imbricational enamel formation spans that were 15% shorter than those of Upper Paleolithic-Mesolithic modern humans. These authors also found *H. antecessor* and *H. heidelbergensis* to have a greater mean number of perikymata than Neandertals, but less than Upper Paleolithic-Mesolithic modern humans (Ramirez Rozzi and Bermúdez de Castro, 2004). The data from the three modern human populations presented in this study put this 15% difference into context: population differences in the mean number of perikymata for each tooth type within our own species are equivalent to or greater than this. Indeed, actual imbricational enamel formation times in modern humans are quite variable for anterior teeth, as recently shown by Reid and Dean (2006), who calculated imbricational as well as cuspal enamel formation times in the Newcastle and Southern African samples used in this study. These authors found that the greatest difference in imbricational enamel

Figure 2. Hypothetical periodicity distribution for Neandertals assuming imbricational enamel formation times equivalent to those of Southern Africans (Column 1) and assuming imbricational enamel formation times **15%** shorter than those of Southern Africans (Column 2). Southern African periodicity distributions (actual values) are plotted on the same graphs for comparison.

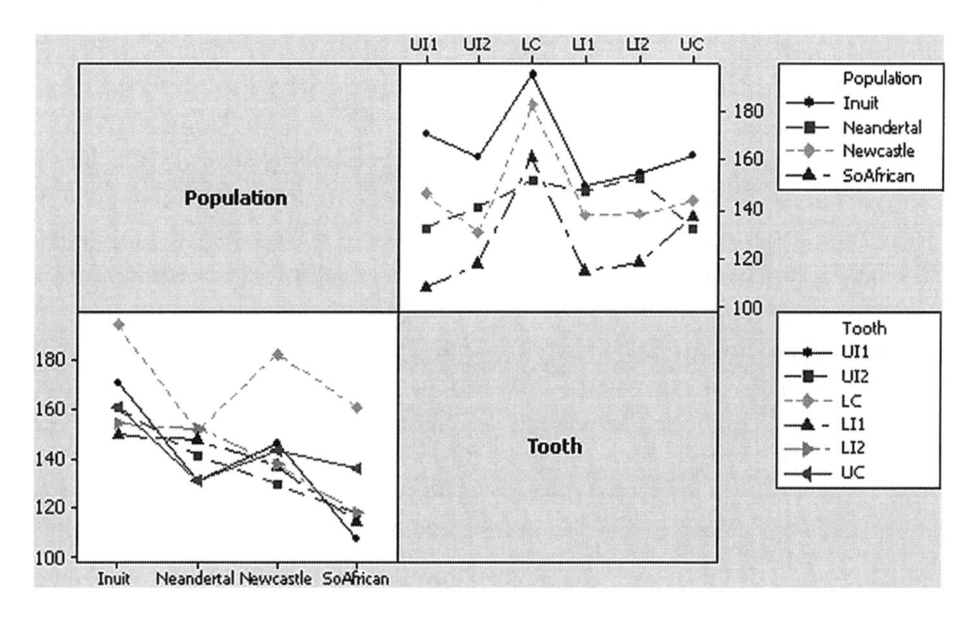

Figure 4. Plot of perikymata/striae numbers by population and tooth type, adjusting for crown height.

formation times between these two samples was for the lower canine. Imbricational enamel took, on average, 3.8 years to form in the Southern African sample versus 4.7 years in the Northern European sample (equivalent to our Newcastle sample). This is a 20% difference in lower canine imbricational enamel formation times for these two modern human populations.

If a common average periodicity is assumed across all samples in this study, then Neandertals would have been forming their teeth over a longer period of time than a sample of modern Southern Africans, during about the same period of time as a sample from Newcastle, and over a shorter time period than a sample of Inuit. When the negative correlation between periodicity and total perikymata/striae numbers is applied to perikymata numbers in Neandertals, our analysis indicates that the lower incisors of Neandertals are very likely to have been growing over a longer time period than those of Southern Africans, while for canines and upper incisors, Neandertal and Southern African imbricational enamel growth may be comparable. Finally, we show that Neandertal teeth are highly unlikely to have been growing in 15% less time than those of Southern Africans. Thus, if the duration of imbricational enamel formation is correlated with the length of childhood growth periods, then these results suggest that Neandertals shared the prolonged growth periods of modern humans.

It is not clear, however, if and to what extent imbricational enamel formation times in anterior teeth are correlated with somatic growth. The relationship between dental and somatic growth has been most clearly established in the high correlations that exist between various primate life history variables (such as age at weaning) and the age at which the first permanent molar erupts (Smith, 1989; Smith, 1991). Dean et al. (2001) have noted that in modern humans and *Paranthropus*, the lower canine completes enamel formation at about the same time that the first permanent molar emerges. If this was also the case for Neandertals, whose eruption patterns were similar to those of modern humans (Tompkins, 1996), then our data would indicate that the time of Neandertal first molar emergence was not different from that of modern Southern Africans.

It has been suggested that if Neandertals suffered high adult mortality rates, then they might be expected to have had abbreviated periods of childhood growth (Trinkaus and Tompkins, 1990; Ramirez Rozzi and Bermúdez de Castro, 2004). Adult mortality rates directly select for the timing of maturation across mammals; a higher risk of dying selects for rapid maturation (Stearns, 1992; Charnov, 1993; Kelley, 2002). However, Smith (2004) notes that if Neandertals had accelerated life histories, then this would leave them with a "peculiar" relationship between brain size and maturation, "two variables that are rarely out of step." Because large brains appear to require extended periods of childhood growth (Mann, 1972; Sacher, 1975; Gould, 1977; Martin, 1983; Bogin, 1997; Leigh and Park, 1998; Leigh, 2001; Crews and Gerber, 2003), the presence of large brains in Neandertals suggests that their adult mortality risks were not high enough to have prevented them from evolving prolonged growth periods.

While our results indicate that imbricational enamel formation times in anterior teeth may not differentiate Neandertals from modern humans, they also reveal some potentially unique features of Neandertal enamel growth. Ramirez Rozzi and Bermúdez de Castro (2004) argued that one of these unique features is a rapid extension rate indicated by lower densities of perikymata, relative to modern humans, in the last three deciles of growth. Our data show that this is not a unique feature of Neandertal teeth: modern South Africans have even lower densities of perikymata in these regions, contributing to their generally lower total perikymata numbers. However, visual inspection of the growth curves across the deciles is suggestive of a unique pattern of enamel growth in Neandertals as evidenced by their generally straighter curves in comparison to those of the modern human groups. Interestingly, Ramirez Rozzi and Bermúdez de Castro (2004) obtain straighter growth curves, resembling those of the Neandertals, for *H. heidelbergensis* and *H. antecessor*. Their Upper Paleolithic-Mesolithic curve is more curvilinear, resembling our modern human curves (Ramirez Rozzi and Bermúdez de Castro, 2004). An additional apparently unique growth pattern is exhibited by Neandertals in their pattern of perikymata numbers across anterior teeth: the modern human samples appear to have curves that parallel each other, while Neandertals appear to show a divergent pattern (Figure 4). Relative to modern humans, Neandertals therefore appear to have greater uniformity in growth within as well as among teeth.

Conclusions

The aim of this study was to compare aspects of imbricational enamel growth in the anterior teeth of Neandertals with those of three modern human population samples. In terms of its potential implications for Neandertal life history and taxonomic status, the most significant of these is the duration of imbricational enamel formation. A recent study by Ramirez Rozzi and Bermúdez de Castro (2004) suggested that Neandertals had periods of imbricational enamel formation abbreviated by 15% relative to modern humans, a difference which they argued means that Neandertals grew up more quickly than modern humans and supports the designation of Neandertals as a separate species. Their study, however, lacked a broad modern human comparative sample, reflecting the range of variation in imbricational enamel formation times that exists across modern human populations. The present study, by comparing Neandertals to modern human samples from diverse regions (Point Hope, Alaska; Newcastle-upon-Tyne, England; Southern Africa), finds that Neandertals do not appear to differ from some modern human groups in the length of time their anterior tooth imbricational enamel takes to form. Indeed, our data strongly indicate that imbricational enamel in the lower incisors of Neandertals takes longer to form than it does in Southern Africans, while for the other anterior teeth, Southern Africans and Neandertals appear to have comparable imbricational enamel formation times.

We do find evidence, however, that imbricational enamel formation in Neandertal anterior teeth may be different from modern humans in two ways: in the shape of their enamel growth curves across deciles from cusp tip to cervix and in their pattern of mean perikymata numbers across anterior tooth types. While these data are suggestive of Neandertal divergence from modern humans in these respects, they require additional investigation and statistical evaluation.

Acknowledgments

We thank Shara Bailey and Jean-Jacques Hublin for inviting us to present this paper in Leipzig and to contribute to this volume and the anonymous reviewers who provided valuable feedback and insight. This work was supported by a Leakey Foundation grant and a College of Social and Behavioral Sciences grant to DG-S from The Ohio State University. Krapina replicas were made with the support of a Leakey foundation grant to CSL. The authors are grateful to the following people and institutions for access to their collections: Kevin Kuykendall and Cynthia Reid at the Medical School of the University of the Witwatersrand, Jakov Radovčić of the Croatian Natural History Museum, Christopher Stringer and Robert Kruszynski of the Natural History Museum of London, Yoel Rak and Alon Barash at Tel Aviv University, Almutt Hoffman of the Museum für Vor-Und Frühgeschichte Archäeologie Europas, Ivana Jarasova of the Anthropos Institute, Phillipe Mennecier at the Musée de l'Homme in Paris, and Ian Tattersall and Ken Mowbray at the American Museum of Natural History. Thanks are also due Dale Hutchinson for loaning the Krapina replicas, Dan Steinberg, for his extensive support, Pamela Walton for preparation of all histological slides used in this study, Cathy Cooke and James Patrick Bell for making epoxy replicas, and to Cameron Begg and Hank Colijn of the Center for Electron Optics of The Ohio State University. Christopher Dean, Gary Schwartz, Rebecca Ferrell, Bruce Floyd, and Tanya Smith provided helpful comments on the MS. Finally, the following people gave advice on the collections and their help in this regard is also greatly appreciated: Shara Bailey, Jeff Schwartz, and Trent Holliday.

References

Aiello L., Dean C., 2002. *An Introduction to Human Evolutionary Anatomy*. Academic Press, London.

Beynon, A.D., Wood, B.A., 1987. Patterns and rates of enamel growth in the molar teeth of early hominids. Nature 326, 493–496.

Bogin, B., 1997. Evolutionary hypotheses for human childhood. Yearbook of Physical Anthropology 40, 63–89.

Bromage, T.G., 1991. Enamel incremental periodicity in the pig-tailed macaque: a polychrome fluorescent labeling study of dental hard tissues. American Journal of Physical Anthropology 86, 205–214.

Bromage, T.G., Dean M.C., 1985. Re-evaluation of the age at death of immature fossil hominids. Nature 317, 525–527.

Charnov, E.L., 1993. *Life History Invariants: Some Explorations of Symmetry in Evolutionary Ecology*. Oxford University. Press, Oxford.

Crews, D.E., Gerber, L.M., 2003. Reconstructing life history of hominids and humans. Collection Anthropology 27, 7–22.

Dean, M.C., 1987. The dental development status of six East African juvenile fossil hominids. Journal of Human Evolution 16, 197–213.

Dean, M.C., Beynon, A.D., Thackery, J.F., Macho, G.A., 1993. Histological reconstruction of dental development and age at death of a juvenile *Paranthropus robustus* specimen, SK

63, from Swartkrans, South Africa. American Journal of Physical Anthropology 91, 401–419.

Dean, C., Leakey, M.G., Reid, D.J., Shrenk, F., Schwartz, G.T., Stringer, C. & Walker, A., 2001. Growth processes in teeth distinguish modern humans from *Homo erectus* and earlier hominins. Nature 414, 628–631.

Dean, M.C., Reid, D.J., 2001a. Anterior tooth formation in *Australopithecus* and *Paranthropus*. In: Brook, A. (Ed.), Dental Morphology. University. Sheffield, Sheffield, pp. 135–143.

Dean, M.C., Reid D.J., 2001b. Perikymata spacing and distribution on hominid anterior teeth. American Journal of Physical Anthropology 116, 209–215.

Dean M.C., Stringer, C.B., Bromage T.G., 1986. Age at death of the Neanderthal child from Devil's Tower, Gibraltar and the implications for studies of general growth and development in Neanderthals. American Journal of Physical Anthropology 70, 301–309.

Fitzgerald, C.M., 1998. Do enamel microstructures have regular time dependency? Journal of Human Evolution 35, 371–386.

Gould, S.J., 1977. *Ontogeny and Phylogeny*. Harvard Univ. Press, Cambridge.

Guatelli-Steinberg, D., Reid, D.J., Bishop, T.A., Larsen, C.S., 2005. Anterior tooth growth periods in Neandertals were comparable to those of modern humans. Proceedings of the National Academy of Sciences of the USA 102, 14197–14202.

Hillson, S., 1996. *Dental Anthropology*. Cambridge University Press, Cambridge.

Kelley, J., 2002. Life-history evolution in Miocene and extant apes. In: McNamara, K.J., Minugh-Purvis N (Eds.), Human Evolution through Developmental Change. Johns Hopkins Univ. Press, Baltimore, pp. 223–248.

Kelley, J., Smith, T., 2003. Age at first molar emergence in early Miocene *Afropithecus turkanensis* and life-history evolution in the Hominoidea. Journal of Human Evolution 44, 307–329.

Kutner M.H., Nachtsheim C.J., Neter, J., 2004. *Applied Linear Regression Models*, 4th Ed. McGraw Hill, Boston, New York.

Leigh, S.R., 2001. Evolution of human growth. Evolutionary Anthropology 20, 223–236.

Leigh, S.R., Park, P.B., 1998. Evolution of human growth prolongation. American Journal of Physical Anthropology 107, 331–350.

Mann, A.E., 1972. Hominid and cultural origins. Man 7, 379–386.

Mann, A.E., Monge, J.M., Lampl, M., 1991. Investigation into the relationship between perikymata counts and crown formation times. American Journal of Physical Anthropology 86, 175—188.

Martin, R.D., 1983. *Human Brain Evolution in an Ecological Context*. American Museum of Natural History, New York.

McKown, T.D., Keith, A., 1939. *The Stone Age of Mount Carmel II: The Fossil Human Remains from the Levallois- Mousterian*. Clarendon Press, Oxford.

Quam, R.M., Smith, F.H., 1998. A reassessment of the Tabun C2 mandible. In: Takeru, A., Aoki, A., Bar-Yosef, O. (Eds.), Neanderthals and Modern Humans in West Asia. Plenum Press, New York, pp. 405–421.

Rak, Y., 1998. Does any Mousterian cave present evidence of two hominid species? In: Takeru, A., Aoki, A., Bar-Yosef, O. (Eds.), Neanderthals and Modern Humans in West Asia. Plenum Press, New York, pp. 353–377.

Rak, Y., Ginzburg, A., Geffen, E., 2002. Does *Homo neanderthalensis* play a role in modern human ancestry? The mandibular evidence. American Journal of Physical Anthropology 119, 199–204.

Ramirez Rozzi, F.V., 1993. Microstructure et développement de l'émail dentaire du Néandertalien de Zafarraya, Espagne. Temps de formation et hypocalcification de l'émail dentaire. Comptes rendus de l'Académie des Sciences Paris 316, 1635–1642.

Ramirez-Rozzi, F.V., Bermúdez de Castro J.M., 2004. Surprisingly rapid growth in Neanderthals. Nature 428, 936–939.

Reid D.J., Dean M.C., 2006. Variation in modern human enamel formation times. Journal of Human Evolution 50, 329–346.

Reid, D.J., Ferrell R.J., 2006. The relationship between total striae of Retzius number and periodicity in imbricational enamel formation. Journal of Human Evolution 50, 195–202.

Ruff, C., Trinkaus, E., Holliday, T., 1997. Body mass and encephalization in Pleistocene *Homo*. Nature 387, 173–176.

Rupert, G.M., 1981. *Simultaneous Statistical Inference*. Springer-Verlag, New York, Heidelberg and Berlin.

Sacher, G.A., 1975. Maturation and longevity in relation to cranial capacity in hominid evolution.

In: Tuttle R.H. (Ed.), Primate Functional Morphology and Evolution. Mouton, The Hague, pp. 417–441.

Schwartz, J.H., Tattersall, I., 2000. The human chin revisited: what it is and who has it? Journal of Human Evolution 38, 367–409.

Schwartz, J.H., Brauer, J., Gordon-Larsen, P., 1995. Brief communication: Tigaran (Point Hope, Alaska) tooth drilling. American Journal of Physical Anthropology 97, 77–82.

Schwartz, G.T., Samonds, K.E., Godfrey, L.R., Jungers, W.L., Simons, E., 2002. Dental microstructure and life history in subfossil Malagasy lemurs. Proceedings of the National Academy of Sciences of the USA 99, 6124–6129.

Smith, B.H., 1989. Dental development as a measure of life history in primates. Evolution 43, 683–688.

Smith, B.H., 1991. Dental development and the evolution of life history in Hominidae. American Journal of Physical Anthropology 86, 157–174.

Smith, B.H., 2004. The paleontology of growth and development. Evolutionary Anthropology 13, 239–241.

Smith, B.H., Tompkins, R.L., 1995. Toward a life history of the Hominidae. Annual Review of Anthropology 24, 257–279.

Smith, T.M., Dean, M.C., Kelley, J., Marin, L.B., Reid, D.J., Schwartz, G.T., 2003. Molar crown formation in Miocene hominoids: a preliminary synthesis. American Journal of Physical Anthropology Suppl. 36, 196.

Smith, T.M., Reid, D.J., Dean, M.C., Olejniczak, A.J., Ferrell, R.J., Martin, L.B., 2007. New perspectives on chimpanzee and human molar development. In: Bailey, S., Hublin, J.J. (Eds.), Dental Perspectives on Human Evolution: State of the Art Research in Dental Anthropology. Springer, Dordrecht. pp. 211–227.

Stearns, S.C., 1992. *The Evolution of Life Histories*. Oxford Univ. Press, Oxford.

Stefan, V., Trinkaus, E., 1998. Discrete trait and dental morphometric affinities of the Tabun 2 mandible. Journal of Human Evolution 34, 443–468.

Tompkins, R.L., 1996. Relative dental development of Upper Pleistocene hominids compared to human population variation. American Journal of Physical Anthropology 99, 103–118.

Trinkaus, E., 1983. *The Shanidar Neanderthals*. Academic Press, New York.

Trinkaus, E., 1987. The Neanderthal face: evolutionary and functional perspectives on a recent hominid face. Journal of Human Evolution 16, 429–443.

Trinkaus, E., 1993. Comment. Current Anthropology 34, 620–622.

Trinkaus, E., Tompkins, R.L., 1990. The Neandertal life cycle: the possibility, probability, and perceptibility of contrasts with recent humans. In: De Rousseau, C.J. (Ed.), Primate Life History and Evolution, Wiley Liss, New York, pp. 153–180.

Wolpoff, M.H., 1979. The Krapina dental remains. American Journal of Physical Anthropology 50, 67–114.

PART III

DENTAL DEVELOPMENT

1. Introduction

B.A. WOOD
The George Washington University
CASHP
Department of Anthropology
2110 G Street, NW
Washington, DC 20052, USA
bernardawood@gmail.com

Introduction

A decade has past since I last reviewed the various ways in which studies of dental development can contribute to what was then still called hominid paleobiology (Wood, 1996) A little more than eighty years have passed but Adolph Schultz's (1924) seminal contribution is still largely ignored and his paper is seldom cited. Nonetheless, its concluding sentence is worth quoting verbatim once again "With these few and scattered observations on the relation of the growth of the primates to man's evolution... it is hoped to have at least stimulated further investigations and thought – perhaps criticism – in this promising field" (*ibid.*, p. 163). It perhaps took longer than Schultz anticipated for researchers to appreciate the prescience of his prediction and even now we are only just beginning to realize how promising the field of comparative primate growth and development is. Indeed, a decade ago my own appreciation of the potential importance of dental development for improving our understanding of human evolutionary history at many levels was very incomplete.

We hear a good deal about "intelligent design", but for the biologist the dentition is a prime example of "fortunate design". This is because millions of years of evolution have crafted in the form of the dentition a morphological complex that provides opportunities to study both the processes at work in evolution and the patterns that result from those processes. They can be studied at the level of the molecule, the cell, in more than one cell of the same type, in a single tissue, and in more than one tissue in a single tooth (e.g., Shellis, 1984, 1998; Dean et al., 1993; Dean, 1995; Dean and Scandrett, 1996; Antoine et al., 2001; Kangas et al., 2004; Lucas, 2004; Schwartz, 2005). Added to this good fortune is the benefit that in mammals including primates the phenotype of this morphological complex comes in the form of discrete meristic morphological packages called teeth. Individual teeth can be studied at each of the first five levels listed above, as can several teeth of the same tooth type and teeth of more than one type. But a tooth is not an independent entity. The functional potential of a tooth cannot be realized unless it works in concert with other teeth within a functionally integrated

S.E. Bailey and J.-J. Hublin (Eds.), Dental Perspectives on Human Evolution, 231–235.

masticatory system (Lucas, 2004) so the hard tissues of the dentition can be studied as an integrated unit and the dentition can be combined with other hard and soft tissues of the masticatory system as an integrated unit. But there is more good fortune. Unlike bone, dental tissues grow in an orderly and incremental fashion and they are not subjected to the remodeling that bedevils attempts to use bone microstructure to recover information about the growth and development of the bony skeleton. Dental microstructure can potentially preserve evidence of the cellular activity involved in the growth and development of teeth in exquisite detail. This potential has been realized, for in some studies the daily activity of single cells has been tracked for hundreds of days. The final piece of good fortune is the fact that teeth are made of the densest tissues in the body. This factor, combined with their compact shape, results in teeth and jaws being particularly well represented in the hominin fossil record.

Until relatively recently, the data extracted from the dental evidence for hominin evolution was essentially limited to variables that described the macrostructure of the crowns of teeth that had completed their ontogeny. These data were either simple linear measurements that described (albeit relatively crudely) the overall size and shape of the crowns of the teeth, or they were non-metrical observations that drew attention to details of the shape of the tooth crowns. The four papers in the Growth and Development section of this volume illustrate some of the ways in which the ontogeny of isolated teeth and of dentitions can help deepen our understanding of human evolution.

If we restrict ourselves to the extant taxa within the African ape clade, we might be tempted to think that there are significant constraints on the relative sizes of the crowns of the anterior (incisors and canine) and posterior (premolars and molars) teeth. Fortunately the hominin fossil record shows us that these constraints are apparent and not real. The hominin taxa many researchers include within the genus *Paranthropus* are assigned to a separate genus in part because they combine small incisor and canine crowns with massive postcanine crowns. This is not a size-related change because it is a departure from allometric trends within and among Old World primate taxa (Wood and Stack, 1980). This suggests the type of developmental independence that is implied in dental field theories of the type put forward by Butler (1939) and refined by Dahlberg (1945). One of the most significant recent advances in dental science has been the quantitative genetic studies of Leslea Hlusko and colleagues. Their seminal contribution was to realize the importance of (and then very effectively exploit) the resource represented by the *Papio hamadryas* colony at the Southwest National Primate Research Center (SNPRC) in San Antonio. Mating is controlled within the colony and there are meticulous records of the pedigree of these animals. Hlusko and Mahaney (2007) mined this resource (along with a pedigreed mouse colony) to test the independence of linear measurements of incisors and molars (in the mice) and premolars and molars (in the baboons). Their study is preliminary, but the results suggest that incisor and molar size are independent and that premolar and molar size are partially independent.

The second and third of the four papers deal with the timing and the sequence of events in dental development. Monge et al. (2007) base their study on radiographs of 170 children from two dental clinics in Pennsylvania. The chronological age of each child was not estimated from radiographs, but was known from the clinical records. Each radiograph was assessed on the basis of the later stages of M_1 calcification and on M_2 calcification. The study shows that when compared to two "industry standard" reference samples, those of Moorrees et al. (1963) and Demirjian et al. (1973), this contemporary sample of children from Pennsylvania is developing their teeth earlier. What this shows is that modern human dental developmental schedules are remarkably plastic. The problem it poses is what developmental schedule should be used to represent modern humans when assessing the developmental schedule of a fossil hominin taxon? Monge et al. remind the reader that Zihlman et al. (2004) have already "rattled the bars" of comparative dental developmental studies by showing that wild chimpanzees have longer developmental schedules than the captive chimpanzees whose schedules have heretofore been used as the comparator when assessing the developmental schedules of fossil hominins. Perhaps the developmental schedules of aboriginal populations of modern humans and wild chimpanzees were more similar than the existing comparators suggest? Let us hope that dental clinics for the Hadza and the Ache are set up soon.

Braga and Heuze (2007) use a creditably large (2089 children) data set to explore whether, at least as far as the sequence of dental development is concerned, the whole is simply a sum of its parts, or whether a comprehensive dental developmental schedule contains more information than is contained in the developmental schedules of each tooth or each tooth type? What is the effect, if any, of the interactions between the teeth in the developing mandibular dentition? Can one tooth type, say the anterior teeth, serve as a proxy for the dentition as a whole? The results of their study have implications for the analysis of the dental developmental schedule of fossil hominins for researchers often have to make do with data from just one or two teeth, from which they are tempted to extrapolate to the developmental schedule of the whole dentition. To cut another elegant analytical story short Braga and Heuze suggest that the developmental schedules of tooth types can best be thought of as semi-independent "modules", much like those used for dental morphology. Effectively, they have posted the equivalent of a "health warning" on a packet of cigarettes; anterior teeth should not be used as a proxy for the dentition as a whole, or for other tooth types. Paleoanthropologists take note.

The fourth paper in this series, by Smith et al., focuses on the ontogeny of individual teeth. One of the disadvantages of relying on the morphology of the crown surface for information about taxonomy and phylogeny is that wear soon removes this information. Fortunately just as CT has enabled researchers to capture information about the internal structure of bones (e.g., the temporal bone, see Spoor and Zonneveld, 1998), the relatively recent availability of microCT is enabling researchers to capture morphological information about the internal structure of teeth. MicroCT produces virtual slices through a tooth crown, and enamel and dentine are sufficiently dissimilar structurally and chemically (Schroeder, 1991) that the boundary between the dentine and the enamel can be detected (e.g., Skinner, 2005). With suitable software the images from all the slices through a tooth crown can be integrated to produce a virtual solid rendering of the outer surface of the dentine (DEJ), which is of course the same as the inner surface of the enamel. Smith et al. (2007) go further and suggest that the spatial geometry of the DEJ can be used to reconstruct the order of cusp initiation and coalescence. A previous study had looked at 24 dm_2 germs at different stages of development (Avishai et al., 2004). Within that sample, although the order of cusp initiation was the same, the order in which the cusps coalesced differed. Specifically they found that while the entoconid was the fourth cusp to be initiated, it was the last to coalesce. In the present paper Smith et al. look at three teeth, two (a dm_2 and an M_1) from the jaw of one individual and the third (an M_1) from the jaw of a second individual. They found that the differences in the development of the dm_2 and the M_1s were greater than the differences in the development of the two M_1s. The title of their paper suggests that the authors have generated a "computer model of dental development". It would have been more accurate to say the authors have presented a method that uses computers to help reconstruct the ontogeny of the enamel cap.

Comparative studies of primate growth and development have come a long way since 1924, and a fair distance since 1996. This volume captures a snapshot of what we must hope will be an ongoing, dare I say developmental, process.

References

Avishai, G., Muller, R., Gabet, Y., Bab, I., Zilberman, U., Smith, P., 2004. New approach to quantifying developmental variation in the dentition using serial microtomographic imaging. Microscopy Research and Techniques 65, 263–269.

Butler, P.M., 1939. Studies of the mammalian dentition – differentiation of the postcanine dentition. Proceedings of the Zoological Society of London Series B 109, 1–36.

Braga, J., Heuze, Y., 2007. Quantifying variation in human dental development sequences: An EVO-DEVO perspective. In: Bailey, S.E., Hublin, J.-J. (Eds.), Dental Perspectives on Human Evolution: State of the Art Research in Dental Paleoanthropology. Springer, Dordrecht. pp. 231–235.

Dahlberg, A.A., 1945. The changing dentition of Man. Journal of the American Dental Association 32, 676–690.

Dean, M.C., 1995. The nature and periodicity of incremental lines in primate dentine and their relationship to periradicular bands in OH 16 (*Homo habilis*): In: Moggi-Cecchi, J. (Ed.), Aspects of Dental Biology: Palaeontology, Anthropology and Evolution. International Institute for the Study of Man, Florence, pp. 239–265.

Dean, M.C., Beynon, A.D., Thackeray, J.F., Macho, G.A., 1993. Histological reconstruction of dental development and age at death of a juvenile *Paranthropus robustus* specimen, SK 63, from Swartkrans, South Africa. American Journal of Physical Anthropology 91, 401–419.

Dean, M.C., Scandrett, A.E., 1996. The relation between long-period incremental markings in dentine and daily cross-striations in enamel in human teeth. Archives of Oral Biology 41, 233–241.

Demirjian, A., Goldstein, H., Tanner, J.M., 1973. A new system of dental age assessment. Human Biology 45, 211–227.

Hlusko, L.J., Mahaney, M.C., 2007. Of mice and monkeys: quantitative genetic analyses of size variation along the dental arcade. In: Bailey, S.E., Hublin, J.-J. (Eds.), Dental Perspectives on Human Evolution: State of the Art Research in Dental Paleoanthropology. Springer, Dordrecht. pp. 231–235.

Kangas, A.T., Evans, A.R., Thesleff, I., Jernvall, J., 2004. Nonindependence of mammalian dental characters. Nature 432, 211–214.

Lucas, P., 2004. *Dental Functional Morphology: How Teeth Work*. Cambridge University Press, Cambridge.

Moorrees, C.F.A., Fanning, E.A., Hunt, E.E., 1963. Age variation of formation stages for ten permanent teeth. Journal of Dental Research 42, 1490–1502.

Monge, J., Mann, A., Stout, A., Roger, J., Wadenya, R., 2007. Dental calcification stages of the permanent M1 and M2 in U.S. children of African American and European American ancestry born in the 1990s. In: Bailey, S.E., Hublin, J.-J. (Eds.), Dental Perspectives on Human Evolution: State of the Art Research in Dental Paleoanthropology. Springer, Dordrecht. pp. 231–235.

Schroeder, H.E., 1991. *Oral Structural Biology*. Georg Thieme Verlag, Stuttgart.

Schultz, A.H., 1924. Growth studies on primates bearing upon Man's evolution. American Journal of Physical Anthropology 7, 149–164.

Schwartz, G.T., 2005. Developmental processes and canine dimorphism in primate evolution. Journal of Human Evolution 48, 97–103.

Schwartz, G.T., Godfrey, L.R., Mahoney, P., 2007. Inferring primate growth, development and life history from dental microstructure: the case of the extinct Malagasy lemur, *Megaladapis*. In: Bailey, S.E., Hublin, J.-J. (Eds.), Dental Perspectives on Human Evolution: State of the Art Research in Dental Paleoanthropology. Springer, Dordrecht. pp. 231–235.

Shellis, R.P., 1984. Variations in growth of the enamel crown in human teeth and a possible relationship between growth and enamel structure. Archives of Oral Biology 29, 697–705.

Shellis, R.P., 1998. Utilization of periodic markings in enamel to obtain information on tooth growth. Journal of Human Evolution 35, 387–400.

Skinner, M.M., Kapadia, R., 2005. An evaluation of microCT for assessing the 3D concordance of dental trait expression between the dentin-enamel junction and the outer enamel surface of modern human molars. American Journal of Physical Anthropology Supplement 40, 191–192.

Smith, P., Müller, R., Gabet, Y., Avishai, G., 2007. A computerized model for reconstruction of dental ontogeny: a new tool for studying evolutionary trends in the dentition. In: Bailey, S.E., Hublin, J.-J. (Eds.), Dental Perspectives on Human Evolution: State of the Art Research in Dental Paleoanthropology. Springer, Dordrecht. pp. 231–235.

Spoor, F., Zonneveld, F., 1998. Comparative review of the human bony labyrinth. Yearbook of Physical Anthropology 41, 211–251.

Wood, B.A., 1996. Hominin paleobiology – have studies of comparative development come of age? American Journal of Physical Anthropology 99, 9–15.

Wood, B.A., Stack, C.G., 1980. Does allometry explain the differences between "gracile" and "robust" australopithecines? American Journal of Physical Anthropology 52, 55–62.

Zihlman, A., Bolter, D., Boesch, C. 2004. Wild chimpanzee dentition and its implications for assessing life history in immature hominin fossils. Proceedings of the National Academy of Sciences of the USA 101, 10541–10543.

2. Of mice and monkeys: Quantitative genetic analyses of size variation along the dental arcade

L.J. HLUSKO
Department of Integrative Biology
University of California
3060 Valley Life Sciences Bldg
Berkeley, CA 94720 USA
hlusko@berkeley.edu

M.C. MAHANEY
Southwest National Primate
Research Center and the Department
of Genetics, Southwest Foundation
for Biomedical Research,
P.O. Box 760549, San Antonio, TX
mmahaney@sfbrgenetics.org

Keywords: quantitative genetics, dental fields, field theory, dental patterning, *Papio hamadryas, Mus*

Abstract

We present preliminary results from quantitative genetic analyses of tooth size variation in two outbred pedigreed populations, baboons and mice. These analyses were designed to test the dental field theory as proposed by Butler (1939), that there are three fields within the dentition: incisor, canine, and molar. Specifically we estimated the genetic correlation between pairs of linear size measurements. Results from the baboon analyses suggest that there may also be a premolar field that is only partially independent of the molar field proposed by Butler (1939). Analyses of the mouse data indicate that for mice, size variation in the incisors appears to be genetically independent of molar size. If the field theory is correct, future analyses on incisor data for the baboons will return similar results of genetic independence. Circumstantial evidence from the fossil record suggests that there will be at least some degree of independence between the anterior and postcanine dentitions of primates.

Introduction

Huxley and de Beer (1934) were among the first embryologists to formalize the concept of the gradient-field in relation to the development of the animal body plan. They defined it as:

"...a region throughout which some agency is at work in a co-ordinated way, resulting in the establishment of an equilibrium within the area of the field.

S.E. Bailey and J.-J. Hublin (Eds.), Dental Perspectives on Human Evolution, 237–245.

A quantitative alteration in the intensity of operations of the agency in any one part of the field will alter the equilibrium as a whole. A field is thus a unitary system, which can be altered or deformed as a whole; it is not a mosaic in which single portions can be removed or substituted by other without exerting any effect on the rest of the system." (1934: 276).

In 1939, Butler evoked this concept to explain the dental pattern of mammals, i.e., the number and morphology of incisors, canines, premolars, and molars. In his morphogenic field theory of dental development, Butler suggests that each tooth primordium is equivalently pluripotent, possessing the potential to develop into any type of tooth in the dentition. Determination of the ultimate form of each tooth is a function of the tooth primordium's exposure to morphogens and the nature and concentration of these morphogen(s), both of which are related to the tooth placode's location within a developmental field. Butler infers that morphogenic fields are distinct: each perfused with a characteristic combination of morphogens and, consequently, one should be able to identify different morphogenic fields within a dental arch based on tooth morphology. By this logic, Butler identifies three distinct dental morphogenic fields in the mammalian dentition: the molar, canine, and incisor fields. In this scheme, incisor size and shape would be developmentally independent of molar size and shape; however, because premolars are within the molar field, the development of their size and shape is correlated with that of molars but independent of incisors, for example (Butler, 1939).

The last 15 years have seen a dramatic increase in our understanding of the genetic mechanisms needed to form teeth and pattern the overall dentition, primarily in rodent models. From this work we know that the patterning of the mouse dentition, or dental formula, first appears histologically at embryonic day 11 (E11) when a region of the mouse oral epithelium thickens to form the dental lamina (for more details see Weiss et al., 1998; Jernvall and Thesleff, 2000; Stock, 2001; Tucker and Sharpe, 2004).

At this dental lamina stage of development, tooth position and fate is induced by the oral epithelium. The dental pattern, or formula that is already determined by this stage of development has been hypothesized to result from one of two possible mechanisms. The first is an odontogenetic combinatorial code (Cobourne and Sharpe, 2003). This is similar in concept to the *Hox* gene patterning of the vertebral column, although *Hox* genes are not expressed in tooth development and therefore do not similarly regulate dental patterning (James et al., 2002). Other regulatory genes including members of the *Barx, Dlx, Msx*, and *Pitx* gene families have been implicated (Cobourne and Sharpe, 2003). Restricted expression of two members of the *Dlx* family to the more caudal region of the developing branchial arch may be important for determining the maxillary versus mandibular jaws (Depew et al., 2002).

The second dental patterning mechanism proposed is that of a reaction diffusion process. Weiss et al. (1998) attribute the periodicity of the dentition to quantitative interactions of diffusible signaling factors. This idea is based on Savart's and Chladni's nineteenth century recognition that different wave lengths mechanically interact to form different interference patterns, easily visualized as patterns formed by sound waves moving through powder on violin plates (see Weiss et al., 1998 for more details). Bateson (1894) applied this concept to serially homologous traits, such as the dentition, and coined the term "meristic variation". Alan Turing (1952) later proposed that chemical interactions could similarly result in wave-like patterns as substances, or morphogens, interact in a *reaction diffusion* process.

To date, there have been no convincing data that refute or support the validity of either of these models. We have not yet been able to successfully test these hypotheses partly because of the derived and reduced dentition of the model animal used in gene expression studies: mice. Mice lack deciduous dentitions, permanent premolars and canines. Therefore, only two tooth types are present – molars in the distal region of the jaw and incisors in the anterior region with a large diastema in between – and consequently, hypotheses about the molecular mechanisms that pattern the dental arcade remain relatively untestable.

Butler's (1939) concept of morphogenic fields is relevant today, as it provides an alternative to the two molecularly-driven hypotheses, or rather it is a combination of the two models. Here, we propose a method through which Butler's morphogenic field hypothesis can be tested using quantitative genetic analyses. Specifically, Butler's model predicts that genetic correlations will be higher within fields than between them. We present preliminary results from two sets of quantitative genetic analyses of tooth size variation, one on a pedigreed population of baboons, and the other a pedigreed population of mice.

Material and Methods

Study Population – Papio hamadryas

Mesiodistal length and mesial buccolingual widths of all maxillary molars and premolars were collected from dental casts made for 630 baboons. These animals are part of a captive, pedigreed breeding colony of baboons, *Papio hamadryas,* housed at the Southwest National Primate Research Center (SNPRC) in San Antonio, Texas. Taxonomy follows Jolly (1993). The colony is maintained in pedigrees with all mating opportunities controlled.

Genetic management of the colony was started over 20 years ago and allows for data collection from non-inbred animals. All non-founder animals in this study resulted from matings that were random with respect to dental, skeletal, and developmental phenotype. The female to male sex ratio is approximately 2:1. The animals from which data were collected are distributed across eleven extended pedigrees that are 3–5 generations deep. The mean number of animals with data per pedigree was 44, and these individuals typically occupied the lower two or three generations of each pedigree.

The Institutional Animal Care and Use Committee, in accordance with the established guidelines. (National Research Council, 1996), approved all procedures related to the treatment of the baboons during the conduct of this study.

Study Population – Mus sp.

Length of the maxillary first molar, incisor, and molar (M^{1-3}) row were collected for 222 individuals that are part of a large pedigreed collection of skeletonized mice made by Richard D. Sage between 1977 and 2002, currently housed at the University of California Berkeley's Museum of Vertebrate Zoology.

This collection is unusual in that it is outbred and founded with at least 7 species of wild-caught *Mus* rather than inbred, homozygous lab strains. The collection is quite diverse at the subgeneric and species levels. Three of the four subgenera are represented: *Coelomys* (shrew mice), *Mus* (house and rice field mice), and *Nannomys* (African pygmy mice). Taxonomy follows Nowak (1991). Table 1 summarizes the taxonomic composition.

These mice are from stocks of wild mice collected in the 1970s by R.D. Sage and his colleagues. The number of founders for each species ranges between 2 and 16. Taxonomy became difficult as various species were crossed, and therefore the designations shown in Table 1 represent the best estimation of each individual's closest taxonomic affinity based on ancestry from the founders.

In total, pedigree data for 299 mice were used to reconstruct the pedigrees, although

Table 1. Taxonomic composition of the Mus sample used in these quantitative genetic analyses

Subgenus	*Species*	*Total # of individuals*
Coelomys	*pahari*	11
Mus ($n = 196$)	*caroli*	23
	cervicolor	92
	cooki	51
	musculus	23
	spicilegus	7
Nannomys	*minuotoides*	4
Hybrids	–	11
	TOTAL	222

dental phenotype data were only collected for 222 animals. Of the 299, 155 are female and 144 are male. For analytical purposes the pedigrees were divided by litter, with approximately 47 pedigrees for each analysis, and an average of 6 individuals per pedigree.

Analytical Methods

All pedigree data management and preparation was facilitated through use of the computer package PEDSYS (Dyke, 1996). Statistical genetic analyses were performed using a maximum likelihood based variance decomposition approach implemented in the computer package SOLAR (Almasy and Blangero, 1998). The phenotypic covariance for each trait within a pedigree is modeled as

$$\Omega = 2\Phi\sigma_G^2 + I\sigma_E^2 \qquad (1)$$

where Φ is a matrix of kinship coefficients for all relative pairs in a pedigree, σ_G^2 is the additive genetic variance, I is an identity matrix (composed of ones along the diagonal and zeros for all off diagonal elements), and σ_E^2 is the environmental variance. Because the components of the phenotypic variance are additive, such that

$$\sigma_P^2 = \sigma_G^2 + \sigma_E^2 \qquad (2)$$

we estimated heritability, or the proportion of the phenotypic variance attributable to additive genetic effects, as

$$h^2 = \frac{\sigma_G^2}{\sigma_P^2} \qquad (3)$$

Phenotypic variance attributable to nongenetic factors is estimated as $e^2 = 1 - h^2$.

Using extensions to univariate genetic analyses that encompass the multivariate state (Hopper and Mathews, 1982; Lange and Boehnke, 1983; Boehnke et al., 1987), we modeled the multivariate phenotype of an individual as a linear function of the measurements on the individual's traits, the means of these traits in the population, the covariates and their regression coefficients, plus the additive genetic values and random environmental deviations, as well as the genetic and environmental correlations between them (for more detailed explanation see Mahaney et al., 1995). We obtained these two correlations from additive genetic and random environmental variance-covariance matrices. Respectively, the additive genetic correlation (ρ_G) and the environmental correlation (ρ_E) are estimates of the additive effects of shared genes (i.e., pleiotropy) and shared environmental (i.e., unmeasured and nongenetic) factors on the variance in a pair of traits. Because the genetic and environmental components of the phenotypic correlation matrix are additive, like those of the corresponding variance-covariance matrix, we use the maximum likelihood estimates of these two correlations to calculate a total phenotypic correlation between two traits, ρ_P, in a way that accommodates the nonindependence between data obtained from relatives as

$$\rho_P = \sqrt{h_1^2}\sqrt{h_2^2}\ \rho_G + \sqrt{(1-h_1^2)}\sqrt{(1-h_2^2)}\ \rho_E. \qquad (4)$$

Significance of the maximum likelihood estimates for heritability and other parameters was assessed by means of likelihood ratio tests (see Hlusko et al., 2004a, b for details).

Results

Papio hamadryas

Analyses of the baboon maxillary length and width data (Table 2) show that the shared genetic affects between premolar measurements are not significantly different from models constraining the genetic correlation to one. The same degree of shared genetic affects is found when data from pairs of first, second, and third molars are analyzed. However, when premolars are compared to molars, the shared genetic affects are estimated to be between 0.63 and 0.50, and significantly different from models constraining the genetic correlations to one (complete pleiotropy) or zero (no pleiotropy).

Mus sp.

Estimates of the shared genetic affects for the mouse data are presented in Table 3. The genetic correlation between incisor and first molar length is not significantly different from the model in which we constrain this correlation to zero. Our analysis of the mesiodistal length of the incisor compared to the overall mesiodistal length of the molar series also returns a genetic correlation not significantly different from zero. However, when the length

*Table 2. Results from bivariate quantitative genetic analyses of the length and widths of baboon maxillary premolars and molars**

Lengths	$h^2_{mesial\ tooth}$	$h^2_{distal\ tooth}$	*n*	ρ_G	$P(\rho_G = 0)$	$P(\lvert\rho_G\rvert = 1)$
$RP^3{:}RP^4$	0.26	0.73	403	0.98	<0.0001	0.46
$RP^4{:}RM^1$	0.72	0.66	537	0.63	<0.0001	<0.0001
$RM^1{:}RM^2$	0.69	0.71	547	0.99	<0.0001	0.40
$RM^2{:}RM^3$	0.79	0.31	541	0.95	0.0003	0.39
$RM^1{:}RM^3$	0.66	0.42	495	0.65	0.01	0.05
Widths	$h^2_{mesial\ tooth}$	$h^2_{distal\ tooth}$	n	ρ_G	$P(\rho_G=0)$	$P(\lvert\rho_G\rvert=1)$
$RP^3{:}RP^4$	0.64	0.66	423	1	<0.0001	na
$RP^4{:}RM^1$	0.61	0.65	536	0.50	0.006	<0.0001
$RM^1{:}RM^2$	0.66	0.59	543	0.85	<0.0001	0.0007
$RM^2{:}RM^3$	0.57	0.48	552	0.90	<0.0001	0.10
$RM^1{:}RM^3$	0.72	0.54	535	0.78	<0.0001	0.003

* h^2 = residual heritability; L = left, R = right; superscript # = maxillary tooth row position; P = premolar; M = molar; length = mesiodistal diameter; width = crown width (mesial buccolingual diameter)

*Table 3. Results from the bivariate quantitative genetic analyses of the length of mice incisors and molars**

Lengths	$h^2_{mesial\ tooth}$	$h^2_{distal\ tooth}$	*n*	ρ_G	$P(\rho_G = 0)$	$P(\lvert\rho_G\rvert = 1)$
$I^1{:}M^1$	0.32	0.42	214	0.25	0.32	0.0009
$I^1{:}M^{1-3}$	0.32	0.38	214	0.22	0.24	0.00008
$M^1{:}M^{1-3}$	0.42	0.38	214	0.93	<0.0001	0.16

* h^2 = residual heritability; L = left, R = right; superscript # = maxillary tooth row position; I = incisor; M = molar; length = mesiodistal diameter

of the first molar is compared to the entire molar row length, the shared genetic affects are estimated as not significantly different from the model in which we constrain the genetic correlation to one.

Discussion

In this paper we present preliminary results from quantitative genetic analyses of dental size variation in an outbred pedigreed population of baboons and an outbred pedigreed population of mice. As is often the case with exploratory research, the results presented here beg more questions than they answer. But in so doing, they demonstrate the potential usefulness of such a comparative approach for refining hypotheses about the genes responsible for the mechanisms that pattern the mammalian dentition.

Before discussing these results, we first address the potential caveats inherent to the analyses presented here. Genetic correlations can result from either pleiotropy or gametic phase disequilibrium between genes affecting different traits. In the case of pleiotropy, the same gene or set of genes influences variation in more than one phenotype. In the case of gametic phase disequilibrium, alleles at two or more loci with similar effects on more than one trait exhibit non-random association (Lynch and Walsh, 1998). The degree of gametic phase disequilibrium is a function of a population's genetic history and demography: e.g., it will be lower in outbred populations with many unrelated founders as recombination exerts its effects each generation, higher in populations undergoing rapid expansion from a small number of founders and those resulting from recent admixture. Given a conducive set of population characteristics, the likelihood of genetic correlation between two traits being due to gametic phase disequilibrium is higher for simple traits, with monogenic (or nearly so) inheritance. However, if variation in a pair of traits is attributable to the effects of multiple alleles at multiple loci – as well as to the effects of multiple environmental factors and interactions between them – gametic phase disequilibrium is not likely to be a major contributor to the genetic correlation (Lynch and Walsh, 1998).

Therefore, we are cautiously confident that significant additive genetic correlations estimated in our analyses of data on pairs of complex, multifactorial dental measures from our non-inbred, extended pedigrees of baboons are primarily indicative of pleiotropy rather than gametic phase disequilibrium. Ongoing and planned whole genome screens and linkage disequilibrium analyses in this population will help confirm this. However, we have less confidence concerning the unambiguous interpretation of significant genetic correlations estimated in our analyses of data from the first and second generation hybrids of mouse species. We are currently investigating whether or not a model of linkage disequilibrium or pleiotropy better fits with the genetic correlations found for these mouse data. These analytical tests are beyond the scope of this chapter and will be published elsewhere in the near future. In this paper, we will focus on the lack of a genetic correlation rather than the presence, a result that is of more relevance to the discussion of Butler's field theory (1939).

One hypothesis derived from the field theory is that the incisor field would be genetically independent of the molar field (Butler, 1939). This indeed is what we find in our analyses of the mouse data. The mesiodistal length of the maxillary incisor appears to be genetically independent of the mesiodistal length of both the first molar and the entire length of the molar row.

The second hypothesis inherent to Butler's theory is that variation in the premolars and molars would be due to the effects of the same gene or genes. That is, they would exhibit genetic correlations because their development, occurring in the same morphogenic field (the molar field), is influenced by the same morphogens whose molecular and biochemical characteristics and concentrations are influenced by the same gene(s). The results of some of our analyses of data for the maxillary postcanine teeth in these pedigreed baboons do accord with this prediction in that size variation in the premolars is influenced by overlapping but not identical genetic affects as is size variation in the molars.

However, not all of our analyses of data from the baboon produce results consistent with Butler's theory. If there are only three dental fields (incisor, canine, and molar), then the degree of genetic correlation among the premolars and molars would be similar since they are part of the same molar field. This is not what we find. Rather, the shared genetic affects are greatest within tooth classes (premolars versus molars) and reduced but not independent between classes. Therefore, these preliminary analyses suggest that there may be a fourth dental morphogenetic field that corresponds to the premolars.

Dahlberg applied the field theory to his study of the human dentition (1945). He proposed different fields depending on which phenotype or trait was under consideration. For example, he argued that *form* was divided into four fields: incisors, canine, premolars, and molars. However, Dahlberg's focus on only one extant species (*Homo sapiens*), led him to state that, "They [a particular set of teeth] are, as a rule, either all small, all large or at some stage in between ..." (1945: 688). Therefore, he proposed that size was uniformly influenced across the dentition. Our results suggest that Dahlberg's concept of four dental fields for form, rather than just three as proposed by Butler (1939), may also apply to size variation.

The genetic relationships between the tooth classes may also be variable, with some teeth more genetically independent of others. For example, incisors may be relatively more genetically independent from molars than are premolars. A key test that remains is to analyze incisor data for the baboons. This work is underway.

In the meantime, suggestive evidence lends support to the hypothesis that the genetic independence we see between mouse incisors and molars may be similarly found in primates. Across primate taxa, the most variable aspect of the dentition is the incisor region. There are numerous dramatic variations in the incisors relative to the rest of the dentition, for example, within the hominids *Australopithecus boisei* has reduced anterior teeth relative to those of *A. garhi* (Asfaw et al., 1999), in the Old World monkeys *Theropithecus oswaldi leakeyi* has a reduced anterior dentition relative to its phyletic ancestor *Theropithecus oswaldi darti* (Leakey, 1993), lemurs (the tooth comb), Oligocene primates from Egypt (*Parapithecus*' extreme reduction of the maxillary incisors and loss of the mandibular incisors), and last but not least, *Daubentonia*'s extremely derived large robust continuously growing rodent-like incisors.

Of course these evolutionary trends may result from different selective forces operating on the anterior and posterior aspects of the dentition. However, the key question that underlies all of evolutionary biology does not concern adaptation or even phylogeny, but rather "what are the characters upon which the process of evolution occurs?" Character definition is in many ways the most fundamental question to be addressed in evolutionary biology (Lewontin, 2001). Does the incisor region represent a character distinct from the postcanine dentition? Are the premolars and molars correlated but not distinct characters? In this paper we present two sets of quantitative genetic analyses of dental variation that are part of an endeavor to better understand the mammalian dental phenotype from the perspective of the genetic architecture, and thereby to identify the phenotypic characters upon which selection operates.

Acknowledgments

Many thanks to S. Bailey, J.-J. Hublin and the Max Planck Institute for hosting the conference that resulted in this volume. We thank K.D. Carey, K. Rice and the Veterinary Staff of the Southwest Foundation for Biomedical Research and the Southwest National Primate Research Center and J. Cheverud (Washington University) for access to baboon specimens. The University of California's Museum of Vertebrate Zoology and Richard D. Sage generously provided access to the mouse specimens, and many thanks to R.D. Sage for producing the collection. L. Buchanan, T. Cannistraro, L. Holder, J. Irwin, A. Liberatore, M.-L. Maas, and D. Pillie assisted with the dental data collection for the baboons. L. Broughton, S. Deldar, N. Do, T. Koh, N. Reeder, and N. Wu collected the mouse data. This material is based upon work supported by the National Science Foundation under Award No. BCS-0130277, the University of Illinois at Urbana-Champaign Research Board, and the University of California's Committee on Research (Junior Faculty Research Grant). NIH/NCRR P51 RR013986 supports the Southwest National Primate Research Center and NIH/NHLBI P01 HL028972 provides support for the pedigreed baboon breeding colony from which data were obtained. We also thank A. Walker, K. Weiss, and J. Rogers for project support and development.

References

Almasy L, Blangero J., 1998. Multipoint quantitative-trait linkage analysis in general pedigrees. American Journal of Human Genetics 62, 1198–1211.

Asfaw, B., White, T., Lovejoy, C.O., Latimer, B., Simpson, S., Suwa, G., 1999. *Australopithecus garhi*: a new species of early hominid from Ethiopia. Science 284, 629–635.

Bateson, W., 1894. *Material for the Study of Variation, Treated with Special Regard to Discontinuity in the Origin of Species*. Macmillan, London.

Boehnke, M., Moll, P.P., Kottke, B.A., Weidman, W.H., 1987. Partitioning the variability of fasting plasma glucose levels in pedigrees. Genetic and environmental factors. American Journal of Epidemiology 125, 679–689.

Butler, P.M., 1939. Studies of the mammalian dentition – differentiation of the postcanine dentition. Proceedings of the Zoological Society of London, Series B 109, 1–36.

Cobourne, M.T., Sharpe, P.T., 2003. Tooth and jaw: molecular mechanisms of patterning in the first branchial arch. Archives of Oral Biology 48, 1–14.

Dahlberg, A.A., 1945. The changing dentition of man. Journal of the American Dental Association 32, 676–690.

Depew, M.J., Lufkin, T., Rubenstein, J.L., 2002. Specification of jaw subdivisions by *Dlx* genes. Science 298, 381–385.

Dyke, B., 1996. *PEDSYS: A Pedigree Database Management System Users Manual*. Population Genetics Laboratory, Department of Genetics, Southwest Foundation for Biomedical Research, San Antonio, TX

Hlusko, L.J., Maas, M.L., Mahaney, M.C., 2004a. Statistical genetics of molar cusp patterning in pedigreed baboons: Implications for primate dental development and evolution. Molecular and Developmental Evolution (Journal of Experimental Zoology) 302B, 268–283.

Hlusko, L.J., Suwa, G., Kono, R., Mahaney, M. C., 2004b. Genetics and the evolution of primate enamel thickness: a baboon model. American Journal of Physical Anthropology 124, 223–233.

Hopper, J.L., Mathews, J.D., 1982. Extensions to multivariate normal models for pedigree analysis. Annals of Human Genetics 46, 373–383.

Huxley, J.S., de Beer, G.R., 1934. *The Elements of Experimental Embryology*. Cambridge University Press, Cambridge.

James, C.T., Ohazama A., Tucker A.S., and Sharpe P.T., 2002. Tooth development is independent of a *Hox* patterning programme. Developmental Dynamics 225, 332–335.

Jernvall, J., Thesleff, I., 2000. Reiterative signaling and patterning during mammalian tooth morphogenesis. Mechanisms of Development 92, 19–29.

Jolly, C.J., 1993. Species, subspecies, and baboon systematics. In: Kimbel, W.H., Martin, L.B. (Eds.), Species, Species Concepts, and Primate Evolution. Plenum Press, New York, pp. 67–107.

Lange, K., Boehnke, M., 1983. Extensions to pedigree analysis. IV. Covariance components models for multivariate traits. American Journal of Medical Genetics 14, 513–524.

Leakey, M.G., 1993. Evolution of *Theropithecus* in the Turkana Basin. In: Jablonski, N.G. (Ed.), *Theropithecus*: The Rise and Fall of a Primate Genus. Cambridge University Press, New York, pp. 85–123.

Lewontin, R., 2001. Foreward. In: Wagner, G.P. (Ed.), The Character Concept in Evolutionary Biology. Academic Press, New York, pp. xvii–xxiii.

Lynch, M., Walsh, B., 1998. *Genetics and Analysis of Quantitative Traits*. Sinauer Associates, Inc., Sunderland, MA.

Mahaney, M.C., Blangero, J., Comuzzie, A.G., VandeBerg, J.L., Stern, M.P., MacCluer, J.W., 1995. Plasma HDL cholesterol, triglycerides, and adiposity. A quantitative genetic test of the conjoint trait hypothesis in the San Antonio Family Heart Study. Circulation 92, 3240–3248.

National Research Council., 1996. *Guide for Care and Use of Laboratory Animals*. National Academy of Sciences, Washington, D.C.

Nowak, R.M., 1991. *Walker's Mammals of the World*, 5th Edition, Volume II. The Johns Hopkins University Press, Baltimore, MD.

Stock, D.W., 2001. The genetic basis of modularity in the development and evolution of the vertebrate dentition. Philosophical Transactions Royal Society London B 356, 1633–1653.

Tucker, A., Sharpe, P., 2004. The cutting edge of mammalian development; how the embryo makes teeth. Nature Reviews Genetics 5, 499–508.

Turing, A., 1952. The chemical basis of morphogenesis. Philosophical Transactions of the Royal Society B 237, 37–72.

Weiss, K.M., Stock, D.W., Zhao, Z., 1998. Dynamic interactions and the evolutionary genetics of dental patterning. Critical Reviews in Oral Biology and Medicine 9, 369–398.

3. Quantifying variation in human dental development sequences: An EVO-DEVO perspective

J. BRAGA
Laboratoire d'Anthropologie Biologique
Université Paul Sabatier (Toulouse 3), FRE 2960
39 allées Jules Guesde, 31000 Toulouse, France
braga@cict.fr

Y. HEUZE
Laboratoire d'Anthropologie Biologique
Université Paul Sabatier (Toulouse 3), FRE 2960
39 allées Jules Guesde, 31000 Toulouse, France
heuze@cict.fr

Keywords: dental development, dental evolution, EVO-DEVO, modularity, human evolution

Abstract

The present paper describes a novel analytical approach to provide a comprehensive description of the complex interactions that exist between the growing permanent mandibular teeth (excluding the third molars), and to quantify variability in sequences of key events during crown and root formation, independent of chronological age. Importantly, our method integrates the fundamental concept of modularity and rejects the old statistical fallacy of analyzing data on the assumption that it contains no information beyond that revealed on a tooth-by-tooth analysis. Indeed, interactions between growing teeth may also contain some information, which enables developmental or evolutionary information to be uncovered. Our training sample is based upon cross-sectional standardized panoramic radiographs of the teeth of a total of 2089 children (1206 girls and 883 boys) of different geographic origins (mainly Western Europe, Southern Iran, and Ivory Coast). We observe that, in extant humans sampled so far, the relative development of the permanent incisors is more plastic and varies more than for other teeth. Therefore, we consider that the quantification of possible variations between onsets, durations and rates of development of different teeth in any given child, within a large sample, is a prerequisite to the analysis of fossil hominids. In particular, we seriously question the assumption that the anterior teeth can serve as a reliable substitute for the other permanent teeth, and in particular for interpretations on somatic maturity and brain size. Our hypothesis of modularity in dental development and our method derived from this concept can serve as a basis for identifying and studying patterns of dental growth and, importantly, for comparisons between extant populations, and/or fossil species. These studies do not need to be hedged with age assessments of unknown accuracy and reliability levels (particularly in fossils), or the assumption of independence between growing teeth.

S.E. Bailey and J.-J. Hublin (Eds.), Dental Perspectives on Human Evolution, 247–261.

Introduction

The aim of our cross-sectional analysis of subadult radiographs is to examine, using an appropriate method and samples from different extant human populations, the variation in the relative sequences (or patterns) of key events during crown and root formation. To date, investigators have focused on the timescale (chronology) of dental development with published median or mean ages of emergence for permanent teeth in different modern human populations (e.g., Kuykendall, 1992; Liversidge, 2003) (in some studies, emergence is often confused with eruption; see Marks and Cahill, 1993 for a definition of the latter term). The debates have focused on the existence of real differences between samples, or on biases due to sampling, rating effects, and statistical procedures (Smith, 1991). However, sequence and timing represent two separate aspects of variation. For example, the first permanent molars of two individuals can mineralize at the same time but in different sequences relative to the other teeth. Compared to variation in timing, variation in patterns is much less documented in the literature. The question of variability in patterns is fundamental to the study of dental developmental processes as a possible foundation for morphological changes during human evolution. For permanent teeth, data are available for variation in sequences of emergence (e.g., Garn et al., 1973; Smith and Garn, 1987; Nonaka et al., 1990) and for differences in sequences of development revealed only between pairs of teeth (e.g., Fanning and Moorrees, 1969; Tompkins, 1996; Liversidge and Speechly, 2001). Variability in sequences of dental development in extant humans is acknowledged by some scholars (Mann et al., 1987; Smith, 1989) but the problem has not properly addressed so far, due to lack of definition, concepts and methods (see below). Little is known of possible shifts in the sequences of formation of teeth that might be due to sex, age, jaw, and geographic origin, among other factors. It is generally assumed that "the pattern of development should be more robust than is age of stage appearance" (Smith, 1989: 77) and that there is little variability in relative sequences of key events during tooth growth. The comparative analyses of the sequences of dental development observed in fossil hominids have led to opposite views regarding their "modern human-like" or "ape-like" status (Smith, 1986, 1989, 1994; Mann et al., 1987; Lampl et al., 1993; Conroy and Kuykendall, 1995). Discrepancies are due to the lack of studies of variability in patterns of dental development both within extant humans and between the two chimpanzee species. In addition to the problem of quantifying variability and the controversy over which standard is the most appropriate, other limitations on the study of dental developmental patterns are mainly methodological: (i) first, the understanding of patterns of developmental relationships should not be age related because non-adult dental age assessment is another complex problem with factors influencing its quality (accuracy and reliability) (Ritz-Timme et al., 2000; Braga et al., 2005); (ii) second, developing teeth do not mineralize randomly with respect to one another. We do not expect partial correlations between the rates of tooth formation to approach zero when controlled for age. Indeed, teeth are topographically, developmentally and functionally associated with each other. Teeth essentially grow as a unit (see below) and should be statistically considered as dependent units which grow within a developmental module (Figure 1).

Figure 1. Modularity in dental development with 126 combinations (numbered as follows, in bold) derived from any sequence comprising 7 permanent teeth.

The growing teeth represent a developmental module, with stable or labile interactions (to be quantified) between its hierarchical units (a tooth or a group of teeth)

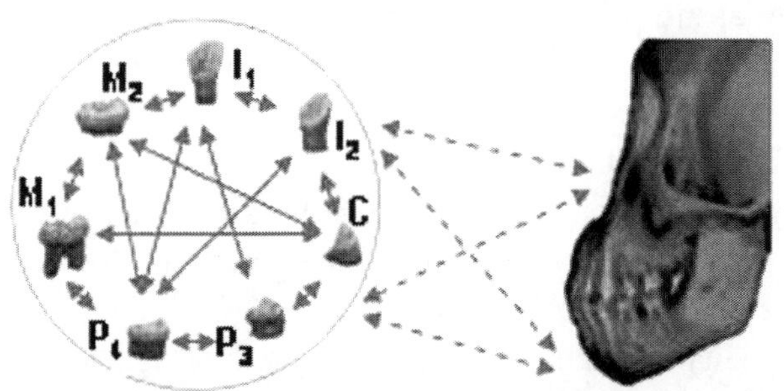

126 combinations (numbered as follows, in bold) derived from a sequence comprising 7 teeth

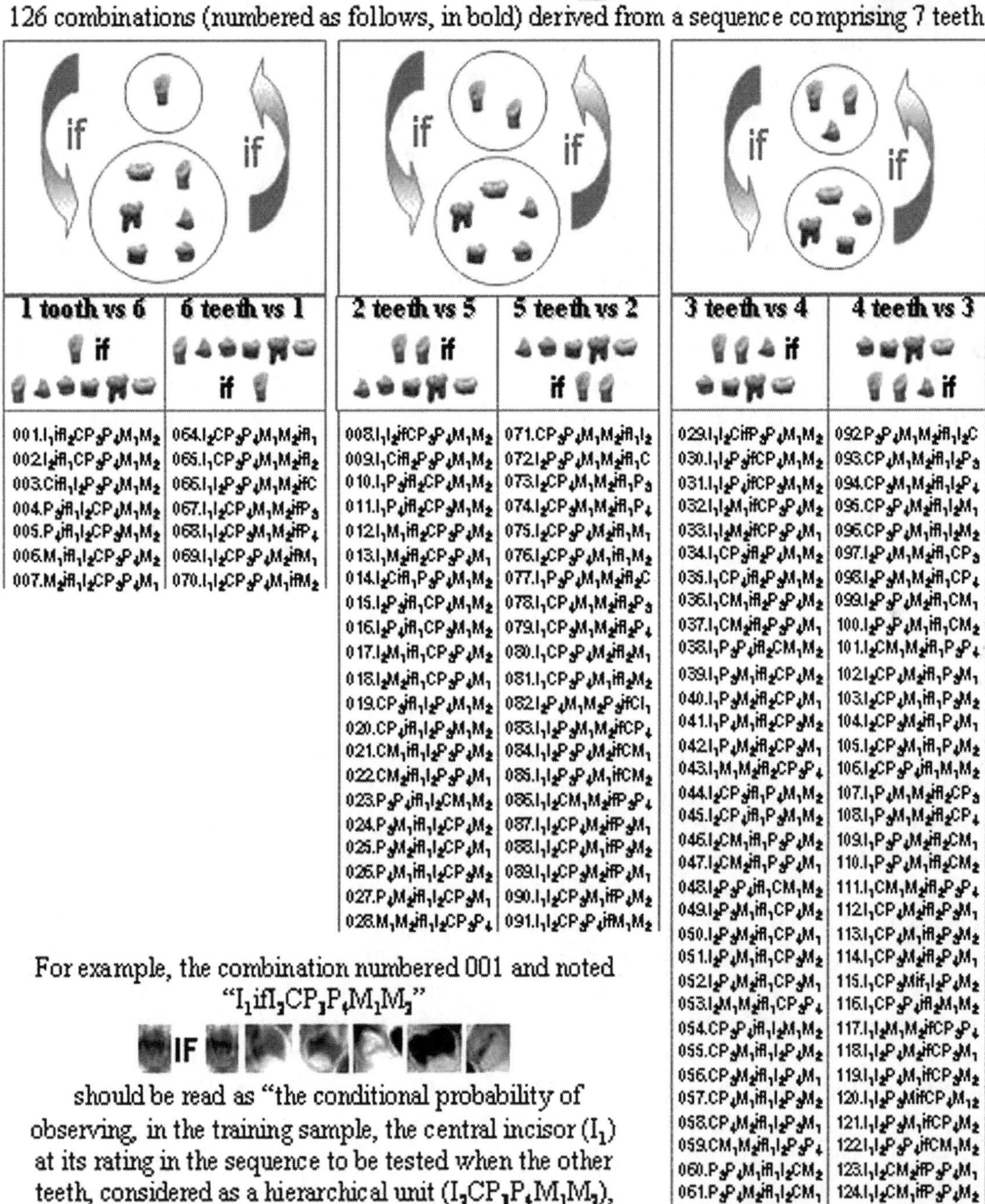

For example, the combination numbered 001 and noted "$I_1ifI_2CP_3P_4M_1M_2$"

should be read as "the conditional probability of observing, in the training sample, the central incisor (I_1) at its rating in the sequence to be tested when the other teeth, considered as a hierarchical unit ($I_2CP_3P_4M_1M_2$), are observed at the ratings seen in the same sequence."

Any assumption of independence (in a statistical sense) between developing teeth is a fallacy. Age should be estimated not only from as many available teeth as is possible, but also by considering teeth as dependent units (Braga et al., 2005); this can be done even in the case of fragmentary remains (Heuzé, 2004); and (iii) third, usually more than two teeth are represented in sequences of dental development. The intricate relationships existing in the set of all teeth need to be examined for an accurate, reliable and comprehensive outline of dental developmental patterns.

The dental developmental status of an individual corresponds to the degree of various growth processes: the formation of crypts, the mineralization of crowns and roots, and their eruption status. Any combination of offset, duration and rate between developing teeth results in a sequence of dental development. A sequence of dental development is a series of discrete events, from the first evidence of mineralization to the closure of the root apex in all teeth (i.e., based on ratings of tooth formation made from panoramic radiographs) (Figure 1). It affects the growing teeth during pre-eruptive (movements of the growing tooth germs within the alveolar process before root formation), pre- functional eruptive (pre-emergent and post-emergent) and functional eruptive phases, when teeth move relative to each other and relative to the developing jaws. A dental mineralization sequence then corresponds to growing elementary units (teeth) embedded within the jaws, observed at a given developmental stage of an individual. Teeth are connected by many interactions that channel variation in sequences of dental development in non-random ways, and that lead to morphological patterns of co-variation. Moreover, even if minor, teeth also exhibit some degree of interaction with their respective jaws. We still have not clearly identified these interactions; however, experimental and clinical observations have demonstrated that tooth development and alveolar bone growth are interdependent (Marks and Cahill, 1987; Wise et al., 2002). There is no unifying analytical approach that provides a description of the dynamics of dental development. It is, therefore, important to identify within and between species both the more conservative and more plastic patterns of co-variation during dental growth. Indeed, conservative interactions among developing teeth arise primarily by strong, direct connections between their developmental pathways. In this case, developing teeth will be considered to co-vary as a unit during evolution, and represent an integrated module (Olson and Miller, 1958). On the contrary, plastic patterns of interactions arise primarily in parallel but separate developmental pathways. Therefore, analysis of the complex interactions existing between growing teeth may well represent a new field in evolutionary developmental biology (EVO-DEVO). Are developmental changes in a set of teeth accompanied by correlative changes in other teeth in extant humans and chimpanzees? How did morphological patterns of co-variation evolve in humans, and which changes in interactions between growing teeth may have produced evolutionary changes? To contribute to answering this question, we hypothesize that the developing dentition represents a developmental module with its several discernible properties defined by Raff (1996): (i) it has "an autonomous, genetically discrete organization"; (ii) it is composed of "hierarchical units and may in turn be parts of larger hierarchical entities"; (iii) it has a physical location within the developing system; (iv) it exhibits varying degrees of connectivity to other modules; (v) it undergoes temporal transformations. Indeed, a sequence of dental development in an individual is not necessarily stable through time. For example, in a given individual examined at two successive stages A and B, of his somatic development; teeth A1 and A2 may show, at stage A, the same degree of mineralization whereas, at stage B, the same teeth, B1 and B2, may be shifted and then show two different degrees of mineralization. As a cross-sectional assessment, this study does not examine the temporal variability of a sequence in an individual. This is also the case in studies applied to fossil hominids and those based on radiographic stages.

This paper describes a novel analytical approach: (i) to provide a comprehensive description of the complex interactions existing between the growing permanent teeth (third molars excepted); and (ii) to quantify variability in sequences of dental development among extant humans at the gross anatomical level and independently of chronological age. Importantly, our method integrates the fundamental concept of modularity by rejecting the old fallacy of analyzing data on the assumption that it contains no information beyond that revealed on a tooth-by-tooth analysis. On the contrary, we believe that the phylogenetic and taxonomic significance of dental developmental patterns arises from the interactions between growing teeth, for it is these that may contain the information which enables developmental or evolutionary information to be uncovered.

Materials and Methods

Materials

The training sample is based on cross-sectional standardized panoramic radiographs of the teeth of 2089 children (1206 girls and 883 boys) aged 36 to 192 months. The geographic composition of the sample is presented in Table 1. These radiographs were taken when indicated by treatment need, and later digitally photographed for subsequent studies in our department (University of Bordeaux 1). Importantly, because magnification may vary and distortions are not uncommon, and also because of the absence of growth standards for tooth length in children from various geographic populations, we decided to assess mineralization radiographically by using fractions of crown and root formed (see above). We selected the children who were clinically free of anomalies in tooth numbers, size or shape. None of the children had undergone orthodontic treatment at the time of the selected radiograph. Our study was ethically approved by a French institutional board (Commission Nationale de l'Informatique et des Libertés); no regulation existed in Iran or in Ivory Coast. Our study includes the left mandibular permanent teeth except third molars, owing to the well-known practical difficulties in the radiographic visualization of the maxillary teeth. We exclude the third molar from our study for several reasons. First, this tooth is often not observed in our sample, and, second, the absence of this developing tooth may or may not correspond to its agenesis. Owing to the high variability (still to be measured across extant human populations) in the timing of the third molar development (Mincer et al., 1993), we cannot determine whether an absence of the germ corresponds to an agenesis or to a "delay" in the mineralization of this tooth.

Radiology has the disadvantage of detecting mineralization with some delay as compared to histology. However, the principal strength of radiology is that it allows a much better assessment of developmental variability: it allows large samples of individuals of varying geographic origin to be observed and any given child will be represented by each tooth category (i.e., incisors, canine, premolars, and molars). Indeed, when microstrutural studies have been conducted on relatively large samples (e.g., Yuan, 2000) (still smaller samples than in radiographic studies), all the teeth of a dentition have never been analysed.

*Table 1. Geographic composition of the sample. (*some of them being, at least in part, of European origin)*

Number of individuals		*Geographic origin*	*Age range (in months)*		*Parental Origin*
Girls	*Boys*		*Girls*	*Boys*	
242	214	South of France	54 to 192	47 to 189	Four grand parents from Western Europe
392	291	South of France	42 to 192	43 to 192	Unknown, or mixed, geographic origin
249	135	Southern Iran	69 to 191	77 to 190	Iran
323	244	Ivory Coast	36 to 181	45 to 190	Ivory Coast *

Rating Dental Development

Each tooth was assigned a developmental status from the first evidence of crown mineralization to the closure of the root apex, by using the height-stages (from A to H) system described by Demirjian et al. (1973) with its precise descriptive points, radiographs and line drawings. We believe that this rating method, with its relatively small number of formation stages, is more accurate than the 14- stage system developed by Moorrees et al. (1963a) for the permanent teeth. Reducing the number of stages decreases the possibilities of inter- and intra-observer errors. These errors also decrease with increased training in scoring. Some scholars argue that rating dental development by using fractions of crown and root formed is largely subjective, because we do not know the total length of the future completed crown or root. However, when quantifying mineralization from linear measurements taken from radiographs (even if not considering the problem of magnification and distortion; see above), we also do not know the total length of the future completed crown and root. So how can we handle the problem of variability in crown and root length in extant human populations, and how do we take it into account to assess dental development? Moreover, inter- and intra-observer errors also exist when taking measurements (only automatic, operator independent, methods are purely non-subjective). One of us (YH) rated the crown and root mineralization in the total training sample. The other author rated dental development in a fraction of our sample. Disagreement occurred in less than 5% and, at the most, by one stage (usually, at the end of the incisor root formation) (4.2%).

Decomposing a Sequence Into Combinations

Our method aims to quantify patterns of dental development and is based on the hypothesis that any sequence represents a developmental module composed of its hierarchical units (i.e., isolated teeth or groups of teeth) (Figure 1), which exhibit varying degrees of interactions. Indeed, any method aiming to measure patterns yet ignoring this aspect in statistical analyses by assuming independence between teeth is rooted on a circular fallacy. For example, in Smith's (1989) method, a dental mean ape is assigned to each developing tooth of an individual from average standards of human and great ape development. This tooth-by-tooth age assessment method obviously assumes independence between teeth because average standards for each tooth are derived from different subsamples of individuals with differing age distributions (e.g., the average standard for the stage observed on the canine is not derived from the same individuals than the average standard observed on the first molar). Indeed, Moorrees et al. (1963a) developed the human standards used by Smith (1989) and mentioned this important limitation on the assessment of dental maturation by acknowledging "possible variation between rates of development of different teeth in a given child". As previously stated by Braga et al. (2005), the variability across age classes is more likely higher for a given radiographic stage observed on an isolated tooth than for a given mineralization sequence from which this isolated tooth is derived. In other words, individuals will be distributed in a larger age range when considering a single tooth with its radiographic stage than when considering the same tooth within a given sequence. As a consequence, we simply argue that Smith's (1989) age based method produces results which largely depend on the independence assumption. However, this method has the advantage that it is easy to use with a simple program and with only stages of dental development as input. Our aim is also to devise an easily used program, but one that is not based on age assessments and an assumption of independence between growing teeth. Inversely, we use the concept of modularity.

Degrees of interactions between isolated teeth, or groups of teeth, can be measured by decomposing any sequence into two subsets with no element in common (Figure 1). Decomposition begins because any sequence comprising 7 developing permanent teeth can be represented by a rearrangement of its seven elements, where none are lost, added, or changed, but distributed into two subsets (each comprising one tooth or more, and noted "sub.k" and "sub.q"; where $k+q=n$ and $n=7$), which are combined by a conditional probability. Any sequence of dental development with seven teeth can then be decomposed in a finite number of combinations (as opposed to permutations because in combinations the order of the teeth in each subset is not important, e.g., a subset comprising the canine and the lateral incisor will be identical to the subset comprising the lateral incisor and the canine). Such combinations correspond to conditional probabilities (Figure 1) of observing one subset at radiographic stages (a prior event) as we condition on the attained developmental status of the other subset (a posterior event). The total number of combinations derived from 7 teeth by grouping into two subsets, is 126 (Figure 1). This number is given by the so-called binomial coefficient. Importantly, this decomposition approach is suitable for use with fragmented remains (e.g., fossil hominids). In this case, the total number of combinations will decrease according to the number of preserved teeth.

Calculating Conditional Probabilities

The probability of observing, in our training sample, each of the 126 combinations derived from any complete (7 teeth) sequence of dental development, is calculated using Bayes's rule of conditional probability, with teeth being considered as statistically dependent units (as opposed to "Naive Bayes", with units are considered as independent). The Bayesian approach has recently been used in non-adult dental age assessment from ratings of crown and root formation, considered as statistically dependent units (Braga et al., 2005). The conditional probability of observing the development status of "subset q" (noted "sub.q"), as we condition on the developmental status of "subset k" (noted "sub.k"), is written as follows:

$$P(\text{sub.q}/\text{sub.k}) = \frac{P(\text{sub.k}/\text{sub.q}) \times P_{\text{prior}}(\text{sub.q})}{[P(\text{sub.k}/\text{sub.q}) \times P_{\text{prior}}(\text{sub.q})] + [P(\text{sub.k}/\sout{\text{sub.q}}) \times P_{\text{prior}}(\sout{\text{sub.q}})]}$$

The term P(sub.k/sub.q) represents, for subset q, the observed proportion of individuals evincing, in the training sample, subset k. The term P(sub.k/~~sub.q~~) represents, for all subsets differing from subset q, the observed proportion of individuals evincing subset k (see Braga et al., 2005; for more details about the computations). Crucial to this approach is the selection of appropriate prior probabilities ("priors"). A comparison of conditional probabilities from different trials requires the assumption that individuals are derived from samples of balanced distributions of radiographic stages. Training samples never exhibit unbiased distributions of radiographic stages, because they are constructed by "availability" sampling. Consequently, in absence of any specific prior knowledge, we assume that each mineralization stage has the same prior probability to be observed (a so-called "uninformative prior"). The Bayesian procedure is then effective in removing part of the factors of confusion due to sampling bias. As the uniform priors appear in both numerator and denominator, by factorization, they have no influence on the calculation of the conditional probabilities.

Bootstrapping and Mean Conditional Probabilities

Because of finite training sample size, it is possible that some dental mineralization sequences in extant human populations will not be sampled. Bootstrapping is a way of testing the reliability of our training sample, i.e., the degree to which adding or removing individuals in the training sample changes the estimate of variability in sequences of dental development. Therefore, we generated 1000 pseudoreplicate datasets by randomly resampling with replacement 1000 individuals from our training sample (bootstrapping). Consequently, instead of deriving one conditional probability from our total training sample, we obtained distribution of 1000 randomized conditional probabilities for each of the 126 combinations. The mean value of each distribution is then used as a measure of the likelihood of each of the 126 combinations (Figure 2). If this mean probability is higher or equal to 0.75 it is considered to be likely. Inversely, if the mean probability is lower or equal to 0.25 the combination is considered to be unlikely (Figure 2). A mean probability ranging between 0.25 and 0.75 is considered to be not informative (Figure 2).

Let us now consider the example given in Figure 2. Using Demirjian et al.'s (1973) rating system for all permanent teeth (except the third molar), from mesial to distal, the sequence tested is written as follows: G, F, F, E, E, G, E. In this example, 4 combinations (numbered 22, 58, 85 and 121) out of 126, are unlikely, with their mean probabilities falling under 0.25. As shown in Figure 1, the combinations numbered 22, 58, 85 and 121 correspond, respectively, to the following conditional probabilities: CM_2 if $I_1I_2P_3P_4M_1$, CP_4M_2 if $I_1I_2P_3M_1$, $I_1I_2P_3P_4M_1$ if CM_2, $I_1I_2P_3M_1$ if CP_4M_2. Therefore, from these combinations, we can conclude that, in this sequence, the developmental rating(s) of the canine and/or the second premolar and/or the second molar deviate significantly from what we observe in our training sample. When we now examine the combinations corresponding to the developmental status of each of these three teeth taken separately, versus of all other permanent teeth (i.e., C if $I_1I_2P_3P_4M_1M_2$, P_4 if $I_1I_2CP_3M_1M_2$, M_2 if $I_1I_2CP_3P_4M_1$, $I_1I_2P_3P_4M_1M_2$ if C, $I_1I_2CP_3M_1M_2$ if P_4, and $I_1I_2CP_3P_4M_1$ if M_2; corresponding, respectively, to numbers 3, 5, 7, 66, 68 and 70), we observe that the lowest probabilities are for the relative dental development of the second molar. Therefore, we conclude that this tooth shows an abnormal developmental status in this sequence.

Figure 2. Example of a sequence of dental development with each of its 126 combinations correspondingto a conditional probability. For each combination, we calculate the mean of 1000 probabilities obtained after 1000 resamplings with replacement.

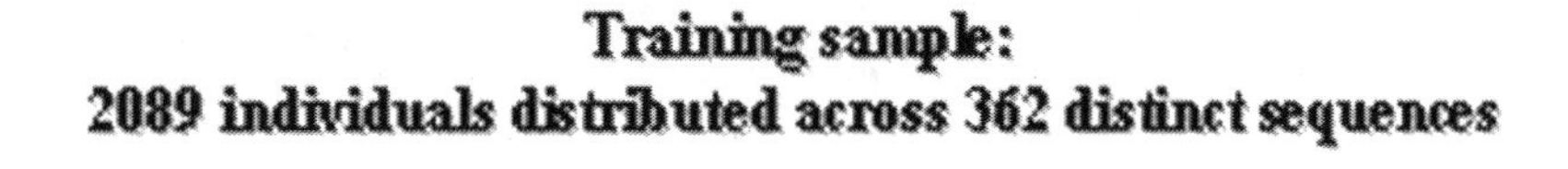

163 individuals with their M1s before or at stage F, distributed across 99 distinct sequences (grey color) (63 singles, 22 double … etc).
1926 individuals with their M1s at stages G or H, distributed across 263 distinct sequences (black color) (117 single, 38 double … etc).

Occurrence level

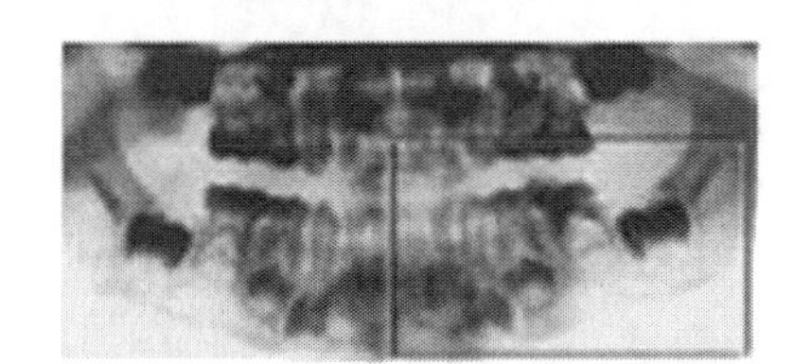

For each individual to be tested with this training sample

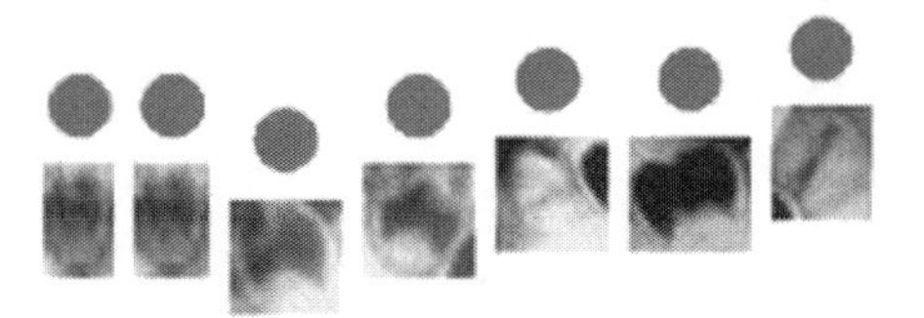

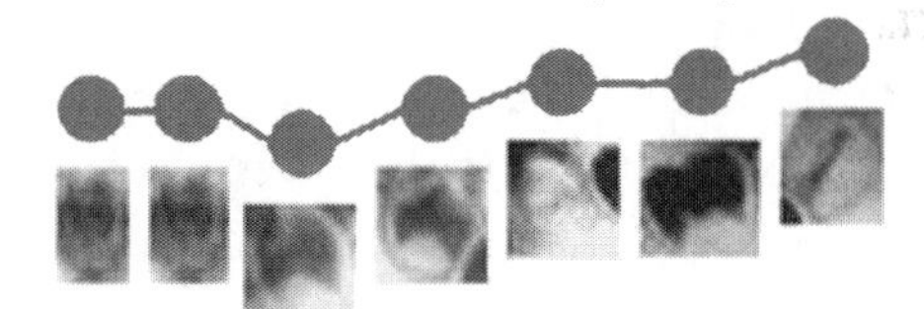

Independence assumption

Dependent approach
The sequence is decomposed into 126 combinations; bold numbers, see Figure 1).
For each of the 126 combinations (represented below, from right to left), we calculate the mean of 1000 probabilities obtained after 1000 resamplings with replacement.

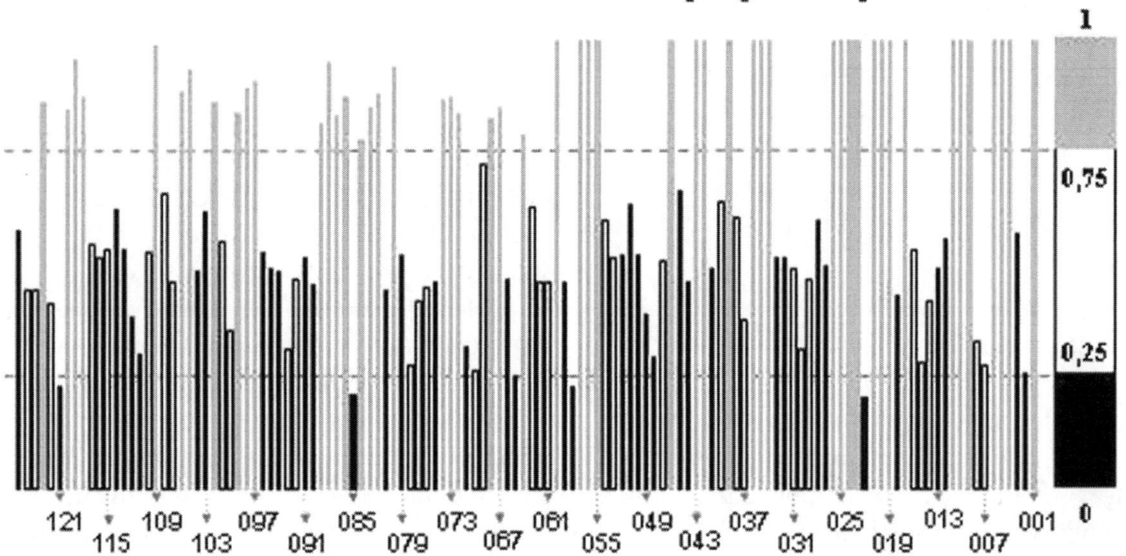

In order to automate the calculations, Dr. F. Houët (from our department) created a Microsoft Excel 2000® Visual Basic macro to determine the relative frequencies of subsets q and k, and to apply Bayes's rule of conditional probability. This program is easy to use, with only stages of dental development as input. We will make it accessible through a web-site in the future.

Testing Sequences and Cross-Validation

After devising our method, which decomposes any sequence into 126 combinations (Figure 1), and which estimates the generalization error (using bootstrapping) for each of the corresponding 126 conditional probabilities (Figure 2), we need to test each sequence of dental development by using a leave-one-out cross- validation. We test sequences of dental development rather than individuals because most individuals share a common sequence, as illustrated in Figure 2 (see above). After cross-validation, the generalization error of our model can be assessed with a representation of the likelihood and the variability of each of the 126 combinations. For each combination, the likelihood and its variability are represented, respectively, by the mean conditional probability and its coefficient of variation, calculated across all sequences that have been tested (Figure 3). This method can then help us to identify the more conservative and the more plastic combinations, or morphological patterns of covariation during dental growth, in our training sample.

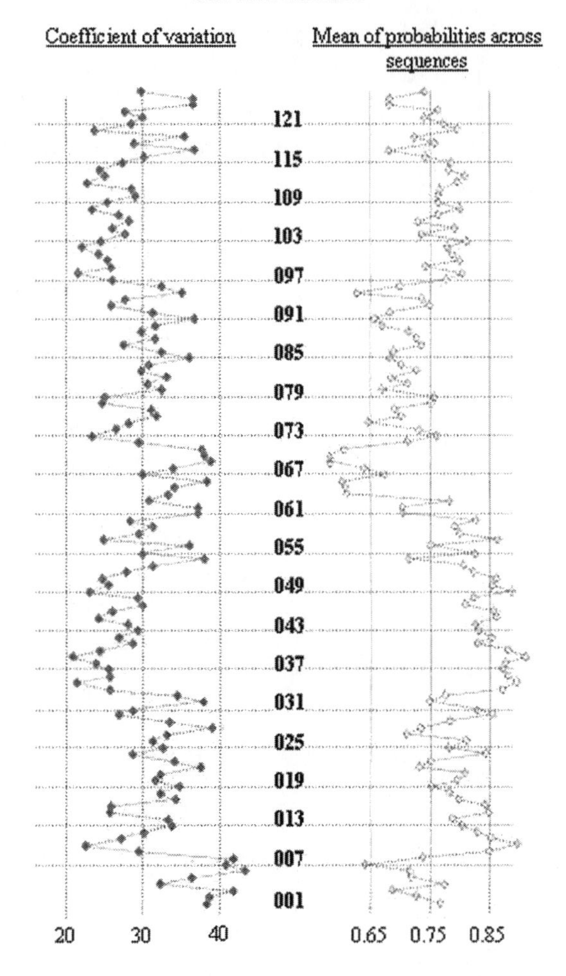

Figure 3. Leave-one-out cross validation in our training sample: 182 multiple sequences (see Figure 2) have been tested.

Results

In our training sample, 1926 individuals out of 2089 had their first permanent molars at stages G or H (roots completed or almost completed; see Demirjian et al., 1973). These 1926 individuals do not share the same developmental sequence. Instead, they are distributed across 263 different sequences that need to be tested. As shown in Figure 2, some of these 263 sequences with the M1 at stages G or H (117), are represented only once in our training sample. This is the lowest occurrence level for a sequence. These sequences are called "single" as opposed to "multiple". Each single sequence corresponds only to one individual. Each multiple sequence corresponds to two or more individuals. Multiple sequences demonstrate different occurrence levels. Nevertheless, as the occurrence level increases, the number of distinct sequences showing this occurrence level, decreases rapidly in the training sample (Figure 2). In our training sample, when we consider only the 1926 individuals with their first permanent molars at stages G or H, there are 117 distinct single sequences (sequences occurring once, each representing only one individual), 38 distinct double sequences (sequences occurring twice, each representing two individuals, 16 distinct triple sequences, and so on (see Figure 2). When we now consider, in our training sample, the 163 individuals with their first permanent molars before or at stage F, we observe the same trend: as the occurrence level increases, the number of distinct sequences showing this occurrence level decreases rapidly.

We now examine which of the 126 combinations are plastic or conserved across the 182 multiple sequences that were tested. In Figure 3, the combinations are numbered and ordered as in Figures 1 and 2, with their corresponding mean probabilities (ranging from 0.58 to 0.91) and coefficients of variation (ranging from 20.8 to 43.9). Assuming a normal distribution of all 126 mean probabilities, 95% are within two standard deviations (within 0.62 and 0.90). The following six combinations fall outside this range: $I_2CP_3P_4M_1M_2$ if I_1, $I_1CP_3P_4M_1M_2$ if I_2, $I_1I_2P_3P_4M_1M_2$ if C, $I_1I_2CP_3P_4M_2$ if M_1, $I_1I_2CP_3P_4M_1$ if M_2, $CP_3P_4M_1M_2$ if I_1I_2 (respectively, numbers 64 to 66 and 69 to 71) (Figure 3). Five out of six of these combinations represent the developmental status of six permanent teeth given the developmental rating of one isolated tooth. From these results, the premolars do not seem to deviate significantly from the other teeth during dental growth. Interestingly, the sixth combination falling outside the 95% confidence interval of the distribution of all mean probability values corresponds to the combination numbered 71: $CP_3P_4M_1M_2$ if I_1I_2 (Figure 1). This means that, in any sequence that was tested, the probability of observing the permanent incisors at their relative developmental status, in the training sample, is low. The corresponding low mean probability is associated with the second higher coefficient of variation (41.9) (the highest – 43.9 – is for combination number 6 which is associated with a mean probability falling inside the 95% confidence interval). In other words, these results mean that the mandibular incisors are generally developmentally more plastic compared to other permanent mandibular teeth.

Discussion

Robert (2004: 129) proposes "... a broad interpretation of evo-devo, one according to which the development of whole organisms and the evolution of their modular parts are deemed the primary analysands, rather than as secondary to the epigenetic expression of purely or primarily genetic potential." Robert (2004: 129) goes on to write: "Such is the force of taking development seriously. A developing system is clearly organized, but in a systemic way; that is, the interrelations between its parts are structured into causal, generative systems with ontogenetic control dispersed throughout these systems". In this evo-devoist perspective, we believe that our results are important. Using radiographic studies (Smith, 1989; Tompkins, 1996), scholars have asserted that there is "a modern human-like sequence of dental development" that was already present in *Homo erectus* and Neanderthals (Dean et al., 2001; Ramirez Rozzi and Bermúdez de Castro, 2004). It is assumed that "a modern human-like sequence of dental development", as opposed to various human-like sequences, characterizes extant humans and their Upper and Middle Pleistocene fossil relatives. This assumed 'modern human-like sequence" is then used to compare fossil hominids represented by single, or few, isolated teeth and to understand tooth (especially enamel) growth processes (Dean et al., 2001; Ramirez Rozzi and Bermúdez de Castro, 2004). In particular, this assumption has been used in microanatomical (or histological) analyses to argue that, the anterior teeth can serve as a reliable substitute for the other permanent teeth (e.g., the molars). It has also been used for interpretations of somatic maturation and brain size. However, even if microanatomical studies now allow the study of small samples of children and the quantification of variation in rates of enamel formation of a given tooth (usually, the permanent incisors), we still have not qualified possible variability between rates of development of different teeth in a given child.

The results of our basic analytical approach (Figure 3) suggest that the developmental relationships between, on the one hand, the permanent incisors (one subset), and on the other hand, the canine, premolars and molars (another subset), are the most statistically plastic in extant humans. Therefore, these two subsets of teeth may well represent two hierarchical units within the same developmental module. Obviously, before applying our method to fossil hominids (preferably in single fossil hominids rather than in fossil samples mixing different periods, populations, or even species), our results need to be confirmed by further testing in extant human populations. We should add that the incisor developmental instability and variation demonstrated in this study may arise primarily by parallel variation of separate developmental pathways and simultaneous effects by an extrinsic source of variation. This source could be the developmental constraints due to the growing premaxilla or incisive bone, which may represent another modular unit. In this case, the developing canine, premolars and molars could be considered to co-vary as a hierarchical unit during evolution, arising primarily by strong connections between developmental pathways.

This study aimed to assess variation in the relative sequences (or patterns) of key events during formation of crowns and roots from samples representing different extant human populations, by using a method that does not assume tooth independence. Our training sample is biased because early radiographic stages (from A to F), especially for early developing permanent teeth (incisors and first molar), are underrepresented (Figure 2). Indeed, a high proportion of individuals with their first permanent molars before or at stage F, is represented by single dental mineralization sequence (63 individuals out of 163; i.e., 38.7%) (Figure 2). For individuals with their first permanent molars at stages G or H, this proportion decreases considerably (117 individuals out of 1926; i.e., 6%) (Figure 2). A temporary limitation of our method, which does not assume independence, is that no computations can be made for those individuals represented by unique dental development sequences. For these unique sequences, a "Naïve Bayes" approach can easily be developed but, again, will not be biologically realistic (because of the independence assumption) (Heuzé, 2004; Braga et al., 2005). Therefore, in future studies, we need to sample more individuals with their first molars before or at stage F. More sampling will also allow us to quantify possible sexual dimorphism and/or geographic variation of sequences of dental development across extant human populations.

So far, our results are limited to extant humans with their very limited lower facial prognathism coupled with reduced jaw elevator muscles (when compared with chimpanzees). An interpretation of our results in a totally different context of facial morphology and growth would be misleading. Importantly, we still need data on dental development in the pygmy chimpanzee and close attention to the interrelationships between jaw and lower/upper tooth growth in both chimpanzee species in which the patterns of closure of the premaxillary (or incisive) suture are totally different (Braga, 1998) but also in extant humans. In this context, we suggest that it would be premature to make direct comparisons, using our method, between extant humans and chimpanzees, or even, between fossil hominids and chimpanzees, in which the premaxilla (containing the alveolar ridge of the four upper incisors), acting as a stabilizing element within the facial skeleton, plays a totally different role in both craniofacial growth and morphology.

Finally, more testing is needed to determine how stable a crown and root-formation sequence is through ontogeny. In other words, would our results be the same using a different sample with a much younger age distribution? Indeed, Moorrees et al. (1963a, b) have demonstrated that later formation stages vary more in their timing than early stages and later-forming teeth vary more in their timing than early forming teeth. However, timing and pattern are two separate things and this trend for timing and its implications for developmental patterns need to be confirmed. This question cannot be addressed simply by examining pairs of developing teeth and testing whether the stage of development of one given tooth increases when another given tooth may increase its range of calcification. This kind of simplistic observation may represent a false correlation caused by the influence of a third single tooth or group of teeth. Therefore, as previously stated, all possible combinations between teeth need to be examined. Nevertheless, our results suggest a developmental instability of the incisors and integration of the developing canine, premolars and molars. We regard this result as consistent with findings suggesting that the first permanent molar provides indications of overall somatic maturity and brain size in living primates (Smith, 1989).

Conclusions

Dental development studies fall mainly into two methods of observation, which now co-exist but which succeeded each other historically, the first method leading to the hypothesis of an evolutionarily early appearance of a human-like dental development and the later method giving birth its negation. Indeed, scholars first used radiographs and attempted to focus on dental macrostructural development. Pioneering comparative studies (Mann, 1975; Skinner, 1978; Skinner and Sperber, 1982) suggested that early hominids followed a human-like schedule of dental growth. This hypothesis of an evolutionarily early appearance of a human-like dental development was established on the principal strength of developmental variability in both extant humans and the great apes (Mann et al., 1987). The opposing hypothesis was built up with a different method of observation, which was based on the study of incremental growth features in enamel to estimate age at death of fossil hominids (Bromage and Dean, 1985). These microanatomical observations are based on the record of a precise timescale for enamel developmental events on growth increments. A large number of studies (e.g., Dean, 1987; Beynon et al., 1998; Dean et al., 2001) now consider this latter approach as a more secure model on which to base studies of the paleodemography, growth and maturation of early hominids. This led to the more general consensus that truly human-like dental development emerged relatively late in human evolution because all early hominids documented thus far, including some earliest fossils attributed to *Homo,* did not resemble modern humans in their development. These two views can now be reconciled because the conceptual framework of modularity enables us to build new methods to achieve an agreement. More precisely, modularity and EVO-DEVO (Raff,

1996; Bolker, 2000) offer a new conceptual framework and new perspectives to provide a picture of the variability in sequences of dental development, as well as in enamel and dentine growth processes, as possible foundations for morphological changes during human evolution. Indeed, our hypothesis of modularity in dental development and our method derived from this concept can serve as a basis for identifying and studying patterns of dental growth and, importantly, for comparisons populations or species that do not need to be hedged with age assessments based on a tooth-by-tooth analysis and its fallacies. At this stage, we suggest that our first comprehensive quantification of variability in patterns of development of permanent teeth in three extant human populations opens an EVO-DEVO perspective. Therefore, we aim to contribute to EVO-DEVO in the future by examining modularity/patterning in fossil hominid samples and by further testing (e.g., influence of sex, age and geographic origin) of the conclusions presented here.

Acknowledgments

We would like to express our gratitude to Shara Bailey and Jean-Jacques Hublin for inviting us to Leipzig to contribute to this special edition and their help throughout its preparation. This study was funded by the French Ministry of Research ("Action Concertée Incitative") and the GDR 2152 of the CNRS. We thank Drs. Chabadel, de Brondeau, Garde, de la Talonnière, Roux, Boileau, Chouvin, Hassid, Blanca, Thévenet, Frapier (France), Sonan (Ivory Coast) and Gueramy (Iran), for access to the orthopantomographs. We also thank F. Houët and B. Dutailly for their help in developing the program for the calculations and an anonymous referee who helped us to improve our discussion.

References

Beynon, A.D., Clayton, C.B., Ramirez Rozzi, F.V., Reid, D.J., 1998. Radiographic and histological methodologies in estimating the chronology of crown development in modern human and great apes: a review, with some applications for studies on juvenile hominids. Journal of Human Evolution 35, 351–369.

Bolker, J.A., 2000. Modularity in development and why it matters to evo-devo. American Zoologist 40, 770–776.

Braga, J., 1998. Chimpanzee variation facilitates the interpretation of the incisive suture closure in South African Plio-Pleistocene hominids. American Journal of Physical Anthropology 105, 121–135.

Braga, J., Heuze, Y., Chabadel, O., Sonan, N.K., Gueramy, A., 2005. Non-adult dental age assessment: correspondence analysis and linear regression versus Bayesian predictions. International Journal of Legal Medicine 119, 260–274.

Bromage, T.G., Dean, M.C., 1985. Re-evaluation of the age at death of immature fossil hominids. Nature 317, 525–527.

Conroy, G.C., Kuykendall, K., 1995. Paleopediatrics: or when did human infants really become human? American Journal of Physical Anthropology 98, 121–131.

Dean, M.C., 1987. Growth layers and incremental markings in hard tissues: a review of the literature and some preliminary observations about enamel structure in *Paranthropus boisei*. Journal of Human Evolution 16, 157–172.

Dean, C., Leakey, M., Reid, D., Schrenk, F., Schwartz, G., Stringer, C., Walker, A., 2001. Growth processes in teeth distinguish modern humans from *Homo erectus* and earlier hominins. Nature 414, 628–631.

Demirjian, A., Goldstein, H., Tanner, J.M., 1973. A new system of dental age assessment. Human Biology 45, 211–227.

Fanning, E.A., Moorrees, C.F., 1969. A comparison of permanent mandibular molar formation in Australian aborigines and Caucasoids. Archives of Oral Biology 14, 999–1006.

Garn, S.M., Sandusky, S.T., Nagy, J.M., Trowbridge, F.L., 1973. Negro-Caucasoid differences in permanent tooth emergence at a constant income level. Archives of Oral Biology 18, 609–615.

Heuzé, Y., 2004. Chronologie et étiologie de la maturation macrostructurale des dents définitives. Doctorat de l'Université Bordeaux 1, Inédit.

Kuykendall, K., 1992. Dental development in chimpanzees (*Pan troglodytes*) and implications for dental development patterns in fossil hominids. Ph.D. Dissertation, Washington University, St Louis.

Lampl, J., Monge, J.M., Mann, A.E., 1993. Further observations on a method for estimating hominoid dental developmental patterns. American Journal of Physical Anthropology 90, 113–127.

Liversidge, H.M., 2003. Variation in modern human dental development. In: Thompson, J.L., Krovitz, G.E., Nelson, A.J. (Eds.), Patterns of Growth and Development in the Genus *Homo*. Cambridge University Press, Cambridge, pp. 73–113.

Liversidge, H.M., Speechly, T., 2001. Growth of permanent mandibular teeth of British children aged 4 to 9 years. Annals of Human Biology 28, 256–262.

Mann, A.E., 1975. *Some Paleodemographic Aspects of the South African Australopithecines*. University of Pennsylvania Publications in Anthropology, Number 1, Philadelphia.

Mann, A.E., Lampl, M., Monge, J., 1987. Patterns of ontogeny in human evolution: evidence from dental development. Yearbook of Physical Anthropology 33, 111–150.

Marks, S.C., Cahill, D.R., 1987. Regional control by the dental follicle of alterations in alveolar bone metabolism during tooth eruption. Journal of Oral Pathology 16, 164–169.

Mincer, H.H., Harris, E.F., Berryman, H.E., 1993 The A.B.F.O. study of third molar development and its use as an estimator of chronological age. Journal of Forensic Science 38, 379–390.

Moorrees, C.F., Fanning, E.A., Hunt, E.E., 1963a. Age variation of formation stages for ten permanent teeth. Journal of Dental Research 42, 1490-502.

Moorrees, C.F., Fanning, E.A., Hunt, E.E., 1963b. Formation and resorption of three deciduous teeth in children. American Journal of Physical Anthropology 21, 205–213.

Nonaka, K., Ichiki, A., Miura, T., 1990. Changes in the eruption order of the first permanent tooth and their relation to season of birth in Japan. American Journal of. Physical Anthropology 82, 191–198.

Olson, E., Miller, R., 1958. *Morphological Integration*. University of Chicago Press, Chicago.

Raff, R.A., 1996. *The Shape of Life: Genes, Development and the Evolution of Animal Form*. University of Chicago Press, Chicago.

Ramirez Rozzi, F., Bermúdez de Castro, J.M., 2004. Surprisingly rapid growth in Neanderthals. Nature 428, 936–939.

Ritz-Timme, S., Cattaneo, C., Collins, M.J., Waite, E.R., Schütz, H.W., Kaatsch, H.-J., Borrman, H.I.M., 2000. Age estimation: the state of the art in relation to the specific demands of forensic practice. International Journal of Legal Medicine 113, 129–136.

Robert, J.S., 2004. *Embryology, Epigenesis and Evolution. Taking development seriously*. Cambridge University Press, Cambridge.

Skinner, M.F., 1978. Dental maturation, dental attrition and growth of the skull in fossil hominidae. Ph.D. Dissertation, University of Cambridge.

Skinner, M.F., Sperber, G.H., 1982. *Atlas of Radiographs of Early Man*. Alan R. Liss, New York.

Smith, B.H., 1986. Dental development in *Australopithecus* and early *Homo*. Nature 323, 327–330.

Smith, B.H., 1989. Growth and development and its significance for early hominid behaviour. Ossa 14, 63–96.

Smith, B.H., 1991. Standards of human tooth formation and dental age assessment. In: Kelley, M.A., Larsen, C.S. (Eds.), Advances in Dental Anthropology. Wiley-Liss, New York, pp. 143–168.

Smith, B.H., 1994. Patterns of dental development in *Homo, Australopithecus, Pan*, and *Gorilla*. American Journal of Physical Anthropology 94, 307–325.

Smith, C.B., Garn, S.M., 1987. Polymorphisms in eruption sequence of permanent teeth of American children. American Journal of Physical Anthropology 74, 289–303.

Tompkins, R.L., 1996. Relative dental development in Upper Pleistocene hominids compared to human population variation. American Journal of Physical Anthropology 99, 103–118.

Yuan, M., 2000. Perikymata counts in two modern human sample populations. Ph.D. Dissertation, Columbia University, New York.

Wise, G.E., Frazier-Bowers, S., D'Souza, R.N., 2002. Cellular, molecular, and genetic determinants of tooth eruption. Critical Reviews in Oral Biology and Medicine 13, 323–334.

4. Dental calcification stages of the permanent M1 and M2 in U.S. children of African-American and European-American ancestry born in the 1990s

J. MONGE

Department of Anthropology
Museum of Anthropology and Archaeology
University of Pennsylvania
Philadelphia, PA 19104
jmonge@sas.upenn.edu

A. MANN

Department of Anthropology
Princeton University
Princeton, New Jersey 08544
mann@Princeton.edu

A. STOUT

Temple University School of Dentistry
3223 North Broad Street
Philadelphia, PA 19140
angela.stout@temple.edu

J. ROGÉR

Marquette University School of Dentistry
P.O. Box 1881
Milwaukee, WI 53201
james.roger@mu.edu

R. WADENYA

University of Pennsylvania School
of Dental Medicine
240 South 40th Street
Philadelphia, PA 19104
wadenya@dolphin.upenn.edu

Keywords: Dental development, European-Americans, African-Americans, *Homo sapiens*, chimpanzee, human variation, human evolution

S.E. Bailey and J.-J. Hublin (Eds.), Dental Perspectives on Human Evolution, 263–274.

Abstract

Reported here are the preliminary results of an ongoing study undertaken to determine if there are significant changes from the results obtained in the middle of the 20th Century in the range of variation in dental development in American children of European and African ancestry. Several thousand orthopanoramic radiographs, available from the Dental Clinics of Temple University and the University of Pennsylvania, in Philadelphia, Pennsylvania will eventually be incorporated into this study. The 170 radiographs that have thus far been analyzed document significant changes in the maturation and development of the first and second permanent lower molars. Children in this study have dentitions that are maturing earlier than those in the samples published by Moorrees et al. (1963) and Demirjian et al. (1973). If the results of this preliminary analysis are confirmed by the incorporation of additional teeth and a larger sample of children, as has been reported elsewhere, it will be necessary to reconsider the characterization of dental development in living humans. Confirmation would also require re-evaluation of the range of inherent plasticity in human dental development. These preliminary results, along with the work by Zihlman et al. (2004), suggest that current concepts of what constitutes "normal" dental development in living humans and chimpanzees may have to be reconsidered; there are also marked implications for the way in which the developing dentition in fossil hominins is characterized.

Introduction

This research began in 2002 when dentists at two independent clinical pediatric practices (Temple University and the University of Pennsylvania, Philadelphia, Pennsylvania) remarked that it was difficult if not impossible to use the staging patterns of teeth traditionally reproduced in countless dental reference volumes used in US dental schools. The children in their practices were developing their teeth at significantly younger ages (e.g., Nadler, 1998). Considering the paucity of available data on dental development in US children in the late 20th and early 21st Centuries, the need for a critical evaluation of the efficacy of these standards to the dental growth patterns in the current cohort of children in the United States was apparent. This same question has been addressed by Rousset et al. (2003) and applied to clinical practice in France.

As evidence has accumulated over the past 35 years, it has become increasingly clear that different populations appear to follow different trajectories in the formation and eruption/emergence of individual teeth (Loevy, 1983; Harris and McKee, 1990; Simpson and Kunos 1998; Liversidge et al., 1999; Rousset et al., 2003). Further, data is now available for quantitative analysis of this trend; this results from the now routine procedure for youngsters entering private and public dental clinics to undergo orthopanoramic radiographs, both at the initiation of dental evaluation and at various milestones through the course of dental care. In concert with our clinical colleagues, we have begun to examine and analyze these radiographic materials. This preliminary report details the analysis of 170 American children, 105 of European-ancestry and 65 of African ancestry. The data presented here are cross-sectional and limited to the examination of the first and second lower permanent molars; it is expected that along with the information from a larger sample of children and from other teeth, this data set will be augmented with at least one additional panoramic radiograph as these children continue their dental care.

From an evolutionary perspective, it is of importance to document the range of human variation in the formation and emergence of the dentition. Although in comparison to other maturational events leading to adulthood [e.g., age of menarche (Herman-Giddens et al., 2004), various determinants of skeletal maturation (Eveleth and Tanner, 1990)], the dentition may be one of the least environmentally sensitive developmental complexes, they are subject to environmental influences

(Nadler, 1998). Further, on the basis of what appears to be population differences in the pattern of dental maturation, Olze et al. (2004) has raised the need for more ethnically based dental development tables for use in the forensic aging of unknown specimens. Recent analysis of data from the Fels Longitudinal Study of US girls born in the 1930s and in each consecutive decade until the 1980s, indicates that the observed lowering of age at menarche is not associated with the trend to increasing BMI (body mass index) documented in childhood and adolescence in recent decades (Demerath et al., 2004). Thus, the changes in at least this specific maturational event appear more complex than simple associations to a single factor in the environment.

These observations can be placed in the context of the growing body of evidence documenting significant variations in maturation variables in the non-human primates. For example, it has been demonstrated that both baboons (Phillips-Conroy and Jolly, 1988; Kahumbu and Eley, 1991) and chimpanzees in the wild possess dental maturation patterns significantly distinct from the patterns recorded on captive animals (compare Nissen and Riesen, 1964 to Zihlman et al., 2004). Thus, various lines of evidence suggest the presence of significant variations in human and other primate dental development and it would appear appropriate to re-evaluate the traditional dental maturational schedules that were based on children who matured a half a century ago.

Literature Review

Recently, Liversidge (2003) summarized all the published studies on the eruption/emergence times and calcification sequences for human populations. Although the results of this study document variation in the timing of calcification and eruption/emergence schedules, Liversidge concludes that these differences are not the result of differences in the formation and timing of the dental development, but primarily on differences in the techniques of analysis used on the samples and the ages of the children from which the data were derived.

Although a substantial body of data exists on the eruption/emergence of teeth in a number of African populations (Krumholt et al., 1971; Blankenstein et al., 1990; Tompkins, 1996 and summarized by Liversidge, 2003), remarkably little information has been published for African-American populations. The notable exception is the Garn et al. (1973 and 1975) study in which a comparison was made in emergence schedules between European and African-American populations (from the Ten States Nutrition Survey 1968–1970) with both groups of low socio-economic status. A similar study was performed on British Black and British White children (Lavelle, 1976). Harris and McKee (1990) evaluated the stage of mineralization of the permanent dentition of a sample of African-American and European-American children between 3.5 and 13 years of age. The stages of the developing dentition of this sample, drawn from the middle southern United States (990 children from the University of Tennessee College of Dentistry) were based on the standards established by Moorrees et al. (1963). The Moorrees et al. (1963) data set derives from a combination of the Fels Longitudinal study supplemented by materials from the Forsyth Dental Infirmary at Harvard University. The children whose radiographs were used in this study were born in the early part of the 20th Century; they were of European-American ancestry and from a middle socio-economic background.

Simpson and Kunos (1998) amassed data on dental development in a sample from the Case Western Reserve University School of Dentistry Pediatric Dental Clinic and the Bolton-Brush Growth Study Center. This sample includes children born in a number of decades, with one portion, from the Case Western Dental Clinic, providing radiographs from children born in the middle 1990s. Unfortunately, the ethnicity of the children was not recorded since this information was not part of the personal histories recorded for these patients. Simpson and Kunos (1998) concluded that their analysis of the data showed remarkable variation in the timing of dental development within their sample and further remarked that there is no simple explanation to explain the variation.

Materials and Methods

Children in the present study ranged in age from four to 14 years. For each child, the ethnicity of the patient was recorded as well as the birth date and the date of panoramic radiograph. Orthopanorama radiographs of African-American children came primarily from the Temple University Dental Clinic but a small sub-sample was collected from the University of Pennsylvania Dental Clinic. Data for the European- American sample were obtained from a private dental practice in a more affluent neighborhood in suburban/urban Philadelphia.

The orthopanoramic radiographs were made during the initial visits of these children and are considered a normal part of dental practice. It is planned that a subset of these individuals will be radiographed at approximately 6 years from the initial clinical visit and this data will be added to the present analysis. Figure 1 illustrates one of these panoramic radiographs, a European-American, female, 5.6 years of age at the time of the initial clinical visit.

In order to facilitate more comprehensive comparative analyses, we applied both the Moorrees et al. (1963) and the Demirjian et al. (1973) staging methods to the individual teeth. Each tooth was independently evaluated by two individuals whose background and training were completely distinct. An example of the data entry sheet along with the Moorrees et al. (1963) standards applied to a panoramic radiograph is shown in Figure 2. Of a total of 4950 tooth stages (both permanent and deciduous teeth were staged but not all have been included here), the raters agreed in the application of the Moorrees et al. (1963) stages 88% of the time; in the remaining 12% of teeth, analysis of dental development varied by one stage only. Given the minor differences in the age associated with the reported median by Moorrees between two adjacent stages, except for the time frame differences in the last stages of root development (apex formation, not used in the analysis here), this interobserver difference appears to be of minor consequence. As many researchers have noted, the Moorrees calcification stages divides each portion of dental development into categories that are often very difficult to precisely resolve. Using the Demirjian et al. (1973) stages of crown and root development (from Stage A through Stage H), accuracy in staging tooth development rose to 92%.

Due to the limitations associated with the age at which the children were radiographed, only the later stages of M1 calcification and partial root completion stages of the M2 are included in this study. The full range of children's ages at these stages is represented in this sample. The same pattern of analysis was applied by Liversidge and Speechly (2001). Developmental status of the incisors, canines and premolars are currently being evaluated.

Discussion

Figures 3 and 4 show the distribution of lower M1 and M2 molar calcification stages using the Moorrees et al. (1963) staging method on European Americans; Figures 5 and 6, on African Americans. As Liversidge and Speechly (2001) point out, the limitations of analysis depend directly on the age distribution of the children included in the study and the relative percentage of children who have achieved each of the stages. Thus, in each case, only those tooth stages where at least 50% of the children had attained that

Tooth Development Database

New Record

Patient Number	8
Date of Birth	11/19/1994
Date of Panorex	7/17/2003
Age at Scan	8.66
Ethnicity	Black
Sex	Female

Special Features

Permanent Teeth

Maxillary Arch				Mandibular Arch			
Stage #1	5	Stage #9	11	Stage #17	7	Stage #25	10
Stage #2	9	Stage #10	12	Stage #18	9	Stage #26	10
Stage #3	12	Stage #11	11	Stage #19	13	Stage #27	11
Stage #4	0	Stage #12	11	Stage #20	11	Stage #28	11
Stage #5	0	Stage #13	10	Stage #21	11	Stage #29	10
Stage #6	11	Stage #14	12	Stage #22	11	Stage #30	14
Stage #7	12	Stage #15	10	Stage #23	10	Stage #31	9
Stage #8	11	Stage #16	7	Stage #24	10	Stage #32	6

PrimaryTeeth

Maxillary Arch				Mandibular Arch			
Stage A	0	Stage F	0	Stage K	5	Stage P	0
Stage B	0	Stage G	0	Stage L	0	Stage Q	0
Stage C	0	Stage H	6	Stage M	0	Stage R	0
Stage D	0	Stage I	0	Stage N	0	Stage S	0
Stage E	0	Stage J	9	Stage O	0	Stage T	0

Jaw Scan

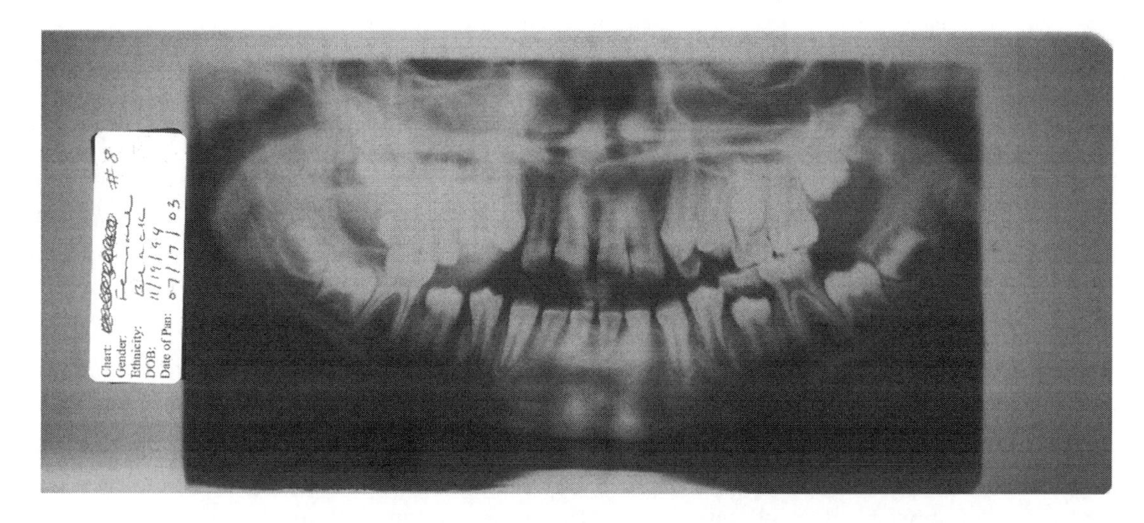

Figure 1. Sample data collection record in FileMakerPro. Moorrees et al. (1963) calcification stages were assessed for both the permanent and the deciduous teeth of each individual. Each panoramic x-ray was imported into the file. All x-rays were either scanned from originals or, as in the case of the most recent cases, output was exclusively produced in digital format. Of the 105 European-American children, 62 were male, 43 female; of the 65 African American children, 32 were male, 33 female.

stage of development were analyzed. Because of the current small sample size, data points on girls and boys have been pooled.

In order to facilitate comparisons beyond those studies using the Moorrees data set, radiographs were also analyzed using the Demirjian et al. (1973) scale. More specifically, we attempted to reproduce the methods of stage analysis presented in the comprehensive study by Liversidge and Speechly (2001) on British children of Bangladeshi or European ancestry. Direct comparisons were

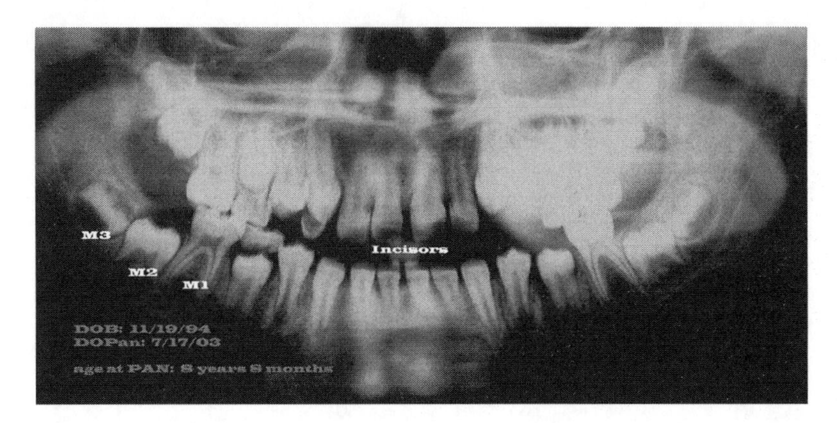

Figure 2. Panoramic radiograph of one child illustrating the overall quality these digitized images. Although it was more difficult to view the developing maxillary dentition, especially the molars, stages were nevertheless determined. All analyses presented here are on the mandibular dentition, where superimposition of the teeth was less apparent than in their maxillary equivalent.

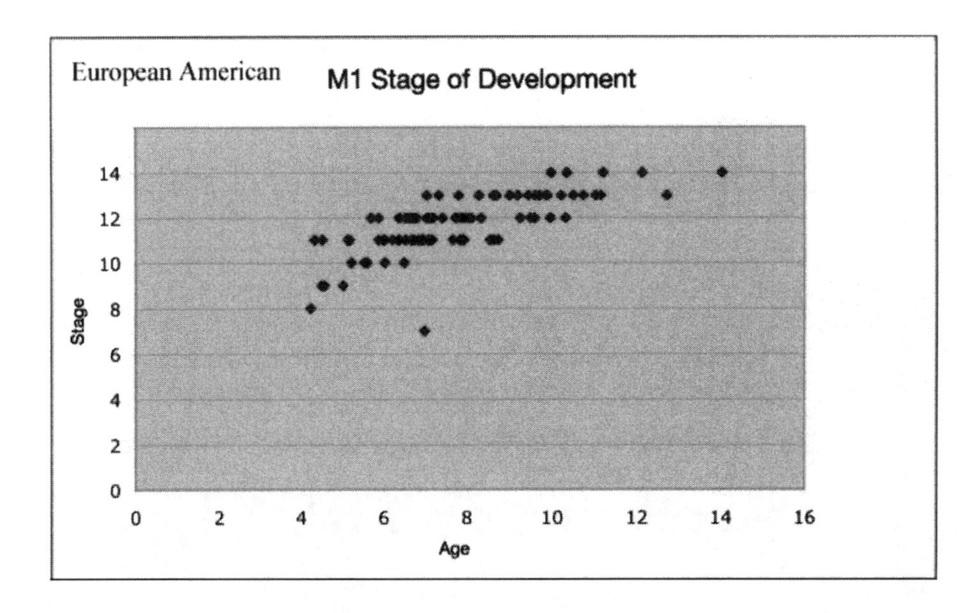

Figure 3. Calcification stages of the lower right M1s (using the stages of Moorrees et al., 1963) applied to the full sample of European-American children.

only possible in M1 Stages F and G; M2, only in Stage E. Mean and standard deviation results are shown in Table 1. Table 2 shows the results of comparison to Moorrees et al. (1963) M1 stage 11 and M2 stage 9. Table 3 illustrates the results of comparisons to the sample of African and European-Americans from the southern part of US (Harris and McKee, 1990) using these same M1 and M2 stages.

As with many other studies (summarized by Liversidge et al., 1999), the original standards produced in the 1960s and 1970s no longer appear to reflect dental development in the world today. Nadler (1998), using the analysis of canine formation at Demirjian Stage G, showed that between the years of 1974 and 1992–1994, there has been a marked decrease in the age of attainment of that Stage by 1.21 years in US boys and 1.52 years in girls. What is remarkable in the present pilot study is the statistically significant decrease in M2 molar development in children of both ethnic groups living in a Northeastern U.S.

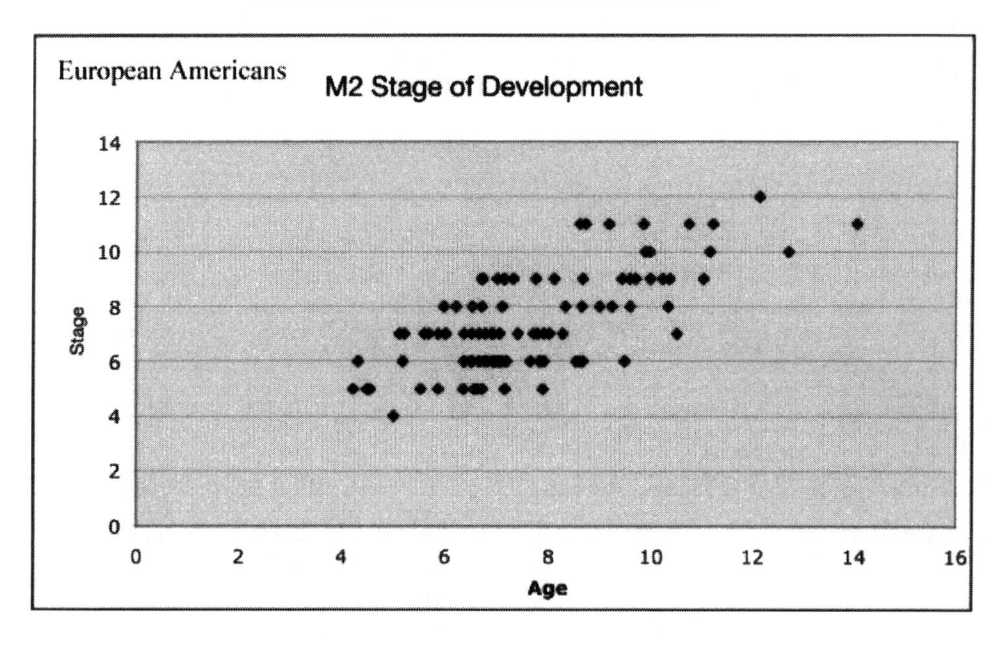

Figure 4. Calcification stages of the lower right M2s (using the stages of Moorrees et al., 1963) applied to the full sample of European-American children.

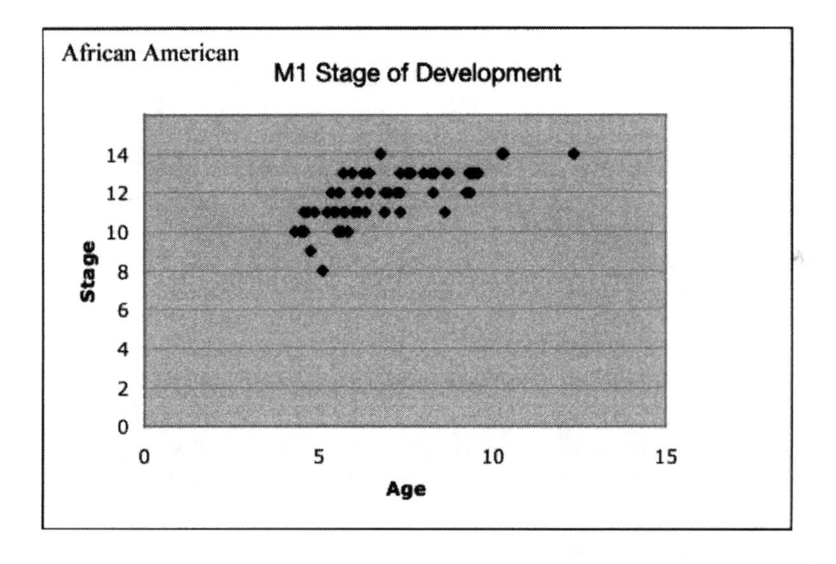

Figure 5. Calcification stages of the lower right M1s (using the stages of Moorrees et al., 1963) applied to the full sample of African-American children.

urban/suburban environment. If and how this is associated with the trend towards decreasing age of certain maturational landmarks, as for example menarche, has yet to be determined, but the decrease in age of this feature is well documented [in the US in 1970 mean age of menarche was recorded as 12.8 years; in the 1990s, as 10 years in American girls of European ancestry, 8 to 9 years in American girls of African ancestry (Nadler, 1998; and see also Herman-Giddens et al., 2004)].

It is possible that the change in the timing of dental maturation may be associated with the documented secular trend in face shape and size in the US and documented by Smith et al. (1986). In fossil hominin specimens, Simpson et al. (1990) considered facial organization as a possible explanatory model for differences

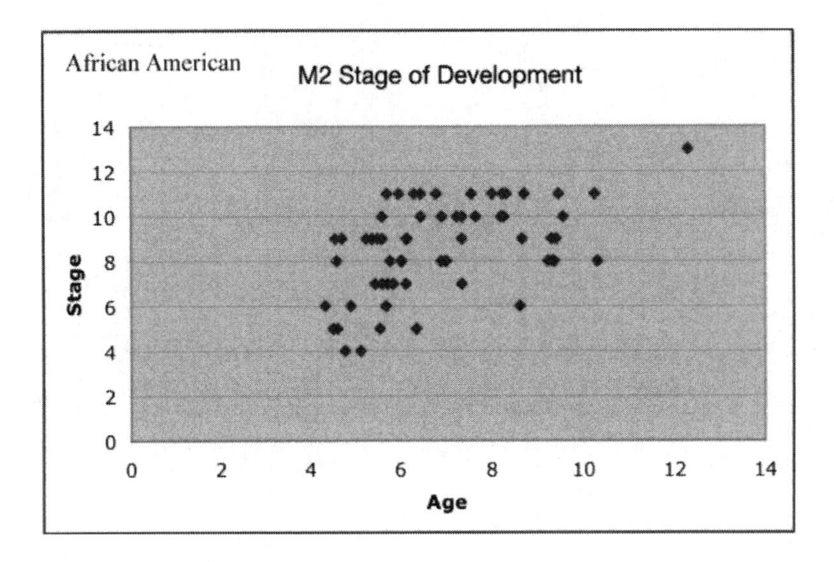

Figure 6. Calcification stages of the lower right M2s (using the stages of Moorrees et al. 1963) applied to the full sample of African American children.

Table 1. Mean age of attainment of mandibular permanent M1 and M2

	E	*F*	*G*
M1		(EA) 5.3 +/− .79 (AA) 4.9 +/− .99 (5.5 +/− .73 female 5.72 +/− .74 male)	(EA) 6.5 +/− .85 (AA) 5.4 +/− 1.09 (6.32 +/− .61 female 6.71 +/− .69 male)
M2	(EA) 6.9 +/− .83 (AA) 5.7 +/− .88 (8.69 +/− .79 female 8.96 +/− .78 male)	(EA) 7.3 +/− 1.06 (AA) 5.9 +/− 1.22 (no data)	

Demirjian et al. (1973) calcification stages applied to the European and African American sample with comparisons to the data presented in Liversidge and Speechly (2001). Statistically significant differences ($p > 01$) exist in each cell between the European and African American samples.
EA = European American sample
AA = African American sample
For comparison, numbers in parentheses at the base of each cell are from Liversidge and Speechly (2001) and divided into female and male mean/standard deviation

in at least pattern differences in the emergence of the developing dentition although Boughner and Dean (2004) have argued that size of the face in baboons and chimpanzees is not associated with at least the initial stages of dental calcification. Adding to the conundrum is the work by Janson et al. (1998) demonstrating that face shape is indeed associated with dental maturation.

Implications

Unfortunately, data on dental calcification in living human populations as well as in chimpanzees are much more limited than is information on emergence times and patterns. Assuming that emergence is indeed broadly equated to dental calcification, as for example, a molar tooth will not generally erupt until at least

Table 2. Mean age of attainment of mandibular permanent M1 and M2 at Stage 11 and 9, respectively, compared to the Moorrees et al. (1963) standards

	STAGE 9	STAGE 11
M1		(EA) 6.81 +/− 1.24 (AA) 5.95 +/− .75 (male 5.5: 4.2–6.7 female 5.2: 4.2–6.6)
M2	(EA) 8.53 +/− 1.49 (AA) 6.91 +/− 1.27 (male 9.1: 7.6–11.5 female 9.0: 7.2–11.1)	

Moorrees, et al. (1963) standards of tooth calcification were extrapolated from the charts produced in their article. These data are presented in parentheses. It appears that for the M1, both the European and African-American sample are at the high end of the range presented in the Moorrees et al. (1963) standards. The M2 data show completely the opposite pattern although the European-American sample is within the low range of variation for both the male and females presented previously by Moorrees et al. (1963).
EA – European Americans
AA – African Americans

Table 3. Mean age of attainment of mandibular permanent M1 and M2 at Stage 11 and 9 respectively

	STAGE 9	STAGE 11
M1		(EA) 6.81 +/− 1.24 (AA) 5.95 +/− .75 (EAM: 7.5 AAM: 7.4 EAF: 7.0 AAF: 6.6)
M2	(EA) 8.53 +/− 1.49 (AA) 6.91 +/− 1.27 (EAM: 10.2 AAM: 9.7 EAF: 9.8 AAF: 9.5)	

Moorrees et al. (1963) standards of tooth calcification compared between two different studies for Stage 11 M1 and Stage 9 M2. All data derives from mandibular teeth. Data in parentheses in each cell, are from Harris and McKee (1990) derived from radiographs of children from the Middle Southern States USA. Harris and McKee (1990) do not report on the decade from which the x-rays were performed on these children. However, it appears that at least a 10 to 15 year gap must exist between the decade of this study and the one undertaken by Harris and McKee (1990). There appears to be a reduction in the time frame of calcification of the molars between both the African and European ancestry Americans as represented in these two studies. Preliminary inspection of the data appears to show that this delay is more pronounced in the M2 than in the M1.
EA – European Americans
AA – African Americans
(from the Harris and McKee (1990) study:
EAM – European American Male
AAM – African American Male
EAF – European American Female)
AAF – African American Female

half of root formation is complete, then there is value in using this comparative database.

Thus, calcification and eruption/emergence of the teeth in human populations is more varied than has been previously thought; the origins of this variation remain unknown. It also appears likely that there has been a significant lowering of the age at which these maturational milestones appear in very recent decades; the plasticity inherent in the genetic systems underlying dental maturation is only now becoming apparent. The data also suggests that as with other dental features (for example, in perikymata numbers (Mann and Monge, in preparation), African populations show the greatest variation, emphasizing the results of studies of genetic diversity in African peoples. Both ends of the range of human variation are present in African populations. African- American children appear to coalesce to the lowest end of that range of variation.

In addition to this emerging information on variation in human populations, there is also accumulating data on chimpanzee dental maturation. This data also indicates significant differences in the developmental timing of chimpanzee teeth based on radiographic analysis (Anemone et al., 1991; Kuykendall, 1996; Anemone et al., 1996; Anemone, 2002). Histological studies on enamel formation reported by Beynon et al. (1998) and Reid

Table 4. Data on the emergence schedules of a mixed sample of maxillary teeth of wild chimpanzees

	WILD CHIMPANZEES	*CAPTIVE CHIMPANZEES 10–90% range*
dC	</– 1.5	0.8–1.4
M1	4.1 (2.6 < × </– 4.9)	2.7–4.1
I1	6.3–8.4 (5.7 < × </– 10.2)	4.7–6.5
I2	7.4–8.6 (6.5 < × </– 10.2)	5.3–6.9
M2	8.2–8.4 (8.2 × </– 10.2)	5.3–7.3
C	10.1–10.8 (8.5 < × </– 14.2)	7.9
M3	12.4 (10.8 < × </– 14.2)	10.5

All information is from Zihlman et al. (2004). For a further summary of data on captive chimpanzee emergence timing, see Kuykendall et al. (1992). For the most part, wild chimpanzees appear to show delayed pattern of dental maturation over their captive counterparts. In some cases, wild animal emergence timing is outside the maximum limit of the recorded range of emergence in captive animals.

et al. (1998), appear to disagree with radiographic studies although Kuykendall (2002) has attempted to reconcile these differences. Dental eruption/emergence studies have been presented by Nissen and Riesen (1964), Kuykendall et al. (1992) and Zihlman et al. (2004). Most studies of chimpanzee dental development have been undertaken on captive born animals. The work of Zihlman et al. (2004), on a combined sample of *Pan troglodytes verus* and *P. troglodytes schweinfurthii*, significantly changes the range of emergence of the permanent dentition in these animals. Comparisons of the results of these studies are presented in Table 4. It now seems clear that the variation between chimps and humans in the timing of dental development must now be re-evaluated.

Finally, in this context, the accumulating data on the range of variation in dental maturation in living humans over the last half century requires some consideration of how we use modern human maturation standards in the reconstruction of the biology of extinct hominins. Numerous studies (i.e., Loevy, 1983; Nadler, 1998; Liversidge et al., 1999; Rousett et al., 2003) have relied on the standards of dental maturation that now appear somewhat obsolete on the basis of new data. How is this emerging data set to be integrated into human evolutionary studies? What standards are to be used in estimating the age at death of earlier hominins? More generally, the new data tends to narrow the difference in dental maturation times between humans and chimpanzees, and thus reduces the distinctiveness of the prolonged human developmental period. This is especially the case when the preliminary data presented here is viewed in concert with the recent work of Zihlman et al. (2004) on the dental development of wild chimpanzees of known age. The accumulation of additional data on dental development in populations of humans and other hominoids will help to resolve these questions. In the meantime, the implications of the present studies strongly suggest that drawing inferences about the pattern of growth and development in extinct hominins that are based on dental data should be done with great caution.

Acknowledgments

We would like to express our appreciation to the organizers of the Conference on Dental Evolution, Drs. Shara Bailey and Jean-Jacques Hublin for their kind invitation to participate in what proved to be a stimulating and collegial meeting. We would also like to thank the staff of the Max Planck Institute for Evolutionary Anthropology (Department of Human Evolution), most especially Diana Carstens, Claudia Feige, Silke Streiber and Allison Cleveland for all their work in the organization of the conference and their kindness and hospitality during our stay in Leipzig.

References

Anemone, R.L., 2002. Dental development and life history in hominid evolution. In: Minugh-Purvis, N., McNamara, K.J. (Eds.), Human Evolution through Developmental Change. Johns Hopkins University Press, Baltimore, Maryland, pp. 249–280.

Anemone, R.L., Mooney, M.P., Siegel, M.I., 1996. Longitudinal study of dental development in chimpanzees of known chronological age: implications for understanding the age at death of Plio- Pleistocene hominids. American Journal of Physical Anthropology 99, 119–133.

Anemone, R.L., Watts, E.S., Swindler, D.R., 1991. Dental development of known-age chimpanzees, *Pan troglodytes* (Primates, Pongidae). American Journal of Physical Anthropology 86, 229–241.

Beynon, A.D., Clayton, C.B., Ramirez Rossi, F.V., Reid, D.J., 1998. Radiographic and histological methodologies in estimating the chronology of crown development in modern humans and great apes: a review, with some applications for studies on juvenile hominids. Journal of Human Evolution 35, 351–370.

Blankenstein, R., Cleaton-Jones, P.E., Luk, K.M., Fatti, L.P., 1990. The onset of eruption of the permanent dentition amongst South African black children. Archives of Oral Biology 35, 225–228.

Boughner, J.C., Dean, M.C., 2004. Does space in the jaw influence the timing of molar crown initiation? A model using baboons (*Papio anubis*) and great apes (*Pan troglodytes, Pan paniscus*). Journal of Human Evolution 46, 255–277.

Demerath, E.W., Towne, B., Chumlea, W.C., Sun, S.S., Czerwinski, S.A., Remsberg, K.E., Siervogel, R.M., 2004. Recent decline in age at menarche: The Fels Longitudinal Study. American Journal of Human Biology 16, 453–457.

Demirjian, A., Goldstein, H. and Tanner, J.M., 1973. A new system of dental age assessment. Human Biology 45, 211–227.

Eveleth, P.B., Tanner, J.M., 1990. *Worldwide Variation in Human Growth*, 2nd ed. Cambridge University Press, Cambridge.

Garn, S.M., Sandusky, S.T., Nagy, J.M., Trowbridge, F.L., 1973. Negro-Caucasoid differences in permanent tooth emergence at a constant income level. Archives of Oral Biology 18, 609–615.

Garn, S.M., Nagy, J.M., Sandusky, S.T., Trowbridge, F., 1975. Economic impact on tooth emergence. American Journal of Physical Anthropology 39, 233–238.

Harris, E.F., McKee, J.H., 1990. Tooth mineralization standards for blacks and whites from the Middle Southern United States. Journal of Forensic Science 35, 859–872.

Herman-Giddens, M.E., Kaplowitz, P.B., Wasserman, R., 2004. Navigating the recent articles on girls puberty in pediatrics: what do we know and where do we go from here? Pediatrics 113, 911–917.

Janson, G.R.P., Rodrigues Martins, D., Tavano, O., Dainesi, E.A., 1998. Dental maturation in subjects with extreme vertical facial types. European Orthodontics 20, 73–78.

Kahumbu, P., Eley, R.M., 1991. Teeth emergence in wild olive baboons in Kenya and formulation of a dental schedule for aging wild baboon populations. American Journal of Primatology 23, 1– 9.

Krumholt, L., Roed-Petersen, B., Pindborg, J.J., 1971. Eruption times of the permanent teeth in 622 Ugandan children. Archives of Oral Biology 16, 1281–1288.

Kuykendall, K.L., 1996. Dental development in chimpanzees (*Pan troglodytes*): the timing of tooth calcification stages. American Journal of Physical Anthropology 99, 135–157.

Kuykendall, K.L., 2002. An assessment of radiographic and histological standards of dental development in chimpanzees. In: Minugh Purvis, N. and McNamara, K.J. (Eds.), Human Evolution through Developmental Change. Johns Hopkins University Press, Baltimore, Maryland, pp. 281–304.

Kuykendall, K.L., Mahoney, C.J., Conroy, G.C., 1992. Probit and survival analysis of tooth emergence ages in a mixed-longitudinal sample of chimpanzees (*Pan troglodytes*). American Journal of Physical Anthropology 89, 379–399.

Lavelle, C.L.B., 1976. Study of tooth emergence in British Blacks and Whites. Journal of Dental Research 55, 1128.

Liversidge, H. M., 2003. Variation in modern human dental development. In: Thompson, J.L., Krovitz, G.E., Nelson, A.J. (Eds.), Patterns of Growth and Development in the Genus *Homo*. Cambridge University Press, Cambridge, pp. 73–113.

Liversidge, H.M., Speechly, T., 2001. Growth of permanent mandibular teeth of British children aged 4 to 9 years. Annals of Human Biology 28, 256–262.

Liversidge, H.M., Speechly, T., Hector, M.P., 1999. Dental maturation in British children: are Demirjian standards applicable?. International Journal of Pediatric Dentistry 9, 263– 269.

Loevy, H.T., 1983. Maturation of permanent teeth in Black and Latino children. Acta Ondontologica Pediatrica 4, 49–62.

Moorrees, C.F.A., Fanning, E.A., Hunt, E.E., 1963. Age variation of formation stages for ten permanent teeth. Journal of Dental Research 42, 1490–1502.

Nadler, G.L., 1998. Earlier dental maturation: fact or fiction? Angle Orthodontist 68, 535–538.

Nissen, H.W., Riesen, A.H., 1964. The eruption of the permanent dentition of chimpanzees. American Journal of Physical Anthropology 22, 285–294.

Olze, A., Schmeling, A., Taniguchi, M., Maeda, H. van Niekerk, P., Wernecke, K.D., Geserick, G., 2004. Forensic age estimation in living subjects: the ethnic factor in wisdom tooth mineralization. International Journal of Legal Medicine 118, 170–173.

Phillips-Conroy, J.E., Jolly, C.J., 1988. Dental eruption schedules of wild and captive baboons. American Journal of Primatology 15, 17–29.

Reid, D.J., Schwartz, G.T., Dean, C., Chandrasekera, M.S., 1998. A histological reconstruction of dental development in the common chimpanzee (*Pan troglodytes*). Journal of Human Evolution 35, 427–448.

Rousset, M.M., Boualam, N., Delfosse, C., Roberts, W.E., 2003. Emergence of permanent teeth: secular trends and variance in a modern sample. Journal of Dentistry for Children 70, 208–214.

Simpson, S.W., Kunos, C.A., 1998. A radiographic study of the development of the human mandibular dentition. Journal of Human Evolution 35, 479–505.

Simpson, S.W., Lovejoy, C.O., Meindl, R.S., 1990. Hominoid dental maturation. Journal of Human Evolution 19, 285–297.

Smith, B.H., Garn, S.M., Hunter, W.S., 1986. Secular trends in face size. Angle Orthodontist 56, 196–204.

Tompkins, R.L., 1996. Relative dental development of Upper Pleistocene hominids compared to human population variation. American Journal of Physical Anthropology 99, 103–118.

Zihlman, A., Bolter, D., Boesch, C., 2004. Wild chimpanzee dentition and its implications for assessing life history in immature hominin fossils. Proceedings of the National Academy of Sciences of the USA 101, 10541–10543.

5. A computerized model for reconstruction of dental ontogeny: A new tool for studying evolutionary trends in the dentition

P. SMITH
Laboratory of Bio-Anthropology and Ancient DNA,
Hadassah Faculty of Dental Medicine
Hebrew University of Jerusalem
POB 12272, 91120 Jerusalem, Israel
pat@cc.huji.ac.il

R. MÜLLER
Institute for Biomedical Engineering
Swiss Federal Institute of Technology (ETH)
University of Zürich, Zürich, Switzerland
ralph.mueller@ethz.ch

Y. GABET
Bone Laboratory,
Institute of Dental Sciences
Faculty of Dental Medicine
Hebrew University, Jerusalem, Israel
gabet@md.huji.ac.il

G. AVISHAI
Laboratory of Bio-Anthropology and Ancient DNA,
Hadassah Faculty of Dental Medicine
Hebrew University of Jerusalem
POB 12272, 91120 Jerusalem, Israel
gavishai@md.huji.ac.il

Keywords: evolutionary-development, tooth development, morphogenesis, microCT, DEJ, morphometry, enamel thickness

Abstract

All hominid molars show the same sequence of cusp initiation, but differ in the later stages of development. The topography of the dentin-enamel junction (DEJ) represents the outcome of differential growth between cusps. Since the cusps grow in an orderly sequence from tip to base (defined by the plane of coalescence with adjacent cusps), quantification of cusp volume and relationships can be used to reconstruct successive stages in development and their contribution to the morphometry of the crown surface. Their volume and spatial relationships at the DEJ enable us to partition cell proliferation in relation to cusp initiation, while quantification of the amount and distribution of enamel overlying the DEJ provides the necessary discrimination between cell proliferation and cell function, expressed in enamel matrix apposition. We have developed a three-dimensional

S.E. Bailey and J.-J. Hublin (Eds.), Dental Perspectives on Human Evolution, 275–288.

computerized model of a lower molar tooth that enables us to identify and quantify the different stages of tooth development defined above. The model is based on serial micro-computed tomography (microCT) images of human teeth that provide accurate quantification of the outer and inner enamel and dentin boundaries of individual cusps. We have used this model to compare morphogenesis in the lower second deciduous molar and lower first permanent molar. Spatial relationships of the cusps, expressed by the topography of the DEJ showed that shape differences were established in the early stages of morphogenesis by differential proliferation within the developing tooth germ and that cusp size and proportions were modified at the crown surface by enamel apposition. Reduction of the hypoconulid in the permanent molar shown at the DEJ was largely masked by the exceptional thickness of enamel on this cusp. We propose that this model provides a novel contribution to the identification of ontogenetic trajectories and their contribution to evolutionary trends in tooth size, shape and enamel thickness.

Introduction

All hominid lower molars are variants of the 5-Y Dryopithecine molar pattern (Gregory and Hellman, 1926; Dahlberg, 1945). The observed variation in tooth size, cusp number and outline (steep versus rounded, large versus small, groove pattern and enamel thickness) that has traditionally been used in phylogenetic studies has however, recently been questioned on the grounds of homoplasy (Collard and Wood, 2000; Finarelli and Clyde, 2004). Enamel thickness, in particular, varies independently of tooth size. In a recent study of enamel thickness variation of inbred baboons Hlusko (2004) proposed that rapid change in this feature could occur under appropriate selective pressures. Thus, variation in enamel thickness may account for much of the homoplasy shown at the crown surface. As it is the last tissue to be completed, enamel modifies the template of the tooth laid down during morphogenesis. For this reason, examination of the earlier stages of tooth development, represented in the completed tooth by the dentin-enamel junction (DEJ), has long been considered a more reliable representation of the underlying phenotype in phylogenetic studies (Kraus, 1952; Korenhof, 1960; 1979; Smith et al., 1997; Sasaki and Kanazawa, 1999; Smith et al., 2000).

While we cannot directly study developmental processes in fossil teeth, the sequential pattern of tooth development means that many early developmental stages are conserved in the structure of completed teeth and can be identified using appropriate imaging systems. The meristic variation shown along the molar tooth row of modern humans, whose developmental pattern is well established, provides an excellent model for examining the extent to which ontogenetic pathways are modified in teeth of different size and rates of development.

The mandibular molars develop along a mesio-distal gradient in which their size and complexity increase and decrease differentially. The size changes are associated with mesio-distal gradients in the duration of development and enamel thickness as well as in overall tooth size and cuspal relationships. All molars show the same order of cusp initiation (protoconid>metaconid>hypoconid> entoconid> hypoconulid), but differ in subsequent amount and rate of growth (Kraus and Jordan, 1965; Butler, 1967, 1999; Winkler et al., 1991; Avishai et al., 2004). In large toothed hominids, the distal cusps are larger relative to other cusps than in small toothed hominids, with the differences most pronounced in the M2 and M3. The cusps develop as infoldings of the inner enamel epithelium initiated by the enamel knots. They develop apically from tip to base. Differentiation of the cells of the inner enamel epithelium into ameloblasts and underlying cells of the dental papilla and odontoblasts follows the same sequence. Cusp size and shape is therefore a function of the temporal and spatial organization of cell proliferation and differentiation, from initiation through the spread of bio-mineralization of individual cusps in a staggered fashion, to coalescence with adjacent cusps along the DEJ (Butler, 1999). This means that the spatial relationships expressed in the topography of the DEJ reflect the outcome of morphodynamic processes in which growth and differentiation proceed simultaneously within the developing tooth germ.

In modern humans there is only a lag of some two weeks between initiation of the dm_2 and M_1, but size differences between them are apparent by the cap stage (Kraus and Jordan, 1965). Identification of different growth trajectories expressed in the final size and form of these teeth may help us better understand the developmental basis of phenotypic variation and its significance for reconstructing phylogenies. In order to achieve this we have developed a model that enables us to reconstruct growth trajectories from the topography of the DEJ in fully formed teeth using 3D microCT imaging techniques.

Our approach differs conceptually from that of previous studies of the DEJ, in two important aspects. Firstly our model does not damage the tooth in any way and so is applicable to rare fossil and pathological specimens. It can also be easily replicated, facilitating evaluation of its precision and accuracy. Secondly our model provides accurate quantification of the size and shape of individual cusps at the DEJ and OES. This enables us to differentiate between two components of size and shape: (i) cell proliferation defined by the DEJ and (ii) ameloblast function defined by the amount of enamel formed. We have used the spatial relationships of cusp tips and coalescence points shown at the DEJ to reconstruct the order of cusp initiation and coalescence. We have used cusp volume to quantify the amount and pattern of cell proliferation taking place between these events and their distribution within the developing tooth germ. We propose that this model will enable us to identify the source and extent of developmental plasticity expressed at the crown surface and so evaluate the phylogenetic significance of variation in crown form.

The Model

The model has been developed as part of a long-term project designed to identify the onset and extent of developmental variation in the early stages of tooth formation. The starting point of our study is the observation that cusps are initiated sequentially and grow apically through infolding of the inner enamel epithelium (Jernvall et al., 2000). Biomineralization begins when cells of the inner enamel epithelium and underlying layer of cells of the dental papilla differentiate into ameloblasts and odontoblasts that form enamel and dentin. This process begins at the cusp tips and spreads apically along the slopes of the cusps. Cusp growth ceases when adjacent cusps coalesce. Until this occurs individual cusps move apart from one another through continued cell division at their rapidly expanding bases. The extent of this growth is expressed at the DEJ by the distance between cusp tips as well as in their height and volume. This means that the spatial geometry of the DEJ can be used to reconstruct the order of cusp initiation and coalescence. For any one cusp its volume, measured from tip to base, represents the extent of cell proliferation and when related to height provides an estimate of cusp shape (i.e., thin or squat).

The accuracy of the model is dependent on the ability of the microCT system to provide accurate images of enamel and dentin boundaries; to compile them for three-dimensional reconstruction and quantification, and on the ability of the operator to correctly identify landmarks used in the three-dimensional reconstructions. Since the process is non-destructive its accuracy can be evaluated through comparison of results obtained through repeated scans. Technical considerations during scanning include the selection of standards used to differentiate between air-enamel and enamel-dentin boundaries, slice thickness and programs used in reconstruction and subsequent calculations (Avishai et al., 2004). In our case, the protocol used for scanning and subsequent reconstructions and analyses was developed by R.M. and all landmarks were located by G.A. directly from the microCT workstation. The combined measurement error calculated from reconstructions derived from repeated scans was less than 1.2 %.

The technique was first applied by Avishai et al. (2004) to obtain serial microCT scans of 24 lower dm_2 tooth germs in different stages of development. They found that the order of cusp initiation was constant but differed from the order of cusp coalescence. Although the entoconid was the fourth cusp initiated it was the last to coalesce, with its base located more apically than those of other cusps.

In this study we have applied the same model to compare morphogenesis in the second deciduous molar (dm_2) and first permanent molar (M_1). These teeth were chosen since they show the greatest morphological resemblance of any two teeth of the molar sequence in modern humans (Dahlberg, 1945). They are also initiated within a short time of each other, but differ in size, enamel/dentin proportions and growth rates (Butler, 1968). Crown development in the M1 takes more than three times as long as that of the dm2 (Kraus and Jordan, 1965). The two teeth resemble one another in the expression of accessory cusps such as Cusp 6 and Cusp 7, but in small toothed individuals the hypoconulid may be reduced or absent in the M_1 even though present in the smaller dm_2.

In order to identify the developmental processes associated with such morphometric variation we have examined three unworn teeth obtained from archaeological collections from Israel. Two of these teeth – a dm_2 and a small M_1 (identified as M_1a) were taken from one mandible. The third tooth (M_1b) was a first permanent molar from a large-toothed individual. Use of these three teeth enabled us to compare growth in two teeth (dm_2 and M_1a) developing in the same jaw as well as in two teeth of the same tooth class but different size (M_1a and M_1b) derived from non-related individuals.

We propose that this approach to the reconstruction of ontogenetic pathways may help to elucidate the role played by tooth type as opposed to more generalized constraints of space and position and so contribute to our understanding of the basis of variation expressed in tooth size and crown pattern.

Materials and Methods

The three teeth used in this study were selected from the archaeological collections of the Hebrew University. They were selected from mandibles of children with unworn teeth showing no overt signs of diagenesis or developmental anomalies. Two specimens were selected - one with small-teeth and one with large teeth. From the small toothed individual the dm2 and adjacent M_1a were carefully removed and were scanned together with the isolated M_1b from the second jaw.

Scanning

The three specimens were scanned in a microCT system (μCT40, Scanco Medical, Bassersdorf, Switzerland). This apparatus contains a microfocus X-ray tube (with a focal spot of 5 μm), a turn-table system for mounting and rotating the specimens and a two-dimensional CCD detector connected to a computer. The original desktop system was developed at the Institute for Biomedical Engineering, the University of Zurich, Switzerland and described by Rüegsegger et al. (1996) and Müller et al. (1998).

Recent applications for this device include: modeling of root canal anatomy (Peters et al., 2000), quantification of peri-radicular bone loss (von Stechow et al., 2003) and structural and mechanical properties of fracture callus (Gabet et al., 2004). For this project we used a specially designed protocol developed by R.M. that provided excellent discrimination between enamel and dentin and enabled us to carry out all necessary calculations and reconstructions directly from the scanner. This also enabled us to access the entire data set throughout our calculations and check reconstructions against the original scans. It also obviated errors caused by transferring data to other work stations.

The radio-opacity of the Scanco scanner is graded automatically on a scale of 0 to 1000; 0 is completely radio-lucent (e.g., air), 1000 is completely radio-opaque (e.g., metal).

A known phenomenon of the radiographic procedure is "Scattering". Scattering appears as a gradient of radio-opacity along the margins of an X-ray absorbing object. It is caused by a dispersion of the electrons hitting the radio-opaque object. This scattering is recorded on the radiographic sensor – be it a radiographic film or the digital sensor of the microCT. To minimize this effect we used the following values:

Enamel:	Inner Value – 500
	Outer Value – 30
Dentin:	Inner Value – 400
	Outer Value – 500

These values were chosen after initial scans showed that 500 is the average radio-opacity of enamel and 400 is the average for dentin. The outer value for the analysis of enamel was 30 (and not 0) to allow for annulment of dust particles and other residues on the surface of the teeth.

For scanning, the three teeth were placed in a specialized holding tube of the microCT, one above the other. They were inverted so that the enamel cusp tips of the protoconid, metaconid and hypoconid, the first three cusps initiated, formed a horizontal plane at a right angle to the long axis of the tube. A solid radiolucent foam sponge was placed between the teeth to hold them in place and facilitate correct orientation. After arranging the teeth in the holding tube, they were inserted into the microCT scanning chamber according to the manufacturer's instructions.

The first stage in scanning was to obtain a preview scan of each of the specimens in the scanning tube. The preview scan took about one minute for each sample and provided a low-resolution image of the tooth, which allowed for definitions of the upper and lower border of each scan. After defining the borders of scanning for each of the three specimens, they were queued for scanning. The scanning was done overnight with a resolution of 16 microns. Each tooth took from 3.5 to 5 hours depending on the size of the tooth. All measurement parameters were saved in a control file, used for 3D reconstructions.

Spatial Analysis

The scanning procedure resulted in a series of 16 micron slices for each tooth. For each cusp the following landmarks were defined and located on the three-dimensional X, Y, Z grid:

1. Enamel tip – the X, Y location of the tip of the cusp at the highest Z slice showing enamel from the specific cusp.
2. Dentin tip – the X, Y location of the tip of dentin at the highest Z slice showing dentin in the specific cusp.
3. Coalescence of cusps – first and second coalescence points.
4. Contour of enamel at the OES.
5. Contour of dentin at the DEJ.

All measurements were taken by G.A. The measurement error, from either scanning or identification of contours and points, was calculated by re-measuring teeth six weeks later. The calculated differences were always less than 1.2 % and not statistically significant. Pearson correlation coefficient between measurements ranged from 0.97 to 1.00. Three-dimensional reconstructions of the tooth crown as assessed by microCT were visualized using an extended Marching Cubes algorithm (Müller et al., 1994).

Cusp height was calculated along the Z (vertical) axis where:

Enamel tip is the location of the tip of the cusp at the highest Z slice showing enamel.

Dentin tip is the location of dentin at the highest Z slice showing dentin.

Point of Fusion is the location of coalescence of dentin of adjacent cusps (each cusp has two points of fusion).

Base of cusp is the virtual plane constructed from the line connecting the points of fusion of a cusp with two adjacent cusps,

Crown Height is the Z axis measurement from the enamel tip of the mesio-buccal cusp to the most apical mesio-buccal point of the crown of the tooth, defined by the presence of enamel. This arbitrary definition was chosen since the cervical enamel boundary is notoriously variable in degree of curvature.

Dentin Cusp Height or DCH is the average Z axis measurement between the dentin tip and the defined base of the cusp.

Enamel height or EH is the Z axis measurement between the enamel tip and the dentin tip of each cusp.

Figure 1 gives a graphical representation of these points and measurements.

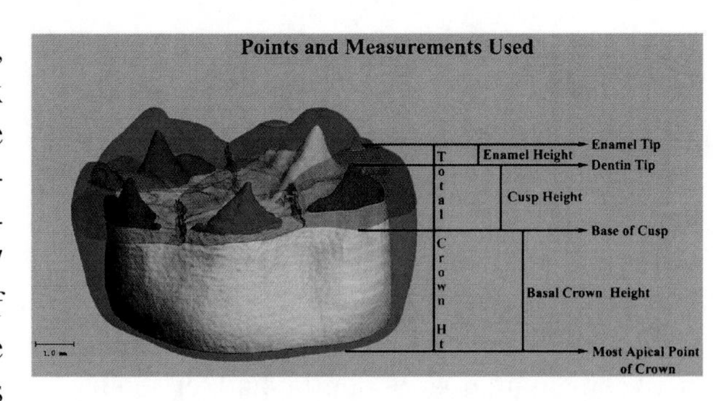

Figure 1. Computerized reconstruction of dm_2, showing landmarks used in measurements.

Defining Cusp Boundaries

Heights of cusp tips were defined by the Z axis coordinates. The XY coordinates were automatically recorded by the curser. The base of each cusp was defined as the most apical border of the dentin and enamel located by the first point of fusion with adjacent cusps. Since each cusp has two points of fusion with adjacent cusps, which are at different heights along the Z axis, the base of the cusp was arbitrarily defined by an angled plane constructed by uniting fusion points. These points are those recorded in the analysis of cusp height.

Contouring was performed on each slice using a contouring program (UCT Evaluation, Scanco Medical, Bassersdorf, Switzerland) with an "iteration" function, which automatically advances or reverses the slices and fits the contour for the last slice to the new slice. This function is able to deduce whether the area contoured has grown or diminished in size. The digital management of iteration copies the contour from the previous slice to the next slice and refits the area copied according to the predefined inner and outer radio-opacity values. A standard method and sequence for contouring cusps and the entire tooth was used following the protocol used by Avishai et al. (2004).

Contouring Each Cusp

Enamel was contoured first. A random slice in which the cusp was not fused with any of the other cusps was chosen as the first slice contoured. A contour was drawn around the outer border of the cusp and fitted to the predefined values (inner – 500, outer – 30). The contour was iterated occlusally until the tip of the cusp was found. Returning to the first slice contoured, the contour was iterated apically until the dentin of the cusp made contact with that of an adjacent cusp (Figure 2). The slice in which the dentin of the cusp united with that of the second adjacent cusp was found, and used to define a contour including only the border of the cusp. These two contours were merged by "morphing" of intervening contours to create a virtual diagonal base for each cusp. The DEJ of each cusp was contoured using the method described for the OES.

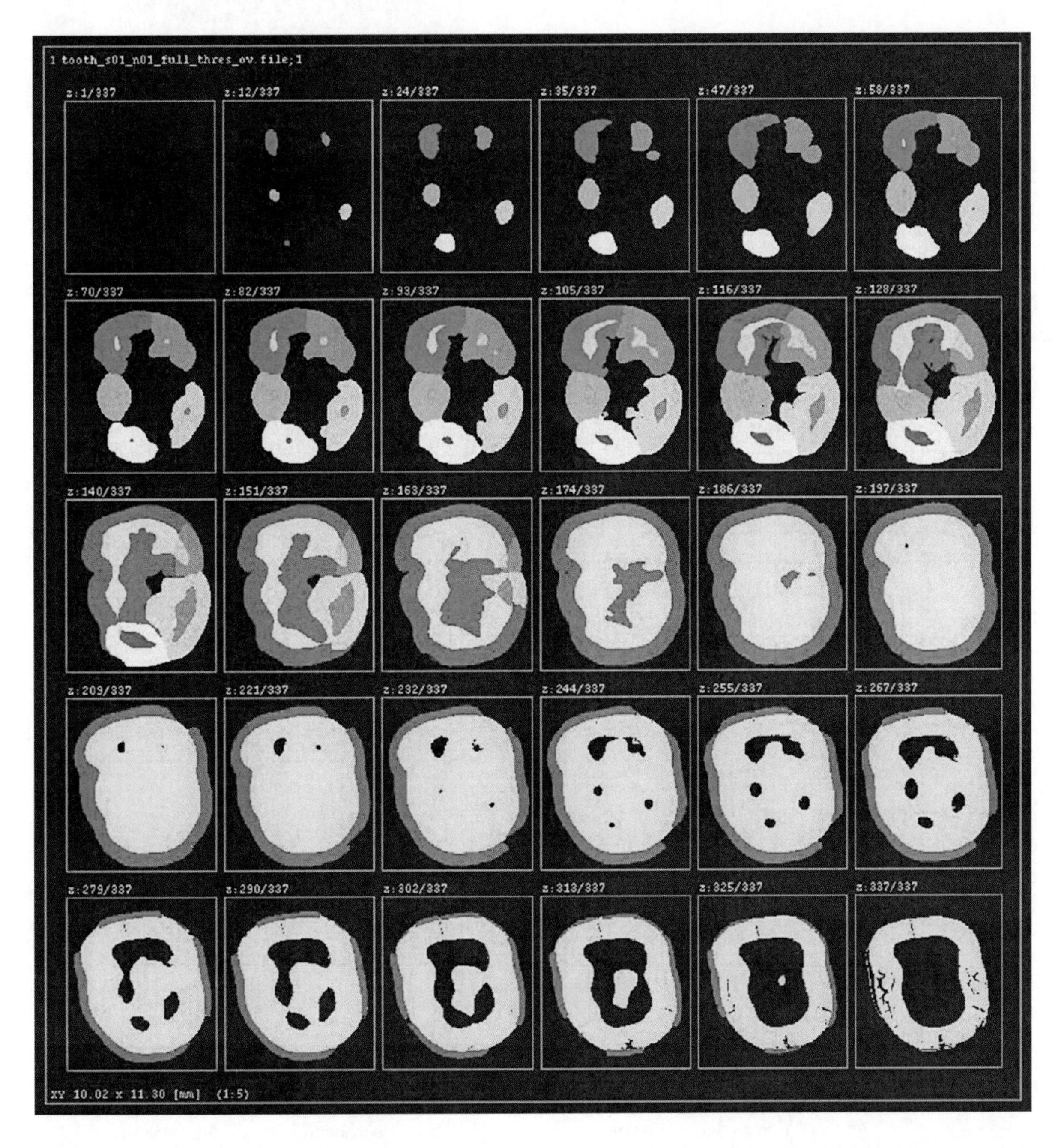

Figure 2. Selected slices from scans of dm2, showing height differences in points of coalescence of dentin and enamel. Note the extreme difference between enamel and dentin area.

A total of 12 sets of contours for each tooth were saved in 12 separate files for later use. All analyses were carried out using the work station of the microCT. This enabled us to access all data and refer to original scans for verification of locations used in analyses.

Results

The maximum horizontal contour recorded at the DEJ was used to estimate size relations of the three teeth. It was least in the deciduous tooth followed by the paired M_1a and was largest in the non-related M_1b. (Table 1, Figure 3).

Variation in cusp formation development by vertical relationships between cusp tips and coalescence is illustrated in Figure 4 where the length of each bar represents the amount of vertical growth of each cusp (from tip to base) and relative height of each bar depicts the extent of differences in growth between initiation and coalescence events. Height differences between successive cusp tips were larger in the deciduous tooth than in the permanent teeth, indicating more growth between initiation of successive cusps in the deciduous molar. Height differences between successive coalescence points were also greater in the deciduous tooth than in the permanent teeth and show a different sequence.

In the dm_2 the protoconid, metaconid and hypoconid showed a similar plane of coalescence that was more coronally located than that of the entoconid and metaconid (Figure 4). In the permanent teeth the plane of coalescence of all cusps was similar. Since the order of cusp initiation shown by the location of cusp tips, was identical in all three teeth, the amount of independent growth in the distal cusps of the permanent teeth was obviously less than in the dm_2 and is reflected in their altered height relations. Notably the hypoconulid of the larger unrelated M_1b deviated less from the dm_2 blueprint than that of the smaller M_1a.

Table 1. Axial cross-sectional area at maximal height of dentin contour (mm^2)

Tooth	*Dentin Area*	*Enamel Area*	*Total Area*
dm_2	55.46	16.16	72.14
M_1a	65.43	17.97	83.4
M_1b	77.90	22.04	99.94

Note: Dentin area includes the pulp complex.

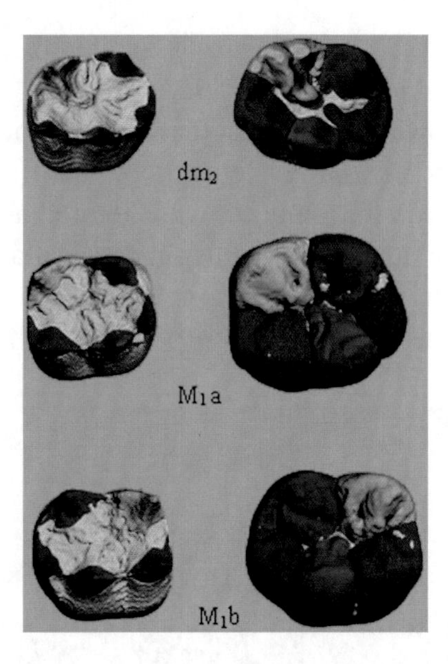

Figure 3. Computerized reconstructions of the DEJ (left) and OES (right), with cusps outlined, showing the difference in appearance and proportions of cusps at the two surfaces.

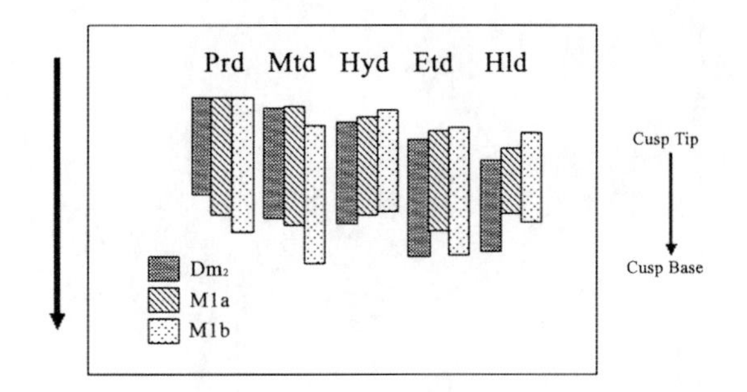

Figure 4. Height differences between cusp tips and coalescence points in the three teeth studied. Note: the marked step between dm_2 cusps. Key: Prd=Protoconid, Mtd=Metaconid, Hyd=Hypoconid, Etd=Entoconid, Hld=Hypoconulid.

These differences in the order of initiation and coalescence were expressed in cusp height and volume (Figure 5). Although the dm_2 was the smallest tooth, it had the tallest hypoconulid. In all three teeth, dentin cusp height was greater in lingual cusps than in buccal cusps, but the rank order of cusp height differed. In the dm_2 the fourth cusp to be initiated, the entoconid, was the tallest and the hypoconid, the third cusp initiated, was the shortest. In both M_1a and M_1b the second cusp initiated, the metaconid, was the tallest and the fifth cusp initiated, the hypoconulid, was the smallest. However, the extent of reduction of these two cusps differed. In the M1a, both entoconid and hypoconulid were shorter than homologous cusps in the associated dm_2, while in the larger unrelated M_1b, only the hypoconulid was shorter than that of the dm_2. Since cusps expand towards their base, these relations were expressed in cusp volume as well as height (Figure 6).

Enamel cusp height in both the deciduous and permanent molars increased distally, with

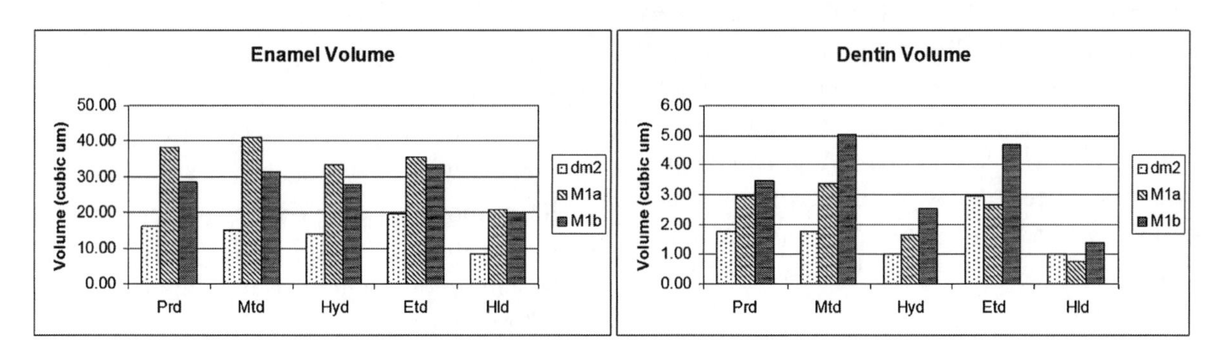

Figure 5. Enamel and dentin volume, Key: Prd=Protoconid, Mtd=Metaconid, Hyd=Hypoconid, Etd=Entoconid, Hld=Hypoconulid.

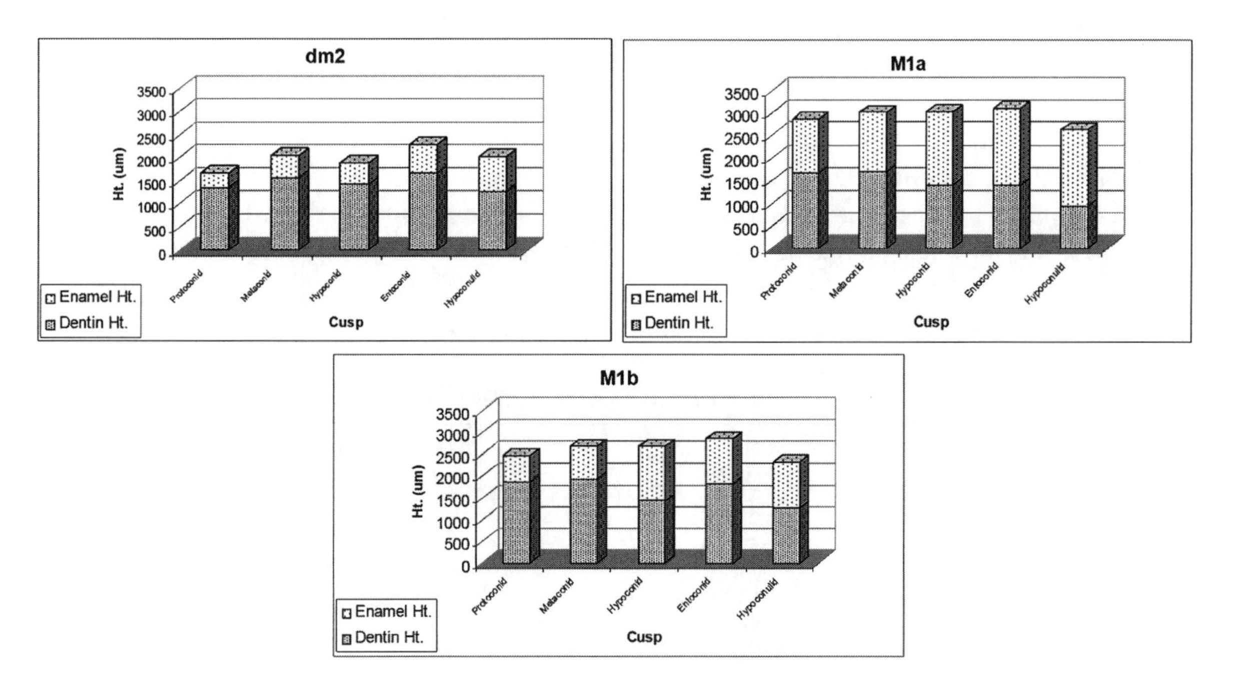

Figure 6. Enamel and dentin height, Key: Prd=Protoconid, Mtd=Metaconid, Hyd=Hypoconid, Etd=Entoconid, Hld=Hypoconulid.

the increase most pronounced in the permanent teeth (Figure 6). This means that the contribution of enamel to total crown height differed between permanent and deciduous teeth and between cusps. The smaller of the two permanent teeth (M_1a) had the thickest enamel and this contributed most of the total height of the hypoconulid. The same pattern was observed for enamel cusp volume (Figure 5).

The three-dimensional relationship of cusp tips at the DEJ and OES provides further evidence of differences in growth patterns (Table 2, Figure 3). The three teeth differ in occlusal outline. The smaller dm_2 is relatively longer (mesio-distally) with a more pentagoid outline than the more rectangular permanent molars because of the greater distal extension of the hypoconulid and entoconid. Once again the differences between the deciduous and permanent teeth are most pronounced in the smaller M_1a with the smallest entoconid and hypoconulid.

Table 3 shows how enamel thickness modifies cuspal relations expressed at the DEJ. In the dm_2 dentin area at the base of the cusp follows the same sequence as dentin cusp height: entoconid>metaconid>protoconid> hypoconid> hypoconulid. In M_1a dentin area at the base of the cusps was: metaconid>entoconid>protoconid> hypoconid> hypoconulid. Dentin area at the base of cusps in the larger M_1b showed the same order of decreasing size as the dm_2,

Table 3. Basal cusp area (mm²)

Area at DEJ					
Cusp	Prd	Mtd	Hyd	Etd	Hld
dm_2	3.60	5.43	3.27	5.55	2.62
M_1 a	4.89	6.52	2.70	6.02	1.72
M_1 b	5.85	8.72	4.12	9.72	3.15
Area at OES					
Cusp	Prd	Mtd	Hyd	Etd	Hld
dm_2	13.59	13.70	12.42	12.06	14.73
M_1 a	21.63	21.23	15.01	18.90	11.77
M_1 b	20.66	22.04	17.06	20.96	17.33

Key: Prd=Protoconid, Mtd=Metaconid, Hyd=Hypoconid, Etd=Entoconid, Hld=Hypoconulid.

but in both the M_1a and M_1b, the basal area of the hypoconulid was only 1/3 that of the entoconid, whereas in the dm_2, hypoconulid area was 1/2 of the entoconid.

The proportions of basal cusp areas to one another at the DEJ and the OES differed only slightly in the dm_2, but in both permanent teeth, cuspal relations at the DEJ differed markedly from those seen on the crown surface, because of the thicker enamel. In M_1a, hypoconulid : entoconid ratios were 1:4 at the DEJ and 1:2 at the OES. In M1b, hypoconulid: entoconid ratios were 1:3 at the DEJ and 1:2 at the OES.

Discussion

Measurements taken at the DEJ demonstrate that all three teeth studied here show the

Table 2. Distance between dentin tips: Euclidian distance between cusp tips (mm)

	Distance between cusp tips at DEJ									
	Prd-Mtd	Prd-Hyd	Prd-Etd	Prd-Hld	Mtd-Hyd	Mtd-Etd	Mtd-Hld	Hyd-Etd	Hyd-Hld	Etd-Hld
dm_2	3.51	3.58	6.46	6.41	5	4.71	6.8	4.64	2.92	3.94
M_1a	4.31	4.56	6.97	6.85	6.48	5.32	7.67	4.88	2.5	4.04
M_1b	4.15	4.18	7.26	7.24	6.27	5.55	8.07	5.44	3.26	4.5
	Distance between cusp tips at OES									
	Prd-Mtd	Prd-Hyd	Prd-Etd	Prd-Hld	Mtd-Hyd	Mtd-Etd	Mtd-Hld	Hyd-Etd	Hyd-Hld	Etd-Hld
dm_2	3.65	3.93	7.12	7.18	5.65	5.2	7.67	5.6	3.45	4.75
M_1a	5.28	4.78	7.45	7.16	7.4	5.99	8.68	5.02	2.51	4.31
M_1b	4.85	3.64	7.45	6.89	6.35	5.67	8.42	5.58	3.29	4.97

Key: Prd=Protoconid, Mtd=Metaconid, Hyd=Hypoconid, Etd=Entoconid, Hld=Hypoconulid.

same order of cusp initiation, expressed by the relative height of cusp tips, but they differ in the amount of growth between initiation events. Height differences between cusps were greatest in the dm_2 and least in M_1a. In each tooth the order of coalescence differed from that of initiation. In the dm_2 the third cusp initiated, the hypoconid, coalesced with the protoconid at approximately the same time as the earlier appearing metaconid and protoconid. The hypoconid and entoconid were initiated later and coalesced with one another shortly after the other cusps, their coalescence points with the hypoconid and entoconid were shifted apically, so that they grew longer (or faster) as an independent unit with the entoconid-metaconid cusps coalescing last. Consequently the entoconid is the tallest cusp. Similar results were reported by Avishai et al. (2004) for 24 developing dm_2 tooth germs, suggesting that this sequence is characteristic of the dm_2.

Cusp relationships in the two permanent teeth differed from those in the deciduous tooth, both in terms of amount of growth between initiation of successive cusps and in the order of their coalescence. In the permanent teeth, height differences between dentin cusp tips were minimal except for the hypoconulid. Initiation of this cusp was delayed relative to that of the other cusps, and it coalesced earlier with the hypoconid resulting in a very short cusp. While both permanent teeth showed reduced growth of the two distal cusps, this was much more pronounced in the smaller M_1a with an extremely reduced hypoconulid at the DEJ. Neither M_1a nor M_1b showed the independent growth phase of the entoconid and hypoconulid observed in the dm_2. However in all three teeth, initiation and coalescence of the hypoconid, traditionally considered one of the talonid cusps was integrated with that of the modified trigonid in which the paraconid, as in all hominids, was lacking. The differences observed at the DEJ between the three teeth studied here were modified at the OES by inter-cusp differences in enamel thickness. This increased distally in all three teeth with the greatest increase found in the permanent teeth affecting basal crown area more than intercusp distances.

We found that in both permanent molars the distribution of cell proliferation, expressed by cusp volume and spatial relationships at the DEJ, shows a marked mesio-distal gradient resulting in a small hypoconulid as well as in altered size relations and position of other cusps. However, our study has also shown that cusp pattern seen at the OES does not accurately reflect the true extent of morphodynamic changes in the pattern of growth and differentiation. The increased enamel thickness of the distal cusps masks these changes to some extent since the smallest cusp, the hypoconulid, has the thickest enamel.

Developmentally larger teeth have been attributed to up-regulation of epithelial cell proliferation either in the bud or enamel cap stage, with emphasis placed on the role of the enamel knots in defining cuspal relations (Jernvall et al., 2000; Zhao et al., 2000; McCollum and Sharpe, 2001). In the human dm2 cusp initiation and biomineralization of the protoconid begin in the rapidly growing tooth germ before the other cusps are morphologically defined (Kraus and Jordan, 1965). This is reflected in our model by the step like differences in height of the dentin cusp tips, which reflect continued rapid growth until all cusps are initiated. In the modern human M_1 tooth germ all soft tissue cusps are defined before biomineralization begins (Kraus and Jordan, 1965) and this is associated with reduced growth of the distal cusps. The mesiodistal gradient shown by the successive phases of cell proliferation and differentiation in the deciduous tooth is curtailed in the permanent tooth, where the rate of differentiation outstrips that of proliferation, even though the process of crown formation takes much longer in the permanent tooth than in the deciduous tooth. Thus, most of the observed variation is related to differential growth of the two distal cusps.

In the M_1 tooth germs are larger than those of dm_2 before cusp initiation but grow more slowly thereafter, as shown by the relation of vertical to horizontal growth trajectories reflected by distance between cusp tips. Euclidean distances between cusp tips at the DEJ in the dm_2 are similar to the vertical distances between them, indicating continued rapid cell proliferation of the tooth germ from initiation to coalescence of all cusps. In the M_1 intercusp tip distances vary and height differences are small. In the permanent molars the early coalescence of the hypoconulid with the hypoconid and entoconid tilts it inward so that enamel cusp tips of the distal cusps are closer together than dentin cusp tips (Table 3). This finding agrees with that expected from direct observation of tooth germs of different stages of development.

Cusps increase in breadth towards their base. However, in the two permanent teeth studied here, thicker enamel on the distal cusps means that this may not be accurately reflected in cuspal proportions and groove patterns at the crown surface. In permanent molars the thicker enamel on the distal cusps, especially on the hypoconulid, leads to this cusp being relatively larger on the enamel surface than on the DEJ, as illustrated in Figure 3 and Table 3. As shown here, enamel thickness increases from mesial to distal cusps of the same tooth and varies between teeth. In the thinner enameled dm_2 cuspal proportions shown at the DEJ are only minimally affected by this uneven distribution of enamel. In the two permanent teeth the marked increase in enamel thickness of the distal cusps partially compensates for their small size at the DEJ. An increase in enamel thickness of distal cusps of upper molars has been previously shown by Macho (1995) and in lower molars by Grine (2005) from examination of ground sections. The volumetric analyses carried out here from the computerized model, confirm their findings and suggest some degree of independence, or alternately compensation, of regulating factors associated with enamel matrix apposition and those operating during morphogenesis.

So far no specific factors have been found that regulate formation of specific cusps or specific teeth within the molars (Jernvall et al., 2000; Zhao et al., 2000). McCullum and Sharpe (2001) have emphasized the importance of space constraints in partitioning tooth classes and Shimizu et al. (2004) have identified quantitative trait loci (QTLs) that affect first molar size in recombinant inbred mice strains. In modern humans similar size constraints may exist and differentially affect the slow growing permanent molars.

Since crown form is not solely the result of the morphodynamics of cell proliferation and differentiation, enamel thickness needs to be taken into account in assessing the phylogenetic significance of crown form in comparing species that differ in this feature.

Conclusions

We present here a computerized model for reconstructing early developmental processes from serial microCT scans of complete teeth. The method is non-destructive and so is applicable for examination of even rare fossil specimens and is accurate to within several microns. The results indicate that the morphodynamic process of tooth development, where growth and differentiation proceed simultaneously in different locations within the developing tooth germ, is reflected in the spatial relationship of cusps at the DEJ. The extent to which this is modified at the OES, depends on enamel thickness and the extent to which it varies over the crown. Variation in development is the mechanism for speciation and evolution.

Our model provides a unique approach to the reconstruction of ontogenetic processes reflected in the fully formed tooth and enables us to distinguish between early and late components of variation in crown form associated with phenotypic diversity. The data presented here suggest that small alterations in the mesiodistal gradient of differentiation relative to proliferation can produce a major change in crown size and proportions.

Acknowledgments

This study was supported by Grant No. 032-5302 from the Israel Science Foundation.

R.M. was supported by Swiss National Science Foundation (SNF, FP 620-58097.99). We would like to thank Prof. J.J. Hublin and Dr. S. Bailey for the invitation extended to P.S. to participate in the May 2005 conference on "Dental Perspectives on Human Evolution".

References

Avishai, G., Muller, R., Gabet, Y., Bab, I., Zilberman, U., Smith, P., 2004. New approach to quantifying developmental variation in the dentition using serial microtomographic imaging. Microscopy Research Techniques 65(6), 263–269.

Butler, P. M., 1967. Comparison of the development of the second deciduous molar and first permanent molar in man. Archives of Oral Biology 12(11), 1245–1260.

Butler, P. M., 1968. Growth of the human second lower deciduous molar. Archives of Oral Biology 13(6), 671–682.

Butler, P.M., 1999. The relation of cusp development and calcification to growth. In: Mayhall, J.T., Heikkinen, T. (Eds.), Dental Morphology 1998: Proceedings of the 11th International Symposium on Dental Morphology. Oulo University Press, Oulu, Finland, pp. 26–32.

Collard, M., Wood, B., 2000. How reliable are human phylogenetic hypotheses? Proceedings of the National Academy of Sciences of the USA 97(9), 5003–5006.

Dahlberg, A., 1945. The changing dentition of man. Journal of the American Dental Association 32, 676–690.

Finarelli, J.A., Clyde, W.C., 2004. Reassessing hominoid phylogeny: evaluating congruence in the morphological and temporal data. Paleobiology 30(4), 614–651.

Gabet, Y., Muller, R., Regev, E., Sela, J., Shteyer, A., Salisbury, K., Chorev, M., Bab, I., 2004. Osteogenic growth peptide modulates fracture callus structural and mechanical properties. Bone 3(1), 65–73.

Gregory, W.K., Hellman, M., 1926. The Dentition of *Drypopithecus* and the Origin of Man. Anthropology Papers of the American Museum of Natural History New York 28, 1–123.

Grine, F., 2005. Enamel thickness of deciduous and permanent molars in modern *Homo sapiens*. American Journal of Physical Anthropology 126(1), 14–31.

Hlusko, L.J., 2004. Integrating the genotype and phenotype in hominid paleontology. Proceedings of the National Academy of Sciences of the USA 101(9), 2653–2657.

Jernvall, J., 2000. Linking development with generation of novelty in mammalian teeth. Proceedings of the National Academy of Sciences of the USA 97(6), 2641–2645.

Jernvall, J., Keranen, S.V., Thesleff, I., 2000. Evolutionary modification of development in mammalian teeth: quantifying gene expression patterns and topography. Proceedings of the National Academy of Sciences of the USA 97(26), 14444–14448.

Korenhof, C.A., 1960. Morphogenetic aspects of the human upper molar. A comparative study of the enamel and dentin surfaces and their relationship to the crown pattern of fossil and recent primates. Academisch Proefschrift, Uitg. mij Neerlandia, Utrecht.

Korenhof, C.A., 1979. The evolution of the lower molar pattern and remnants of the trigonid crests in man. Nederlandisch. Tijdschr Tandheelkd 86 (Suppl 17), 6–31.

Kraus, B.S., 1952. Morphological relationships between enamel and dentin surfaces of lower first molar tooth. Journal of Dental Research 31, 248–256.

Kraus, B.S., Jordan, R.E., 1965. *The Human Dentition Before Birth*. Lea & Febiger, Philadelphia.

Macho, G.A., 1995. The significance of hominid enamel thickness for phylogenetic and life-history reconstruction. In: Moggi-Cecchi, J. (Ed.), Aspects of Dental Biology: Paleontology, Anthropology and Evolution. Angelo Pontecorbi, Florence, pp. 51–68.

McCollum, M., Sharpe, P.T., 2001. Evolution and development of teeth. Journal of Anatomy 199 (Pt 1–2), 153–159.

Müller, R., Hildebrand, T., Rüegsegger, P., 1994. Non-invasive bone biopsy: A new method to analyse and display the three-dimensional structure of trabecular bone. Physics in Medicine and Biology 39, 145–164.

Müller, R., Van Campenhout, H., Van Damme, B., Van Der Perre, G., Dequeker, J., Hildebrand, T., Rüegsegger, P., 1998. Morphometric analysis of human bone biopsies: a quantitative structural comparison of histological sections and micro-computed tomography. Bone 23(1), 59–66.

Peters, O. A., Laib, A., Ruegsegger, P., Barbakow, F., 2000. Three-dimensional analysis of root canal geometry by high-resolution computed tomography. Journal of Dental Research 79(6), 1405–1409.

Rüegsegger, P., Koller, B., Müller, R., 1996. A microtomographic system for the nondestructive evaluation of bone architecture. Calcified Tissue International 58, 24–29.

Sasaki, K., Kanazawa, E., 1999. Morphological traits on the dentino-enamel junction of lower deciduous molar series. In: Mayhall, J., Heikkinen, T. (Eds.), Dental Morphology 1998: Proceedings of the 11th International Symposium on Dental Morphology. Oulo University Press, Oulu, Finland, pp. 167–178.

Shimizu, T., Oikawa, H., Han, J., Kurose, E., Maeda, T., 2004. Genetic analysis of crown size in the first molars using SMXA recombinant inbred mouse strains. Journal of Dental Research 83(1), 45–49.

Smith, P., Gomorri, J. M., Spitz, S., Becker, J., 1997. Model for the examination of evolutionary trends in tooth development. American Journal of Physical Anthropology 102(2), 283–294.

Smith, P., Gomori, J., Shaked, R., Haydenblit, R. and Joskowicz, L., 2000. A computerized approach to reconstruction of growth patterns in hominid molar teeth. In: Mayhall, J., Heikkinen, T. (Eds.), Proceedings of the 11th International Symposium on Dental morphology. Oulo University Press, Oulo, Finland, pp. 388–397.

von Stechow, D., Balto, K., Stashenko, P., Muller, R., 2003. Three-dimensional quantitation of periradicular bone destruction by micro-computed tomography. Journal of Endodonty 29(4), 252–256.

Winkler, L.A., Schwartz, J.H., Swindler, D.R., 1991. Aspects of dental development in the orangutan prior to eruption of the the permanent dentition. American Journal of Physical Anthropology 86(2), 255–273.

Zhao, Z., Weiss, K.M. and Stock, D.W., 2000. Development and evolution of dentition patterns and their genetic basis. In: Teaford, M.F., Smith, M.M., Ferguson, M.W.J. (Eds.), Development Function and Evolution of Teeth. Cambridge University Press, Cambridge, pp. 152–172.

PART IV

DENTITION AND DIET

1. Introduction

F.E. GRINE
Departments of Anthropology & Anatomical Sciences
Stony Brook University
Stony Brook
NY 11794-4364 USA
frederick.grine@stonybrook.edu

Teeth convey information about a number of interesting and potentially significant biological phenomena. Because the development of the dentition is closely related to that of the individual possessing it, teeth may reveal aspects of somatic development and life history (Dean et al., 1986, 2001; Smith, 1989; Smith et al., 1994, 1995; Bermúdez de Castro et al., 1999; Kelley and Smith, 2003; Ramirez Rozzi and Bermúdez de Castro, 2004; Nargolwalla et al., 2005). These revelations, in turn, may be of some importance to questions relating to the taxonomy of extinct hominins.

By virtue of differences in size and/or morphology, teeth might also be informative to the biological distinctiveness of the organisms that possess them (Rosenberger et al., 1991; Suwa et al., 1996; Ramirez Rozzi, 1998; Carrasco, 2000; Hlusko, 2002; Bailey, 2004, 2005; Scott and Lockwood, 2004; Bailey and Lynch, 2005). Such features may say something about population affinities (Irish and Guatelli-Steinberg, 2003; Manabe et al., 2003). They also may reflect phylogenetic relations (Stringer et al., 1997; Strait et al., 1997; Strait and Grine, 2004; Kimbel et al., 2006), although recent work on the genetics of dental development and morphology suggests that these laudable goals may be rather more complex tasks than has been realized heretofore (Weiss et al., 1998; Jernvall and Jung, 2000; Salazar-Ciudad and Jernvall, 2002; Kangas et al., 2003, 2004; Line, 2003; Hlusko et al., 2004).

Teeth also might be informative to the levels of systemic physiological stress experienced by recent as well as fossil populations (Oglivie et al., 1989; Skinner, 1996; Dirks et al., 2002; Cunha et al., 2004; Skinner and Hopwood, 2004; Guatelli-Steinberg et al., 2004; Corruccini et al., 2005).

Ultimately, however, the fact that teeth can be related to somatic development and species life-history, that their developmental histories might reflect systemic physiological stress, and that their morphological attributes are taxonomically and potentially phylogenetically informative are all merely fortuitous happenstances. This is because the only reason teeth exist is as part of an individual's trophic apparatus. This is no less true for humans and their close relatives than for any other order of dentate gnathostomes. Indeed, it is precisely because teeth serve a vital trophic function that they have proven to be such difficult characters with which to deal in attempts to disentangle the complicated form-function-phylogeny embranglement.

S.E. Bailey and J.-J. Hublin (Eds.), Dental Perspectives on Human Evolution, 291–302.

Teeth exist to process food, and this is their biological significance. At the same time, diet is central to a species' ecology and behavior. The seasonal availability of items comprising the dietary repertoire may impact attributes such as mobility patterns, population size and social organization. The distribution and mechanical properties of food items may selectively affect postcranial and cranial morphologies related to their procurement, ingestion and mastication. It is, therefore, quite understandable that considerable effort has been expended to elucidate the dietary proclivities of our extinct hominin relatives.

Paleodietary reconstructions have been attempted from a number of disparate sources of information, including (1) the archaeological record, (2) biomechanical models of cranial and mandibular morphology, (3) tooth size and morphology, including enamel thickness and structure, (4) isotope chemistry, and (5) dental microwear. Each of these approaches has its own particular strengths, and each is beset with its own attendant weaknesses.

The archaeological record is relevant only if there is one. At present, lithic flakes and/or cores that are identifiable as having been purposefully manufactured, and animal bones with stone tool cutmarks are known only as far back as some 2.5 Myr ago (Kimbel et al., 1996; de Heinzelin et al., 1999; Domínquez-Rodrigo et al., 2005). This leaves well over half of the hominin fossil record devoid of any durable artefacts that might be informative of dietary habits.

Even when there is an archaeological record, its relevance depends upon the attribution of particular artifact assemblages to a particular taxon. Unfortunately, it is not always clear who was responsible for archaeological debris (e.g., Brain et al., 1988; Susman, 1988). Thus, in both eastern and southern Africa, for example, lithic (Oldowan and Developed Oldowan) assemblages are known from sites and/or temporal horizons between about 2.5 Myr and 1.5 Myr that contain the remains of two (or more) hominin species.

Even if the question of species attribution is left aside, the interpretation of archaeological remains vis-à-vis diet is not always straightforward. Thus, for example, the bone tools from the Early Pleistocene deposits at Swartkrans, South Africa, have been interpreted by some authorities (Brain and Shipman, 1993) as evidence for digging-up edible bulbs such as those of grass stars (*Hypoxis costata*) and lilies (*Scilla marginata*). Other workers (Backwell and d'Errico, 2001, 2003), however, have argued that they were used to excavate termite mounds implying a very different food resource. Furthermore, the interpretation of other archaeological traces can be equally contentious, as illustrated by the seemingly endless debate that has surrounded meat-eating versus scavenging by extinct hominins (Binford, 1981; Bunn, 1981, 1983, 1994; Bunn and Kroll, 1986; Shipman, 1986; Marean, 1988; Potts, 1988; Gifford-Gonzales, 1991; Stiner, 1994; Blumenshine, 1995; Stanford and Bunn, 1999; Domínguez-Rodrigo, 2002; Domínguez-Rodrigo and Barba, 2006).

Biomechanical models (Rak, 1983, 1986; Demes and Creel, 1988; Hylander, 1988; Daegling and Grine, 1991; McCollum, 1994; Chen and Grine, 1997; Wood and Aiello, 1998; Hylander and Johnson, 2002; Daegling and Grine, 2007) are appealing because the material properties of dietary items almost certainly influence cranial and mandibular anatomy on an evolutionary timescale. However, the exact relationships between loading environments and particular morphologies are unclear. Indeed, because it is the mechanical properties of ingested items rather than diet *per se* that impacts jaw morphology, even a full appreciation of cranial and mandibular design can provide only limited insight into diet.

Aspects of dental morphology such as molar cusp height and alignment, as well as enamel thickness and structure are appealing for the same reason that biomechanical models of cranial and mandibular design have attracted such interest (Kay, 1981, 1984, 1985; Ungar, 1998, 2004; Ungar et al., 1999; Macho et al., 2005). Morphological attributes may reflect the efficiency by which teeth can process different types of food, but they do not provide direct evidence for diet. The oftentimes imprecise functional relationship between form and diet is also especially problematic when applied to morphologies (or combinations thereof) that are unique to the paleontological record (Ross et al., 2002). In essence, the anatomical record may speak to specific adaptations that occurred in a lineage's history, but these may or may not be germane to what individuals of a particular species actually consumed. Individuals (or, for that matter, species) may not always do what is suggested by the form/function/adaptation relationship that we infer.

The limits of direct analogy from the neontological present to the palaeontological past should be self-evident, although these rarely seem to be acknowledged by most practitioners of the comparative method who would interpret functional morphology. The comparative method is limited by the fact that no extinct species was exactly like a living species, and nor can an extinct taxon be accurately portrayed as a simple theoretical composite of attributes of a number of various living ones.

Nevertheless, details of molar occlusal relief and their attendant wear patterns provide information about historical adaptation, the process of mastication and, ultimately, something about the dietary habits of a species (M'Kirera and Ungar, 2003; Dennis et al., 2004; Ungar, 2004). The contributions by Ungar (2007) and Ulhaas et al. (2007) in the present volume represent two examples of how the analysis of dental topography and gross patterns of molar wear can inform us about potential differences in the diets of extinct hominin species. Both apply new, cutting-edge research methodologies to questions of paleoanthropological significance.

Ungar (2007) here provides a review of dental topographic analysis, which is a comparatively novel tool to be employed in the study of occlusal functional morphology. This landmark-free, three-dimensional approach involves the creation and measurement of digital models of a tooth crown using point cloud data and Geographic Information Systems software. Using this technique, Ungar elegantly demonstrates that living species with different diets have corresponding and predictable differences in the shapes of their molars at comparable stages of tooth wear, and that individuals of a given species evince consistent changes in crown shape as their teeth wear down. These are very important findings, because they provide evidence for the existence of species-specific wear patterns. This, in turn, is a necessary prerequisite for the inference of function from form in worn teeth. Ungar then utilizes this methodology in an analysis of molar topography in the hominins *Australopithecus africanus* and *Paranthropus robustus*, and shows that their occlusal morphologies differ in a manner expected of primates that often eat similar foods, but whose diets differ in terms of the "fallback" foods utilized during periods of stress. The dental topography data that he presents suggest that *P. robustus* probably relied more on hard, brittle items whereas *A. africanus* relied on tougher, more elastic foods when their preferred resources were less available. This represents a significant refinement in the models of trophic differentiation that have been proposed in the past for these species (e.g., Robinson, 1954; Grine, 1981, 1986).

Ulhaas et al. (2007) also employ a three-dimensional approach to compare occlusal surface morphologies of worn molars of the same two extinct taxa examined by

Ungar (2007) – *Australopithecus africanus* and *Paranthropus robustus* – and of an earlier species, *Praeanthropus afarensis* (which is called by them *Australopithecus afarensis*). They employed high resolution optical topometry to measure computer models of second lower molar crowns, and in so doing, document different modes of occlusal wear in these three taxa. They conclude that the molars of *Pr. afarensis* provided a means by which that species could cope with a wide range of food qualities. Echoing the findings of Ungar (2007), Ulhaas et al. conclude that the molars of *P. robustus* suggest an omnivorous generalist that relied on hard objects as fall-back foods.

In contrast to biomechanical/adaptive models of jaw and tooth design, isotope chemistry and microwear preserve epigenetic signals relating to an individual's diet. These latter approaches can tell us something about the sorts of foods that were actually consumed, rather than what a species might have been capable of processing.

Isotope chemistry and trace element analysis are extremely promising avenues of paleodietary research (Sillen and Kavanagh, 1991; Lee-Thorpe et al., 1994; Burton and Wright, 1995; Sillen et al., 1995; Drucker et al., 1999; Sponheimer and Lee-Thorpe, 1999, 2003; Bocherens et al., 2001, 2005; Richards et al., 2001; Lee-Thorpe, 2002; Bocherens and Drucker, 2003; Drucker and Bocherens, 2004; Sponheimer et al., 2005a, b; Drucker and Henry-Gambier, 2005). Recent work has shown that by considering different chemical signals in combination [e.g., strontium/calcium (Sr/Ca) considered together with barium/calcium (Ba/Ca) and stable carbon isotopes ($^{13}C/^{12}C$)], it may be possible to produce a more refined picture of the agents responsible for the signatures trapped by fossil hominin teeth (Sponheimer et al., 2005a, b).

Isotopic studies, however, are beset by their own particular problems. These include diagenetic alteration of tooth and (especially) bone chemistry, and the temporal/preservation limitations imposed by elements (e.g., nitrogen) that are restricted to the organic phases of these hard tissues. In addition, particular food sources can yield different results depending upon the environment from which they come, and different types of foods can yield similar chemical signatures. Trophic levels can also influence interpretations, where, for example, carnivorous and/or insectivorous behaviors result in the ingestion of the chemical components of the foliage eaten by the prey. Thus, while knowledge of environmental parameters and diet may predict isotope ratios in hominin fossils, the reverse is not necessarily the case (Burton and Wright, 1995).

Nevertheless, the exciting results obtained through well-devised studies such as that presented here by Humphrey et al. (2007) demonstrate that isotope chemistry and trace element analysis can address important questions relating to changes in an individual's diet over time (or, at least over that time that dental enamel is being formed). This study has important implications on several fronts. First, through the use of laser ablation of tooth enamel coupled with inductively coupled plasma mass spectrometry, Humphrey et al. (2007) were able to develop a microfocal sampling strategy that could be cross-referenced to incremental growth structures in the enamel cap. Having established an incredibly precise level of resolution (c. 30 μm diameter sample pits), these workers were able to demonstrate that there is a consistent pattern of change in the Sr/Ca ratio across the neonatal line in human deciduous teeth. This is strongly suggestive that incremental growth structures in enamel can be used as a basis for inferring the chronology of marked changes in diet during infancy and childhood. This study, however, also has implications for the interpretative validity of analyses of the chemical composition of tooth enamel that employ bulk sampling approaches. The Sr/Ca ratio at a given point in the tooth crown is influenced by a range of parameters, including the age of initiation of enamel matrix formation at the sampling point, the age of onset and duration of mineralization for the tooth cusp as a whole, dietary intake during the entire period of enamel mineralization, as well as the thickness of the enamel and the duration of matrix secretion. Thus, samples taken from the same position in homologous teeth of children with the same dietary history may produce different Sr/Ca ratios as a result of individual differences in tooth crown development.

Dental microwear has been extensively studied in attempts to elucidate the dietary habits of extinct hominin species (Grine, 1981, 1986; Walker, 1981; Puech and Albertini, 1983; Puech, 1984, 2001; Grine and Kay, 1988; Ryan and Johanson, 1989; Ungar and Grine, 1991; Ryan, 1993; Ungar, 1998; Pérez-Pérez et al., 1999; Teaford et al., 2002; Scott et al., 2005; Ungar et al., 2006). Studies of dental microwear have demonstrated a relationship between occlusal surface texture and the types of dietary items that comprise a species' diet (Walker et al., 1978; Scott et al., 2005). Occlusal molar microwear is capable of distinguishing among broad dietary categories when they correspond to differences in the physical characteristics of their constituent food items (Teaford and Walker, 1984; Teaford, 1985, 1986; Teaford and Glander, 1996; Daegling and Grine, 1999; Nelson et al., 2005). Occlusal microwear is also capable of identifying subtle differences in the diets of closely related species, as well as short-term variations in diet, thus permitting the detection of seasonal and other ecological differences (Teaford, 1986; Teaford and Oyen, 1989; Teaford and Robinson, 1989; Teaford and Glander, 1996).

However, while microwear fabrics may vary with diet, they actually vary with the physical properties of the foods that are chewed. Moreover, the ability of microwear fabrics to indicate subtle, short-term variations in diet means that they potentially fossilize information pertaining only to the last meals consumed before an individual's death. This is the so-called "Last Supper Effect" (Grine, 1986). It is a potentially confounding influence in the analysis of fossil assemblages that accumulated over a prolonged period of time and which may sample different seasons in unequal abundance. Finally, taphonomic artifacts, including environmental erosion or etching, and damage inflicted through the preparation of fossils, have the potential to confound microwear interpretations if they are not properly diagnosed (Teaford, 1988; King et al., 1999). In particular, such artifacts may severely limit available samples, especially where the fossil assemblages comprise surface collections from fluvial channel sand deposits (e.g., the Shungura Formation of southern Ethiopia). Thus, in a recent study of molar microwear in early *Homo*, only 19 of 83 molar crowns from the Plio-Plestocene deposits in East and South Africa were found to be suitable for analysis (Ungar et al., 2006). Taphonomic limitations on available sample sizes exacerbate the "Last Supper Effect."

Notwithstanding these caveats, dental microwear analysis, like the study of isotope chemistry, can provide important information relating to the dietary proclivities of extinct individuals. One important question that microwear has the potential to address relates to change in a species' diet over time, or in response to different environmental or ecological conditions.

In the present volume, Teaford (2007) provides a comprehensive review of the various techniques that have been and are presently being applied in microwear research, and provides us with timely caveats about their limitations. He points out that dental microwear analyses have "come a long way in the past 25 years," which I can only read to be a reference to the very primitive state of the art in 1981 (Grine, 1981)! Teaford (2007) notes that each method of data collection has its strengths and attendant limitations, but that microwear analysis has the potential to give us direct glimpses of the past because it can tell us about how teeth were actually used rather than what they were evolutionarily capable of doing. He also reinforces the observation that different methods may result in data that are not necessarily directly comparable (Grine et al., 2002).

Indeed, it is almost impossible for different workers to obtain the same (or even similar) results when highly subjective methods, such as low-power standard refractive light microscopy (Solounias and Semprebon, 2002; Godfrey et al., 2004; Semprebon et al., 2004), are employed. As Alan Walker pointed out to those who espouse this particular technique, it is incapable of distinguishing microscopic wear features (i.e., puncture/crushing pits) from macroscopic wear features (i.e., dentine exposures) (Walker, pers. comm., in Semprebon et al., 2005)! In contrast, and taking a positive step forward in the development of methodologies that can be successfully applied in the objective and repeatable quantification of dental microwear, Ungar and colleagues (Ungar et al., 2003; Scott et al., 2005) have devised a technique that employs scale-sensitive fractal analysis to provide a true representation of wear features in three dimensions.

In their contribution to the present volume, Estebaranz et al. (2007) undertake an assessment of the ability of such automated three-dimensional methods to provide accurate information about microwear. They employ interferometric microscopy to examine microtopographic relief on the buccal surfaces of molars, and conclude that surface roughness, and the postmortem taphonomic artefacts that can intensify it render suspect the results of fully automated procedures. In particular, they conclude that such techniques still require the researcher to select surfaces that are free of artefacts prior to analysis, something which has been done in some published studies that utilized such automated techniques (e.g., Scott et al., 2005). However, it is worth pointing out that taphonomically altered surfaces are as easily recognized with high magnification confocal light microscopy as by high magnification scanning electron microscopy.

Indeed, it is quite possible that the problems of surface roughness and taphonomic alteration over which Estebaranz et al. (2007) agonize are peculiar to the surface that they examined. That is, because the buccal surface of a molar is not used in mastication, it will retain all of the original surface irregularities (i.e., "roughness"), upon which various antemortem and postmortem scratches and acidic etchings can be superimposed. It is unlikely, however, that scratches induced by antemortem activities related to diet will ever obliterate the original surface texture because they occur so infrequently. This is in contrast to the occlusal surface. Here, the small-scale (and even large-scale) irregularities that constitute surface "roughness" will be obliterated as the tooth is worn. Thus, the original surface texture will be replaced by a microwear fabric (or a taphonomically modified fabric). This means that the original surface "roughness" cannot possibly constitute a problem on the occlusal aspect of the crown in terms of microwear analysis; although postmortem artefacts on the occlusal surface can be potentially problematic, these can be diagnosed (Teaford, 1988). On the other hand, original surface "roughness" will, indeed, be a persistently troublesome feature on the buccal (or lingual) aspect of the crown in terms of automated 3-D microwear analysis, as argued by Estebaranz et al. (2007).

It is only fitting that a volume dedicated to state of art research in dental paleoanthropology should conclude with a section on diet, for that is what teeth are all about. It is also fitting that this section comprises the papers that it does, for these represent the state of art research in functional morphology, isotope chemistry and microwear studies that elucidate the diets of our extinct relatives.

Acknowledgments

I am grateful to Shara Bailey and Jean-Jacques Hublin for the invitation to participate in this conference, and for everything that they did to ensure that I and all of the other participants fully enjoyed out time in Leipzig. I would also like to thank Diana Carstens and Silke Streiber, whose diligent efforts provided for effortless travel to and from the symposium.

References

Backwell, L.R., d'Errico, F., 2001. First evidence of termite-foraging by Swartkrans early hominids. Proceedings of the National Academy of Sciences of the USA 98, 1358–1363.

Backwell, L.R., d'Errico, F., 2003. Additional evidence on the early hominid bone tools from Swartkrans with reference to spatial distribution of lithic and organic artifacts. South African Journal of Science 99, 259–267.

Bailey, S.E., 2004. A morphometric analysis of maxillary molar crowns of Middle-Late Pleistocene hominins. Journal of Human Evolution 47, 183–198.

Bailey, S.E., 2005. Inter- and intraspecific variation in *Pan* tooth crown morphology: implications for Neandertal taxonomy. American Journal of Physical Anthropology Suppl. 40, 68.

Bailey, S.E., Lynch, J.M., 2005. Diagnostic differences in mandibular P4 shape between Neandertals and anatomically modern humans. American Journal of Physical Anthropology 126, 268–277.

Bermúdez de Castro, J.M., Rosas, A., Carbonell, E., Nicolas, M.E., Rodriguez, J., Arsuaga, J.L., 1999. A modern human pattern of dental development in Lower Pleistocene hominids from Atapuerca – TD6 (Spain). Proceedings of the National Academy of Sciences of the USA 96, 4210–4213.

Binford, L.R., 1981. *Bones: Ancient Men and Modern Myths.* Academic, New York.

Blumenshine, R.J., 1995. Percussion marks, tooth marks and the experimental determinations of the timing of hominid and carnivore access to long bones at FLK *Zinjanthropus*, Olduvai Gorge, Tanzania. Journal of Human Evolution 29, 21–51.

Bocherens, H., Drucker, D., 2003. Reconstructing Neandertal diet from 120,000 to 30,000 BP using carbon and nitrogen isotopic abundances. In: Patou-Mathis, M., Bocherens, H. (Eds.), Le Rôle de l'Environment dans les Comportements des Chasseurs-cueilleurs Préhistoriques, BAR International Series 1105, pp. 1–7.

Bocherens, H., Drucker, D.G., Billiou, D., Patou-Mathis, M., Vandermeersch, B., 2005. Isotopic evidence for diet and subsistence pattern of the Saint Césaire I Neanderthal: review and use of a multi-source mixing model. Journal of Human Evolution 49, 71–87.

Bocherens, H., Toussaint, M., Billiou, D., Patou-Mathis, M., Bonjean, D., Otte, M., Mariotti, A., 2001. New isotopic evidence for dietary habits of Neandertals from Belgium. Journal of Human Evolution 40, 497–505.

Brain, C.K., Churcher, C.S., Clark, J.D., Grine, F.E., Shipman, P., Susman, R.L., Turner, A., Watson, V., 1988. New evidence of early hominids, their culture and environment from the Swartkrans Cave, South Africa. South African Journal of Science 84, 828–835.

Brain, C.K., Shipman, P., 1993. The Swartkrans bone tools. In: Brain, C.K. (Ed.), Swartkrans: A Cave's Chronicle of Early Man. Transvaal Museum, Pretoria, pp. 195–215.

Bunn, H.T., 1981. Archaeological evidence for meat-eating by Plio-Pleistocene hominids from Koobi Fora and Olduvai Gorge. Nature 291, 574–577.

Bunn, H.T., 1983. Evidence on the diet and subsistence patterns of Plio-Pleistocene hominids at Koobi Fora, Kenya, and Olduvai Gorge, Tanzania. In: Clutton-Brock, J., Grigson, C. (Eds.), Animals and Archaeology: 1. Hunters and Their Prey. British Archaeological Reports, Oxford, pp. 21–30.

Bunn, H.T., 1994. Early Pleistocene hominid foraging strategies along the ancestral Omo River at Koobi Fora, Kenya. Journal of Human Evolution 27, 247–266.

Bunn, H.T., Kroll, E.M., 1986. Systematic butchery by Plio-Pleistocene hominids at Olduvai Gorge, Tanzania. Current Anthropology 27, 431–452.

Burton, J.H., Wright, L.E., 1995. Nonlinearity in the relationship between bone Sr/Ca and diet: paleodietary implications. American Journal of Physical Anthropology 96, 273–282.

Carrasco, M.A., 2000. Species discrimination and morphological relationships of kangaroo rats (*Dipodomys*) based on their dentition. Journal of Mammalogy 81, 107–122.

Chen, X., Grine, F.E., 1997. The effect of cortical thickness on mandibular torsional strength of South African early hominids. Journal of Human Evolution 32, A6.

Corruccini, R.S., Townsend, G.C., Schwerdt, W., 2005. Correspondence between enamel hypoplasia and odontometric bilateral asymmetry in Australian twins. American Journal of Physical Anthropology 126, 177–182.

Cunha, E., Ramirez Rozzi, F.R., Bermúdez de Castro, J.M., Martinon-Torres, M., Wasterlain, S.N., Sarmiento, S., 2004. Enamel hypoplasias and physiological stress in the Sima de los Huesos Middle Pleistocene hominins. American Journal of Physical Anthropology 125, 220–231.

Daegling, D.J., Grine, F.E., 1991. Compact bone distribution and biomechanics of early hominid mandibles. American Journal of Physical Anthropology 86, 321–339.

Daegling, D.J., Grine, F.E., 1999. Terrestrial foraging and dental microwear in *Papio ursinus*. Primates 40, 559–572.

Daegling, D.J., Grine, F.E., 2007. Mandibular biomechanics and the paleontological evidence for the evolution of human diet. In: Ungar, P.S. (Ed.), The Evolution of Hominin Diets: the Known, the Unknown and the Unknowable. Oxford University Press, New York, pp. 77–105.

Dean, M.C., Leakey, M.G., Reid, D., Schrenk, F., Schwartz, G.T., Stringer, C.B., Walker, A., 2001. Growth processes in teeth distinguish modern humans from *Homo erectus* and earlier hominins. Nature 414, 628–631.

Dean, M.C., Stringer, C.B., Bromage, T.G., 1986. Age at death of the Neanderthal child from Devil's Tower, Gibralter and the implications for studies of general growth and development in Neanderthals. American Journal of Physical Anthropology 70, 301–309.

de Heinzelin, J., Clark, J.D., White, T.D., Hart, W., Renne, P., WoldeGabriel, G., Beyene, Y., Vrba, E., 1999. Environment and behavior of 2.5-million year-old Bouri hominids. Science 284, 625–629.

Demes, B., Creel, N., 1988. Bite force, diet, and cranial morphology of fossil hominids. Journal of Human Evolution 17, 657–670.

Dennis, J.C., Ungar, P.S., Teaford, M.F., Glander, K.E., 2004. Dental topography and molar wear in *Alouatta palliata* from Costa Rica. American Journal of Physical Anthropology 125, 152–161.

Dirks, W., Reid, D.J., Jolly, C.J., Phillips-Conroy, J.E., Brett, F.L., 2002. Out of the mouths of baboons: stress, life history, and dental development in the Awash National Park hybrid zone, Ethiopia. American Journal of Physical Anthropology 118, 239–252.

Domínguez-Rodrigo, M., 2002. Hunting and scavenging by early humans: the state of the debate. Journal of World Prehistory 16, 1–54.

Domínguez-Rodrigo, M., Barba, R., 2006. New estimates of tooth mark and percussion mark frequencies at the FLK Zinj site: the carnivore-hominid carnivore hypothesis falsified. Journal of Human Evolution 50, 170–194.

Domínquez-Rodrigo, M., Pickering, T.R., Semaw, S., Rogers, M.J., 2005. Cutmarked bones from Pliocene archaeological sites at Gona, Afar, Ethiopia: implications for the function of the world's oldest stone tools. Journal of Human Evolution 48, 109–121.

Drucker, D., Bocherens, H., 2004. Carbon and nitrogen stable isotopes as tracers of change in diet breadth during Middle and Upper Palaeolithic in Europe. International Journal of Osteoarchaeology 14, 162–177.

Drucker, D., Bocherens, H., Mariotti, A., Lévêque, F., Vandermeersch, B., Guadelli, J.L., 1999. Conservation des signatures isotopiques du collagene d'os et de dents du Pléistocene supérieur (Saint-Césaire, France): implications pour les recostitutions des régimes alimentaires des Néandertalians. Bulletins et mémoires de la Société d'anthropologie de Paris 11, 289–305.

Drucker, D.G., Henry-Gambier, D., 2005. Determination of the dietary habits of a Magdalenian woman from Saint-Germain-la-Rivière in southwestern France using stable isotopes. Journal of Human Evolution 49, 19–35.

Estebaranz, F., Galbany, J., Martinez, L.M.,Pérez-Pérez, A., 2007. Interferometric microscopy applied to the study of buccal enamelmicrowear. In: Bailey, S., Hublin, J.-J. (Eds.), Dental Perspectives on Human Evolution: State of the Art Research in Dental Paleoanthropology. Springer, Dordrecht, pp. 291–302.

Gifford-Gonzales, D.P., 1991. Bones are not enough: analogues, knowledge, and interpretive strategies in zooarchaeology. Journal Anthropology and Archaeology 10, 215–254.

Godfrey, L.R., Semprebon, G.M., Jungers, W.L., Sutherland, M.R., Simons, E.L., Solounias, N., 2004. Dental use wear in extinct lemurs:

evidence of diet and niche differentiation. Journal of Human Evolution 47, 145–169.

Grine, F.E., 1981. Trophic differences between "gracile" and "robust" australopithecines: a scanning electron microscope analysis of occlusal events. South African Journal of Science 77, 203–230.

Grine, F.E., 1986. Dental evidence for dietary differences in *Australopithecus* and *Paranthropus*: a quantitative analysis of permanent molar microwear. Journal of Human Evolution 15, 783–822.

Grine, F.E., 2004. Geographic variation in human tooth enamel thickness does not support Neandertal involvement in the ancestry of modern Europeans. South African Journal of Science 100, 389–394.

Grine, F.E., Kay, R.F., 1988. Early hominid diets from quantitative image analysis of dental microwear. Nature 333, 765–768.

Grine, F.E., Ungar, P.S., Teaford, M.F., 2002. Error rates in dental microwear quantification using scanning electron microscopy. Scanning 24, 144–153.

Guatelli-Steinberg, D., 2004. Analysis and significance of linear enamel hypoplasia in Plio-Pleistocene hominins. American Journal of Physical Anthropology 123, 199–215.

Guatelli-Steinberg, D., Larsen, C.S., Hutchinson, D.L., 2004. Prevalence and the duration of linear enamel hypoplasia: a comparative study of Neandertals and Inuit foragers. Journal of Human Evolution 47, 65–84.

Hlusko, L.J., 2002. Identifying metameric variation in extant hominoid and fossil hominid mandibular molars. American Journal of Physical Anthropology 118, 86–97.

Hlusko, L.J., Maas, M.L., Mahaney, M.C., 2004. Statistical genetics of molar cusp patterning in pedigreed baboons: Implications for primate dental development and evolution. Journal of Experimental Zoology B 302B, 268–283.

Humphrey, L.T., Dean, M.C., Jeffries, T.E., 2007. An evaluation of changes in strontium/calcium ratios across the neonatal line in human deciduous teeth. In: Bailey, S., Hublin, J.-J. (Eds.), Dental Perspectives on Human Evolution: State of the Art Research in Dental Paleoanthropology. Springer, Dordrecht, pp. 291–302.

Hylander, W.L., 1988. Implications of in vivo experiments for interpreting the functional significance of "robust" australopithecine jaws. In: Grine, F.E. (Ed.), Evolutionary History of the "Robust" Australopithecines. Aldine de Gruyter, New York, pp. 55–83.

Hylander, W.L., Johnson, K.R., 2002. Functional morphology and in vivo bone strain patterns in the craniofacial region of primates: beware of biomechanical stories about fossil bones. In: Plavcan, J.M., Kay, R.F., Jungers, W.L., van Schaik, C. (Eds.), Reconstructing Behavior in the Primate Fossil Record. Plenum, New York, pp. 43–72.

Irish, J.D., Guatelli-Steinberg, D., 2003. Ancient teeth and modern human origins: an expanded comparison of African Plio-Pleistocene and recent world dental samples. Journal of Human Evolution 45, 113–144.

Jernvall, J., Jung, H.S., 2000. Genotype, phenotype, and developmental biology of molar tooth characters. Yearbook of Physical Anthropology 43, 171–190.

Kangas, A.T., Mustonen, T., Mikkola, M.L., Thesleff, I., Jernvall, J., 2003. A single-gene mutation leads to changes in several dental characters in mouse. Integrative and Comparative Biology 43, 1027.

Kangas, A.T., Evans, A.R., Thesleff, I., Jernvall, J., 2004. Nonindependence of mammalian dental characters. Nature 432, 211–214.

Kay, R.F., 1981. The nut-crackers: a new theory of the adaptations of the Ramapithecinae. American Journal of Physical Anthropology 55, 141–151.

Kay, R.F., 1984. On the use of anatomical features to infer foraging behavior in extinct primates. In: Rodman, P.S., Cant, J.G.H. (Eds.), Adaptations for Foraging in Nonhuman Primates. Columbia University Press, New York, pp. 21–53.

Kay, R.F., 1985. Dental evidence for the diet of *Australopithecus*. Annual Review of Anthropology 14, 315–341.

Kelley, J., Smith, T.M., 2003. Age at first molar emergence in early Miocene *Afropithecus turkanensis* and life-history evolution in the Hominoidea. Journal of Human Evolution 44, 307–329.

Kimbel, W.H., Lockwood, C.A., Ward, C.V., Leakey, M.G., Johanson, D.C., 2006. Was *Australopithecus anamensis* ancestral to *A. afarensis*? A case of anagenesis in the hominin fossil record. Journal of Human Evolution 51, 134–152.

Kimbel, W.H., Walter, R.C., Johanson, D.C., Reed, K.E., Aronson, J.L., Assefa, Z., Marean, C.W., Eck, G.G., Bobe, R., Hovers, E., Rak, Y., Vondra, C., Yemane, T., York, D., Chen, Y., Evensen, N.M., Smith, P.E., 1996. Late Pliocene

Homo and Oldowan tools from the Hadar formation (Kada Hadar Member), Ethiopia. Journal of Human Evolution 31,549–561.

King, T., Andrews, P., Boz, B., 1999. Effect of taphonomic processes on dental microwear. American Journal of Physical Anthropology 108, 359–373.

Lee-Thorpe, J.A., 2002. Hominid dietary niches from proxy chemical indicators in fossils: the Swartkrans example. In: Ungar, P.S., Teaford, M.F. (Eds.), Human Diet: Its Origin and Evolution. Bergin and Garvey, Westport, CT, pp. 123–142.

Lee-Thorpe, J.A., van der Merwe, N.J., Brain, C.K., 1994. Diet of *Australopithecus robustus* at Swartkrans from stable carbon isotopic analysis. Journal of Human Evolution 27, 361–372.

Line, S.R.P., 2003. Variation of tooth number in mammalian dentition: connecting genetics, development, and evolution. Evolution and Development 5, 295–304.

Macho, G., Shimizu, D., Jiang, Y., Spears, I.R., 2005. *Australopithecus anamensis*: a finite-element approach to studying the functional adaptations of extinct hominins. The Anatomical Record Part A 283, 310–318.

Manabe, Y., Oyamada, J., Kitagawa, Y., Rokutanda, A., Kato, K., Matsushita, T., 2003. Dental morphology of the Dawenkou Neolithic population in North China: implications for the origin and distribution of Sinodonty. Journal of Human Evolution 45, 369–380.

Marean, C.W., 1988. A critique of the evidence for scavenging by Neandertals and early modern humans: new data from Kobeh Cave (Zagros Mountains, Iran) and Die Kelders Cave 1 Layer 10 (South Africa). Journal of Human Evolution 35, 111–136.

McCollum, M.A., 1994. Mechanical and spatial determinants of *Paranthropus* facial form. American Journal of Physical Anthropology 93, 259–273.

M'Kirera, F., Ungar, P.S., 2003. Occlusal relief changes with molar wear in *Pan troglodytes troglodytes* and *Gorilla gorilla gorilla*. American Journal of Primatology 60, 31–41.

Nargolwalla, M.C., Begun, D.R., Dean, M.C., Reid, D.J., Kordos, L., 2005. Dental development and life history in *Anapithecus hernyaki*. Journal of Human Evolution 49, 99–121.

Nelson, S., Badgley, C., Zakem, E., 2005. Microwear in modern squirrels in relation to diet. Palaeontologia Electronica 8, 14A (1–8).

Oglivie M., Curran B., Trinkaus, E., 1989. Incidence and patterning of dental enamel hypoplasia among the Neandertals. American Journal of Physical Anthropology 79, 25–41.

Pérez-Pérez, A., Bermúdez de Castro, J.M., Arsuaga, J.L., 1999. Nonocclusal dental microwear analysis of 3000,000-year-old *Homo heidelbergensis* teeth from Sima de los Huesos (Sierra de Atapuerca, Spain). American Journal of Physical Anthropology 108, 433–457.

Potts, R., 1988. *Early Hominid Activities at Olduvai.* Aldine, New York.

Puech, P.-F., 1984. Acidic food choice in *Homo habilis* at Olduvai. Current Anthropology 25, 349–350.

Puech, P.-F., 2001. Dental microwear in Neandertal highlight a closed cultural trajectory. Journal of Dental Research 80, 224.

Puech, P.-F., Albertini, H., 1983. Usure des dentes chez *Australopithecus afarensis*: examen au microscope e complexe canine supérieure/première prémolaire inférieure. Comptes Rendus de l'Académie des Sciences. Série II 296, 817–822.

Rak, Y., 1983. *The Australopithecine Face.* Academic Press, New York.

Rak, Y., 1986. The Neanderthal: a new look at an old face. Journal of Human Evolution 15, 151–164.

Ramirez Rozzi, F., 1998. Can enamel microstructure be used to establish the presence of different species of Plio-Pleistocene hominids from Omo, Ethiopia? Journal of Human Evolution 35, 543–576.

Ramirez Rozzi, F., Bermúdez de Castro, J.M., 2004. Surprisingly rapid growth in Neanderthals. Nature 428, 936–939.

Richards, M.P., Pettitt, P.B., Stiner, M.C., Trinkaus, E., 2001. Stable isotope evidence for increasing dietary breadth in the European mid upper Palaeolithic. Proceedings of the National Academy of Sciences of the USA 98, 6528–6532.

Robinson, J.T., 1954. Prehominid dentition and hominid evolution. Evolution 8, 324–334.

Rosenberger, A.L., Hartwig, W.C., Takai, M., Setoguchi, T., Shigehara, N., 1991. Dental variability in *Saimiri* and the taxonomic status of *Neosaimiri fieldsi*, an early squirrel monkey from La Venta, Colombia. International Journal of Primatology 12, 291–301.

Ross, C.F., Lockwood, C.A., Flaegle, J.G., Jungers, W.L., 2002. Adaptation and behavior in the primate fossil record. In: Plavcan, J.M., Kay, R.F., Jungers, W.L., van Schaik, C. (Eds.), Reconstructing Behavior in the Primate Fossil Record. Plenum, New York, pp.1–41.

Ryan, A.S., 1993. Anterior dental microwear in Late Pleistocene human fossils. American Journal of Physical Anthropology S16, 171.

Ryan, A.S., Johanson, D.C., 1989. Anterior dental microwear in *Australopithecus afarensis*. Journal of Human Evolution 18, 235–268.

Salazar-Ciudad, I., Jernvall, J., 2002. A gene network model accounting for development and evolution of mammalian teeth. Proceedings of the National Academy of Sciences of the USA 99, 8116–8120.

Scott, J.E., Lockwood, C.A., 2004. Patterns of tooth crown size and shape in great apes and humans and species recognition in the hominid fossil record. American Journal of Physical Anthropology 125, 303–319.

Scott, R.S., Bergstrom, T.S., Brown, C.A., Grine, F.E., Teaford, M.F., Walker, A., Ungar, P.S., 2005. Dental microwear texture analysis reflects diets of living primates and fossil hominins. Nature 436, 693–695.

Semprebon, G.M., Godfrey, L.R., Solounias, N., Sutherland, M.R., Jungers, W.L., 2004. Can low-magnification stereomicroscopy reveal diet? Journal of Human Evolution 47, 115–144.

Semprebon, G.M., Godfrey, L.R., Solounias, N., Sutherland, M.R., Jungers, W.L., Simons, E.L., 2005. Erratum. Journal of Human Evolution 49, 662–663.

Shipman, P., 1986. Scavenging or hunting in early hominids: theoretical framework and tests. American Anthropologist 88, 27–43.

Sillen, A., Kavanagh, M., 1991. Strontium and paleodietary research: a review. Yearbook of Physical Anthropology 25, 67–90.

Sillen, A., Hall, G., Armstrong, R., 1995. Strontium-calcium ratios (Sr/Ca) and strontium isotope ratios (87Sr/86Sr) of *Australopithecus robustus* and *Homo* sp. from Swartkrans. Journal of Human Evolution 28, 277–285.

Skinner, M.F., 1996. Developmental stresses in immature hominines from Late Pleistocene Eurasia: evidence from enamel hypoplasia. Journal of Archaeological Science 23, 833–852.

Skinner, M.F., Hopwood, D., 2004. Hypothesis for the causes and periodicity of repetitive linear enamel hypoplasia in large, wild African (*Pan troglodytes* and *Gorilla gorilla*) and Asian (*Pongo pygmaeus*) apes. American Journal of Physical Anthropology 123, 216–235.

Smith, B.H., 1989. Dental development as a measure of life history in primates. Evolution 43, 683–688.

Smith, B.H., Crummett, T.L., Brandt, K.L., 1994. Ages of eruption of primate teeth: a compendium for aging individuals and comparing life histories. Yearbook of Physical Anthropology 37,177–231.

Smith, R.J., Gannon, P.J., Smith, B.H., 1995. Ontogeny of australopithecines and early *Homo*: evidence from cranial capacity and dental eruption. Journal of Human Evolution 29, 155–168.

Solounias, N., Semprebon, G.M., 2002. Advances in the reconstruction of ungulate ecomorphology with application to early fossil equids. American Museum Novitates 3366, 1–49.

Sponheimer, M., Lee-Thorpe, J.A., 1999. Isotopic evidence for the diet of an early hominid, *Australopithecus africanus*. Science 283, 368–370.

Sponheimer, M., Lee-Thorpe, J.A., 2003. Differential resource utilization by extant great apes and australopithecines: towards solving the C-4 conundrum. Comparative Biochemistry and Physiology A 136, 27–34.

Sponheimer, M., de Ruiter, D., Lee-Thorpe, J.A., Späth, A., 2005a. Sr/Ca and early hominin diets revisited: new data from modern and fossil tooth enamel. Journal of Human Evolution 48, 147–156.

Sponheimer, M., Lee-Thorpe, J.A., de Ruiter, D., Codron, D., Codron, J., Baugh, A.T., Thackeray, F., 2005b. Hominins, sedges, and termites: new carbon isotope data from the Sterkfontein Valley and Kruger National Park. Journal of Human Evolution 48, 301–312.

Stanford, C.B., Bunn, H.T., 1999. Meat eating and hominid evolution. Current Anthropology 40, 726–728.

Stiner, M., 1994. *Honor Among Thieves: a Zooarchaeological Study of Neandertal Ecology*. Princeton University Press, Princeton.

Strait, D.S., Grine, F.E., 2004. Inferring hominoid and early hominid phylogeny using craniodental characters: the role of fossil taxa. Journal of Human Evolution 47, 399–452.

Strait, D.S., Grine, F.E., Moniz, M.A., 1997. A reappraisal of early hominind phylogeny. Journal of Human Evolution 32, 17–82.

Stringer, C.B., Humphrey, L.T., Compton, T., 1997. Cladistic analysis of dental traits in recent humans using a fossil outgroup. Journal of Human Evolution 32, 389–402.

Susman, R.L., 1988. Hand of *Paranthropus robustus* from Member 1, Swartkrans: fossil evidence for tool behavior. Science 240, 781–784.

Suwa, G., White, T.D., Howell, F.C., 1996. Mandibular postcanine dentition from the Shungura Formation, Ethiopia: crown morphology, taxonomic allocations, and Plio-Pleistocene hominid evolution. American Journal of Physical Anthropology 101, 247–282.

Teaford, M.F., 1985. Molar microwear and diet in the genus *Cebus*. American Journal of Physical Anthropology 66, 363–370.

Teaford, M.F., 1986. Dental microwear and diet in two species of *Colobus*. In: Else, J., Lee, P. (Eds.), Primate Ecology and Conservation. Cambridge University Press, Cambridge, pp. 63–66.

Teaford, M.F., 1988. Scanning electron microscope diagnosis of wear patterns versus artifacts on fossil teeth. Scanning Microscopy 2, 1167–1175.

Teaford, M.F., 2007. Dental microwear and palaeoanthropology: cautions and possibilities. In: Bailey, S., Hublin, J.-J. (Eds.), Dental Perspectives on Human Evolution: State of the Art Research in Dental Paleoanthropology. Springer, Dordrecht, pp. 291–302.

Teaford, M.F., Glander, K.E., 1996. Dental microwear in a wild population of mantled howlers (*Alouatta palliata*). In: Norconk, M., Rosenberger, A., Garber, P. (Eds.), Adaptive Radiations of Neotropical Primates. Plenum, New York, pp. 433–449.

Teaford, M.F., Oyen, D., 1989. *In vivo* and *in vitro* turnover in dental microwear. American Journal of Physical Anthropology 80, 447–460.

Teaford, M.F., Robinson, J.G., 1989. Seasonal or ecological zone differences in diet and molar microwear in *Cebus nigrivittatus*. American Journal of Physical Anthropology 80, 391–401.

Teaford, M.F., Walker, A.C., 1984. Quantitative differences in dental microwear between primate species with different diets and a comment on the presumed diet of *Sivapithecus*. American Journal of Physical Anthropology 64, 191–200.

Teaford, M.F., Ungar, P.S., Grine, F.E., 2002. Paleontological evidence for the diets of African Plio-Pleistocene hominins with special reference to early *Homo*. In: Ungar, P.S., Teaford, M.F. (Eds.), Human Diet: Its Origins and Evolution. Bergin and Garvey, Westport, CT, pp. 143–166.

Ulhaas, L., Kullmer, O., Schrenk, F., 2007. Tooth wear diversity in early hominid molars – a case study. In: Bailey, S., Hublin, J.-J. (Eds.), Dental Perspectives on Human Evolution: State of the Art Research in Dental Paleoanthropology. Springer, Dordrecht, pp. 291–302.

Ungar, P.S., 1998. Dental allometry, morphology and wear as evidence for diet in fossil primates. Evolutionary Anthropology 6, 205–217.

Ungar, P.S., 2004. Dental topography and diets of *Australopithecus afarensis* and early *Homo*. Journal of Human Evolution 46, 605–622.

Ungar, P.S., 2007. Dental topography and human evolution. In: Bailey, S., Hublin, J.-J. (Eds.), Dental Perspectives on Human Evolution: State of the Art Research in Dental Paleoanthropology. Springer, Dordrecht, pp. 291–302.

Ungar, P.S., Grine, F.E., 1991. Incisor size and wear in *Australopithecus africanus* and *Paranthropus robustus*. Journal of Human Evolution 20, 313–340.

Ungar, P.S., Brown, C.A., Bergstrom, T.S., Walker, A., 2003. Quantification of dental microwear by tandem scanning confocal microscopy and scale sensitive fractal analyses. Scanning 25, 185–193.

Ungar, P.S., Grine, F.E., Teaford, M.F., El Zaatari, S., 2006. Dental microwear and diets of African early *Homo*. Journal of Human Evolution 50, 78–95.

Ungar, P.S., Teaford, M.F., Grine, F.E., 1999. A preliminary study of molar occlusal relief in *Australopithecus africanus* and *Paranthropus robustus*. American Journal of Physical Anthropology S28, 269.

Walker, A.C., 1981. Diet and teeth: dietary hypotheses and human evolution. Philosophical Transactions of the Royal Society of London B 292, 57–64.

Walker, A.C., Teaford, M.F., 1989. Inferences from quantitative analysis of dental microwear. Folia Primatologica 53, 177–189.

Walker, A.C., Hoeck, H.N., Perez, L., 1978. Microwear of mammalian teeth as an indicator of diet. Science 201, 808–810.

Weiss, K.M., Stock, D.W., Zhao, Z., 1998. Dynamic interactions and the evolutionary genetics of dental patterning. Critical Reviews in Oral Biology and Medicine 9, 369–398.

Wood, B.A., Aiello, L.C., 1998. Taxonomic and functional implications of mandibular scaling in early hominins. American Journal of Physical Anthropology 105, 523–538.

2. An evaluation of changes in strontium/calcium ratios across the neonatal line in human deciduous teeth

L.T. HUMPHREY
Department of Palaeontology
The Natural History Museum
Cromwell Road
London SW7 5BD, UK
l.humphrey@nhm.ac.uk

M.C. DEAN
Evolutionary Anatomy Unit
Department of Anatomy and Developmental Biology
University College London
London WC1E 6BT, UK
Chris.dean@ucl.ac.uk

T.E. JEFFRIES
Department of Mineralogy
The Natural History Museum
Cromwell Road
London SW7 5BD, UK
t.jeffries@nhm.ac.uk

Keywords: enamel, trace element, intra-tooth analysis, strontium, calcium, neonatal line, birth, dietary reconstruction

Abstract

Analysis of human tooth enamel using laser ablation inductively coupled plasma mass spectrometry (LA-ICP-MS) provides a basis for systematic evaluation of variation in the chemical composition of enamel in relation to tooth crown geometry. Analysis of thin sections allows a sampling strategy that can be cross-referenced to incremental growth structures in tooth enamel. Strontium and calcium are incorporated into developing teeth in a manner that reflects changing physiological concentrations in the body. Strontium/calcium (Sr/Ca) ratios are expected to decease at birth in breastfed infants, because the mammary gland exerts a greater activating effect on calcium transfer than the placenta. However, Sr/Ca ratios should increase at birth in infants fed on a formula derived from cow's milk. Changes in Sr/Ca ratios across the neonatal line in five out of six deciduous teeth from children of known mode of feeding within the first few months after birth conform to the predicted direction of change, indicating that changes in physiological concentrations of strontium and calcium resulting from a dietary shift during the secretory stage of enamel formation may not be completely overwhelmed during enamel maturation. Implications for the reconstruction of longitudinal records of infant diet from tooth enamel are discussed.

S.E. Bailey and J.-J. Hublin (Eds.), Dental Perspectives on Human Evolution, 303–319.

Introduction

Over the last decade a number of studies have sought to overcome the limitations of traditional studies of paleodiet, seasonal mobility and migration by undertaking discrete analysis of tissues that form at different stages of an individual's development (e.g., Sealy et al., 1995; Wright and Schwarcz, 1998; Balasse et al., 2001; Fuller et al., 2003). This study will focus on the potential of tooth enamel for the reconstruction of individual profiles of dietary change during the first few years of life. Tooth enamel offers many advantages for this type of study. First, the age of onset and duration of enamel formation varies between teeth, such that different teeth preserve a record of different periods of development (Hillson, 1996). Second, since tooth enamel is not remodeled, it retains a durable record of dietary intake during the period of enamel formation. Finally, enamel is an ideal matrix for chemical analysis because it is more resistant to diagenetic change than other biological tissues (Lee-Thorpe and van der Merwe, 1991; Koch et al., 1997; Budd et al., 2000).

Traditional studies of intra-tooth variation use a sampling strategy that involves drilling enamel from an intact tooth in an ordered series of horizontal bands, running from the tip to the cervix of the crown. Sampling procedures such as this will result in a set of samples that is, broadly speaking, chronologically ordered, but which does not respect either the direction of enamel secretion or subsequent mineralization (Balasse, 2003). Techniques now exist for analysing the chemical composition of very small quantities of enamel, providing a basis for discrete multiple sampling investigations of variation in trace element and isotope composition through the tooth crown (Wurster et al., 1999; Kang et al., 2004; Webb et al., 2005; Zazzo et al., 2005). Incremental growth features in the tooth are laid down during the secretory (matrix formation) stage of enamel formation, and preserve a permanent record of the timing of the onset of enamel crystal growth at a given point (Dean, 1987; Boyde, 1989). The pattern and timing of subsequent mineralization will determine whether these incremental growth features are relevant to the development of micro-sampling strategies and their interpretation within a detailed chronological framework. An area of particular interest for life history reconstructions is whether it is possible to use incremental growth structures as a basis for inferring the chronology of changes in diet during infancy and childhood.

Ideas concerning the pattern and process of progressive mineralization of developing enamel have changed considerably over the last 150 years (early work is summarized by Crabb and Darling, 1962). There is now a consensus that enamel formation proceeds via a series of distinct stages, which can be defined on the basis of changes in the structure and function of ameloblasts and their affect on enamel composition (Robinson et al., 1997; Smith 1998). The secretory stage of enamel formation involves the formation of an organic matrix, which starts in the region of the dentine horn and continues outwards through the thickness of the cuspal enamel and simultaneously along the enamel dentine junction to the cervix (Boyde, 1989). Enamel crystals seed within the organic matrix almost immediately after it is laid down and quickly elongate in the wake of retreating ameloblasts to produce thin enamel ribbons extending the full thickness of the enamel (Fincham and Simmer, 1997; Robinson et al., 1997). Following completion of enamel matrix secretion – which may take more than a year in some human molar cusps – ameloblasts undergo a morphological and functional transition and enter the maturation phase (Smith, 1998). This stage involves selective removal of water and organic materials together with addition of more mineral ions. As a result, there is a substantial increase in the width and breadth of the existing apatite crystallites that then grow into the spaces previously occupied by water and organic material. The maturation stage transforms the partially mineralized immature enamel into a highly mineralized and durable tissue (Boyde, 1989; Smith, 1998).

The extent to which enamel secretion and maturation represent distinct and discontinuous phases of enamel formation is nevertheless debatable. There is some evidence to suggest that limited growth in width and thickness of crystallites occurs during the secretory and transition stages of enamel formation (Robinson et al., 1997). Glick (1979) measured a linear increase in crystal size as a function of distance from the ameloblast Tomes' process in human and rat enamel. Kerebel and colleagues (1979) also measured gradual increase in the width and thickness of enamel crystallites in samples of human fetal (secretory) enamel with increasing distance from the enamel surface, and interpreted their data as a logarithmic growth curve. The calcium concentration of developing human molar enamel has been shown to increase swiftly in an approximately linear fashion with distance from the enamel surface (Rosser et al., 1967).

The enamel matrix proteins produced by secretory ameloblasts have an inhibitory effect on crystal growth and act to regulate both growth and orientation (Aoba, 1996; Fincham et al., 1999; Du et al., 2005). The substantial growth of enamel crystals in width and thickness, which characterizes the maturation stage, is made possible by the processing and degradation of growth-inhibitory enamel matrix proteins by enamel proteinases. Serine proteases are expressed predominantly during the transition and maturation stages of enamel formation, but have also been identified in the deeper (earliest forming) layers of secretory stage of porcine enamel (Robinson et al., 1997; Simmer and Hu, 2002). The early presence of these products indicates that some breakdown and removal of enamel protein already occurs during the secretory stage and may underlie the increased crystal width and thickness found in the deeper layers of secretory stage tissue of thicker enamels (Eisenmann, 1989; Robinson et al., 1997).

The mineral laid down during the secretory stage of enamel formation is not replaced during maturation (Boyde, 1997), and the chemical composition of this part of the enamel crystallite (if it could be isolated) would respect the chronological framework inferred from the incremental growth markers in enamel. At the end of the matrix secretion stage enamel contains about 30% mineral by weight and 10% by volume, whereas mature enamel contains about 95% mineral by weight and 70% by volume implying that the weight of mineral in immature enamel is only about 14% of that in mature enamel (Smith, 1998). The substantial increase in mineral content brought about during enamel mineralization will alter the chemical composition of the enamel at any given sampling point, resulting in a time-averaged signal. This means that discrete sampling strategies, whether carried out in traditional horizontal fashion or following a strategy that follows the pattern and direction of matrix secretion may preserve a certain amount of chronological information but do not represent discrete or well defined amounts of time (Passey and Cerling, 2002; Balasse, 2003). The rate and pattern of subsequent mineralization in different parts of the tooth could significantly affect interpretations of the chemical composition of tooth enamel within a chronological framework. At best this process would dampen but not overwhelm evidence of dietary changes during the secretory stage of enamel formation. At worst the additional mineral deposited during maturation could completely obliterate any information relating to mineral deposition during the secretory stage and could introduce patterns of variation that cannot be interpreted in a time resolved manner.

Available evidence concerning the direction and timing of matrix protein resorption and subsequent mineralization is not sufficient to resolve which of these scenarios is more likely. Suga (1982, 1989) argued that enamel maturation is accomplished in several successive stages, and that the pattern and direction of increasing mineralization are different in the matrix formation and maturation stages. The first phase of maturation increases mineralization from the enamel surface towards the inner layer and the second phase progresses from the innermost enamel towards the surface. Finally, a sharp increase in mineralization occurs in the narrow subsurface layer, which becomes the most heavily mineralized area of enamel. Successive waves of maturation vary in pattern and rate and do not necessarily respect the direction of matrix secretion (Suga, 1982, 1989). Other research has produced contradictory results with some evidence implying that protein degradation and maturation starts at the enamel dentine junction and progresses outwards (see discussion following Robinson et al., 1997).

In recent years, as researchers have sought to optimize enamel microsampling strategies and data interpretation, several different approaches have been used to explore the effect of the pattern and timing of enamel mineralization on the interpretation of intra-tooth variation in chemical composition (Passey and Cerling, 2002; Balasse, 2003; Hoppe et al., 2004). Zazzo and colleagues (2005) demonstrated that the degree of isotopic damping based on micromilling of samples parallel to the appositional front was lower than that based on conventional sampling procedures, but that sampling of the thin layer of highly mineralized enamel close to the enamel dentine junction resulted in the lowest level of isotopic damping.

Here we use new approach to explore whether chemical changes associated with dietary shifts during the secretory stage of enamel formation are overwhelmed during the maturation stage, based on changes in the strontium/calcium (Sr/Ca) ratio across the neonatal line. The rationale for evaluating changes in Sr/Ca ratio is explained below. Laser ablation inductively coupled plasma mass spectrometry (LA-ICP-MS) allows enamel samples to be analysed directly from the solid state (Jeffries et al., 1998; Jeffries, 2004). The analysis of thin longitudinal sections enables the position of each sampling point to be defined in relation to incremental growth features in enamel. The neonatal line is an accentuated line that reflects physiological disturbances associated with the birth process, and is present in all deciduous teeth in humans (Schour, 1936; Whittaker and Richard, 1978; Skinner and Dupras, 1993; Smith and Avishai, 2005). The neonatal line separates enamel initiated prior to birth from enamel initiated after birth. We evaluated changes in Sr/Ca ratios across the neonatal line in order to determine: (a) whether there are consistent changes across the neonatal line within a tooth; (b) whether we can differentiate between infants fed on breast milk or infant formula based on the direction of change in Sr/Ca ratios across the neonatal line; and (c) whether there is a delay in the uptake of sufficient mineral to show the postnatal signal.

The behaviour of strontium and calcium in biological systems has attracted considerable research interest due to concerns over the effects of radioactive Sr-90 on human health (Wasserman, 1963; Rivera and Harley, 1965). It has also attracted the attention of researchers interested in reconstructing the weaning process in past human populations (Sillen and Smith, 1984; Hühne-Osterloh and Grupe, 1989; Mays, 2003). Sr/Ca ratios are useful for detecting dietary change during infancy and early childhood because human milk has a very low Sr/Ca ratio compared to most foods, particularly those given during weaning. Strontium and calcium are incorporated into developing teeth in a manner that reflects their changing physiological concentrations in the body. The physiological processes that cause major differences in strontium and calcium behaviors are those that involve the passage of these ions across membranes under metabolic control. These are gastrointestinal absorption, renal excretion, lactation and placental transfer (Comar, 1963). This paper will focus primarily on the implications of the last two of these processes, the transfer of strontium and calcium across the placenta and mammary gland, for changing physiological levels of strontium and calcium in the perinatal period.

A summary of the major parameters that influence physiological levels of strontium and calcium is given in Table 1. Prior to birth, the developing fetus obtains all of its dietary strontium and calcium directly from its mother. The Sr/Ca ratio of tissues forming in the fetus depends on four factors: the mother's dietary intake of calcium and strontium; differential transfer of strontium and calcium across the mother's digestive tract; release of strontium and calcium from the mother's skeleton; and differential transfer of strontium and calcium across the placenta. While exclusively breast-fed, an infant continues to obtain all of its dietary strontium and calcium from its mother. The Sr/Ca ratio in the tissues of a breast-fed infant is affected by the following factors: the mother's dietary intake of calcium and strontium; differential transfer of strontium and calcium across the mother's digestive tract; release of strontium and calcium from the mother's skeleton; differential transfer of strontium and calcium across the mammary gland; and differential transfer of strontium and calcium across the infant's digestive tract.

The mother's dietary intake and the transfer of strontium and calcium across her digestive tract (parameters A and B) are assumed to remain fairly constant during pregnancy

Table 1. Summary of the main parameters influencing physiological levels of strontium and calcium in the fetus, breastfed infant and formula fed infant

Parameter		Prior to birth	During exclusive breastfeeding	During exclusive formula feeding
A	Mother's dietary intake	Yes	Yes	/
B	Differential transfer of strontium and calcium across the mother's digestive tract	Yes	Yes	/
C	Release of strontium and calcium from the mother's skeleton	Yes	Yes	/
D	Differential transfer of strontium and calcium across the placenta	Yes	/	/
E	Differential transfer of strontium and calcium across the mammary gland	/	Yes	/
F	Differential transfer of strontium and calcium across the infant's digestive tract	/	Yes	Yes
G	Infant's independent dietary intake	/	/	Yes

and lactation. The release of strontium and calcium from the mother's skeleton (parameter C) may increase during pregnancy and lactation in line with the infant's requirements. This parameter may exhibit considerable variation between individuals and even between successive pregnancies. Although there are no good data to model the effect of this parameter on physiological changes in strontium and calcium in the mother and infant, there is no theoretical expectation of a sudden shift at birth. The major difference between the fetal stage and early postnatal period in a breastfed infant is the shift from the placenta to the mammary gland as means of transfer of strontium and calcium from the mother to infant (parameters D and E).

The Observed Ratio (OR) describes the differential movement of strontium and calcium across physiological barriers and is defined as:

$$OR_{sample-precursor} = \frac{Sr/Ca \text{ sample}}{Sr/Ca \text{ precursor}} \quad (1)$$

The fetus-diet Observed Ratio ($OR_{fetus/diet}$), which describes the relationship between the Sr/Ca ratio in the mother's diet and the Sr/Ca ratio in the fetus, is estimated at 0.16 in late pregnancy based on an analysis of stable strontium in bones of the newborn (Comar, 1963). The $OR_{milk/diet}$ is estimated at 0.10, based on a study of dietary Sr-90 and calcium secretion into breast milk (Comar, 1963). This difference is expected to result in a marked reduction in the Sr/Ca ratio of tissues forming immediately after birth in exclusively breastfed infants.

Several more recent studies have evaluated trace element concentration in maternal sera, and corresponding umbilical cord sera and/or colostrum samples in order to investigate the impact of the human placenta and mammary gland on the transfer of trace elements from the mother to the baby (Krachler et al., 1999; Rossipal, 2000). Results of these studies demonstrate that both the placenta and the mammary gland can exert an activating, inhibiting or gradient mode of action for selected trace elements. The concentration of Ca in umbilical cord sera is approximately 120% of that of the maternal sera but there is an almost perfect correlation between umbilical cord and maternal sera for strontium (Krachler et al., 1999). The concentration of Ca in colostrum is approximately 222% of that of the maternal sera (Rossipal, 2000). The results demonstrate that there is an active transport mechanism for calcium from the mother to the newborn across the placenta and mammary gland and that the mammary gland exerts a greater activating effect on calcium transfer than the placenta. The transfer of strontium across both the placenta and mammary gland follows concentration gradients indicating neither an activating or inhibiting effect.

The extent to which differential transport of strontium and calcium across the placenta (parameter D) may vary through pregnancy is not known for humans. Experiments performed in rodents to evaluate the extent of the placental barrier against Sr in comparison to Ca showed a maximum of Sr/Ca discrimination early in pregnancy (von Zallinger and Tempel, 1998). At the end of pregnancy Sr and Ca were transported across the placenta in almost equal proportions. The relative transportation of strontium and calcium across the mammary gland (parameter E) is tentatively assumed to remain constant during the period of exclusive breastfeeding since there is some evidence to suggest that the strontium concentration of human milk remains constant throughout lactation (Anderson, 1993; Perrone et al., 1993; Rossipal and Krachler, 1998). Longitudinal analyses of the trace element composition of human milk during the first 5–6 months of lactation found no significant change in strontium concentration (Anderson, 1993; Perrone et al., 1994).

The Sr/Ca ratio of tissues forming in an exclusively formula fed infant are independent of the mother, and are effected solely by the infant's own dietary intake (parameter G) and differential transfer of strontium and calcium across the infant's digestive tract (Parameter F). The Sr/Ca ratio of dairy foods is higher than that of human milk. Infants fed on milk substitute derived from cow's milk are expected to exhibit a moderate increase in Sr/Ca ratio of tissues forming after birth, achieving an equilibrium at about 50% above the level in late pregnancy (Comar, 1963).

After birth, the gastro-intestinal system of the infant (parameter F) will exert an influence on physiological levels of strontium and calcium in all infants regardless of mode of feeding. The ability of the gastro-intestinal system to discriminate against strontium changes with age in humans and other mammals that have been studied (McClellan, 1964; Rivera and Harley, 1965; Sugihira and Suzuki, 1991). The gastro-intestinal tracts of human infants discriminate poorly against strontium (Lough et al., 1963). The ability to discriminate against strontium develops during the first few years of life, and strontium levels will fall as discrimination is established (Comar, 1963; Rivera and Harley, 1965).

Materials and Methods

The sample comprises deciduous canines or second molars from six children. Information concerning mode of feeding in the first few months after birth was provided retrospectively by the mothers. Three of the children were exclusively breastfed for at least two months after birth. Three of the children were predominantly or exclusively fed on infant formulas derived from cow's milk from after birth or shortly thereafter. Details of the teeth analysed and diet are given in Table 2.

Each tooth was sectioned longitudinally using a slow speed rotating Isomet (Buehler) diamond saw. One cut face was polished with 3-micron aluminum oxide power. The polished block face was fixed with epoxy resin adhesive to a glass microscope slide under pressure and a mesiobuccal section, approximately 100 microns thick was cut from the block together with the slide. The section was lapped planoparallel and polished, in a Buehler slide holder. Using polarized light, a low power (x80) digital photographic record of the prism paths in the mesiobuccal aspects of the crown and of the major long-period incremental chronology of the enamel markings was made. The neonatal line was used to identify the portion of enamel formed postnatally and also as a baseline for establishing a chronological framework for sampling within the enamel crown.

Laser ablation inductively coupled plasma mass spectrometry (LA-ICP-MS) was used to determine Sr/Ca ratios. The LA-ICP-MS facility at the Natural History Museum comprises a New Wave Research UP213

Table 2. Details of dietary history for each child and teeth analysed

Child	*Mode of feeding after birth*	*Deciduous tooth analysed*
A	Exclusive and prolonged breastfeeding	upper second molar
B	Exclusive and prolonged breastfeeding	lower canine
C	Exclusive and prolonged breastfeeding	lower second molar
D	Exclusive formula feeding	lower canine
E	Breastfeeding attempted. Exclusive formula feeding within a few days of birth	upper canine
F	Breastfeeding attempted. Formula introduced as predominant mode of feeding within 10 days of birth.	upper canine

aperture imaged frequency quintupled Nd:YAG laser ablation accessory operating at 213nm coupled to a Thermo Elemental PlasmaQuad 3 quadrupoled based ICP-MS with enhanced sensitivity (s-option) interface. A mixed He:Ar sample carrier gas was used throughout. Analytical and instrumental conditions are presented in Table 3. Discrete 30 μm diameter areas of the sectioned tooth were sampled during analysis. At this resolution, the instrumentation provides a detection limit for strontium in teeth at approximately 5 ppb. For each analysis data were collected for 120 seconds. During the first ca 60 seconds of collection, background data in the form of gas blank and electronic noise were acquired. The laser was then fired at the sample and data acquired for a further ca 60 seconds. Background data were subsequently subtracted from the ablation signal in the tooth. Data were collected in discrete runs of 20 analyses, beginning and ending with two analyses of the National Institute of Standards Technology (NIST) standard reference material SRM NIST612. Initial data processing and reduction, including selection and integration of background and ablated signal intervals were performed off-line using LAMTRACE, a Lotus™ 123 macro-based spreadsheet created by Simon Jackson, Macquarie University, Australia. Further data processing including normalization of Sr/Ca ratios based on the measured ratio in NIST612 was performed using Microsoft Excel™ spreadsheet software. The normalization process corrects data for instrumental mass bias, to provide an internally consistent data set. Sampling points were evenly distributed along a series of trajectories running from the enamel dentine junction (EDJ) to the enamel surface and spaced at regular intervals along the tooth cusp (Figure 1).

Following laser ablation, a higher power (x25 objective) digital montage of the enamel containing the ablation pits was constructed. All ablation points in the crown were chronologically aged with respect to birth by either tracking daily counts along prisms to each ablation point and extending down through the enamel along long-period incremental markings and continuing the counts out along new prism paths to each ablation point in turn in the chronological sequence, or, when these were not easily visible, by calculating

Table 3. Analytical and instrumental operating conditions

ICP-MS		
Forward power		1350 W
Gas flows	Coolant	Ar: 13 l min^{-1}
	Auxiliary	Ar: 0.9 l min^{-1}
	Sample transport	He: 0.8 l min^{-1}
		Ar: 0.8 l min^{-1}
Laser		
Wavelength		213 nm
Pulse width (FWHM)		3 ns
Pulse energy		0.05 mJ per pulse
Energy density		3 J cm^{-2}
Energy distribution		Homogenized, flat beam, aperture imaged
Repetition rate		10 Hz
Spot size diameter		30 μm
Scanning mode		Peak jumping, 1 point per peak
Analysis protocol		
Acquisition mode		Time resolved analysis
Duration of analysis		120s, (ca 60s background, 60s ablated signal)
Dwell times		10 ms all masses

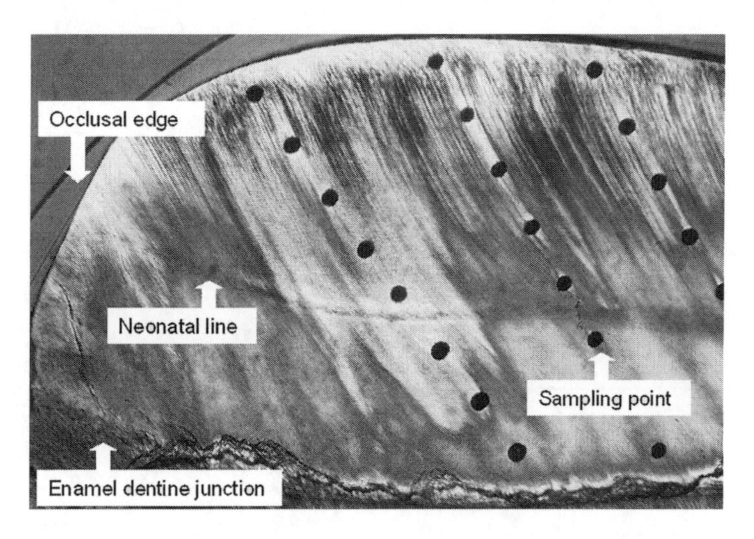

Figure 1. Deciduous cuspal enamel showing prism trajectories, neonatal line and ablation pits. Ablation pits are 30 microns diameter.

the number of days along linear thickness of cuspal enamel using polynomial equations derived from more than 20 human deciduous teeth (Dean et al., 2001).

Results

In all teeth there is a negative correlation between the Sr/Ca ratio determined for each sampling point and the minimum distance from the enamel surface. Typically the highest ratios occur at innermost sampling point on each trajectory and the lowest values occur closest to the enamel surface. There is a clear tendency for Sr/Ca ratios to decrease between successive ablation points along individual trajectories running from the enamel dentine junction to the enamel surface (Figure 2). The amount of decrease between successive ablation points is not constant within or between trajectories, and some trajectories show a temporary reversal of the underlying trend.

For child A strontium/calcium ratios were determined for six rows of ablation points that included sampling points on either side of the neonatal line (Figure 2a). On each of these trajectories a particularly pronounced reduction in Sr/Ca ratio occurs between the outermost sampling point that lies inside the neonatal line and the innermost sampling points that falls outside the neonatal line. These points sample enamel that initiates formation in the last 5 weeks before birth and the first 3 weeks after birth. The seventh row of ablation points and subsequent rows (not shown), which only sample enamel that starts to form at least 10 days after birth, do not show this pattern. Three of the trajectories also exhibit an increasing trend in Sr/Ca ratios close to the enamel dentine junction. These increases occur at 57 to 30 days before birth (first trajectory), 96 to 60 days before birth (second trajectory) and 85 to 51 days before birth (third trajectory).

For Child B, strontium/calcium ratios were determined for four rows of ablation points that included sampling points on either side of the neonatal line (Figure 2b). On each trajectory there is a marked reduction Sr/Ca ratio between the outermost sampling point that lies inside the neonatal line and the innermost sampling points that falls outside the neonatal line. There is also a marked reduction between the first two sampling points on the fifth trajectory, which sample enamel that initiate formation at birth, and

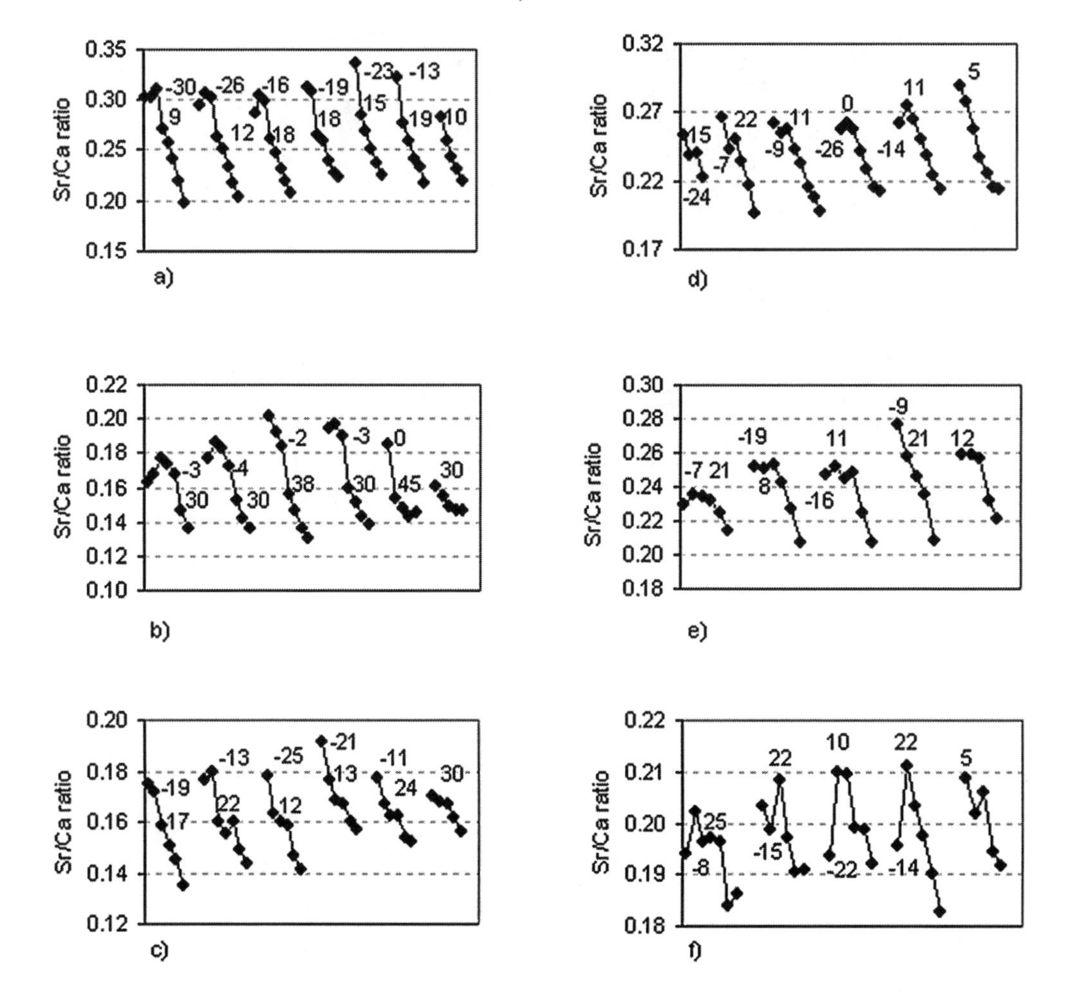

Figure 2. Strontium-calcium ratios for discrete sampling points analysed from thin sections of enamel from 6 children. Details of mode of feeding in the early postnatal period are given in Table 2. Each spot gives the result for a single sampling point. Lines connect the Sr/Ca ratios for series of points ablated along a single trajectory, running from the enamel dentine junction towards the enamel surface. The trajectories are arranged from those closest to the occlusal edge on the left to those closest to the cervix of the tooth on the right. Numbers refer to the number of days from birth of initiation of enamel matrix formation for sampling points on either side of the neonatal line.

at 45 days after birth. There is no equivalent reduction in Sr/Ca ratio on any of the other trajectories, which sample enamel that initiates formation after birth. Three of the trajectories also exhibit an increasing trend in Sr/Ca ratios between the sampling points closest to the enamel dentine junction. These increases occur at 128 to 80 days before birth (first trajectory), 120 to 60 days before birth (second trajectory) and 57 to 31 days before birth (fourth trajectory).

For Child C, strontium/calcium ratios were determined for five rows of ablation points that included sampling points on either side of the neonatal line (Figure 2c). On each trajectory there is a marked reduction Sr/Ca ratio between the outermost sampling point that lies inside the neonatal line and the innermost sampling points that falls outside the neonatal line. None of the trajectories sampling enamel that initiates formation after birth shows an equivalent reduction in Sr/Ca ratio. On one of the trajectories there is an increase in Sr/Ca ratio between the two sampling points closest to the neonatal line, which initiate formation at 44 and 13 days before birth.

For Child D, strontium/calcium ratios were determined for five rows of ablation points that included sampling points on either side of the neonatal line (Figure 2d). On four of these trajectories there is an increase in Sr/Ca ratio between the outermost sampling point that lies inside the neonatal line and the innermost sampling points that falls outside the neonatal line. These points sample enamel that initiates formation in the last 4 weeks before birth and the first 4 weeks after birth. On the remaining trajectory there is an increase between the last data point prior to birth and the sampling point that samples enamel that initiates formation at birth (the innermost edge of the ablation pit touches the neonatal line). The sixth row of ablation points and subsequent rows (not shown), which only sample enamel that starts to form at least 5 days after birth, do not show this pattern. None of the trajectories in Child D exhibits a marked reduction in Sr/Ca ratio between pairs of adjacent ablation points equivalent to that seen across the neonatal line in the breastfed infants.

For Child E, strontium/calcium ratios were determined for four rows of ablation points that included sampling points on either side of the neonatal line (Figure 2e). Changes in Sr/Ca ratio between the outermost sampling point that lies inside the neonatal line and the innermost sampling points that falls outside the neonatal line are small and inconsistent. On two of the trajectories (1 and 2) there is almost no change in Sr/Ca ratio, while on the third trajectory there is a small rise and on the fourth trajectory there is a small reduction in Sr/Ca ratio across the neonatal line.

For Child F, strontium/calcium ratios are determined for four rows of ablation points that included sampling points on either side of the neonatal line (Figure 2f). On all four trajectories there is a marked increase in Sr/Ca ratio between the outermost sampling point that lies inside the neonatal line and the innermost sampling points that falls outside the neonatal line. All of the earliest postnatal ablation points sample enamel that initiates formation after the age at which infant formula was introduced into the diet.

Discussion

In these human deciduous teeth there is an underlying trend for strontium/calcium levels to decrease from the enamel dentine junction to the enamel surface. A similar trend has also been observed in human and non-human permanent contemporary and archaeologically derived teeth (unpublished data). The pattern seen here is broadly consistent with the distribution of mineral density in human tooth enamel and, hence, variation in the degree of mineralization in different parts of the tooth crown. Mineral density is lowest in the deepest enamel beneath the tooth cusp and decreases from the enamel dentine junction to the outer enamel in permanent and deciduous human molars (Robinson et al., 1986; Wilson and Beynon, 1989, 1990; Wong et al., 2004). For this study it is the deviations from this underlying trend, and particularly those that occur at the neonatal line, that are of interest.

The primary goal of this study was to determine whether there is a consistent pattern of change in the Sr/Ca ratio across the neonatal line in human deciduous teeth. In each of the breastfed infants (Children A, B and C) there is a larger than expected reduction in Sr/Ca ratio between the outermost sampling point lying inside the neonatal line and the innermost sampling points lying outside the neonatal line on all rows of ablation points, which sample enamel on both sides of the neonatal line. The result is consistent with a marked drop in the Sr/Ca ratio at birth, as would be predicted for a breast fed infant. None of the trajectories of the predominantly formula fed infants exhibits a marked drop equivalent to that observed between adjacent ablation points on either side of the neonatal line in the breastfed infants. This suggests that infants who were exclusively breastfed for an appropriate length of time after birth can be recognized on the basis of a marked reduction in Sr/Ca ratio across the neonatal line. The duration of breastfeeding required to generate this breastfeeding signal will need to be established by further experimental work.

In the two formula fed infants (Children D and F) there is an increase in Sr/Ca ratio between the outermost sampling point, which lies inside the neonatal line and the innermost sampling points, which fall on or outside the neonatal line in all rows of ablation points that sample enamel on both sides of the neonatal line. The result is consistent with an increase in the Sr/Ca ratio at, or shortly after birth in the case of child F, as would be predicted for a formula fed infant. In child E the changes in the Sr/Ca ratio across the neonatal line are inconsistent, and the variation does not relate in an obvious way to the proximity of sampling points to the neonatal line. Child E is the older sibling of child D and for this reason the contrasting results for the two children warrant further exploration. For child E, breastfeeding was briefly attempted while child D was exclusively breastfed, but since none of the ablation points for child E sample enamel that initiated formation in the brief period for which breastfeeding was attempted, this may not explain the difference between the two siblings. Differences in physiological maturity between the two children at the time of birth may be relevant here since child D was born at term whereas child E was born three weeks premature.

Increases in Sr/Ca ratio of a similar magnitude to those occurring between adjacent sampling points on either side of the neonatal line in child D were observed between adjacent sampling points in enamel forming close to the enamel dentine junction on some of the trajectories sampled on the deciduous teeth from both breastfed and formula fed infants. In all cases the increases are between the innermost two or occasionally three sampling points, which sample the deepest and earliest forming enamel. The timing of the increases in Sr/Ca ratio varies between teeth. The rises occur at between 90 and 30 days before birth in Child A, 125 and 60 days before birth in Child B, 44 and 13 days before birth in Child C, 32 and 7 days before birth in Child E and 57 and 31 days before birth in Child F. This pattern may reflect physiological changes during pregnancy. In rodents the placenta is less effective as a barrier against Sr in comparison to Ca in late pregnancy than in early pregnancy (von Zallinger and Tempel, 1998). The rises in Sr/Ca ratio observed here may reflect a similar phenomenon in humans. Alternatively the rises may reflect some other parameter related to the geometry of the tooth and/or the pattern and degree of enamel mineralization in this zone of enamel.

The neonatal line in the deciduous canine of one of the formula fed infants (Child D) lies close to the enamel dentine junction so it is possible that the increases in Sr/Ca ratio between adjacent sampling points on either side of the neonatal line reflect some other parameter that contributes to variation in Sr/Ca ratios in enamel, which just happens to coincide with a known dietary shift in this child. The canine from Child F had a longer period of prenatal enamel formation and exhibits both a prenatal increase in Sr/Ca ratio between the sampling points closest to the EDJ and a separate increase in Sr/Ca ratio between sampling points on either side of the neonatal line, which coincides with the introduction of formula shortly after birth. In view of the ambivalent results recorded for child E, this study does not provide evidence for a consistent increase in Sr/Ca ratio associated with the introduction of formula. However, in this sample, infants who were exclusively or predominantly fed on formula derived from cow's milk could be identified by the absence of the breastfeeding signal.

The consistent but opposite direction of change in Sr/Ca ratio in the teeth of three children who were breastfed for the first few months after birth and the two children who were exclusively or predominantly formula fed in the first few months after birth suggests that the change that occurs at neonatal line reflects variation in physiological levels of strontium and calcium associated with dietary intake rather than some other parameter that changes at birth. This in turn would imply that the shift in Sr/Ca ratio that occurs in association with dietary changes at birth is not completely obliterated by subsequent mineralization, indicating that abrupt transitions in Sr/Ca ratios that occur during the period of matrix formation can still be detected in mature teeth.

Each ablation pit measures approximately 30 microns. Enamel secretion rate in the human deciduous teeth used in this study was consistently close to 4 microns per day and varied little beyond 3.5 to 4.5 microns per day throughout the whole crown (see also Schour and Poncher, 1937). Thus each pit represents about 10 days enamel growth across its maximum diameter. The depth of the ablation pit is limited by the thickness of the section, which is approximately 100 microns and may contain 20 to 25 close-packed prism layers. Any obliquity in the plane of section may result in the enamel sampled, as the laser ablates into the tooth section, being formed slightly earlier or later than is apparent from the microscopic image of the ablation pit. This will depend on the on the geometry of the tooth, the position and orientation of the thin section relative to the dentine tip and midline of the tooth, and the orientation of the enamel prisms, but is likely to be a very small error in such a thin section. The age assigned to each sampling point refers to the age at the onset of enamel matrix formation determined for the innermost point of each ablation point as imaged by polarized light microscopy. We infer that each the Sr/Ca ratio determined for each sample corresponds to the Sr/Ca ratio of enamel that initiates formation over a period of approximately 10 days.

At present it is not possible to resolve exactly how close to the neonatal line the changes in Sr/Ca ratio manifest themselves. This is partly due to sampling factors, including the position of sampling points relative to the neonatal line, the size of the samples and intervals between sampling points. Other relevant parameters include the sharpness of the neonatal line and the length of time taken to establish feeding after birth, particularly in breastfed infants. In the three breastfed children, samples of enamel that initiates formation at or before birth (and therefore extending up to 10 days after birth) yielded high Sr/Ca ratios, and samples of enamel that initiate formation at 9 or more days after birth yield markedly lower Sr/Ca ratios that are indicative of a breast milk signal. In Child D, samples of enamel that initiate formation between 7 and 24 days before birth yield lower Sr/Ca ratios than samples of enamel that initiate formation between birth and 22 days after birth. These preliminary results imply that a postnatal dietary signal can be detected in enamel that initiates mineralization within 2 weeks of birth. A more targeted sampling strategy could improve this resolution and we are currently exploring time resolved methods of data analysis.

The results of this study have implications for other studies of the chemical composition of tooth enamel that use a bulk sampling approach. The Sr/Ca ratio of each part of the enamel is influenced by a range of interacting physiological, developmental and geometric parameters. These include the age of initiation of enamel matrix formation at the sampling point, the age of onset and duration of mineralization for the tooth cusp as a whole, dietary intake during the entire period of enamel mineralization, enamel thickness and the duration of matrix secretion at different position along the tooth crown. Thus even samples taken from the same position in homologous teeth of children of the same dietary history may produce different Sr/Ca ratios as a result of differences in tooth crown development between individuals. Further work is required to fully understand the nature and implications of variation in Sr/Ca ratios in human tooth enamel, and research on children of known dietary history could help elucidate these issues. Prospective recruitment would allow more detailed and reliable information on dietary histories and other relevant parameters to be collected.

Conclusions

Analysis of human tooth enamel using LA-ICP-MS provides a basis for systematic evaluation of variation in the chemical composition of enamel in relation to tooth crown geometry, and is used here to determine Sr/Ca ratios in deciduous teeth from six modern children with known nursing histories. The position of sampling points was cross-referenced to incremental growth structures in tooth enamel, in order to determine the age of initiation of enamel formation at each sampling point and to examine changes in Sr/Ca ratio across the neonatal line in human deciduous teeth. The neonatal line is an accentuated increment that reflects physiological disturbances associated with the birth process, and separates enamel that initiates mineralization prior to birth and enamel that initiates mineralization in the early postnatal period. Physiological levels of strontium were expected to decrease at birth in breastfed infants, primarily because of differential transport of strontium and calcium across the placenta and mammary gland. In contrast strontium levels were expected to increase in an infant fed on a formula derived from cow's milk.

There is an underlying trend for Sr/Ca ratios to decrease between successive sampling points along individual trajectories running from the enamel dentine junction to the enamel surface. Particularly marked reductions in Sr/Ca ratio were observed between successive sampling points on either side of the neonatal line in the teeth of three children who were breastfed for the first few months after birth. This is consistent with the prediction of a decrease in Sr/Ca ratio in breastfed infants. Increases in Sr/Ca ratio were observed between successive sampling points on either side of the neonatal line in deciduous teeth from two of the three children who were fed predominantly on infant formula from birth or shortly after birth, but changes in Sr/Ca ratio in the third formula-fed child were inconsistent. None of the teeth from formula fed children exhibited the consistent marked reduction in Sr/Ca ratios across the neonatal line that characterized the teeth of breastfed children. This difference provides a basis for interpreting nursing behaviour in the early postnatal period from the chemical composition of deciduous teeth.

The results presented here further suggest that a postnatal dietary signal can be detected in enamel that initiates mineralization within two weeks of birth. These results imply that changes in physiological concentrations of strontium and calcium resulting from a dietary shift during the secretory stages of enamel mineralization are reflected in the chemical composition of the immature enamel crystallites and that these changes may not be completely overwhelmed during enamel maturation. This study suggests that incremental growth structures in tooth enamel may provide a basis for interpreting the chronology of dietary transitions during infancy and childhood.

Acknowledgments

We would like to thank Shara Bailey and Jean-Jacques Hublin for the opportunity to participate in this conference. We are grateful to D. Cooper, P. Cooper, B. Cornish, T. Elliot, C. Mayes, R. Garcia-Sanchez, T. Stringer, C. Soligo, and T. Wighton for technical support and advice. We thank Allison Cleveland and two reviewers for helpful comments on an earlier version of this paper.

References

Anderson, R.R., 1993. Longitudinal changes of trace-elements in human milk during the 1st 5 months of lactation. Nutrition Research 13, 499–510.

Aoba, T., 1996. Recent observations on enamel crystal formation during mammalian amelogenesis. Anatomical Record 225, 208–218.

Balasse, M., 2003. Potential biases in sampling design and interpretation of intra-tooth analysis. International Journal of Osteoarchaeology 13, 3–10.

Balasse, M., Bocherens, H., Mariotti, A., Ambrose, S.H., 2001. Detection of dietary changes by intra-tooth carbon and nitrogen isotopic analysis: an experimental study of dentine collagen of cattle (*Bos taurus*). Journal of Archaeological Science 28, 235–245.

Boyde, A., 1989. Enamel. In: Oksche, A., Vollrath, L. (Eds.), Handbook of Microscopic Anatomy, Vol. 6: Teeth. Springer, Berlin, pp. 309–473.

Boyde, A., 1997. Microstructure of enamel. In: Chadwick, D.J., Cardew, G. (Eds.), Dental enamel. Proceedings of the Ciba Foundation Symposium 205. John Wiley, Chichester, pp. 18–31.

Budd, P., Montgomery, J., Barreiro, B., Thomas, R.G., 2000. Differential diagenesis of strontium in archaeological human dental tissues. Applied Geochemistry 15, 687–694.

Comar, C.L., 1963. Some over-all aspects of strontium-calcium discrimination. In: Wasserman, R.H. (Ed.), The Transfer of Calcium and Strontium Across Biological Membranes. Academic Press, New York, pp. 405–419.

Crabb, H.S.M., Darling, A.I., 1962. *The Pattern of Progressive Mineralization in Human Dental Enamel*. Pergamon Press, Oxford.

Dean, C., Leakey, M.G., Reid, D., Schrenk, F., Schwartz, G.T., Stringer, C., Walker, A., 2001. Growth processes in teeth distinguish modern humans from *Homo erectus* and earlier hominins. Nature 414, 628–631.

Dean, M.C., 1987. Growth layers and incremental markings in hard tissues: a review of the literature and some preliminary observations about enamel structure in *Paranthropus boisei*. Journal of Human Evolution 16, 157–172.

Du, C., Falini, G., Fermani, S., Abbott, C., Moradian-Oldak, J., 2005. Supramolecular assembly of amelogenin nanospheres into birefringent microribbons. Science 307, 1450—1454.

Eisenmann, D.R., 1989. Amelogenesis. In: Ten Cate, A.R. (Ed.), Oral Histology: Development, Structure and Function. Third edition. The C.V. Mosby Company, St. Louis, pp. 197–213.

Fincham, A.G., Simmer, J.P., 1997. Amelogenin proteins of developing dental enamel. In: Chadwick, D.J., Cardew, G. (Eds.), Dental enamel. Proceedings of the Ciba Foundation Symposium 205. John Wiley and Sons Inc, Chichester, pp. 118–134.

Fincham, A.G., Moradian-Oldak, J., Simmer, J.P., 1999. The structural biology of developing dental enamel matrix. Journal of Structural Biology 126, 270–299.

Fuller, B.T., Richards, M.P., Mays, S.A., 2003. Stable carbon and nitrogen isotope variations in tooth dentine serial sections from Wharram Percy. Journal of Archaeological Science 30, 1673–1684.

Glick, P.L., 1979. Patterns of enamel maturation. Journal of Dental Research. Special Issue B. 58, 883–892.

Hillson, S., 1996. *Dental Anthropology*. Cambridge University Press, Cambridge.

Hoppe, K.A., Stover, S.M., Pascoe, J.R., Amundson, R., 2004. Tooth enamel biomineralization in extant horses: implications for isotopic microsampling. Paleogeography, Paleoclimatology, Paleoecology 206, 355–365.

Hühne-Osterloh, G., Grupe, G., 1989. Causes of infant mortality in the middle ages revealed by chemical and palaeopathological analyses of skeletal remains. Zeitschrift für Morphologie und Anthropologie. 77, 247–258.

Jeffries, T.E., 2004. Laser ablation inductively coupled plasma mass spectrometry. In: Janssens, K., Van Grieken, R. (Eds.), Wilson and Wilson's Comprehensive Analytical Chemistry, Vol. XLII: Non-destructive Microanalysis of Cultural Heritage Materials. Elsevier, Amsterdam, pp. 313–358.

Jeffries, T.E., Jackson, S.E., Longerich, H.P., 1998. Application of a frequency quintupled Nd: YAG source (λ = 213 nm) for laser ablation inductively coupled plasma mass spectrometric analysis of minerals. Journal of Analytical Atomic Spectrometry 13, 935–940.

Kang, D., Amarasiriwardena, D., Goodman, A.H., 2004. Application of laser ablation-inductively coupled plasma-mass spectrometry (LA-ICP-MS) to investigate trace metal spatial distributions in human tooth enamel and dentine growth layers and pulp. Analytical and Bioanalytical Chemistry 378, 1608–1615.

Kerebel, B., Daculsi, G., Kerebel, L.M., 1979. Ultrastructural studies of enamel crystallites. Journal of Dental Research 58, 844–850.

Koch, P.L., Tuross, N., Fogel, M.L., 1997. The effects of sample treatment and diagenesis on the isotopic integrity of carbonate in biogenic hydroxylapatite. Journal of Archaeological Science 24, 417–429.

Krachler, M., Rossipal, E., Micetic-Turk, D., 1999. Trace element transfer from the mother to the newborn – investigations on triplets of colostrum, maternal and umbilical cord sera. European Journal of Clinical Nutrition 53, 486–494.

Lee-Thorpe, J.A., van der Merwe, N.J., 1991. Aspects of the chemistry of modern and fossil biological apatites. Journal of Archaeological Science 18, 343–35.

Lough, S.A., Rivera, J., Comar, C.L., 1963. Retention of strontium, calcium and phosphorus in human infants. Proceedings of the Society of Experimental Biology and Medicine 112, 631–636.

Mays, S., 2003. Bone strontium: calcium ratios and duration of breastfeeding in a Mediaeval skeletal population. Journal of Archaeological Science 30, 731–741.

McClellan, R.O., 1964. Radiobiology: calcium-strontium discrimination in miniature pigs as related to age. Nature 202, 104–106.

Passey, B.H., Cerling, T.E., 2002. Tooth enamel mineralization in ungulates: implications for recovering a primary isotopic time-series. Geochimica et Cosmochimica Acta 66, 3225–3234.

Perrone, L., Dipalma, L., Ditoro, R., Gialanella, G., Moro, R., 1993. Trace-element content of human-milk during lactation. Journal of Trace Element Electrolytes in Health and Disease 7, 245–247.

Perrone, L., Dipalma, L., Ditoro, R., Gialanella, G., Moro, R., 1994. Interaction of trace-elements in a longitudinal-study of human-milk from full-term and preterm mothers. Biological Trace Element Research 41, 321–330.

Rivera, J., Harley, J.H., 1965. The HASL bone program: 1961–1964. U.S. Atomic Energy Commission Health and Safety Lab Report, 163.

Robinson, C., Kirkham, J., Weatherell, J.A., Strong, M., 1986. *Dental Enamel – A Living Fossil.* BAR International Series 291, 31–55.

Robinson, C., Brookes, S.J., Bonass, W.A., Shore, R.C., Kirkham, J., 1997. Enamel maturation. In: Chadwick, D.J., Cardew, G. (Eds.), Dental enamel. Proceedings of the Ciba Foundation Symposium 205. John Wiley and Sons Inc., Chichester, pp. 118–134.

Rosser, H., Boyde, A., Stewart, A.D.G., 1967. Preliminary observations of the calcium concentration in developing enamel assessed by scanning electron-probe X-ray emission microanalysis. Archives of Oral Biology 12, 431–440.

Rossipal, E., 2000. Investigation on the transport of trace elements across barriers in humans: Studies of placental and mammary transfer. Journal of Trace and Microprobe Techniques 18, 493–497.

Rossipal, E., Krachler, M., 1998. Pattern of trace elements in human milk during the course of lactation. Nutrition Research 18, 11–24.

Schour, I., 1936. Neonatal line in enamel and dentin of human deciduous teeth and first permanent molar. Journal of the American Dental Association 23, 1946–1955.

Schour, I., Poncher, H.G., 1937. Rate of apposition of enamel and dentine, measured by the effect of acute fluorosis. American Journal of Diseases of Children 54, 757–776.

Sealy, J., Armstrong, R., Schrire, C., 1995. Beyond lifetime averages: tracing life histories through isotopic analysis of different calcified tissues from archaeological human skeletons. Antiquity 69, 290–300.

Sillen, A., Smith, P., 1984. Weaning patterns are reflected in strontium-calcium ratios of juvenile skeletons. Journal of Archaeological Science 11, 237–245.

Simmer, J.P., Hu, J.C.C., 2002. Expression, structure, and function of enamel proteinases. Connective Tissue Research 43, 441–449.

Skinner, M., Dupras, T., 1993. Variation in birth timing and location of the neonatal line in human enamel. Journal of Forensic Sciences 38, 1383–1390.

Smith, C.E., 1998. Cellular and chemical events during enamel maturation. Critical Reviews in Oral Biology and Medicine 9, 128–161.

Smith, P., Avishai, G., 2005. The use of dental criteria for estimating postnatal survival in skeletal remains of infants. Journal of Archaeological Science 32, 83–89.

Suga, S., 1982. Progressive mineralization pattern of developing enamel during the maturation stage. Journal of Dental Research 61, 1532–1542.

Suga, S., 1989. Enamel hypomineralization viewed from the pattern of progressive mineralization of human and monkey developing enamel. Advances in Dental Research 3, 188–198.

Sugihira, N., Suzuki, K.T., 1991. Discrimination between strontium and calcium in suckling rats. Biological Trace Element Research 29, 1–10.

Von Zallinger, C., Tempel, K., 1998. Transplacental transfer of radionuclides. A review. Journal of Veterinary Medical Science, Series A Physiology Pathology Clinical Medicine 45, 581–590.

Wasserman, R.H., 1963. *The Transfer of Calcium and Strontium Across Biological Membranes.* Academic Press, New York.

Webb, E., Amarasiriwardena, D., Tauch, S., Green, E.F., Jones, J., Goodman, A.H., 2005. Inductively coupled plasma-mass (ICP-MS) and atomic emission spectrometry (ICP-AES): Versatile analytical techniques to identify the archived elemental information in human teeth. Microchemical Journal 81, 201–208.

Whittaker, D.K., Richard, D., 1978. Scanning electron microscopy of the neonatal line in human enamel. Archives of Oral Biology 23, 45–50.

Wilson, P.R., Beynon, A.D., 1989. Mineralization differences between human deciduous and permanent enamel measured by quantitative microradiography. Archives of Oral Biology 34, 85–88.

Wilson, P.R., Beynon, A.D., 1990. Mineralization levels in pre-and post natal human deciduous molar enamel. Journal of Paedeatric Dentistry 6, 35–39.

Wong, F.S.L., Anderson, P., Fan, H., Davis, G.R., 2004. X-ray microtomographic study of mineral concentration distribution in deciduous enamel. Archives of Oral Biology 49, 937–944.

Wright, L.E., Schwarcz, H.P., 1998. Stable carbon and oxygen isotopes in human tooth enamel: identifying breastfeeding and weaning in prehistory. American Journal of Physical Anthropology 106, 1–18.

Wurster, C.M., Patterson, W.P., Cheatham, M.M., 1999. Advances in micromilling techniques: a new apparatus for acquiring high-resolution oxygen and carbon stable isotope values and major/minor elemental ratios from accretionary carbonate. Computational Geosciences 25, 1159–1166.

Zazzo, A., Balasse, M., Patterson, W.P., Patterson, P., 2005. High-resolution delta C-13 intratooth profiles in bovine enamel: Implications for mineralization pattern and isotopic attenuation. Geochimica et Cosmochimica Acta 69, 3631–3642.

3. Dental topography and human evolution with comments on the diets of *Australopithecus africanus* and *Paranthropus*

P.S. UNGAR
Department of Anthropology
University of Arkansas
Old Main 330
Fayetteville, AR 72701 USA
pungar@uark.edu

Keywords: dental functional morphology, early hominin dietary adaptations

Abstract

Dental functional morphology can inform us on the dietary adaptations of early hominins and other fossil primates. Traditional approaches to understanding dental form-function relationships have relied mostly on unworn teeth for analysis. This has limited our samples and our understanding of how teeth are adapted to wear in a manner that keeps them mechanically efficient for chewing. This paper reviews a relatively new tool for the study of occlusal functional morphology, dental topographic analysis. This landmark-free, three-dimensional approach involves the creation and measurement of digital models of teeth using point cloud data and Geographic Information Systems software. Three examples are presented. First, a study of living great apes is reviewed to show that worn teeth can be included in the study of dental topography, and that species with different diets have corresponding and predictable differences in the shapes of their molars at comparable stages of tooth wear. Second, a longitudinal study of howling monkeys is summarized to demonstrate that different individuals within this species have consistent changes in crown shape as their teeth wear down. This suggests species-specific wear patterns, a necessary prerequisite for the inference of function from form of worn fossil teeth. Third, a new dental topographic analysis is presented for *Australopithecus africanus* and *Paranthropus robustus* to illustrate that this approach can offer insights into the dietary adaptations of early hominins and other fossil primates. Results presented here confirm that the *A. africanus* and *P. robustus* differed in their dietary adaptations. The degree of difference between their occlusal morphologies is on the order expected of species that often eat similar foods, but differ at "crunch" times. Dental topography data suggest that *P. robustus* probably fell back more on hard, brittle items whereas *A. africanus* relied on tougher, more elastic foods when preferred resources were less available.

S.E. Bailey and J.-J. Hublin (Eds.), Dental Perspectives on Human Evolution, 321–343.

Introduction

Studies of living primates show us just how important food choices are to the daily lives of these animals. Diet underlies many of the behavioral and ecological differences we observe, from group size and structure to positional behavior and locomotion (see Fleagle, 1999). It should come as no surprise then that primatologists focus considerable attention on documenting even the subtlest aspects of feeding ecology. Such studies allow us to better understand the ecological "role" of a primate species, and how it adapts to the environment in which it lives.

It follows that researchers would view primate (including human) evolution in terms of changing adaptations reflecting changing food resources or patterns of resource exploitation. Diet is therefore an important key to understanding paleoecology of past primates, including early hominins, and many theories concerning human evolution have centered on the feeding adaptations and subsistence practices of our distant ancestors.

Most studies of early hominin feeding behaviors and adaptations have focused on dental remains, because they are so well-suited to reconstructing diet. First, mammal teeth function to procure and process food. These durable parts of the digestive system are often the only surviving link between an extinct species and its diet. Second, teeth are the most abundant elements in most fossil assemblages, so can provide a rich source of data.

There are two distinct lines of evidence researchers look to when reconstructing diet from dental remains: genetic signals, which tell us something about what a species was adapted to eat, and non-genetic or epigenetic signals, which tell us something about what an animal ate during its lifetime. Evidences of adaptation include enamel thickness and structure, dental allometry (tooth size), and dental morphology (tooth shape). Evidences of foods eaten during life include tooth chemistry (e.g., stable isotope ratios) and dental microwear. A consideration of all of these approaches is beyond the scope of this paper; reviews can be found easily enough in the literature (e.g., Ungar, 2002; Teaford and Ungar, 2006; Teaford, 2007).

Background

This chapter considers one line of evidence for dietary adaptation, dental functional morphology. Researchers have recognized for hundreds of years that tooth form reflects function (e.g., Hunter, 1771; Owen, 1840–1841; Cuvier, 1863; Cope, 1883; Osborn, 1907; Gregory, 1922). The rate of this work has increased dramatically over the past few decades, with an emphasis on mammalian teeth as guides for chewing (e.g., Crompton and Sita-Lumsden, 1970; Kay and Hiiemae, 1974). Dental biomechanists have since viewed molar morphology in terms of mechanical efficiency for particular masticatory movements as the lower teeth come into and out of occlusion with corresponding uppers.

It was in this light that, more than a quarter century ago, Grine (1981) first observed differences between facet inclinations of different early hominins. He noted that *Australopithecus africanus* molars have steeper facets than do those of *Paranthropus robustus* (especially *P. robustus* from Swartkrans). The more inclined facets evinced by *A. africanus* suggested that these hominins engaged in more shearing, where facet faces slide past one another nearly parallel to their planes of contact. In contrast, *P. robustus* had less inclined facets, suggesting to Grine a shallower approach into and out of centric occlusion and more grinding activity (defined by both perpendicular and parallel components to occlusal contact). He proposed that this, along with other lines of evidence, such as dental allometry, microwear, and craniofacial morphology, implied that *P. robustus* ate harder and more fibrous foods than were eaten by *A. africanus*.

At about the same time, dental functional morphologists began working in earnest on unraveling relationships between tooth shape and diet in living primates. While it was known that primates that habitually crush foods have flat molar surfaces, and that those that shear and slice have highly crested teeth (Rosenberger and Kinzey, 1976; Seligsohn and Szalay, 1978), it was soon recognized that a quantitative approach was necessary to better understand these form-function relationships and to use them to infer diet from tooth shape for fossil forms.

The most widely applied measurement has been Kay's (1978, 1984) "shearing quotient" (SQ), a gauge of the shearing potential of a primate molar tooth. The SQ approach involves measurement of mesiodistally oriented crest lengths of unworn molars for a range of closely related species with similar diets. A least-squares regression line is fit to summed crest length and mesiodistal occlusal surface length in logarithmic space. SQs are computed as residuals, or deviations from the regression line as percent differences between observed and expected shearing crest length for a given tooth length. For example, a high SQ value calculated from a cercopithecine regression indicates longer shearing crests than expected of a frugivorous Old World monkey.

Studies by Kay and colleagues (Kay, 1984; Kay and Covert, 1984; Anthony and Kay, 1993; Kay and Ungar, 1997; Meldrum and Kay, 1997) have shown that SQ accurately tracks diet within all major groups of anthropoids. Among closely related species, folivores tend to have high SQs (relatively long shearing crests) and hard-object feeders have low SQs relative to a regression equation for soft-fruit eaters. Indeed, these relationships between SQ and diet in living primates have been used to infer diet from shearing crest lengths of fossil primates from every epoch in which they are well-known (e.g., Kay, 1977; Kay and Simons, 1980; Anthony and Kay, 1993; Strait, 1993; Williams and Covert, 1994; Ungar and Kay, 1995; Kay and Ungar, 1997; Meldrum and Kay, 1997; Ungar et al., 2004; Ungar, 2005).

Shearing quotients have also been calculated for unworn M_2s of South African early hominins (Ungar et al., 1999; Teaford et al., 2002). While available sample sizes are small, all *Paranthropus robustus* specimens examined ($n = 5$) had lower SQ values than the average for any extant hominoid species. Further, the average SQ value for *Australopithecus africanus* ($n = 4$) was lower than that for any extant hominoid, though higher than that for the *P. robustus*. This suggested that both species had flat, blunt molar teeth compared with most extant apes, but that *A. africanus* had more occlusal relief than *P. robustus*, consistent with suggestions made by Grine (1981).

Such studies show the potential of occlusal morphology to tell us something about the diets of early hominins and other fossil primates, but also underscore the limitations of conventional techniques. First, it is difficult to interpret results with such small samples. SQ studies require unworn teeth because measurements are made between cusp tips (which are quickly obliterated by wear) and the notches between them. Because teeth begin to wear as soon as they come into occlusion, few fossil specimens are suitable for SQ analysis. The entire sample of early hominins from South Africa, for example, boasts less than ten unworn M_2s (the teeth most often used in functional studies)!

Another constraint of studying unworn teeth is that this leads to an incomplete picture of the form-function relationship. Wear is a normal phenomenon. Natural selection should also act on worn teeth, favoring morphologies that wear in a manner that keep them mechanically efficient for fracturing foods (Kay, 1981; Teaford, 1983; Ungar and Williamson, 2000). We are missing a lot of information if we exclude worn teeth from our analyses.

There have been some attempts to study functional aspects of tooth form on worn molar teeth. Smith (1999), for example, using a technique modified from Wood and coauthors (1983), measured 2D planimetric areas of individual cusps from captured video images. Relative cusp areas do not change markedly with wear in occlusal view so long as cusp boundaries can be identified. Smith's results suggest that cusp proportions may reflect diet to some degree – e.g., chimpanzees are linked with gibbons, rather than with gorillas.

Even this approach though, is not ideal. First, specimens must still be sufficiently unworn to distinguish individual cusp boundaries. These can go quickly on thin enameled molars, such as those of chimpanzees and gorillas. More importantly, planimetric area studies do not give us information on the third dimension. This is a problem because mastication occurs in a 3D environment, and two teeth with similar projected 2D areas may differ greatly in cusp relief. Cusp relief is critical to the angle of approach of mandibular and maxillary teeth as facets come into occlusion during mastication. This in turn determines the method by which foods are fractured (i.e., shearing or crushing).

Dental Topographic Analysis

What is needed is a landmark-free, three-dimensional approach to characterizing and comparing occlusal morphology on unworn and worn teeth. This is where dental topographic analysis comes in. Three-dimensional point cloud data representing the surface of a tooth are collected using a scanner and analyzed using geographic information systems (GIS) software. The scanner (or other device) used depends on the resolution and work envelope required. The electromagnetic digitizer (Zuccotti et al., 1998), laser scanner (Ungar and Williamson, 2000), and confocal microscope (Jernvall and Selänne, 1999; Jernvall et al., 2000) have all been used for dental topographic analysis with success.

Geographic information systems assemble, store, manipulate, analyze and display geographically referenced information. GIS tools have been developed to model and examine the physical surface of the Earth. The idea behind dental topographic analysis is that teeth can be modeled as landscape surfaces, so that GIS tools can be applied to measure and analyze functional aspects of occlusal form. This is a three step process: (1) scan the surface, (2) model the surface, and (3) analyze the model.

The process can be demonstrated with three examples. The first example reviews studies of extant great ape molars to demonstrate that occlusal topography reflects diet, and that worn teeth can be compared between species. The second example involves a longitudinal study of howling monkeys that shows that individuals within a species wear their teeth down in consistent, predictable ways. This is a necessary prerequisite for functional studies of variably worn teeth of fossil taxa. The final example presents new data on dental topography of *Australopithecus africanus* and *Paranthropus robustus*. Results suggest that these species differed in the efficiency with which they could comminute foods with given fracture properties. Neither species had molar teeth well-suited to shearing extremely tough or ductile foods, but *P. robustus* could probably more efficiently crush brittle items than could *A. africanus*. The degree of difference between the two species is comparable to that between living primates that eat similar foods most of the time, but differ in the fallback resources they exploit.

Dental Topographic Analysis of Extant Hominoids

The first step in assessing the value of dental topographic analysis for paleoanthropology is to consider variably worn teeth of living species with known differences in diet. *Gorilla gorilla gorilla, Pan troglodytes troglodytes*, and *Pongo pygmaeus pygmaeus* are well-suited to this task because of the modest degree to which they differ in the material properties of the foods they consume, and because they share a basic occlusal "bauplan" with one another and with early hominins. Here I review results first presented by Ungar and M'Kirera (2003), and Ungar and Taylor (2005) (see also Teaford and Ungar, 2006; Ungar, 2005).

Central African chimpanzees and western lowland gorillas overlap considerably in their diets where the two taxa are sympatric – both prefer soft, succulent fruits when widely available. These apes do differ though, especially at times of fruit scarcity, when gorillas fall back more on tough, fibrous foods (Tutin et al., 1991; Remis, 1997). Reports put fruit flesh at about 70–80% of the diet for *P. t. troglodytes* and about 45–55% of the diet for *G. g. gorilla* (Williamson et al., 1990; Kuroda, 1992; Nishihara, 1992; Tutin et al., 1997). While Bornean orangutan food preferences depend greatly on seasonal availability, their diet may be considered intermediate between those of the chimpanzee and lowland gorilla, with annual fruit flesh consumption reported to make up about 55–65% of the diet (Rodman, 1977; MacKinnon, 1977) – noting caveats concerning differences in data collection methods (Doran et al., 2002).

Dental topographic analysis of these species was conducted on high-resolution replicas of variably worn, undamaged M_2s of wild-shot individuals. Specimens used include *Pan troglodytes troglodytes* ($n = 54$) and *Gorilla gorilla gorilla* ($n = 47$) from the Cleveland Museum of Natural History and *Pongo pygmaeus pygmaeus* ($n = 51$) from the State Collection of Anthropology and Palaeoanatomy in Munich. Pigmented epoxy casts (Epotek 301, Epoxy Technologies) were poured into polyvinylsiloxane molds (President's Jet, Regular Body, Coltène-Whaledent), allowed to harden, and then coated with Magniflux Spotcheck (SKD-S2 Developer, Illinois Tool Works) to mitigate specimen translucency.

Occlusal surfaces were scanned with lateral and vertical resolutions of 25.4 μm using a Surveyor 500 (Laser Design) laser scanner. Resulting point clouds were saved as ASCII data files and opened as tables in ArcView 3.2 (ESRI) GIS software with Spatial Analyst and 3D Analyst extensions. Occlusal surface models were interpolated by inverse distance weighting, and resulting digital elevation models were cropped to exclude areas below the lowest point of the occlusal basin. While no landmark on a wearing occlusal surface is stable, repeated study suggests that cropping to the basin low point yields reasonably comparable surfaces for analysis. Average slope between adjacent points (surface slope) was then recorded for each specimen, and wear scores were assigned following Scott's (1979) scoring protocol (see M'Kirera and Ungar, 2003; Ungar and M'Kirera, 2003 for details). While other attributes were considered in the original analyses, discussion here is limited to occlusal slope because this variable effectively separates taxa and illustrates well how dental topographic analysis works. Results for other dental topography variables for these taxa, including surface relief and angularity, can be found in M'Kirera and Ungar (2003) and Ungar and M'Kirera (2003).

All data were ranked prior to analysis to mitigate violation of assumptions required of parametric statistical tests. Only specimens at stages of wear represented by all three taxa were considered in the statistical analysis, where wear stages were defined as (1) Scott scores 10–14, (2) Scott scores 15–19, and (3)

Scott scores 20–24 (see Ungar, 2004). Surface slope and occlusal relief were analyzed using two-way ANOVAs with taxon and wear score as the factors. This allowed assessment of the effects of taxon and degree of wear on each model, as well as the effects of interactions between the two factors. Bonferroni pairwise multiple comparisons tests were used to determine the sources of significant variation for taxon and wear stage differences.

Results are presented in Figure 1 and Table 1. As expected, more worn molar surfaces of each taxon had less sloping occlusal surfaces. Molar teeth become flatter as they wear. At any given stage of wear, however, gorillas had the steepest slopes, followed by orangutans. Chimpanzees had the least sloping molar cusps. Finally, there was no significant interaction between wear stage and taxon.

This example suggests several things. First, tooth shape changes with wear. As teeth wear down, cusp slope values decline. Second, apes with varying diets differ in the shapes of their teeth in ways that reflect the mechanical properties of foods that they eat. Species adapted to shearing and slicing tough leaves should have steeper sloped cusps than those adapted to crushing and grinding fruit – and at any given stage of wear, they do. Finally, differences between species are of similar magnitude at different stages of wear, as suggested by the lack of interaction between the factors. This means that differences between species remain consistent through the wear sequence. In other words, we can contrast chimps, gorillas, and orangutans at any given wear stage and get the comparable results. This is important because it suggests that species need not be represented by unworn teeth as long as there is a baseline of comparative data for specimens with similar degrees of wear. This dramatically increases both the number of fossil specimens and variety of species that can be analyzed for occlusal functional morphology compared with techniques requiring unworn teeth.

Longitudinal Study of Mantled Howling Monkeys

Dental topographic analyses of museum specimens (such as that presented above) make the assumption that individuals within a species wear their teeth down in similar ways, and thus can be used to construct species-specific wear sequences. This is a necessary assumption if we are to reconstruct diets from variably worn teeth of different individuals, as must be the case for fossil assemblages. Testing this assumption is not an easy task, as it requires a longitudinal study

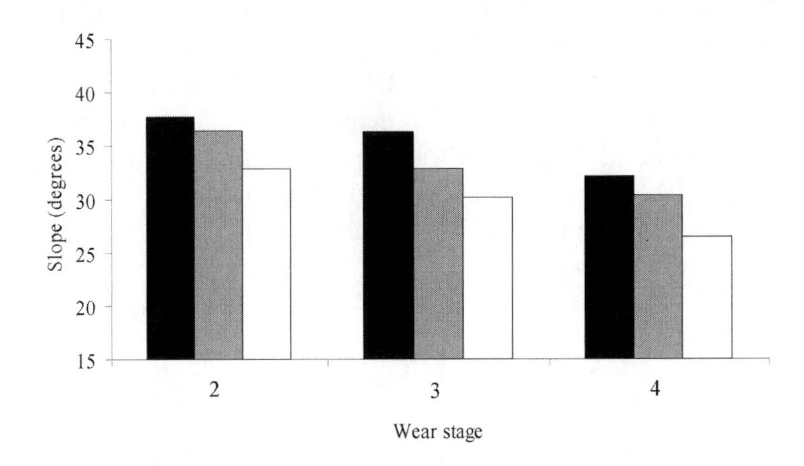

Figure 1. Mean occlusal slopes for extant hominoids at three stages of wear. Black = *Gorilla gorilla gorilla*, grey = *Pongo pygmaeus pygmaeus*, and white = *Pan troglodytes troglodytes*.

Table 1. Occlusal slope comparisons for the extant hominoid study

a. Descriptive statistics

	Gorilla			*Pongo*			*Pan*		
Wear Stage	*Mean*	*sd*	n	*Mean*	*sd*	n	*Mean*	*sd*	n
1	37.75	5.036	7	36.38	2.447	24	32.88	5.859	5
2	36.29	2.665	10	32.86	2.274	16	30.15	5.771	28
3	32.13	5.069	14	30.32	2.170	6	26.48	4.680	18

b. Two factor ANOVA: Occlusal slope data (ranked)

Source	SS	*df*	MS	*F*	*p*
Species	32065	2	16032	11.027	0.000
Wear	47936	2	23968	16.485	0.000
Interaction	34844		871	0.599	0.664
Error	53065	119	1286		

c. Multiple comparison test results

	Gorilla	*Pongo*	*Pan*
Gorilla	0.0		
Pongo	−49.366*	0.0	
Pan	−45.960*	−26.105*	0.0

*$p < 0.05$

examining changes in morphology of a single tooth over time, and comparing such changes between individuals. This was the goal of work recently completed by Dennis et al. (2004) for *Alouatta palliata* from Hacienda La Pacifica, in Guanacaste Province, Costa Rica.

La Pacifica is well-suited to this sort of analysis. First, the howling monkeys there have been studied nearly continuously for the past three decades (Glander, 1978; Clarke and Glander, 1984; Glander et al., 1987; Clarke et al., 1998; Clarke et al., 2002; Teaford et al., 2006), with one focus of this research involving collection of dental impressions for tooth wear analysis (Teaford and Glander, 1991, 1996). Second, groups can be distinguished into different microhabitats, allowing us to assess the effects of this variable on tooth wear patterns. Dennis et al. (2004) focused on Groups 7 and 19 from the gallery forest lining the Corobici River and Groups 1 and 33, which live in drier forest patches subject to more pronounced seasonal changes in resources (Teaford and Glander, 1996).

In total, 14 individuals were used in this study. These were captured and released repeatedly at 2, 4 and 7 year intervals following initial capture whenever possible. Animals were darted and dental impressions were taken following Glander et al. (1991) and Teaford and Glander (1996). High resolution epoxy casts were prepared as described above for the extant ape study. M_2 replicas were scanned by laser scanner at 25.4 μm lateral and vertical resolutions, and resulting x, y, z coordinates were opened as tables in ArcView 3.2 (ESRI) GIS software. As with the hominoid dental topographic analysis, monkey molar surfaces were cropped to include only data above the lowest point on the central basin of the occlusal table, and average surface slope was recorded for each individual. Given missing values resulting from the vagaries of darting primates in the wild, non-normal distributions, and other limitations of the dataset, signed ranks tests were used to assess changes in slope between initial capture and years 2, 4 and 7. Possible effect of environment was assessed using Wilcoxon's two-sample tests comparing river (7, 19) and nonriver (1, 33) groups. Analyses of other variables, including surface relief and angularity, can be found in the literature (Dennis et al., 2004).

Results are presented in Figure 2 and Table 2. As expected, individual teeth showed a general trend toward decreasing surface slope over time. Howling monkey teeth become flatter as they wear down. A one-tailed test showed significant decreases in occlusal slope between initial capture and years 2, 4 and 7. Further, there was no significant difference in change in slope between river and non-river groups. This suggests minimal effect of microhabitat on changes in occlusal topography with wear (though small samples make this result tentative).

Table 2. Slope data results for Alouatta palliata

	Signed ranks test (individuals)			*Wilcoxon test (habitats)*		
Interval	N	t_s	p	N	t_s	p
2 years	11	25.0	0.024	6,5	22.0	0.178
4 years	6	9.5	0.063	3,3	14.0	0.200
7 years	8	18.0	0.008	4,4	13.0	0.200

Dental topographic analysis of *Alouatta palliata* suggests consistent changes in molar morphology over time for the howlers of La Pacifica. The general trend toward decreasing slope mirrors that suggested for gorillas, chimpanzees, and orangutans in the museum study. It is also notable that these patterns of change with wear hold both for the river

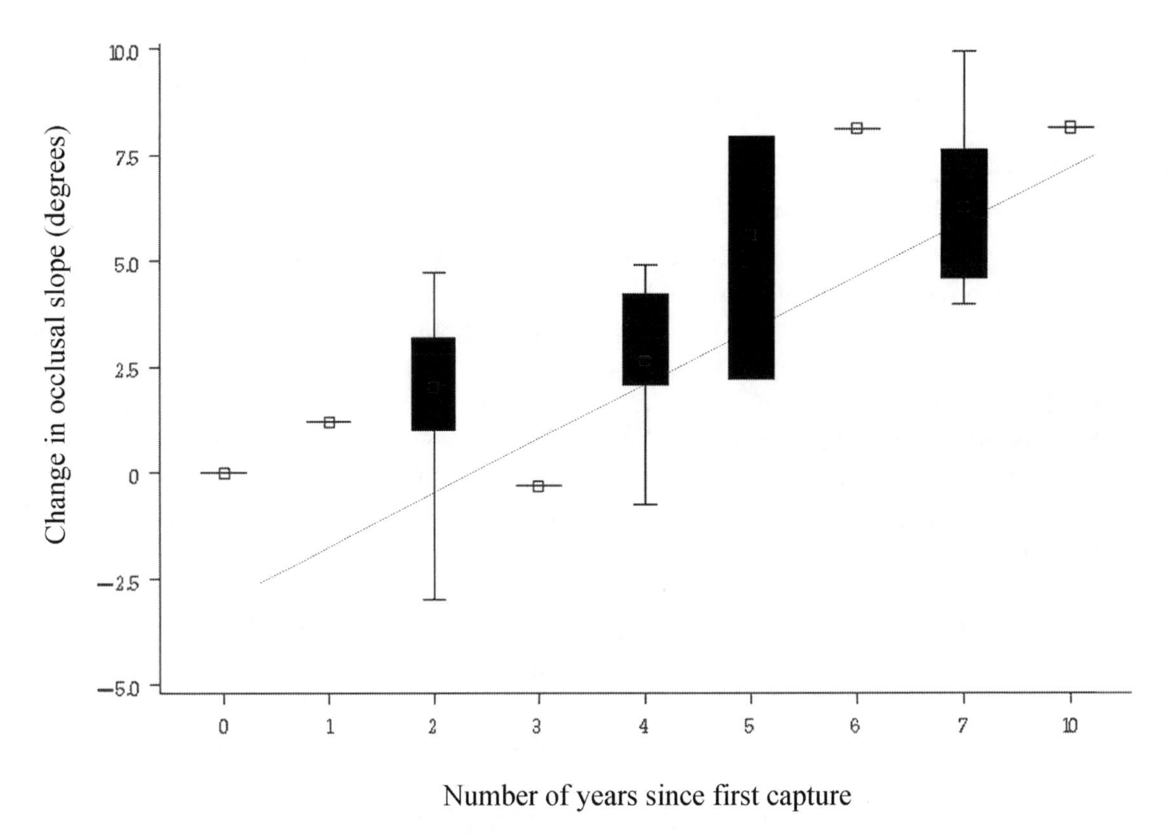

Figure 2. Average change in occlusal slope over a given number of years for howling monkeys from La Pacifica. Center lines = medians, hinges = first and third quantiles, whiskers = 1.5 Hspreads of the hinges. Data from Dennis (2002).

and nonriver groups. While microscopic wear patterns differ between these groups (Teaford and Glander, 1996), it might be that species-specific wear sequences are independent of the subtle dietary differences that characterize conspecifics living in different microhabitats.

The implications of this study for paleobiology are clear. The fact that individual howling monkeys show similar changes in occlusal morphology over time suggests that different individuals at different stages of wear can indeed be used to construct a species wear sequence, at least for this species. This is an important assumption for researchers that wish to use variably worn teeth to reconstruct the diets of early hominins and other fossil primates. It is also a relief to note the lack of evidence for differences in changing occlusal morphology with wear between microhabitats (with sample size caveats). Temporal and geographic control over most hominin assemblages are too limited to expect that individuals found at one site, even within a single deposit, lived in the same place at the same time. If microhabitat differences had a substantive effect on changing occlusal morphology with wear, it would be difficult to compare fossil hominin species using samples of worn teeth.

Dental Form and Diets of *Australopithecus africanus* and *Paranthropus robustus*

The ultimate goal of dental topographic analysis is the reconstruction of the dietary adaptations of fossil forms. The classic contrast between *Australopithecus africanus* and *Paranthropus robustus* provides an excellent example of how this works.

Reconstructing the Diets of South African Australopiths

Researchers have been fixated on the diets of the South African early hominins for more than half a century, ever since Robinson (1954) proposed that morphological differences between *Australopithecus africanus* and *Paranthropus robustus* were related to their dietary adaptations. Robinson (1954, 1963) noted that *P. robustus* had relatively larger cheek teeth and smaller incisors, larger masticatory muscles, and more postcanine enamel chipping than did *A. africanus*. This suggested to him that *Paranthropus* was herbivorous, whereas *Australopithecus* was more omnivorous.

Many subsequent workers have focused on the differences between these hominins in relative tooth sizes (e.g., Tobias, 1967; Groves and Napier, 1968; Jolly, 1970; Pilbeam and Gould, 1974; Szalay, 1975; Kay, 1975a; Wood and Stack, 1980; Peters, 1981; Kay, 1985; Lucas et al., 1985; Wood and Ellis, 1986; Demes and Creel, 1988; McHenry, 1988; Ungar and Grine, 1991; Teaford and Ungar, 2000). While there has been much debate, many would probably now agree that larger molars and smaller incisors in *Paranthropus* are consistent with a diet dominated by larger quantities of lower quality foods requiring less incisal preparation but more mastication when compared with *Australopithecus*. While these results appear to be consistent with craniofacial functional morphology (Du Brul, 1977; Ward and Molnar, 1980; Rak, 1983; Sakka, 1984; Daegling and Grine, 1990), relationships between tooth size and diet remain difficult to understand. For example, while assumptions that molar size relates to food quality and quantity might explain why among platyrrhines, folivores have relatively larger molars than frugivores, it is not immediately clear why among cercopithecoids, folivores tend to have smaller molars than do frugivores. In the end, dental allometry likely reflects many different selective pressures, from food particle size and shape, to mechanical properties and stickiness (Lucas, 2004). We are not yet at the point where we can identify and separate these adaptive signals.

Researchers have also looked to dental microwear to compare the diets of *Australopithecus africanus* and *Paranthropus robustus*. Grine and colleagues (Grine, 1981, 1986; Grine and Kay, 1988; Ungar and Grine, 1991) have demonstrated significant differences in both molar and incisor microwear between these species. Grine (1986) noted, for example, that *P. robustus* molars have more microwear features, especially more pits than do those of *A. africanus*. In contrast, *A. africanus* molars have relatively more striations, and these are longer, thinner, and more homogeneously oriented than are the scratches found in *P. robustus* cheek teeth. This suggested to these authors that *Paranthropus* habitually crushed and ground harder foods, whereas *Australopithecus* more often consumed softer fruits and/or young leaves. Ungar and Grine (1991) also reported incisor microwear differences between these taxa. *Australopithecus africanus* incisors have more microwear on their surfaces than do those of *P. robustus*. This result was interpreted to suggest that the *Australopithecus* used their incisors to process a greater variety of foods, including larger, more abrasive items than did *Paranthropus*.

While there have been many functional studies on dental allometry, microwear, and craniofacial morphology of the South African early hominins, there has been less written on the dental functional morphology of these taxa. Still, the work of Grine (1981) and Ungar et al. (1999), as described above, have suggested differences. Again, *Australopithecus africanus* appears to have more inclined molar facets and relatively higher SQ values on average than does *Paranthropus robustus*. Nevertheless, our understanding of dental-dietary adaptations of these hominins could certainly benefit from further study.

Dental Topographic Analysis of Australopithecus Africanus and Paranthropus Robustus: Materials and Methods

Here I present a new study of dental topography of *Australopithecus africanus* and *Paranthropus robustus*. High-resolution replicas were prepared following the procedure described for the extant hominoid study. Only undamaged M_2s were used to assure comparability of results following usual practice (Williamson et al., 2000; M'Kirera and Ungar, 2003; Ungar and M'Kirera, 2003; Dennis et al., 2004; Ungar, 2004). A total of 33 specimens were suitable for analysis. These included molars of *Australopithecus africanus* ($n = 18$) from Makapansgat and Sterkfontein and *Paranthropus robustus* ($n = 15$) from Drimolen, Swartkrans and Kromdraai (Table 3).

Table 3. Specimens used in the hominin microwear analysis

Australopithecus africanus	Paranthropus robustus
MLD 2	DNH 19
MLD 24	DNH 51
MLD 40	SK 1
Sts 52	SK 10/1648
Sts 7	SK 1587
Stw 109	SK 23
Stw 213	SK 25
Stw 269	SK 37
Stw 308	SK 55
Stw 327	SK 6
Stw 384	SK 843
Stw 404	SKW 5
Stw 412	SKX 4446*
Stw 424	TM 1517
Stw 498	TM 1600
Stw 534	
Stw 537	
Stw 560	

*Schwartz and Tattersall (2003) have recently argued that SKX 4446 is *Homo*. Metric reanalysis of the specimen suggests that it ***is*** appropriately attributed to *Paranthropus* (Grine, 2005). Further, exclusion of this specimen from analysis has no significant affect on the results presented here.

Taxonomic identifications for individual specimens follow those reported in the literature (Grine, 1989; Keyser, 2000; Hlusko, 2004; Moggi Cecchi cited in Hlusko, 2004; Grine, 2005).

Specimens were examined using a PICZA PIX-4 3D piezo scanner (Roland DGA). PICZA scanners collect data much more slowly (it can take several hours to scan a single tooth) and have a lower minimum resolution (50 μm scanning pitch) than the laser scanner, confocal microscope, and 3-D digital photogrammetric system used in previous studies of occlusal topography (Jernvall and Selänne, 1999; Ungar and Williamson, 2000; Kullmer et al., 2004). On the other hand, PICZA scanners are becoming increasingly popular for modeling teeth because of their low price and ease of use (Aguirre de Enriquez et al., 1974; Archer and Sanson, 2002; Okada et al., 2003; Eguchi et al., 2004; King et al., 2005a, b; Martin-de las Heras et al., 2005). These scanners use piezo sensors, which allow scans with much less contact force than possible with conventional touch-probe scanners. Indeed, I have been unable to discern impact marks at resolutions down to 0.18 μm after scanning epoxy replicas using the PICZA scanner.

All specimens were placed on the scanner stage so as to maximize buccolingual and mesiodistal lengths in top view and scanned to a lateral resolution of 50 μm. In other words, depth values were sampled at intervals of 0.05 mm along both x- and y- axes. Resulting data for each specimen were then exported as ASCII files of x, y, z coordinates.

Surfaces were interpolated and analyzed as described above for the extant primates. ASCII data were opened as tables in ArcView 3.2 with Spatial Analyst and 3D Analyst extensions. Each specimen was cropped to the lowest point on the occlusal basin to approximate the functional occlusal table, and average surface slope was calculated for each specimen. Data were ranked and analyzed using two-way ANOVAs with taxon and wear score as the factors. Wear staging followed Scott's (1979) scores as follows: (1) <10, (2) 10–14, (3) 15–19, (4) 20–24.

Specimens with Scott scores above 24 were excluded from statistical analysis because little morphology is preserved beyond that point. The principal interest here is with the functional life of the tooth and adaptive aspects of its morphology. Tooth functional efficiency does ultimately decrease with extreme wear, especially once the occlusal surface is reduced to an enamel rim with a single, large dentin island. Teaford and Glander (1996) noted, for example, a decline in shearing crest length in *Alouatta palliata* with extreme wear, suggesting to them that older individuals might not be processing their food as well as prime adults. Dennis et al. (2004) similarly found decreased occlusal angularity for these primates, indicating less jagged surfaces for shearing and slicing in the most worn specimens. The oldest mantled howling monkeys at La Pacifica do indeed average larger food particles in their stomachs and feces, strongly suggesting a decrease in masticatory efficiency (Teaford, pers. comm.). This echoes King and coauthor's (2005a) suggestion that extreme tooth wear in sifakas can decrease chewing efficiency, nutrient acquisition, and reproductive fitness (see also Lanyon and Sanson, 1986). Thus, the study of dental functional morphology in primates, including early hominins, might best exclude the most worn specimens (again, Scott scores of 25 and above in this case).

Dental Topographic Analysis of Australopithecus Africanus and Paranthropus Robustus: Results and Discussion

Results for the dental topographic analysis of *Australopithecus africanus* and *Paranthropus robustus* are presented in Table 4 and Figures 3–4. As observed for the extant hominoids, there is significant variation in the

Table 4. Occlusal slope comparisons for the early hominin study

a. Descriptive statistics

	Australopithecus			Paranthropus		
Wear	*Mean*	*sd*	n	*Mean*	*sd*	n
1	42.85	2.774	6	35.79	6.03	7
2	36.17	4.581	4	35.61	4.511	2
3	27.89	3.926	4	25.99	2.834	2
4	20.22	–	1	18.14	–	1
5	28.08	4.347	3	16.91	3.113	3

b. Two factor ANOVA: Occlusal slope data (ranked)

Source	SS	df	MS	F	p
Species	174.022	1	174.022	7.571	0.011
Wear stage	1396.931	1	1396.931	60.779	0.000
Interaction	56.817	1	56.817	2.472	0.130
Error	528.631	23	22.984		

model both between species and between wear stages, but no significant interaction between the two factors. *Australopithecus africanus* molars have more sloping occlusal surfaces than do those of *P. robustus*. Further, less worn specimens have more sloping surfaces than more worn specimens. Finally, the lack of interaction between species and wear stage suggests that the difference between *A. africanus* and *P. robustus* remains consistent at comparable degrees of wear.

Steeper occlusal surfaces of *Australopithecus* suggest that this hominin could more efficiently shear and slice tough foods than could *Paranthropus* specimens with comparably worn molars. *Paranthropus robustus*, on the other hand, could more efficiently crush hard, brittle foods than could *A. africanus*. These results are consistent both with SQ results for these taxa (Ungar et al., 1999) and with observations that *A. africanus* molars have more inclined occlusal facets than seen for most *P. robustus* specimens (Grine, 1981). These results also make sense given studies of dental allometry, microwear, and craniofacial morphology (see above).

While earlier studies also identified variation between these South African early hominins, they tell us little about the degree to which they differed from one another in their diets. Grine (1981, 1986) noted that *Paranthropus* may have habitually consumed harder or more fibrous items as their principal dietary source than did *Australopithecus*, or such foods may have constituted a small but critical part of their diets. He reasoned that, given microwear results, the crushing and grinding adaptations of the *P. robustus* molar probably represent a primary specialization to hard and more fibrous vegetable matter.

This brings up an important point. Researchers have often assumed that preferred foods, or at least those commonly eaten, have the greatest selective influence on molar design (e.g., Kay, 1975b). This need not be true. Kinzey (1978) noted for *Callicebus moloch* and *C. torquatus*, for example, that while both are primarily frugivorous, the former have longer shearing crests for slicing leaves and the latter have larger talonid basins for crushing insect chitin. In this case then, dental morphology reflects adaptations not only to primary foods, but to less frequently eaten but still critical ones.

This is, in fact, probably more the rule than the exception for living hominoids. Apes have a penchant for succulent, sugar-rich foods – a legacy of the ancestral catarrhine dietary adaptation (Ross, 2000; Ungar, 2005). Differences in diet between catarrhines often rest largely with the seasonal shift to fallback foods taken when preferred resources are less available (Rogers et al., 1992; Lambert et al., 2004). In these cases, preferred resources are easy to digest, offer a low cost-benefit ratio, and may not result in selective pressures that would tax functional morphology. On the other hand, less desirable but seasonally critical fallback foods might require some morphological specialization (Robinson and Wilson, 1998).

Australopithecus africanus (Stw404)

Paranthropus robustus (SK 23)

Figure 3. Triangulated irregular networks of early hominin M_2s. Both individuals are at Wear stage 2.

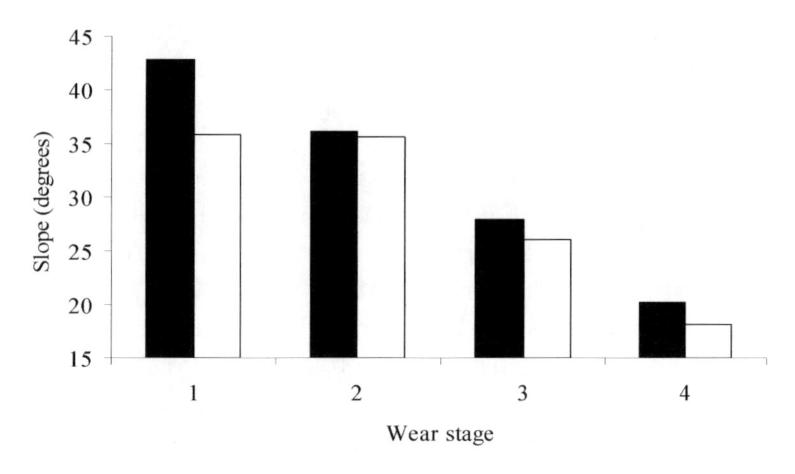

Figure 4. Mean occlusal slopes for early hominins at four wear stages. Black = *Australopithecus africanus*, white = *Paranthropus robustus*.

Sympatric western lowland gorillas and chimpanzees at Lopé, for example, overlap in 60–80% of the plant species they consume (Williamson et al., 1990; Tutin and Fernandez, 1993), with both apes eating soft fruits much of the year. Differences become evident at "crunch times", when preferred fruits are scarce. At such times, gorillas fall back more on leaves and other fibrous plant parts. The same is true for sympatric mountain gorillas and chimpanzees at the Bwindi Impenetrable National Park in Uganda (Stanford and Nkurunungi, 2003). As Remis (2002) has shown, gorillas will choose foods high in non-starch sugars and sugar-to-fiber ratios when such foods are available. The difference between gorilla and chimpanzee occlusal morphology described above then, reflects fallback food choice more than dietary preference *per se*.

It is thus reasonable to suggest that early hominin species preferred the same sorts of foods too – easy to digest fruits rich in simple sugars and low in fiber (Ungar, 2004). If so, it is likely that *A. africanus* and *P. robustus* differed mostly at times of resource stress. This is consistent with Wood and Strait's (2004) recent suggestion that *Paranthropus* may not have been stenotopic, but rather more of a generalist in its diet. As with the gorilla example, the craniodental adaptations may have allowed *P. robustus* more access to mechanically challenging foods (in this case, harder, more brittle ones) when preferred resources were less available.

Kay and Ungar (1997) argued that ranges of variation in occlusal morphology between fossil species may be interpreted in light of ranges between living species with differing diets. While different instruments were used to collect the data for the living apes and the South African australopiths, it is clear that the degree of difference in occlusal slope between *Australopithecus africanus* and *Paranthropus robustus* is no greater than, and probably somewhat less than that between extant *Gorilla gorilla gorilla* and *Pan troglodytes troglodytes* (compare Figures 1 and 4). Since the differences in occlusal morphology between the extant hominoids may be explained by differences in fallback resources rather than preferred foods, the same may be true for the South African australopiths.

Still, the microwear differences remain to be explained. Grine and colleagues did demonstrate significant variation in microwear between the South African australopiths (Grine, 1986; Grine and Kay, 1988; Ungar and Grine, 1991). These suggest real differences in their diets, and underscore the differences between epigenetic and genetic evidence for diet. Dental microwear tells us something about what an individual ate during its lifetime, whereas occlusal morphology tell us something about what a species is adapted to eat (Ungar et al., 2004; Ungar, 2007). These two lines of evidence are complementary, but clearly reflect different things. If preferred foods are not mechanically challenging, they may not have a marked selective influence on molar design. In such a case, molar morphology should be adapted to less preferred foods that are more difficult to fracture, so long as they are important to the survival of the individual.

That said, recent microwear analysis actually does suggest considerable overlap between the diets of *Australopithecus africanus* and *Paranthropus robustus* (Scott et al., 2005). The emphasis of traditional microwear studies has been on demonstrating differences between species. Observer error rates in microwear feature quantification have made it difficult to assess within species variation in microwear patterning (Grine et al., 2002). Microwear texture analysis obviates this problem by offering an objective, repeatable way to characterize microwear surfaces (Ungar et al., 2003; Scott et al., 2005). We can now be confident that slight differences in microwear reflect real differences between surfaces, and not observer error. This allows us to reliably compare distributions of microwear values within and between species.

Microwear texture analysis results for the South African australopiths do indicate significant differences between the "robust" and "gracile" forms (Figure 5). *Paranthropus robustus* has more complex molar microwear surfaces and more variability in degree of complexity. On the other hand, *Australopithecus africanus* has more anisotropic molar microwear surfaces and more variability in degree of anisotropy. Nevertheless, the two species do overlap substantively in their microwear surface textures. This suggests that much of the time "robust" and "gracile" individuals ate foods with similar fracture and/or abrasive properties, but that when they differed, *A. africanus* ate tough foods requiring more shearing/slicing (as indicated by surface texture directionality), and *P. robustus* ate brittle foods requiring more tooth-tooth contact normal to opposing occlusal facets (as indicated by surface complexity). These results may indicate occasional but not everyday differences in diet between the hominins. If so, one would expect the australopiths to have taken foods more and more difficult to fracture as preferred ones became less and less available. It is not surprising then, that we see a "tailing off" of *P. robustus* surfaces with increasing texture complexity (see Figure 5).

In sum then, the microwear texture and dental topography results for the South African australopiths loan themselves to a new model for early hominin diets. These hominins may have eaten foods with similar fracture properties most of the time. Notable differences would have come, however, at "crunch times" when easy to fracture preferred foods were less available. At such times, *Australopithecus africanus* would have been able to efficiently fracture tougher, more elastic foods whereas *Paranthropus robustus* would have been more adept at comminuting harder, more brittle items.

Conclusions and Directions for Future Research

If we are to make the most of dental form-function relationships and their implications for reconstructing the diets of early hominins

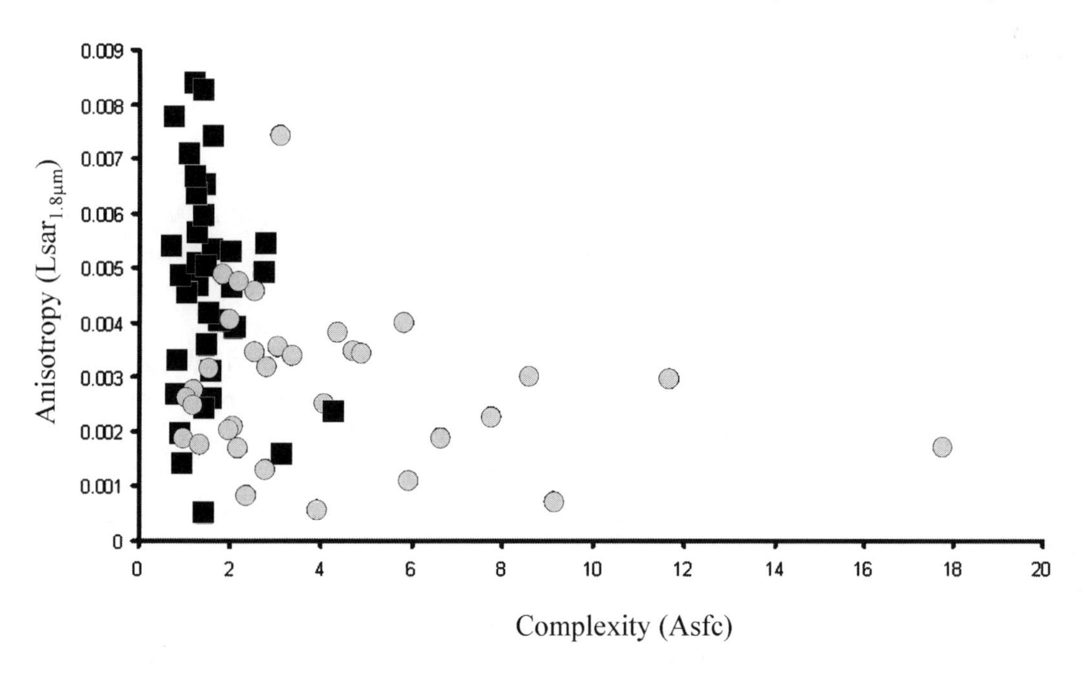

Figure 5. Microwear texture analysis of sampled *Australopithecus africanus* (black squares) and *Paranthropus* robustus (grey circles) surfaces. Each individual analyzed is represented by four surfaces. Data are from Scott et al. (2005).

and other fossil primates, we must be able to include worn teeth in our analyses. This is true both because sample sizes are otherwise too limiting for functional analyses in many cases, and because form-function relationships are complicated by the fact that occlusal form changes with wear. The key to characterizing functional aspects of worn tooth form is a landmark-free, three-dimensional approach, such as dental topographic analysis.

The examples presented here demonstrate the efficacy of dental topographic analysis. The study of living apes shows that species with differing diets differ as expected in the occlusal morphology of their worn teeth. Degree of wear can be controlled for, and species can be compared directly using a two-factor ANOVA model. In this case, species that consume more tough foods, such as leaves, have more sloping occlusal surfaces at a given stage of wear. This allows them to more efficiently shear and slice tough foods as opposing molars come into centric occlusion at steeper angles relative to their contact facets (see Strait, 1993; Spears and Crompton, 1996; Yamashita, 1998; Lucas et al., 2004 for discussions of dental biomechanics and food properties).

The longitudinal study of *Alouatta palliata* molars shows that, at least for this species, individuals wear their teeth in similar ways, with consistent changes in morphology over time. This result is expected given the assumption that teeth have evolved to wear down in a specific manner. If this was not the case, it would be difficult to reconstruct the dietary adaptations of a species using morphology of variably worn teeth. Indeed, all species examined by dental topographic analysis thus far, both living and fossil, show a pattern wherein more worn teeth have predictably less sloping occlusal surfaces – except for the most extremely worn cases (Ungar and Williamson, 2000; M'Kirera and Ungar, 2003; Ungar and M'Kirera, 2003; Dennis et al., 2004; Gembressi, 2004; Ungar, 2004; King et al., 2005a, b). This also suggests consistent changes in morphology with wear.

Finally, the study of *Australopithecus africanus* and *Paranthropus robustus* confirm that dental topographic analysis can be applied to fossils. The difference in number of M_2s available for consideration using this approach ($n = 33$) compared with SQ study ($n = 9$) presents a compelling argument for dental topographic analysis. Such larger samples allow us to go beyond "the species differ" to explore the degree to which they differ, with their implications for the paleoecology of these taxa. In this case, again, while others have noted differences in dental morphology between *A. africanus* and *P. robustus* (e.g., Grine, 1981; Ungar et al., 1999), dental topographic analysis suggests that occlusal topography differences are rather slight, on the order of those expected of two species that differ mostly in the properties of occasional fallback resources taken during times that preferred foods are less available.

Future Directions

Dental topographic analysis should also allow us to go beyond documenting differences between species in dental morphology at a given stage of wear, to deduce how teeth wear in ways that keep them functionally efficient for fracturing the foods they are adapted to. Only one attribute, average surface slope, was presented here but more precision is possible, as other aspects of occlusal morphology likely also reflect function. GIS analyses allow characterizations of many different aspects of topography, from cusp relief, surface area, angularity or jaggedness, aspect, basin volume, "drainage patterns", and just about any other topographic attributes that can be measured (Zuccotti et al., 1998; Ungar and Williamson, 2000). Further, individual features can be isolated with the help of contour lines to examine different parts of the occlusal surface. These and other datasets will allow us get a better handle on how teeth change shape as they wear.

Occlusal angularity provides one example. Blade theory teaches us that jaggedness or serratedness of a cutting surface affects the directions of forces acting on a food item, and can have dramatic effects on fracture efficiency (e.g., Frazzetta, 1988). Primate occlusal surface jaggedness is dictated in part by tooth wear as enamel gives way to softer dentin, and pits develop steeply angled walls. This phenomenon may be comparable, to an extent, to that seen in selenodont ungulate cheek teeth, where complex infoldings and lophs form sharp edges for shearing and grinding tough foods. Perhaps this explains why many primate folivores have thin molar enamel (Kay, 1981).

It is particularly notable then, that occlusal angularity does not change with wear in African apes or howling monkeys in the same manner as other attributes such as occlusal slope and surface relief (M'Kirera and Ungar, 2003; Ungar and M'Kirera, 2003; Dennis et al., 2004). Angularity values do differ between taxa (e.g., gorillas have more jagged occlusal surfaces than do chimpanzees), but do not change with wear until the teeth are extremely worn. It may then be that this is a functional aspect of occlusal morphology that is maintained through the wear sequence by natural selection.

Wear should sculpt teeth in predictable ways in part because of differences in hardness between enamel and dentin. If so, future research should focus on understanding relationships between enamel crown shape and the shape of the underlying dentin cap. New technologies, such as computerized X-ray microtomography, hold the promise of allowing us to work out these relationships (Figure 6; Bjørndal et al., 1999; Olejniczak et al., 2007 and references therein). This in turn should lead to a better

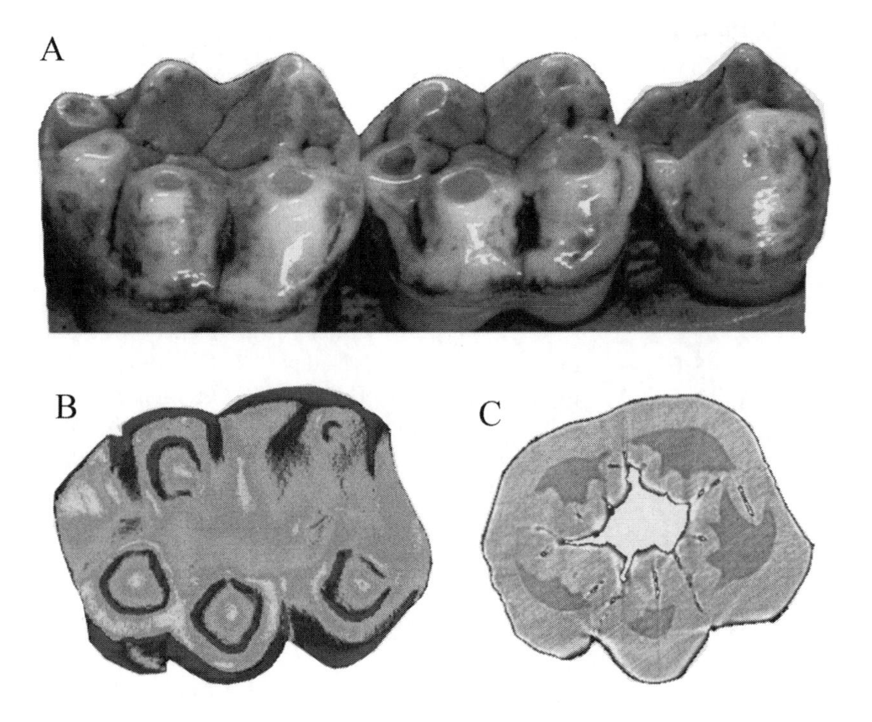

Figure 6. Enamel sculpting, dentin exposure and surface jaggedness. (A) Gorilla tooth with slight dentin exposure; (B) Gorilla occlusal surface slope map (steeper surfaces in darker shades); (C) μCT slice of a Neandertal molar (courtesy of Roberto Macchiarelli).

understanding of tooth form and function, and the complex nature of dental-dietary adaptations.

Acknowledgments

I thank Shara Bailey and Jean-Jacques Hublin for inviting me to participate in this volume. I am grateful to several colleagues for discussions of dental topographic analysis and/or early hominin paleoecology over the years, including John Dennis, Ken Glander, Fred Grine, Rich Kay, Fred Limp, Peter Lucas, Francis M'Kirera, Gildas Merceron, Alejandro Pérez-Pérez, Mike Plavcan, Jerry Rose, Rob Scott, Sarah Taylor, Mark Teaford, Alan Walker, Malcolm Williamson, John Wilson, and Lucy Zuccotti. Howling monkey data were collected with permission from the Board of Directors at Hacienda la Pacifica to Ken Glander and Mark Teaford. Access to museum specimens was granted by curators at the Cleveland Museum of Natural History, the State Collection of Anthropology and Palaeoanatomy in Munich, the Transvaal Museum, and the University of the Witwatersrand. Dental replicas were prepared with the help of Mark Teaford, Fred Grine, Alejandro Pérez-Pérez, and Ken Glander, and much of the data on the extant primates were originally generated by John Dennis, Francis M'Kirera, and Sarah Taylor. The μCT slice in Figure 6 is illustrated courtesy of Roberto Macchiarelli. Research described in this paper was funded in part by the University of Arkansas, the LSB Leakey Foundation, and the US National Science Foundation.

References

Aguirre de Enriquez, E., Blumenberg, B., Collins, D., Delson, E., Howells, W.W., Humphreys, A.J.B., Kress, J.H., Malik, S.C., Partridge, T.C., Poirier, F.E., Raemsch, B.E., Sharma, A., Tobias, P.V., Todd, N.B., Wolpoff, M.H., Zihlman, A., Butzer, K.W., Blumenberg, B., Tuttle, R., 1974. Discussion: recent thinking on human evolution. Current Anthropolgy 15, 398–426.

Anthony, M.R.L., Kay, R.F., 1993. Tooth form and diet in ateline and alouattine primates: reflections on the comparative method. American Journal of Science 293A, 356–382.

Archer, D., Sanson, G., 2002. Form and function of the selenodont molar in southern African ruminants in relation to their feeding habits. Journal of Zoology 257, 13–26.

Bjørndal, L., Carlsen, O., Thuesen, G., Darvann, T., Kreiborg, S., 1999. External and internal macrotomography in 3D-reconstructructed maxillary molars using computerized X-ray microtomography. International Endodontic. Journal 32, 3–9.

Clarke, M.R., Crockett, C.M., Zucker, E.L., Zaldivar, M., 2002. Mantled howler population of Hacienda La Pacifica, Costa Rica, between 1991 and 1998: effects of deforestation. American Journal of Primatology 56, 155–163.

Clarke, M.R., Glander, K.E., 1984. Female reproductive success in a group of free-ranging howling monkeys (*Alouatta palliata*) in Costa Rica. In: Small, M.F. (Eds.), Female Primates: Studies by Female Primatologists. Alan R. Liss, New York, pp. 111–126.

Clarke, M.R., Glander, K.E., Zucker, E.L., 1998. Infant-nonmother interactions of free-ranging mantled howlers (*Alouatta palliata*) in Costa Rica. International Journal of Primatology 19, 451–472.

Cope, E.D., 1883. On the trituberculate type of molar tooth in the Mammalia. Paleontological Bulletin Number 37. Proceedings of the American Philosophical Society 21, 324–326.

Crompton, A.W., Sita-Lumsden, A.G., 1970. Functional significance of therian molar pattern. Nature 227, 197–199.

Cuvier, G., 1863. *The Animal Kingdom.* London.

Daegling, D.J., Grine, F.E., 1990. Biomechanics of australopithecine mandibles from computed tomography. American Journal of Physical Anthropology 81, 211–211.

Demes, B., Creel, N., 1988. Bite force, diet, and cranial morphology of fossil hominids. Journal of Human Evolution 17, 657–670.

Dennis, J.C., 2002. Dental topography of *Alouatta palliata* (the Mantled Howling Monkey). M.A. Thesis, University of Arkansas.

Dennis, J.C., Ungar, P.S., Teaford, M.F., Glander, K.E., 2004. Dental topography and molar wear in *Alouatta palliata* from Costa Rica.

American Journal of Physical Anthropology 125, 152–161.

Doran, D.M., McNeilage, A., Greer, D., Bocian, C., Mehlman, P., Shah, N., 2002. Western lowland gorilla diet and resource availability: new evidence, cross-site comparisons, and reflections on indirect sampling methods. American Journal of Primatology 58, 91–116.

Du Brul, E.L., 1977. Early hominid feeding mechanisms. American Journal of Physical Anthropology 47, 305–320.

Eguchi, S., Townsend, G.C., Richards, L.C., Hughes, T., Kasai, K., 2004. Genetic contribution to dental arch size variation in Australian twins. Archives of Oral Biology 49, 1015–1024.

Fleagle, J.G., 1999. *Primate Adaptation and Evolution* (2nd Edition). Academic Press, New York.

Frazzetta, T.H., 1988. The mechanics of cutting and the form of shark teeth (Chondrichthyes, Elasmobranchii). Zoomorphology 108, 93–107.

Gembressi, V., 2004. Dental wear analysis on the species *Cebus apella* using geographic information systems technology. B.A. Thesis, Stony Brook University.

Glander, K.E., 1978. Howling monkey feeding behavior and plant secondary compounds: a study of strategies. In: Montgomery, G.G. (Eds.), The Ecology of Arboreal Folivores. Smithsonian Institution Press, Washington, D.C., pp. 561–573.

Glander, K.E., Fedigan, L.M., Fedigan, L., Chapman, C., 1991. Field methods for capture and measurement of three monkey species in Costa Rica. Folia Primatologica 57, 70–82.

Glander, K.E., Whitehead, J., Schon, M., Chapman, C., Clarke, M., Milton, K., Pope, T., Estrada, A., Crockett, C., 1987. Howling monkeys: past and present. International Journal of Primatology 8, 403–403.

Gregory, W.K., 1922. The origin and evolution of human dentition. Baltimore.

Grine, F.E., 1981. Trophic differences between 'gracile' and 'robust' australopithecines: a scanning electron microcope analysis of occlusal events. South African Journal of Science 77, 203–230.

Grine, F.E., 1986. Dental evidence for dietary differences in *Australopithecus* and *Paranthropus*: a quantitative analysis of permanent molar microwear. Journal of Human Evolution 15, 783–822.

Grine, F.E., 1989. New hominid fossils from the Swartkrans Formation (1979–1986 excavations): craniodental specimens. American Journal of Physical Anthropology 79, 409–449.

Grine, F.E., 2005. Early *Homo* at Swartkrans, South Africa: a review of the evidence and an evaluation of recently proposed morphs. South African Journal of Science 101, 43–52.

Grine, F.E., Kay, R.F., 1988. Early hominid diets from quantitative image analysis of dental microwear. Nature 333, 765–768.

Grine, F.E., Ungar, P.S., Teaford, M.F., 2002. Error rates in dental microwear quantification using scanning electron microscopy. Scanning 24, 144–153.

Groves, C.P., Napier, J.R., 1968. Dental dimensions and diet in australopithecines. Proc. VIII Int. Cong. Anthrop. Ethnological Science 3, 273–276.

Hlusko, L.J., 2004. Protostylid variation in *Australopithecus*. Journal of Human Evolution 46, 579–594.

Hunter, J., 1771. *The Natural History of Human Teeth*. London.

Jernvall, J., Keranen, S.V.E., Thesleff, I., 2000. Evolutionary modification of development in mammalian teeth: quantifying gene expression patterns and topography. Proceedings of the National Academy of Sciences of the USA 97, 14444–14448.

Jernvall, J., Selänne, L., 1999. Laser confocal microscopy and geographic information systems in the study of dental morphology. Paleontologia Electronica 2, 18 pp.

Jolly, C.J., 1970. The seed-eaters: a new model of hominid differentiation based on a baboon analogy. Man 5, 5–26.

Kay, R.F., 1975a. Allometry and early hominids (comment). Science 189, 63–63.

Kay, R.F., 1975b. Functional adaptations of primate molar teeth. American Journal of Physical Anthropology 43, 195–215.

Kay, R.F., 1977. Evolution of molar occlusion in Cercopithecidae and early catarrhines. American Journal of Physical Anthropology 46, 327–352.

Kay, R.F., 1978. Molar structure and diet in extant Cercopithecidae. In: Butler, P.M., Joysey, K.A. (Eds.), The Development, Function, and Evolution of Teeth. Academic Press, New York, pp. 309–339.

Kay, R.F., 1981. The nut-crackers: a new theory of the adaptations of the Ramapithecinae. American Journal of Physical Anthropology 55, 141–151.

Kay, R.F., 1984. On the use of anatomical features to infer foraging behavior in extinct primates. In: Rodman, P.S., Cant, J.G.H. (Eds.), Adaptations for Foraging in Nonhuman Primates: Contributions to an Organismal Biology of Prosimians, Monkeys and Apes. Columbia University Press, New York, pp. 21–53.

Kay, R.F., 1985. Dental evidence for the diet of *Australopithecus*. Annual Review of Anthropology 14, 315–341.

Kay, R.F., Covert, H.H., 1984. Anatomy and behavior of extinct primates. In: Chivers, D.J., Wood, B.A., Bilsborough, A. (Eds.), Food Acquisition and Processing in Primates. Plenum Press, New York, pp. 467–508.

Kay, R.F., Hiiemae, K.M., 1974. Jaw movement and tooth use in recent and fossil primates. American Journal of Physical Anthropology 40, 227–256.

Kay, R.F., Simons, E.L., 1980. The ecology of Oligocene African Anthropoidea. International Journal of Primatology 1, 21–37.

Kay, R.F., Ungar, P.S., 1997. Dental evidence for diet in some Miocene catarrhines with comments on the effects of phylogeny on the interpretation of adaptation. In: Begun, D.R., Ward, C., Rose, M. (Eds.), Function, Phylogeny and Fossils: Miocene Hominoids and Great Ape and Human Origins. Plenum Press, New York, pp. 131–151.

Keyser, A.W., 2000. The Drimolen skull: The most complete australopithecine cranium and mandible to date. South African Journal of Science 96, 189–193.

King, S.J., Blanco, M.B., Godfrey, L.R., 2005a. Dietary reconstruction of *Archeolemur* using dental topographic analysis. American Journal of Physical Anthropology Suppl. 40, 133.

King, S.J., Arrigo-Nelson, S.J., Pochron, S.T., Semprebon, G.M., Godfrey, L.R., Wright, P.C., Jernvall, J., 2005b. Dental senescence in a long-lived primate links infaut survival to rainfall. Proceedings of the National Academy of science of the USA 102, 16579-16583.

Kinzey, W.G., 1978. Feeding behavior and molar features in two species of titi monkey. In: Chivers, D.J., Herbert, J. (Eds.), Recent Advances in Primatology, Volume 1, Behavior. Academic Press, New York, pp. 373–385.

Kullmer, O., Engel, K., Ulhaas, L., Winzen, O., Schrenk, F., 2004. Occlusal fingerprint analysis (OFA): quantifying individual wear pattern of tooth crowns using optical 3-D topometry. American Journal of Physical Anthropology Suppl. 38, 130.

Kuroda, S., 1992. Ecological interspecies relationships between gorillas and chimpanzees in the Ndoki-Nouabale Reserve, Northern Congo. In: Itoigawa, N., Sugiyama, Y., Sackett, G.P., Thompson, R.K.R. (Eds.), Topics in Primatology, Volume 2, Behavior, Ecology and Conservation. University of Tokyo Press, Tokyo, pp. 385–394.

Lambert, J.E., Chapman, C.A., Wrangham, R.W., Conklin-Brittain, N.L., 2004. The hardness of cercopithecine foods: implications for the critical function of enamel thickness in exploiting fallback foods. American Journal of Physical Anthropology 125, 363–368.

Lanyon, J.M., Sanson, G.D., 1986. Koala (*Phascolarctos cinereus*) dentition and nutrition. II. Implications of tooth wear in nutrition. Journal of Zoology 209, 169–181.

Lucas, P.W., 2004. *Dental Functional Morphology: How Teeth Work.* Cambridge University Press, New York.

Lucas, P.W., Corlett, R.T., Luke, D.A., 1985. Plio-Pleistocene hominid diets: an approach combining masticatory and ecological analysis. Journal of Human Evolution 14, 187–202.

Lucas, P.W., Prinz, J.F., Agrawal, K.R., Bruce, I.C., 2004. Food texture and its effect on ingestion, mastication and swallowing. Journal of Texture. 35, 159–170.

M'Kirera, F., Ungar, P.S., 2003. Occlusal relief changes with molar wear in *Pan troglodytes troglodytes* and *Gorilla gorilla gorilla*. American Journal of Primatology 60, 31–41.

MacKinnon, J., 1977. A comparative ecology of Asian apes. Primates 18, 747–772.

Martin-de las Heras, S., Valenzuela, A., Ogayar, C., Valverde, A.J., Torres, J.C., 2005. Computer-based production of comparison overlays from 3D-scanned dental casts for bite mark analysis. Journal of Forensic Science 50, 127–133.

McHenry, H.J., 1988. New estimates of body weight in early hominids and their significance to encephalization and megadontia in 'robust' australopithecines. In: Grine, F.E. (Ed.), Evolutionary History of the Robust Australopithecines. Aldine, New York, pp. 133–148.

Meldrum, D.J., Kay, R.F., 1997. *Nuciruptor rubricae*, a new pitheciin seed predator from the Miocene of Colombia. American Journal of Physical Anthropology 102, 407–427.

Nishihara, T., 1992. A preliminary report on the feeding habits of western lowland gorillas (*Gorilla gorilla gorilla*) in the Ndoki Forest, Northern Congo. In: Itoigawa, N., Sugiyama, Y., Sackett, G.P., Thompson, R.K.R. (Eds.), Topics in Primatology, Volume 2, Behavior, Ecology and Conservation. University of Tokyo Press, Tokyo, pp. 225–240.

Okada, M., Sadaki, N., Kaihara, Y., Okasa, R., Amano, H., Miura, K., Kozai, K., 2003. Oral findings in Noonan syndrome: report of a case. Journal of Oral Sciences 45, 117–121.

Olejniczak, A.J., Grine, F.E., Martin, L.B., 2007. Micro-computed tomography of primate molars: methodological aspects of three-dimensional data collection. In: Bailey, S.E., Hublin, J.-J. (Eds.), Dental Perspectives on Human Evolution: State of the Art Research in Dental Paleoanthropology. Springer, Dordrecht.

Osborn, H.F., 1907. *Evolution of Mammalian Molar Teeth to and from the Triangular Type.* The MacMillan Company, New York.

Owen, R., 1840–1841. *Odontography.* Hippolyte Bailliere, London.

Peters, C.R., 1981. Robust vs. gracile early hominid masticatory capabilities: the advantages of the megadonts. In: Mai, L.L., Shanklin, E., Sussman, R.W. (Eds.), The Perception of Human Evolution. University of California Press, Los Angeles, pp. 161–181.

Pilbeam, D., Gould, S.J., 1974. Size and scaling in human evolution. Science 186, 892–901.

Rak, Y., 1983. *The Australopithecine Face.* Academic Press, New York.

Remis, M.J., 1997. Western lowland gorillas (*Gorilla gorilla gorilla*) as seasonal frugivores: use of variable resources. American Journal of Primatology 43, 87–109.

Remis, M.J., 2002. Food preferences among captive western gorillas (*Gorilla gorilla gorilla*) and chimpanzees (*Pan troglodytes*). International Journal of Primatology 23, 231–249.

Robinson, B.W., Wilson, D.S., 1998. Optimal foraging, specialization, and a solution to Liem's paradox. American Naturalist 151, 223–235.

Robinson, J.T., 1954. Prehominid dentition and hominid evolution. Evolution 8, 324–334.

Robinson, J.T., 1963. Adaptive radiation in the australopithecines and the origin of man. In: Howell, F.C., Bourliere, F. (Eds.), African Ecology and Human Evolution. Aldine de Gruyter, Chicago, pp. 385–416.

Rodman, P.S., 1977. Feeding behaviour of orangutans of the Kutai Nature Reserve, East Kalimantan. In: Clutton-Brock, T.H. (Eds.), Primate Ecology: Studies of Feeding and Ranging Behaviour in Lemurs, Monkeys and Apes. Academic Press, London, pp. 383–413.

Rogers, M.E., Maisels, F., Williamson, E.A., Tutin, C.E., Fernandez, M., 1992. Nutritional aspects of gorilla food choice in the Lopé Reserve, Gabon. In: Itoigawa, N., Sugiyama, Y., Sackett, G.P., Thompson, R.K.R. (Eds.), Topics in Primatology, Volume 2, Behavior, Ecology and Conservation. University of Tokyo, Tokyo, pp. 267–281.

Rosenberger, A.L., Kinzey, W.G., 1976. Functional patterns of molar occlusion in platyrrhine primates. American Journal of Physical Anthropology 45, 281–297.

Ross, C.F., 2000. Into the light: the origin of Anthropoidea. Annual Review of Anthropology 29, 147–194.

Sakka, M., 1984. Cranial morphology and masticatory adaptations. In: Chivers, D.J., Wood, B.A., Bilsborough, A. (Eds.), Food Acquisition and Processing in Primates. Plenum, New York, pp. 415–427.

Scott, E.C., 1979. Dental wear scoring technique. American Journal of Physical Anthropology 51, 213–217.

Scott, R.S., Ungar, P.S., Bergstrom, T.S., Brown, C.A., Grine, F.E., Teaford, M.F., Walker, A., 2005. Dental microwear texture analysis reflects diets of living primates and fossil hominins. Nature 436, 693–695.

Seligsohn, D., Szalay, F.S., 1978. Relationship between natural selection and dental morphology: tooth function and diet in *Lepilemur* and *Hapalemur*. In: Butler, P.M. Joysey, K.A. (Eds.), Development, Function and Evolution of Teeth. Academic Press, New York, pp. 289–307.

Smith, E., 1999. A functional analysis of molar morphometrics in living and fossil hominoids using 2-D digitized images. Ph.D. Dissertation, University of Toronto.

Spears, I.R., Crompton, R.H., 1996. The mechanical significance of the occlusal geometry of great ape molars in food breakdown. Journal of Human Evolution 31, 517–535.

Stanford, C.B., Nkurunungi, J.B., 2003. Do wild chimpanzees and mountain gorillas compete for food? American Journal of Physical Anthropology Supplement 36, 198–199.

Strait, S.G., 1993. Molar morphology and food texture among small bodied insectivorous mammals. Journal of Mammalogy 74, 391–402.

Szalay, F.S., 1975. Hunting-scavenging protohominids: a model for hominid origins. Man 10, 420–429.

Teaford, M.F., 1983. Differences in molar wear gradient between adult macaques and langurs. International Journal of Primatology 4, 427–444.

Teaford, M.F., 2007. Micro-computed tomography of primate molars: methodological aspects of dental microwear and paleoanthropology: cautions and possibilities three-dimensional data collection. In: Bailey, S.E., Hublin, J.-J. (Eds.), Dental Perspectives on Human Evolution: State of the Art Research in Dental Paleoanthropology. Springer, Dordrecht.

Teaford, M.F., Glander, K.E., 1991. Dental microwear in live, wild-trapped *Alouatta palliata* from Costa Rica. American Journal of Physical Anthropology 85, 313–319.

Teaford, M.F., Glander, K.E., 1996. Dental microwear and diet in a wild population of mantled howling monkeys (*Alouatta palliata*). In: Norconk, M.A., Rosenberger, A.L., Garber, P.A. (Eds.), Adaptive Radiations of Neotropical Primates. Plenum Press, New York, pp. 433–449.

Teaford, M.F., Lucas, P.W., Ungar, P.S., Glander, K.E., 2006. Mechanical defenses in leaves eaten by Costa Rican *Alouatta palliata*. American Journal of Physical Anthropology 129, 99–104.

Teaford, M.F., Ungar, P.S., 2000. Diet and the evolution of the earliest human ancestors. Proceedings of the National Academy of Sciences of the USA 97, 13506–13511.

Teaford, M.F., Ungar, P.S., 2007. Dental adaptations of African apes. In: Kenke, W., Rothe, W., Tattersall, I. (Eds.), Handbook of Paleoanthropology. Volume 1: Principles, Methods, and Approaches. Springer Verlag, Heidelberg, in press.

Teaford, M.F., Ungar, P.S., Grine, F.E., 2002. Paleontological evidence for the diets of African Plio-Pleistocene hominins with special reference to early *Homo*. In: Ungar, P.S., Teaford, M.F (Eds.), Human Diet: Its Origin and Evolution. Bergin and Garvey, Westport, CT, pp. 143–166.

Tobias, P.V., 1967. *The Cranium and Maxillary Dentition of Australopithecus (Zinjanthropus) boisei*. Cambridge.

Tutin, C.E.G., Fernandez, M., 1993. Composition of the diet of chimpanzees and comparisons with that of sympatric lowland gorillas in the Lopé Reserve, Gabon. American Journal of Primatology 30, 195–211.

Tutin, C.E.G., Fernandez, M., Rogers, M.E., Williamson, E.A., Mcgrew, W.C., 1991. Foraging profiles of sympatric lowland gorillas and chimpanzees in the Lopé Reserve, Gabon. Philosophical Transactions of the Royal Society of London B 334, 179–186.

Tutin, C.E.G., Ham, R.M., White, L.J.T., Harrison, M.J.S., 1997. The primate community of the Lopé Reserve, Gabon: Diets, responses to fruit scarcity, and effects on biomass. American Journal of Primatology 42, 1–24.

Ungar, P.S., 2002. Reconstructing the diets of fossil primates. In: Plavcan, J.M., Kay, R.F., Junger, W.L., van Schaik, C.P. (Eds.), Reconstructing Behavior in the Primate Fossil Record. Kluwer Academic / Plenum Publishers, New York, pp. 261–296.

Ungar, P.S., 2004. Dental topography and diets of *Australopithecus afarensis* and early *Homo*. Journal of Human Evolution 46, 605–622.

Ungar, P.S., 2005. Dental evidence for the diets of fossil primates from Rudabánya, northeastern Hungary with comments on extant primate analogs and "noncompetitive" sympatry. Palaeontographica Italiana 90, 97-111.

Ungar, P.S., 2007. Dental functional morphology: the known, the unknown and the unknowable. In: Ungar, P.S. (Eds.), Early Hominin Diets: The Known, The Unknown and the Unknowable. Oxford University Press, Oxford, pp. 39-55.

Ungar, P.S., Brown, C.A., Bergstrom, T.S., Walkers, A., 2003. Quantification of dental microwear by tandem scanning confocal microscopy and scale-sensitive fractal analyses. Scanning 25, 185–193.

Ungar, P.S., Grine, F.E., 1991. Incisor size and wear in *Australopithecus africanus* and *Paranthropus robustus*. Journal of Human Evolution 20, 313–340.

Ungar, P.S., Kay, R.F., 1995. The dietary adaptations of European Miocene catarrhines. Proceedings of the National Academy of Sciences of the USA 92, 5479–5481.

Ungar, P.S., M'Kirera, F., 2003. A solution to the worn tooth conundrum in primate functional anatomy. Proceedings of the National Academy of Sciences of the USA 100, 3874–3877.

Ungar, P.S., Taylor, S.R., 2005. Dental topographic analysis: tooth wear and function. American

Journal of Physical Anthropology Suppl. 40, 210.

Ungar, P.S., Teaford, M.F., Grine, F.E., 1999. A preliminary study of molar occlusal relief in *Australopithecus africanus* and *Paranthropus robustus*. American Journal of Physical Anthropology Suppl. 28, 269.

Ungar, P.S., Teaford, M.F., Kay, R.F., 2004. Molar microwear and shearing crest development in Miocene catarrhines. Anthropos 42, 21–35.

Ungar, P.S., Williamson, M., 2000. Exploring the effects of tooth wear on functional morphology: a preliminary study using dental topographic analysis. Paleontologica Electronica. 3, 18pp.

Ward, S.C., Molnar, S., 1980. Experimental stress analysis of topographic diversity in early hominid gnathic morphology. American Journal of Physical Anthropology 53, 383–395.

Williams, B.A., Covert, H.H., 1994. New early Eocene anaptomorphine primate (Omomyidae) from the Washakie Basin, Wyoming, with comments on the phylogeny and paleobiology of anaptomorphines. American Journal of Physical Anthropology 93, 323–340.

Williamson, E.A., Tutin, C.E.G., Rogers, M.E., Fernandez, M., 1990. Composition of the diet of lowland gorillas at Lopé in Gabon. American Journal of Primatology 21, 265–277.

Williamson, M.D., Ungar, P.S., Teaford, M.F., Glander, K.E., 2000. Gross wear and molar morphology in *Alouatta palliata*: a preliminary study using dental topographic analysis. American Journal of Physical Anthropology Supplement 30, 323.

Wood, B., Strait, D., 2004. Patterns of resource use in early *Homo* and *Paranthropus*. Journal of Human Evolution 46, 119–162.

Wood, B.A., Abbott, S.A., Graham, S.H., 1983. Analysis of the dental morphology of Plio-Pleistocene hominids. II. Mandibular molars. Study of cusp areas, fissure pattern and cross-sectional shape of the crown. Journal of Anatomy 137, 287–314.

Wood, B.A., Ellis, M., 1986. Evidence for dietary specialization in the "robust" australopithecines. Anthropos 23, 101–124.

Wood, B.A., Stack, C.G., 1980. Does allometry explain the differences between gracile and robust australopithecines? American Journal of Physical Anthropology 52, 55–62.

Yamashita, N., 1998. Functional dental correlates of food properties in five Malagasy lemur species. American Journal of Physical Anthropology 106, 169–188.

Zuccotti, L.F., Williamson, M.D., Limp, W.F., Ungar, P.S., 1998. Technical note: Modeling primate occlusal topography using geographic information systems technology. American Journal of Physical Anthropology 107, 137–142.

4. Dental microwear and Paleoanthropology: Cautions and possibilities

M.F. TEAFORD

Center for Functional Anatomy & Evolution
Johns Hopkins University School of Medicine
1830 East Monument Street, Room 303
Baltimore, MD 21205 USA
mteaford@jhmi.edu

Keywords: dental microwear, diet, australopithecines, primates, scanning electron microscopy, SEM, confocal microscopy

Abstract

Fifty years ago, investigators realized they could gain insights into jaw movement and tooth-use through light-microscope analyses of wear patterns on teeth. Since then, numerous analyses of modern and fossil material have yielded insights into the evolution of tooth use and diet in a wide variety of animals. However, analyses of fossils and archeological material are ultimately dependent on data from three sources, museum samples of modern animals, living animals (in the wild or in the lab), and *in vitro* studies of microwear formation. These analyses are not without their problems. Thus, we are only *beginning* to get a clearer picture of the dental microwear of the early hominins. Initial work suggested qualitative differences in dental microwear between early hominids, but it wasn't until Grine's analyses of the South African australopithecines that we began to see quantitative, statistical evidence of such differences. Recent analyses have (1) reaffirmed earlier suggestions that *Australopithecus afarensis* shows microwear patterns indistinguishable from those of the modern gorilla, and (2) shown that the earliest members of our genus may also be distinguishable from each other on the basis of their molar microwear patterns. While this work hints at the possibilities of moving beyond standard evolutionary-morphological inferences, into inferences of actual differences in tooth use, we still know far too little about the causes of specific microwear patterns, and we know surprisingly little about variations in dental microwear patterns (e.g., between sexes, populations, and species). In the face of such challenges, SEM-analyses may be reaching the limits of their usefulness. Thus, two methods are beginning to catch attention as possible "next steps" in the evolution of dental microwear analyses. One technique involves a return to lower magnification analyses, using qualitative assessments of microwear patterns viewed under a light microscope. The advantages of these analyses are that they are cheap and fast, and may easily distinguish animals with extremely different diets. The disadvantages are that they are still subjective and may not be able to detect subtle dietary differences or artifacts on tooth surfaces. Another technique involves the use of scale-sensitive fractal analyses of data from a confocal microscope. Advantages include the ability to quickly and objectively characterize wear surfaces in 3D over entire wear facets. The main disadvantage lies in the newness of the technique and challenges imposed by developing such cutting edge technology. With the development of new approaches, we may be able to take dental microwear analyses to a new level of inference.

S.E. Bailey and J.-J. Hublin (Eds.), Dental Perspectives on Human Evolution, 345–368.

Introduction

Paleoanthropologists are, in many ways, like forensic scientists who travel through time. They must use any available clues to help decipher what went on, eons ago; and only by considering the total range of evidence can they begin to appreciate the limits of what can be said about past behaviors. Unfortunately, much of the evidence available to paleoanthropologists is not *direct* evidence, in the sense of something visible on a bone or tooth, caused directly by something that happened during the individual's lifetime. For instance, the relative size of certain bones may or may not be indicative of what an animal actually did, as the animal may have, for example, relatively long hindlimbs simply because its ancestors had relatively long hindlimbs. So, when looking through the evidence, paleoanthropologists are constantly forced to evaluate their data, to see what they can, and cannot, say about the hypotheses being tested.

The most common elements in the human fossil record are teeth – largely because they are the most resilient structures in the body. For the most part, they are made of inorganic materials, and they tend to remain intact well after death. Thus, it is perhaps no surprise they have provided many clues about the paleobiology of our ancestors. For instance, analyses of tooth shape have shown that species adapted to eat tough, elastic foods generally have longer molar shearing crests than do species adapted to eat hard and brittle foods (Kay, 1975; Kay and Hylander, 1978; Lucas, 1979, 2004). However, most of these studies have focused on analyses of unworn teeth (see Ungar, 2004, 2007 for a revolutionary new perspective on this topic). Yet, like death and taxes, tooth wear is one of life's inevitabilities. As soon as a tooth reaches occlusion, it begins to wear down. In some cases, such as in guinea pigs, wear even begins *in utero* (Ainamo, 1971; Teaford and Walker, 1983). Its first steps are imperceptible to the naked eye – microscopic scratches and pits nicking the surface. But those microscopic effects add up, leading to the formation of wear facets on the teeth, and eventually dentin exposure, as the overlying enamel is worn away. So, while the shape of unworn teeth can tell us a great deal about what a tooth is *capable* of processing, tooth *wear* can give us insights into how a tooth was actually used. This paper will focus on the evidence provided by microscopic wear patterns on the chewing surfaces of teeth – what is often referred to as dental microwear analyses. This is different from most other analyses of fossils, because it is *direct* evidence of past behavior – ultimately based on microscopic wear caused by food or abrasives on food during an animal's lifetime. As a result, this technique has the potential to yield information about prehistoric diet and tooth use at a unique level of resolution.

Postmortem Wear

One of the first questions that springs to mind in contemplating dental microwear analyses of fossils is: if a tooth has been lying in the ground for thousands or millions of years, how do we know that the wear on it was really caused during the animal's lifetime? Actually, it is surprisingly easy (Teaford, 1988b), because the wear patterns caused during chewing are laid down in regular patterns at specific locations on teeth (see Figure 1a). By contrast, when a tooth is buried in the ground it is subjected to wear at innumerable, unusual locations and angles (see Figure 1b) (Puech et al., 1985; Teaford, 1988b; King et al., 1999b). This so-called postmortem wear is certainly a problem when analyzing fossils – but generally not because we cannot recognize it. Instead, it is a problem because we *can* recognize it, and have to eliminate many specimens from our analyses.

Obviously, the degree of postmortem wear can be a function of many factors, such as the length of time a specimen has been exposed

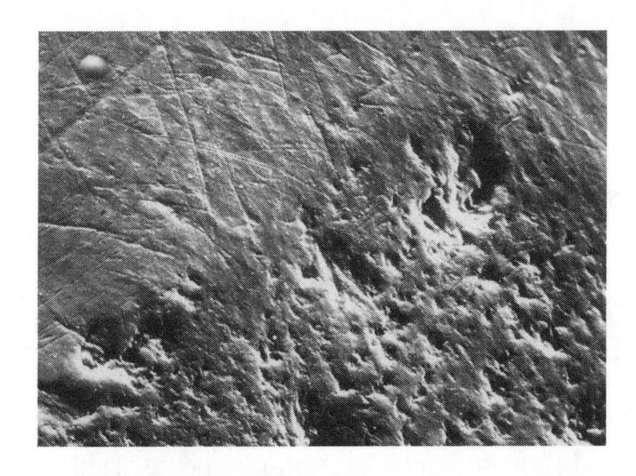

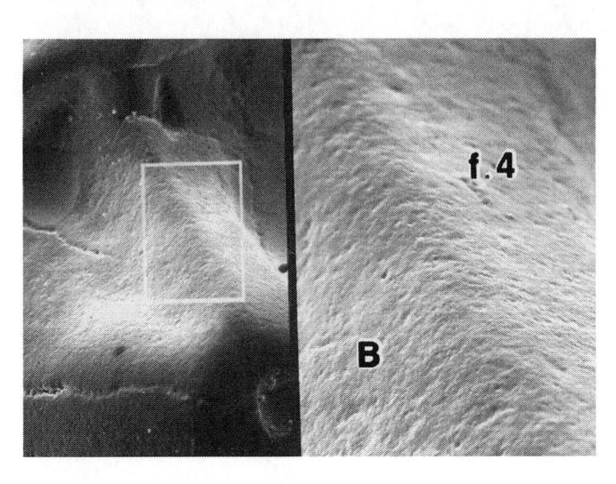

Figure 1. a. SEM micrograph of occlusal and nonocclusal surfaces of a molar of *Cebus apella* (from Teaford, 1988b). The boundary between surfaces curves diagonally across the image so that the lightly scratched nonocclusal surface is in the top third of the micrograph. b. Postmortem abrasive wear on 50 million year old *Cantius* molar (from Teaford, 1988b). B = buccal side of metacone, and f.4 = facet 4 on mesial occlusal aspect of metacone. Thus the boundary between occlusal and nonocclusal surfaces passes diagonally from the top left to the bottom right of the higher magnification image on the right. Identical pitting on both surfaces indicates that both have been subjected to postmortem wear.

to the elements; the presence of destructive acids in the postdepositional environment; whether or not the tooth was excavated or collected on the surface; how it was prepared in cleaning; the types of preservative applied to its surface, etc. As a result, the proportion of fossil specimens useful for dental microwear analyses may vary dramatically from site-to-site – e.g., less than 25 % at Koobi Fora, or more than 60 % at Olduvai (personal observations).

Brief History of Dental Microwear Analyses

Assuming we can recognize the effects of postmortem wear, how has dental microwear traditionally been analyzed? Initial analyses were qualitative in nature and based on light microscope assessments of tooth surfaces. For instance, Butler and Mills (Butler, 1952; Mills, 1955, 1963) noticed characteristic orientations of scratches on the teeth of different mammals, providing the initial evidence for two "phases" of jaw movement in primates. Similarly, Dahlberg and Kinzey (1962) noted the possibility of documenting differences in diet based on (among other things) differences in the amount of microscopic scratching on teeth. Subsequent work by a number of people rekindled interest in the topic (Walker, 1976; Puech, 1977; Rensberger, 1978; Walker et al., 1978; Puech and Prone, 1979; Ryan, 1979; Puech et al., 1980; Walker, 1980, 1981; Grine, 1981; Puech et al., 1981; Ryan, 1981; Rensberger 1982), as workers generally shifted to using the scanning electron microscope, due to its superior depth of focus and resolution of detail. Of course, finer microscopic resolution raised the possibility of finer dietary distinctions – as long as that information could be put to efficient use. This led numerous workers to begin quantifying dental microwear, by counting the incidence of scratches and pits, measuring their length and width, and attempting to measure their orientation, using various forms of computer-controlled digitizers or calipers in conjunction with SEM micrographs or enlarged prints of them (e.g., Fine and Craig, 1981; Gordon, 1982, 1984b, c; Teaford and Walker, 1984;

Teaford, 1985; Grine, 1986; Kelley, 1986; Solounias et al., 1988; Young and Robson, 1987). However, as the number of studies began to grow, it quickly became apparent that many people were using very different methods to measure dental microwear. This presented researchers with an array of methodological difficulties (Covert and Kay, 1981; Gordon, 1982; Gordon and Walker, 1983; Kay and Covert, 1983; Gordon, 1984b, 1988; Teaford, 1988a), some of which are still haunting us (see "cautions" below). In an attempt to standardize techniques, Ungar developed a "semi-automated" procedure for measuring dental microwear (Ungar et al., 1991; Ungar, 1995), a method that is still used by many researchers today. To use it, images need to be stored in a specific digital format so they can be opened by a freeware package known as "Microware" (http://comp.uark.edu/~pungar/). Given the complexity of SEM micrographs, each microwear feature within each micrograph still needs to be identified and "measured" by the researcher, using a mouse and cursor on a computer screen. So, while the technique provides a standardized series of measurements for analysis and stores them in a format readily accessible to most statistical packages, the work is still very time-consuming. The availability of this standardized technique, however, prompted work on an even wider range of taxa. Unfortunately, initial attempts to take analyses one step further, using a combination of image processing and image analysis (Kay, 1987; Grine and Kay, 1988), were limited by the capabilities of the software at the time.

Review of Studies of Modern Material

Given the number of potential avenues of investigation, it is best to review the different approaches before returning to some of the methodological challenges facing dental microwear analyses.

Analyses of Museum Material

Analyses of mammalian teeth from museum collections have always served as a major source of information, by demonstrating correlations between certain diets, or patterns of tooth use, and certain microwear patterns. These correlations depend on which teeth are analyzed, because anterior teeth are used differently than posterior teeth, with the incisors and canines being used to *ingest* food, and the premolars and molars being used to *chew* food once it has been ingested.

Analyses of incisor microwear have yielded two basic conclusions. First, animals that use their incisors very heavily in the ingestion of food show higher densities of incisal microwear features (Ryan, 1981; Kelley, 1986, 1990; Ungar, 1990, 1994). Second, the orientation of striations on the incisors reflects the direction of preferred movement of food (or other items) across the incisors (Walker, 1976; Rose et al., 1981; Ryan, 1981; Ungar, 1994). Thus, for example, the orang-utan, which generally uses its incisors a great deal in preparing food, shows more scratches on its incisors than does the gibbon, and those scratches often run in a more mesiodistal direction, reflecting a tendency to pull branches mesiodistally between the front teeth (Ungar, 1994). Analyses of incisor microwear have also yielded an interesting insight that may be more generally applicable – i.e., that the size of abrasives may be reflected in the size of microscopic scratches on the teeth, and that this, in turn, may be indicative of feeding height in the canopy, as phytoliths in leaves are generally larger than the abrasive particles in clay-based soils. (Ungar, 1990, 1994).

Further back in the mouth, analyses of molar microwear have demonstrated a few more points. Following in the footsteps of the earliest dental microwear analyses, correlations between orientations of jaw movement and scratches on mammalian molars continue to yield insights into chewing in a variety of mammalian species (Gordon, 1984c; Rensberger, 1986; Young and Robson, 1987; Hojo, 1996). More recent work has demonstrated that grazers tend to show more microscopic scratches on their molars as compared to browsers (Solounias and Moelleken, 1992a, b; Solounias and Hayek, 1993; Solounias and Moelleken, 1994; MacFadden et al., 1999), and animals that eat hard objects usually show large pits on their molars, while leaf-eaters tend to have relatively more scratches than pits on their molar enamel (Teaford and Walker, 1984; Teaford, 1988a) (Figure 2). Those "hard objects" can evidently include hard nuts, but also smaller items like insect exoskeletons (Strait, 1993; Silcox and Teaford, 2002). Microwear is also found on the buccal or lingual ("nonocclusal") surfaces of molars, which may give additional indications of the abrasiveness of the diet, the size of food items, or even the degree of terrestriality (Puech, 1977; Lalueza Fox, 1992; Lalueza Fox and Pérez-Pérez, 1993; Ungar and Teaford, 1996). Even more interestingly, museum analyses of molar microwear have yielded glimpses of subtler differences associated with dietary variation. Differences have ranged from those between closely-related genera (Solounias and Hayek, 1993; Teaford, 1993; Daegling and Grine, 1999; Oliveira, 2001), to those between subspecies (e.g., *Gorilla gorilla berengei* vs. *G.g. gorilla*) (King et al., 1999a), to those between populations within the same species (e.g., *Cebus nigrivitattus*) (Teaford and Robinson, 1989). Obviously, such analyses are only as good as the dates and locations of collection for the museum samples, and the published dietary information for those species. For instance, there are very few collections that provide the exact date and precise location of collection for each specimen, with some having little more than "British East Africa" for the location and

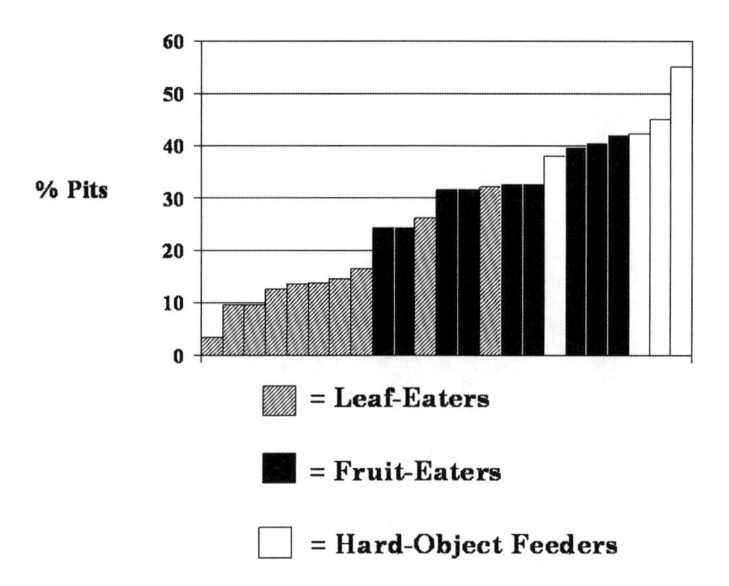

Figure 2. Histogram of the incidence of pitting on molars in primates with different diets (data from Rafferty et al., 2002, Teaford, 1988a, 1993, and Teaford and Runestad, 1992). Leaf-eaters = (left to right in figure) *Gorilla gorilla beringei, Theropithecus gelada, Colobus guereza, Procolobus badius, Nasalis larvatus, Allouatta palliata, Semnopithecus entellus, Alouatta seniculus, Procolobus verus*, and *Presbytis aygula*. Fruit-eaters = (left to right in figure) *Pan troglodytes, Papio cyncocephalus, Ateles belzebuth, Saimiri sciureus, Aotus trivirgatus, Macaca fascicularis, Pithecia pithecia, Mandrillus sphinx*, and *Cercopithecus nictitans*. Hard-object feeders = (left to right in figure) *Chiropotes satanus, Pongo pygmaeus, Cebus apella*, and *Lophocebus albigena*.

the name of the expedition for the date. As a result, finer-resolution studies of diet and dental microwear based on museum samples are relatively rare.

Analyses of Live Primates

Unfortunately, double-checking correlations between dental microwear and diet in live animals, in the lab or in the wild, is not easy either. In fact, keeping animals in a laboratory setting is extremely difficult and expensive. Moreover, since the animals have to be anesthetized to make copies of their teeth, the exact timing and type of anesthesia is often a matter of discussion and debate, as most veterinarians prefer to stick with "tried-&-true" methods (e.g., the use of ketamine administered after 8–12 hours of fasting) which leave the animals rigidly hard to work with, salivating excessively, and with thick organic films built-up on their teeth. As a result, it is perhaps not surprising that the only successful study using laboratory animals to-date is one from the 1980s. Teaford and Oyen (1989a, b) raised a group of vervet monkeys on hard and soft diets to check for the effects of food properties on cranio-facial growth. As the hard diet consisted of monkey chow and apples, and the soft diet water-softened monkey chow and pureed applesauce, you might expect the effects on the teeth to be relatively similar as both diets had the same basic ingredients and were very abrasive. However, there were two surprising differences. First, the incisors of the soft-food animals were more heavily worn than those of the hard-food animals, because the former were routinely rubbing handfuls of food across their incisors, whereas the latter were hardly using their incisors at all. Second, in the molar region, animals on the soft diet showed smaller pits on the occlusal surfaces, perhaps due to adhesive wear caused by repeated tooth-tooth contacts in chewing. The laboratory study also reaffirmed what had been noted in museum studies: that molar facets used for shearing or crushing showed different microwear patterns. Finally, the laboratory study also showed that the turnover in dental microwear could be quite rapid in animals with an abrasive diet, as all of the microwear features in an area sampled by an SEM micrograph would change in 1–2 weeks, depending on whether the animal was raised on the hard or soft diet (Figure 3).

A more feasible option, for studies of live primates, might involve the use of human volunteers fed specific food items. However, regulations concerning the use of human subjects make such work difficult, if external funding is to be sought, and despite the amount of *other* research done on dental patients, surprisingly little work has involved the use of dental microwear (e.g., Morel et al., 1991). A pilot study by Noble and Teaford (1995) using American foods normally thought to be hard or abrasive did reaffirm that few foods in our diet (e.g., popcorn kernels) scratch enamel. From a different perspective, rates of microscopic wear (Teaford and Tylenda, 1991) have also been used to gain insights into dental clinical problems, for instance, monitoring the incidence of tooth-grinding in patients with various symptoms of temporomandibular joint disease (Raphael et al., 2003). Otherwise, remaining work has focused primarily on the wear of dental materials, where the presence/absence of specific materials has, for instance, been shown to change rates of wear of certain dental restorative materials (e.g., Turssi et al., 2005; Wu et al., 2005), where the use of certain dental clinical procedures has been shown to cause certain types of microscopic wear (e.g., Plagmann et al., 1989; Östman-Andersson et al., 1993) and various forms of "microabrasion" have been shown to aid in the whitening of teeth (Allen et al., 2004; Chafaie, 2004; Bezerra et al., 2005).

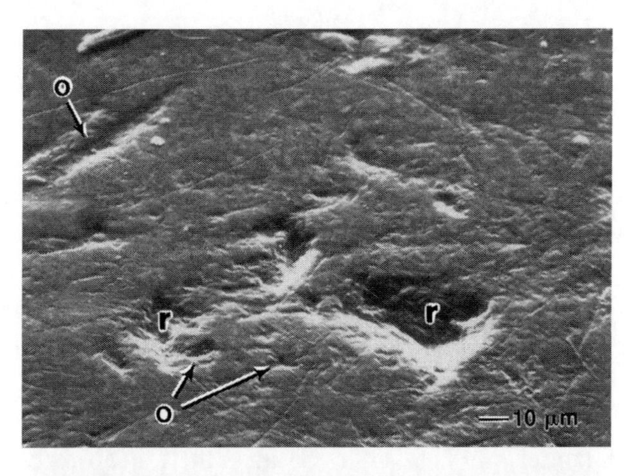

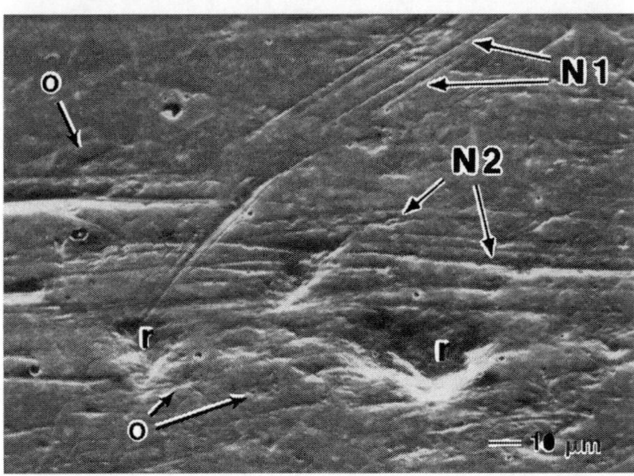

Figure 3. Changes in dental microwear over a 3-day period in a laboratory vervet monkey fed monkey chow and apples. Top = baseline micrograph. Bottom = follow-up micrograph of same surface, after three days. ("R" = reference features in both micrographs, "O" = features nearly obliterated between baseline and follow-up, and "N" = new features appearing on follow-up micrograph) (from Teaford and Oyen, 1989b).

Of course, laboratory studies of living animals are limited in how they can change diets, and most laboratory diets are not nearly as diverse as diets in the wild, where seasonal, geographic, and annual differences in diet have the potential to have a huge impact of dental microwear patterns. Thus, work with animals in the wild is a potential goldmine of information, as demonstrated by the pioneering study of Walker et al. (1978) on hyraxes, where skulls were collected directly from the same area in which behavioral observations were recorded. Unfortunately, while studies of living primates in the wild have been attempted a number of times, they have usually met with little success. Primates often live in forested habitats where they are hard to see, and even harder to catch. Even in open habitats (e.g., baboons in the East African savanna), the work is difficult.

Thus far, only two studies have consistently yielded high-quality copies of primate teeth in the wild. The first is the on-going study at La Pacifica in the Guanacaste region of Costa Rica (Teaford and Glander, 1991; Ungar et al., 1995; Teaford and Glander, 1996; Dennis et al., 2004). There, howling monkeys (*Alouatta palliata*) are regularly observed, captured, and released in a dry tropical forest setting. That work has certainly verified some of the standard correlations from museum analyses (e.g., leaf-eating and scratches on teeth). It has also given us glimpses of other complicating factors. For instance, the amount of molar microwear may vary from season to season, and between riverine and nonriverine microhabitats (Teaford and Glander, 1996). The studies at La Pacifica have also shown that tooth wear generally proceeds at a rapid pace in the wild – at about 8–10 times the pace of that in U.S. dental patients (Teaford and Glander, 1991). This has led to the idea of the "Last Supper" phenomenon (Grine, 1986) – i.e., that, in some situations, dental microwear may only record the effects of the most-recently eaten foods on the teeth, although some investigators feel that microwear on the sides of the teeth may show far slower turnover (Pérez-Pérez et al., 1994).

Recently, a second long-term study has begun to yield high resolution casts of primate teeth in the wild (Nystrom et al., 2004). The study populations, from the anubis-hamadryas hybrid zone of Awash National Park, Ethiopia, have been the focus of multidisciplinary work for over thirty years (e.g., Nagel, 1973; Phillips-Conroy, 1978; Sugawara, 1979; Phillips-Conroy and Jolly, 1986; Phillips-Conroy et al., 1991; Szmulewicz et al., 1999; Phillips-Conroy et al., 2000; Dirks et al., 2002) and have yielded fascinating insights into the behavioral, ecological, and anatomical ramifications of species hybridization in the wild. The precise timing of dental microwear analyses in this case (before the heavy onset of new leaves and grasses in this seasonal environment), allowed Nystrom et al. (2004) to implicate "small-caliber environmental grit" as the main cause of the observed microwear patterns, which included no significant differences between the sexes, age groups, or different troops.

Studies such as these make us look very carefully at the specific causes of dental microwear, and what can and cannot be documented in changes in dental microwear patterns. In fact, many foods are not hard enough to scratch teeth (Lucas, 1991), and, in modern human diets, exceptionally few foods could be expected to scratch teeth, because the foods are so clean, cooked and processed. Still, without such processing, some foods (e.g., certain leaves), include abrasives which can cause striations on teeth (Lucas and Teaford, 1995; Danielson and Reinhard, 1998; Reinhard et al., 1999; Gügel et al., 2001; Teaford et al., 2006) (Figure 4a). Others include acids, which can etch the teeth (Figure 4b) (Puech, 1984b; Puech et al., 1986; Teaford, 1988a, 1994; Ungar, 1994; King et al., 1999b). In addition, as suggested by many authors (e.g., Puech, 1986a; Teaford, 1988a; Pastor 1992, 1993; Teaford, 1994; Ungar, 1994; Ungar et al., 1995; Nystrom et al., 2004), many microwear patterns might be caused by what might be termed the *indirect* effects of food on dental microwear. For instance, certain cooking procedures or methods of food preparation (e.g., cooking food directly within the ashes of a campfire) may introduce abrasives into foods, causing a high incidence of microscopic scratches on teeth – scratches not caused by the foods themselves, but by the methods with which they were prepared (Pastor, 1992, 1993; Teaford and Lytle, 1996). Similarly, animals may also eat soft foods, and still show many scratches on their teeth – if the food is coated with abrasives (e.g., earthworms coated with dirt) (Silcox and Teaford, 2002). Finally, if an animal has a soft but tough diet, tooth-on-tooth wear can yield characteristic microwear patterns as enamel edges penetrate the food and grind past each other yielding a high incidence of small pits on their teeth – pits probably caused by the adhesive wear of enamel on enamel (Puech et al., 1981; Walker, 1984; Puech, 1984a, 1986a; Puech et al., 1986; Radlanski and Jäger, 1989; Teaford and Runestad, 1992; Rafferty et al., 2002).

Figure 4. a. SEM micrograph of long silica "trichomes" on leaf routinely eaten by *Alouatta palliata* in Costa Rica (from Teaford et al., 2006). b. Chemical wear of a molar of *Alouatta palliata*. Top = baseline micrograph. Bottom = follow-up micrograph after 2–3 second exposure to a 30 % solution of phosphoric acid (note the removal of smaller microwear features) (from Teaford, 1994).

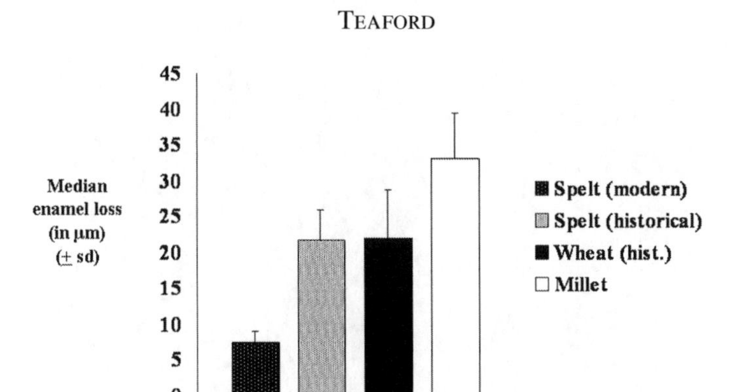

Figure 5. Differences in the amount of enamel lost through in vitro abrasion of enamel by different cereal grains (data from Gügel et al., 2001).

In Vitro Laboratory Studies

So, if studies of living primates are so difficult, why not do experimental studies of dental enamel abraded by different foods? Early studies showed that substances like acids could have a profound effect on enamel surfaces (Mannerberg, 1960; Boyde, 1964). Still, experimental work has proceeded in fits and spurts. Some studies have demonstrated that the orientation of scratches on a tooth's surface can indeed reflect the orientation of tooth-food-tooth movements (e.g., Ryan, 1979; Teaford and Walker, 1983; Gordon, 1984c; Walker, 1984; Teaford and Byrd, 1989; Morel et al., 1991). Other studies have shown that certain agents, such as wind-blown sand, or various acids, can leave characteristic microwear patterns on teeth (Puech and Prone, 1979; Puech et al., 1980, 1981; Gordon, 1984a; Puech et al., 1985; Puech, 1986a; Rensberger and Krentz, 1988; King et al., 1999b). However, there have been surprisingly few controlled studies of the wear patterns caused by different types of foods.

Peters (1982) used standard physical property-testing equipment while examining the effects of a range of African foods on dental microwear, ultimately showing that few foods could actually scratch enamel, with extraneous abrasives being one of the prime culprits instead (see also Puech et al., 1986). Only with more detailed analyses did subsequent work (e.g., Gügel et al., 2001) begin to demonstrate the effects of specific foods on microwear patterns (e.g., "cereal-specific" microwear related to phytolith content in certain grains) (Figure 5).

Analyses of Paleontological Samples

When dental microwear analyses are aimed at the past, they often raise more questions than they answer, largely because they give new and different glimpses of the intricacies of previous behavior. Of course, interpretations of results are dependent on our knowledge of present-day correlations between diet and dental microwear. Thus, while we often have significant differences in dental microwear between teeth from different sites or time periods, the exact meaning of those differences may be subject to discussion and debate until better data are available for modern species.

Paleontological analyses have included a wide variety of animals, ranging from rodents (Rensberger, 1978, 1982), horses (MacFadden et al., 1999), and ungulates (Solounias and Hayek, 1993; Solounias and Moelleken, 1992a, b, 1994; Solounias and Semprebon, 2002), to carnivores (Van Valkenburgh et al., 1990), tyrannosaurids (Schubert and Ungar, 2005), and conodonts (Purnell, 1995). But within the primates, dental microwear analyses have also led to some major insights.

Analyses of Miocene hominoid material have helped document an impressive array of dietary adaptations in the early apes (Teaford and Walker, 1984; Ungar, 1996; King et al., 1999a). By contrast, analyses of Plio-Pleistocene cercopithecoid material have documented a surprisingly limited array of dietary adaptations in East Africa (Lucas and Teaford, 1994; Leakey et al., 2003), but a larger array in South Africa (El-Zaatari et al., 2005), while also yielding insights into the degree of terrestriality in some species (Ungar and Teaford, 1996). Molar microwear analyses have also helped to document the effects of phylogenetic constraints in fossil apes by documenting similar functions in taxa which have undergone shifts in molar morphology through time (Ungar et al., 2004).

As might be expected, analyses of human ancestors have focused on whichever fossils are available. For the anterior teeth, qualitative studies have suggested similarities between early hominin incisor wear and that observed on modern primates that routinely employ a great deal of incisal preparation (Puech and Albertini, 1984). Quantitative analyses of *Australopithecus afarensis* suggested incisal microwear similarities with those documented for lowland gorillas or perhaps savanna baboons (Ryan and Johanson, 1989). More detailed analyses of *Paranthropus robustus* and *Australopithecus africanus* (Ungar and Grine, 1991) showed great variabilitiy within each species in most standard microwear measurements. However, the greater density of features on the incisors of *A. africanus* helped to show that this species placed a higher emphasis on incisal preparation than in *P. robustus*.

In the molar region, qualitative analyses have raised many possibilities that have been often repeated in the literature. For instance, the robust australopithecines (e.g., *Paranthropus*) have been characterized as indistinguishable from modern chimpanzees or orang-utans (Walker, 1981), perhaps with more abrasive molar wear than in the gracile australopithecines (e.g., *Australopithecus*) (Puech et al., 1985; Puech, 1986b; Puech et al., 1986). By contrast, *Homo habilis* has been characterized as using high occlusal pressures, but on foods that can chemically etch the enamel (Puech et al., 1983; Puech, 1986b).

Quantitative analyses have begun to refine these interpretations, from many different perspectives. Studies of non-occlusal microwear have focused primarily on more recent, European taxa, such as the Neanderthals, together with specimens now attributed to *Homo heidelbergensis*. Initial analyses portrayed the Neanderthals as more carnivorous than their immediate predecessors, or subsequent *Homo sapiens* (Lalueza Fox and Pérez-Pérez, 1993; Lalueza et al., 1996). However, subsequent work has raised the possibility of sexual differences in diet in *Homo heidelbergensis* (Pérez-Pérez et al., 1999), and a more heterogeneous diet for the Neanderthals, with a shift in food processing in the Upper Paleolithic (Pérez-Pérez et al., 2003).

Quantitative analyses of fossil hominin occlusal microwear began with Grine's pioneering work on the South African australopithecines, where *Paranthropus robustus* was shown to exhibit more microwear and more pitting on its molars than did *Australopithecus africanus* (Grine, 1981, 1986, 1987; Grine and Kay, 1988; Kay and Grine, 1989). This leant a lot of support to Robinson's ideas of dietary differences among the australopithecines, with the so-called robust forms consuming harder foods that required more variable grinding movements in chewing. Recent work has taken analyses a step further by incorporating samples of australopithecines and early *Homo* from East and South Africa (Ungar et al., 2001; Teaford et al., 2002b). The work is still being completed, but initial results gave further credence to Ryan and Johanson's (1989) idea of similarities between

Australopithecus afarensis and lowland gorillas, this time for the molars (Teaford et al., 2002b; Grine et al., 2006). In conjunction with other morphological data for the australopithecine grade of human evolution (Teaford and Ungar, 2000; Teaford et al., 2002a), they also helped to make the distinction between dental *capabilities* and dental *use*, as the capability to process certain foods may well have been of critical importance in certain situations. Meanwhile, analyses of early *Homo* have begun to help sort through the variable assemblage that now encompasses early *Homo*, with *Homo erectus/ergaster* showing a higher incidence of pitting on its molars than that found in *Homo habilis* (Ungar et al., 2006), suggesting the consumption of tougher or harder food items by the former group, again, as a possible critical fallback food (Figure 6).

As for more recent human populations, the transition from hunting-gathering to agriculture has left a complex signal in the microwear record, depending on which populations are examined, in which habitats, etc. (Bullington, 1991; Pastor, 1992; Pastor and Johnston, 1992; Schmidt, 2001; Teaford, 1991, 2002; Teaford et al., 2001). Once the change to agriculture was made, human diets did not simply stay stagnant. Some became more homogeneous, as evidenced by fairly uniform microwear patterns, while others became more variable (Molleson and Jones, 1991). Some cereal diets left characteristic microwear patterns remarkably similar to those documented in laboratory studies (Gügel et al., 2001). Some changes in food processing, most notably the boiling of foods, led to a marked decrease in the amount of microwear at some sites (Molleson et al., 1993). The net effect, however, is that, with the advent of food preparation (in particular, cooking), the effects of food on human teeth changed dramatically, leaving modern nonhuman primates as perhaps the best modern analogues for analyses of the earliest hominins.

Cautions

While at first glance, it might seem that we know a great deal about dental microwear and diet, in reality, all we have are tiny windows into a complex world. Studies of living primates have really only been carried out on two species (*Alouatta palliata* and *Papio hamadryas*), in two habitats (the dry tropical forest of Costa Rica and the thornbush and savanna grassland of Ethiopia). Those settings certainly have their inherent complexities (e.g., dramatic seasonal changes in rainfall

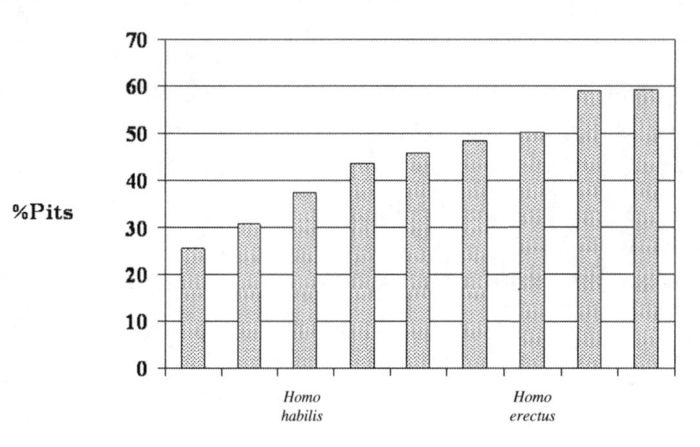

Figure 6. Histogram of the incidence of pitting on molars of *Homo habilis* and *Homo erectus* and extant primates (data from Ungar et al., 2006). From left to right, comparative samples include *Gorilla gorilla, Homo sapiens* (Arikara), *Homo habilis, Pan troglodytes, Homo sapiens* (Aleut), *Papio cynocephalus, Homo erectus, Cebus apella, Lophocebus albigena.*

and resource availability), but how representative are they of all the other ecological zones in the world? Would dental microwear patterns differ for primates in other habitats? Undoubtedly. How might other species share a habitat with either of these species, and how would that be reflected in differences in dental microwear? What is the magnitude of seasonal, annual, geographic, and interspecific differences in dental microwear for other species elsewhere in the world? How does the incidence of dental microwear relate to specific abrasives and foods in the wild? Clearly, a massive amount of work has yet to be done on live primates in the wild if we are to use that information to help interpret results from fossil samples. Unfortunately, that work may need to be done quickly, as major sources of information for dental microwear analysis may be vanishing before our eyes, as huge tracts of the earth's environment disappear or are damaged beyond repair. In the process, habitats and organisms of crucial importance for future microwear interpretations may be lost, effectively leaving certain questions unanswerable.

Meanwhile, laboratory studies have barely begun to sort through the intricacies of dental microwear formation. As noted earlier, the effects of specific food items have yet to be documented in any systematic fashion, and the effects of foods naturally consumed in the wild have yet to be examined in any detail, with the work from Costa Rica and Ethiopia giving us just a teasing glimpse of possibilities. As primate diets are normally quite variable, and as dental microwear features can change quite quickly (Teaford and Oyen, 1989b; Teaford and Glander, 1991), what are the effects of different food items on overall microwear patterns within a specific diet? Will items that are abrasive, hard, or acidic effectively swamp other microwear patterns? Will the so-called "Last Supper" phenomenon vary between species or populations? Again, much more work needs to be done.

Analyses of museum material have probably been pushed closer to their limits than studies of living animals, but only because there are relatively few museums where the associated collection data is of sufficient detail to aid the documentation of geographic or seasonal differences in diet and dental microwear. Moreover, there are virtually no collections of primate material for which dietary information has been collected before the animals were collected. Thus, virtually all studies of museum samples are limited in their resolution by the lack of associated dietary information for the animals in question. Similarly, analyses of archeological and paleontological material are limited by the size and extent of collections (and sometimes, in the case of fossils, access to them), and by the associated information for those collections (e.g., geological information, paleoecological interpretations, presence of associated cultural remains, etc.).

Still, when fossils are found, our current methods may lead to answers, or questions, requiring a different resolution of details than what we had anticipated. Thus, for instance, they force us to consider questions about subtle intraspecific differences in diet – questions that most analyses of museum material cannot begin to answer. Similarly, "what you see is what you get" in fossils, in terms of dental samples. So preservation may be poor, postmortem wear may be rampant, and, for some collections, we may even have an overabundance of certain tooth types for which we have no analyses of modern material (noting that most analyses to-date have focused on incisors or molars). In the face of such problems and possibilities, innumerable questions still need to be answered. For instance, what is the relationship between the biomechanical demands of processing certain foods and the generation of microwear patterns? Can the structural capabilities of bone, or variations in enamel properties (e.g., Cuy et al.,

2002), be correlated with variations in microwear pattern within and between jaws? For that matter, what *is* the relationship between microwear patterns between upper and lower jaws? We know their patterns are roughly similar (Teaford and Walker, 1984), but can analyses of upper-lower microwear integration shed new light on jaw movements and food processing in mammals? What about the microwear of other dental materials like dentin? Until now, investigators have shied away from it, mainly because it is hard to clean without introducing artificial microwear patterns. However, since it is softer than enamel, might it be an indicator of even subtler diet distinctions?

Despite all these questions, when all is said and done, the biggest challenge facing dental microwear analyses is a methodological one. Standard scanning electron microscope analyses are difficult, costly, time-consuming, and (most notably) subjective, in that the "measurement" of certain microwear features depends on the recognition of "landmarks" that may be defined differently by different researchers. Even with the use of semi-automated, computerized, digitizing routines (Ungar et al., 1991; Ungar, 1995), inter-observer error rates are often unacceptably high, ranging from 3 to 13% depending on which measurements, of which types of features, are being measured (Grine et al., 2002). Thus, measurements computed by different researchers should probably not be compared directly, leaving researchers few alternatives but to either have one person do all the measuring (e.g., Organ et al., 2005), or average the measurements, computed for the same specimens, by a number of individuals (e.g., Ungar et al., 2006). When this is all coupled with the fact that most analyses to-date have used relatively small samples (even for species with variable diets), the net effect is that dental microwear analyses have barely begun to live up to their potential.

Recently, workers have begun to address this issue through the use of two new approaches – lower magnification work by light microscopy (Solounias and Semprebon, 2002; Semprebon et al., 2004), and a higher magnification combination of confocal microscopy and scale sensitive fractal analysis (Ungar et al., 2003; Scott et al., 2005). In the former, epoxy casts are viewed at a magnification of 35X while a fiber optic light source is used to direct light obliquely across the cast. Features are then quickly counted as "pits" and "scratches," and also grouped into various size categories (e.g., "fine" versus "hypercoarse" scratches). Its potential advantages include the fact that it is much quicker than standard SEM analyses, making possible the use of larger sample sizes. Also, because the analysis is done at lower magnifications, larger surface areas are covered, giving more representative coverage of the tooth. Third, published tests of inter-observer error hint at better replicability than standard SEM-based analyses (Semprebon et al., 2004), although published measures of error rates are not presented in a form that is comparable between studies. However, the technique is not without its drawbacks either. First, it requires significant training to master – i.e., it is not the sort of thing where you can merely pick up a dissecting microscope and go measure teeth! Second, inter-observer error rates have yet to be reported for the categorization of features into different sizes. As many of the diet differences reported to-date (e.g., Godfrey et al., 2004) depend upon the accurate identification of features of different sizes, how readily can those measures be generated by different observers? Moreover, because the technique works at low magnifications, it may only yield information on gross dietary categorizations. The fact that data have been combined for shearing and crushing facets (Semprebon et al., 2004), with no evident differences between those facets, suggests this may be the case. Until further work is done, claims such as those by Godfrey et al. (2004) that there were no dietary differences between species of *Megaladapis* should be viewed extremely cautiously, as they may be nothing more than a reflection of the limitations of the technique. Finally, the low magnification technique may only be able to detect the most obvious effects of postmortem wear. Thus, since we already know that at some fossil sites (e.g., Koobi Fora) over 75% of specimens are not usable due to postmortem wear, can this technique successfully recognize postmortem wear? Only further work will tell.

As for the confocal technique, once again, an epoxy cast is examined, but this time at a series of higher magnifications using a white light confocal microscope. Resultant x, y, and z coordinates can be used to create "photosimulations" of the surface, or 3D models of the surface (Figure 7). Of course, even though the resolution of the system in the z-dimension is outstanding (in fractions of microns), the resultant maps of tooth surfaces are nothing more than pretty pictures without some form of analysis. What makes this system uniquely useful is that scale-sensitive fractal analyses have been used to characterize the wear surfaces. These analyses are based on the assumption that the apparent area of a rough surface (and the apparent length of a profile from a rough surface) will change with the scale of measurement. Thus, for a relatively smooth surface, a limited number of large patches may accurately characterize the surface area, whereas for a rough surface, a much larger number of small patches may be needed to accurately characterize surface area. So, if you systematically vary the scale of measurement, and thus the size of the patches, and plot them against changes in relative area, you can use the slope of that plot as a measure of the complexity of that surface. Similarly, if the orientation of profiles across the surface is changed systematically, a measure of the degree of difference (or "anisotropy") of the orientation of surface features can be calculated. The net effect is that the entire wear surface can be analyzed or characterized rather than treating each pit or scratch as a "feature" to be measured. As those analyses are completed at a series of different scales, they will provide a more objective picture of which magnifications are most useful for making dietary distinctions. Thus, it would seem to be the closest thing available to putting a specimen in and getting useful numbers out. Its advantages include speed, as large numbers of specimens can be processed quickly. But it is also objective and thus repeatable, in that the only subjective component is in the choice of which specimens to use. Also, unlike all previous types of analyses, it includes measures of height or depth, thus opening new possibilities for analysis. Finally, due to the scale-sensitive nature of its analyses, it effectively covers a wide range of magnifications to objectively determine at which resolution relevant dietary/functional distinctions can be made. However, with all that said, it is still a work in progress, and as a result, there are still some disadvantages. First, by anthropological standards, white light confocal microscopes are rare and expensive. Second, new analytical software is still being developed, so most workers still do not have access to the software, and those who do are still determining which fractal analyses will be most useful. Third, a database for future interpretations is still being gathered and comparisons with data generated by previous techniques are still being completed. Fourth, some postmortem wear seems to be detectable by an absence of detail in the objective measurements or characterizations, but it still may be dependent on visual inspection (or, ironically, SEM examination) to determine which surfaces of fossil teeth are suitable for use. Initial results are promising (Scott et al., 2005), yielding insights into diet variability in the South African australop-

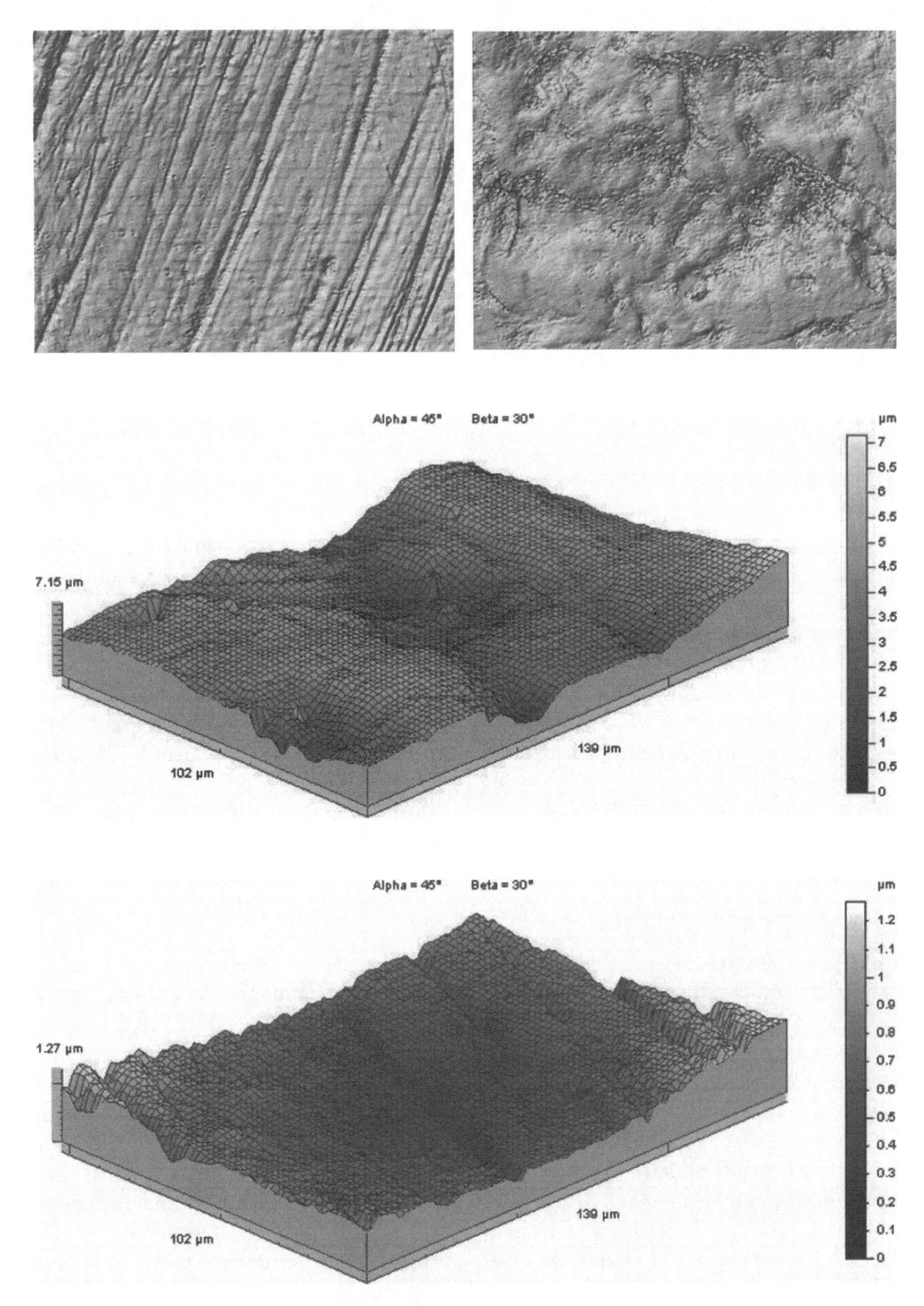

Figure 7. 3D models and "photosimulations" of the teeth of *Cebus apella* (top right image and middle image) and *Alouatta palliata* (top left and bottom image) derived from 3D coordinate data from a confocal microscope. (Note the dramatic difference in the scales of the 3D maps for each specimen).

ithecines, including the possible importance of critical fallback foods in their diets. With the added capability of providing data in three dimensions, it also raises the possibility of answering innumerable new questions concerning topics like the volume of tooth loss in different wear regimes, and the depth of enamel removed by certain abrasives or bite forces, thus giving hope for even better correlations between tooth use and wear patterns.

Conclusions

Dental microwear analyses have come a long way in the past 25 years. While some pieces of information may remain invisible, we always need to be open to new opportunities, as the effects of some foods, or the means of documenting them, may be hard to anticipate. Each method of data collection, and each piece of evidence, has its strengths and weaknesses. Dental microwear analysis is certainly no exception, as it definitely has its limitations. But it also has the potential to give us direct glimpses of the past. As such, it can tell us about how teeth were actually used rather than what they were evolutionarily capable of doing. So we need to better understand its strengths and weaknesses. New methods raise new hopes of doing so. Of course, in the long run, the picture we are trying to decipher is incredibly complicated. So we also need to consider every piece of evidence, be it dental microwear, or otherwise. With a little luck and foresight, we will have the good fortune to contribute to a better understanding of the origin and evolution of human diet, among many other things!

Acknowledgments

I wish to offer my deepest thanks to Professor Jean-Jacques Hublin and Dr. Shara Bailey for inviting me to participate in this conference. They, and their organizational team, put together a wonderful conference enjoyed by all. Allison Cleveland and Shara Bailey deserve special thanks for their efforts (and patience!) in dealing with innumerable questions about the papers and manuscripts. Similarly, Diana Carstens and Silke Streiber worked wonders in handling the travel arrangements. I would also like to thank all the participants at the conference for their stimulating discussions, and I would like to thank the National Science Foundation for its generous support of my research through the years. Much of the work on early hominin diets stems from on-going research with Fred Grine and Peter Ungar, who have given invaluable feedback on many of the thoughts and ideas in this paper (and helped me gain a new appreciation for pilsner beer in the process). I also wish to thank Rob Scott and Peter Ungar for their work in creating Figure 7 and allowing me to reproduce it here.

References

Ainamo, J., 1971. Prenatal occlusal wear in guinea pig molars. Scandinavian Journal of Dental Research 79, 69–71.

Allen, K., Agosta, C., Estafan, D., 2004. Using microabrasive material to remove fluorosis stains. Journal of the American Dental Association 135, 726.

Bezerra, A.C., Leal, S.C., Otero, S.A., Gravina, D.B., Cruvinel, V.R., Ayrton de Toledo, O., 2005. Enamel opacities removed using two different acids, an *in vivo* comparison. Journal of Clinical Pediatric Dentistry 29, 147–150.

Boyde, A., 1964. The structure and development of mammalian enamel. Ph.D. Dissertation, University of London.

Bullington, J., 1991. Deciduous dental microwear of prehistoric juveniles from the lower Illinois River valley. American Journal of Physical Anthropology 84, 59–73.

Butler, P.M., 1952. The milk molars of Perissodactyla, with remarks on molar occlusion. Proceedings of the Zoological Society of London 121, 777–817.

Chafaie, A., 2004. Minimally invasive aesthetic treatment for discolored and fractured teeth in adolescents: a case report. Practical Proceedings of Aesthetic Dentistry 16, 319–324.

Covert, H.H., Kay, R.F., 1981. Dental microwear and diet: implications for determining the feeding behaviors of extinct primates, with a comment on the dietary pattern of *Sivapithecus*. American Journal of Physical Anthropology 55, 331–336.

Cuy, J.L, Mann, A.B., Livi, K.J., Teaford, M.F., Weihs, T.P., 2002. Nanoindentation mapping of the mechanical properties of molar enamel. Archives of Oral Biology 47, 281–291.

Daegling, D.J., Grine, F.E., 1999. Terrestrial foraging and dental microwear in *Papio ursinus*. Primates 40, 559–572.

Dahlberg, A.A., Kinzey, W.G., 1962. Etude microscopique de l'abrasion et de l'attrition sur la surface des dents. Bulletin du Groupement International pour la Recherche Scientifique en Stomatologie et Odontologie (Bruxelles) 5, 242–251.

Danielson, D.R., Reinhard, K.J., 1998. Human dental microwear caused by calcium oxalate phytoliths in prehistoric diet of the lower Pecos region, Texas. American Journal of Physical Anthropology 107, 297–304

Dennis, J.C., Ungar, P.S., Teaford, M.F., Glander, K.E., 2004. Dental topography and molar wear in *Alouatta palliata* from Costa Rica. American Journal of Physical Anthropology 125, 152–161.

Dirks, W., Reid, D.J., Jolly, C.J., Phillips-Conroy, J.E., Brett, F.L., 2002. Out of the mouths of baboons: stress, life history, and dental development in the Awash National Park hybrid zone, Ethiopia. American Journal of Physical Anthropology 118, 239–252.

El-Zaatari, S., Grine, F.E., Teaford, M.F., Smith, H.F., 2005. Molar microwear and dietary reconstructions of fossil Cercopithecoidea from the Plio-Pleistocene deposits of South Africa. Journal of Human Evolution 49, 180–205.

Fine, D., Craig, G.T., 1981. Buccal surface wear of human premolar and molar teeth: a potential indicator of dietary and social differentiation. Journal of Human Evolution 10, 335–344.

Godfrey, L.R., Semprebon, G.M., Jungers, W.L., Sutherland, M.R., Simons, E.L., Solounias, N., 2004. Dental use wear in extinct lemurs: evidence of diet and niche differentiation. Journal of Human Evolution 47, 145–170.

Gordon, K.D., 1982. A study of microwear on chimpanzee molars: implications for dental microwear analysis. American Journal of Physical Anthropology 59, 195–215.

Gordon, K.D., 1984a. Taphonomy of dental microwear, II. American Journal of Physical Anthropology 63, 164–165.

Gordon, K.D., 1984b. Hominoid dental microwear: complications in the use of microwear analysis to detect diet. Journal of Dental Research 63, 1043–1046.

Gordon, K.D., 1984c. The assessment of jaw movement direction from dental microwear. American Journal of Physical Anthropology 63, 77–84.

Gordon, K.D., 1988. A review of methodology and quantification in dental microwear analysis. Scanning Microscopy 2, 1139–1147.

Gordon, K.D., Walker, A.C., 1983. Playing 'possum: a microwear experiment. American Journal of Physical Anthropology 60, 109–112.

Grine, F.E., 1981. Trophic differences between "gracile" and "robust" australopithecines: a scanning electron microscope analysis of occlusal events. South African Journal of Science 77, 203–230.

Grine, F.E., 1986. Dental evidence for dietary differences in *Australopithecus* and *Paranthropus*: a quantitative analysis of permanent molar microwear. Journal of Human Evolution 15, 783–822.

Grine, F.E., 1987. Quantitative analysis of occlusal microwear in *Australopithecus* and *Paranthropus*. Scanning Microscopy 1, 647–656.

Grine, F.E., Kay, R.F., 1988. Early hominid diets from quantitative image analysis of dental microwear. Nature 333, 765–768.

Grine, F.E., Ungar, P.S., Teaford, M.F., 2002. Error rates in dental microwear quantification using scanning electron microscopy. Scanning 24, 144–153.

Grine F.E., Ungar, P.S., Teaford, M.F., El Zaatari, S., 2006. Molar microwear in *Praeanthropus afarensis*: Evidence for dietary stasis through time and under diverse paleoecological conditions. Journal of Human Evolution 51, 297–319.

Gűgel, I.L., Grupe, G., Kunzelmann, K.H., 2001. Simulation of dental microwear: characteristic traces by opal phytoliths give clues to ancient human dietary behavior. American Journal of Physical Anthropology 114, 124–138.

Hojo, T., 1996. Quantitative analyses of microwear and honing on the sloping crest of the P_3 in female Japanese monkeys (*Macaca fuscata*). Scanning Microscopy 10, 727–736.

Kay, R.F., 1975. The functional adaptations of primate molar teeth. American Journal of Physical Anthropology 42, 195–215.

Kay, R.F., 1987. Analysis of primate dental microwear using image processing techniques. Scanning Microscopy 1, 657–662.

Kay, R.F., Covert, H.H., 1983. True grit: a microwear experiment. American Journal of Physical Anthropology 61, 33–38.

Kay, R.F., Grine, F.E., 1989. Tooth morphology, wear and diet in *Australopithecus* and *Paranthropus* from southern Africa. In: Grine, F.E. (Ed.), The Evolutionary History of the Robust Australopithecines. Aldine de Gryter, New York, pp. 427–444.

Kay, R.F., Hylander, W.L., 1978. The dental structure of mammalian folivores with special reference to primates and Phalangeroidea (Marsupialia). In: Montgomery, G.G. (Ed.), The Biology of Arboreal Folivores. Smithsonian Institution Press, Washington, pp. 173–192.

Kelley, J., 1986. Paleobiology of miocene hominoids. Ph.D. Dissertation, Yale University.

Kelley, J., 1990. Incisor microwear and diet in three species of *Colobus*. Folia Primatologica 55, 73–84.

King, T., Aiello, L.C., Andrews, P., 1999a. Dental microwear of *Griphopithecus alpani*. Journal of Human Evolution 36, 3–31.

King, T., Andrews, P., Boz, B., 1999b. Effect of taphonomic processes on dental microwear. American Journal of Physical Anthropology 108, 359–373.

Lalueza, C., Pérez-Pérez, A., Turbon, D., 1996. Dietary inferences through buccal microwear analysis of middle and upper Pleistocene human fossils. American Journal of Physical Anthropology 100, 367–387.

Lalueza Fox, C., 1992. Dental striation pattern in Andamanese and Veddahs from skull collections of the British Museum. Man in India 72, 377–384.

Lalueza Fox, C., Pérez-Pérez, A., 1993. The diet of the Neanderthal child Gibralter 2 (Devils' Tower) through the study of the vestibular striation pattern. Journal of Human Evolution 24, 29–41.

Leakey, M.G., Teaford, M.F., Ward, C.W., 2003. Cercopithecidae from lothagam. In: Leakey, M.G., Harris, J.M. (Eds.), Lothagam: Dawn of Humanity in Eastern Africa. Columbia University Press, New York, pp. 201–248.

Lucas, P.W., 1979. The dental-dietary adaptations of mammals. Neues Jahrbuch für Geologie und Paläontologie 8, 486–512.

Lucas, P.W., 1991. Fundamental physical properties of fruits and seeds in the diet of Southeast Asian primates. In: Ehara, A., Kimura, T., Takenaka, O., Iwamoto, M. (Eds.), Primatology Today. Elsevier, Amsterdam, pp. 128–152.

Lucas, P.W., 2004. *Dental Functional Morphology: How Teeth Work*. Cambridge University Press, New York.

Lucas, P.W., Teaford, M.F., 1994. The functional morphology of colobine teeth. In: Oates, J., Davies, A.G. (Eds.), Colobine Monkeys: Their Evolutionary Ecology. Cambridge University Press. Cambridge, pp. 173–203.

Lucas, P.W., Teaford, M.F., 1995. Significance of silica in leaves to long-tailed macaques (*Macaca fascicularis*). Folia Primatologica 64, 30–36.

MacFadden, B.J., Solounias, N., Cerling, T.E., 1999. Ancient diets, ecology, and extinction of 5-million-year-old horses from Florida. Science 283, 824–827.

Mannerberg, F., 1960. Appearance of tooth surface as observed in shadowed replicas. Odontologisk Revy 11 Suppl. 6, 114 pp.

Mills, J.R.E., 1955. Ideal dental occlusion in the primates. Dental Practitioner 6, 47–61.

Mills, J.R.E., 1963. Occlusion and malocclusion of the teeth of primates. In: Brothwell, D.R. (Ed.), Dental Anthropology. Pergamon Press, New York, pp. 29–52.

Molleson, T., Jones, K., 1991. Dental evidence for dietary change at Abu Hureyra. Journal of Archaeological Science 18, 525–539.

Molleson, T., Jones, K., Jones, S., 1993. Dietary change and the effects of food preparation on microwear patterns in the Late Neolithic of Abu Hureyra, northern Syria. Journal of Human Evolution 24, 455–468.

Morel, A., Albuisson, E., Woda, A., 1991. A study of human jaw movements deduced from scratches on occlusal wear facets. Archives of Oral Biology 36, 195–202.

Nagel, U., 1973. A comparison of anubis baboons, hamadryas baboons and their hybrids at a species border in Ethiopia. Folia Primatologica 19, 104–165.

Noble, V.E., Teaford, M.F., 1995. Dental microwear in Caucasian American *Homo sapiens*: preliminary results. American Journal of Physical Anthropology Suppl. 20, 162.

Nystrom, P., Phillips-Conroy, J.E., Jolly, C.J., 2004. Dental microwear in anubis and hybrid baboons (*Papio hamadryas*, sensu lato) living in Awash National Park, Ethiopia. American Journal of Physical Anthropology 125, 279–291.

Oliveira, E.V., 2001. Micro-desgaste dentário em alguns Dasypodidae (Mammalia, Xenarthra). Acta Biologica Leopoldensia 23, 83–91.

Organ, J.M., Teaford, M.F., Larsen, C.S., 2005. Dietary inferences from dental occlusal microwear at Mission San Luis de Apalachee. American Journal of Physical Anthropology 128, 801–811.

Östman-Andersson, E., Marcusson, A., Hörstedt, P., 1993. Comparative SEM studies of the enamel surface appearance following the use of glass ionomer cement and a diacrylate resin for bracket bonding. Swedish Dentistry Journal 17, 139–146.

Pastor, R.F., 1992. Dietary adaptations and dental microwear in mesolithic and chalcolithic south Asia. Journal of Human Ecology 2 (spec issue), 215–228.

Pastor, R.F., 1993. Dental microwear among inhabitants of the Indian subcontinent: a quantitative and comparative analysis. Ph.D. Dissertation, University of Oregon.

Pastor, R.F., Johnston, T.L., 1992. Dental microwear and attrition. In: Kennedy, K.A.R. (Ed.), Human Skeletal Remains From Mahadaha: A Gangetic Mesolithic Site. Cornell University Press, Ithaca, pp. 271–304.

Pérez-Pérez, A., Lalueza, C., Turbón, D., 1994. Intraindividual and intragroup variability of buccal tooth striation pattern. American Journal of Physical Anthropology 94, 175–187.

Pérez-Pérez, A., Bermúdez de Castro, J.M., Arsuaga, J.L., 1999. Nonocclusal dental microwear analysis of 300,000-year-old *Homo heidelbergensis* teeth from Sima de los Huesos (Sierra de Atapuerca, Spain). American Journal of Physical Anthropology 108, 433–457.

Pérez-Pérez, A., Espurz, V., Bermúdez de Castro, J.M., de Lumley, M.A., Turbón, D., 2003. Non-occlusal dental microwear variability in a sample of Middle and Late Pleistocene human populations from Europe and the Near East. Journal of Human Evolution 44, 497–513.

Peters, C.R., 1982. Electron-optical microscopic study of incipient dental microdamage from experimental seed and bone crushing. American Journal of Physical Anthropology 57, 283–301.

Phillips-Conroy, J.E., 1978. Dental variability in Ethiopian baboons: an examination of the anubis-hamadryas hybrid zone in the Awash National Park, Ethiopia. Ph.D. Dissertation, New York University.

Phillips-Conroy, J.E., Jolly, C.J., 1986. Changes in the structure of the baboon hybrid zone in the Awash National Park, Ethiopia. American Journal of Physical Anthropology 71, 337–350.

Phillips-Conroy, J.E., Jolly, C.J., Brett, F.L., 1991. Characteristics of hamadryas-like male baboons living in anubis baboon troops in the Awash hybrid zone, Ethiopia. American Journal of Physical Anthropology 86, 353–368.

Phillips-Conroy, J.E., Bergman, T., Jolly, C.J., 2000. Quantitative assessment of occlusal wear and age estimation in Ethiopian and Tanzanian baboons. In: Whitehead, P.F., Jolly, C.J. (Eds.), Old World Monkeys. Cambridge University Press, Cambridge, pp. 321–340.

Plagmann, H.-Chr., Wartenberg, M., Kocher, Th., 1989. Schmelzoberflächenveränderungen nach Zahnsteinentfernung. Deutsche Zahnärztl Zeitschrift 44, 285–288.

Puech, P-F., 1977. Usure dentaire en anthropolgie étude par la technique des répliques. Revue d'Odonto-Stomatologie 6, 51–56.

Puech, P-F., 1984a. A la recherche du menu des premiers hommes. Cahiers Ligures de Préhistoire et de Protohistoire 1, 46–53.

Puech, P-F., 1984b. Acidic food Choice in *Homo habilis* at Olduvai. Current Anthropology 25, 349–350.

Puech, P-F., 1986a. Dental microwear features as an indicator for plant food in early hominids: a preliminary study of enamel. Human Evolution 1, 507–515.

Puech, P-F., 1986b. Tooth microwear in *Homo habilis* at Olduvai. Mèmoires Musee Histoire National, Paris (sèrie C) 53, 399–414.

Puech, P-F., Albertini, H., 1984. Dental microwear and mechanisms in early hominids from Laetoli and Hadar. American Journal of Physical Anthropology 65, 87–91.

Puech, P-F., Cianfarani, F., Albertini, H., 1986. Dental microwear features as an indicator for plant food in early hominids: a preliminary study of enamel. Human Evolution 1, 507–515.

Puech, P-F., Prone A., 1979. Reproduction experimentale des processes d'usure dentaire par abrasion: implications paleoecologique chex l'Homme fossile. Comptes Rendus de l'Academie des Sciences Paris 289, 895–898.

Puech, P-F., Prone, A., Albertini, H., 1981. Reproduction expérimentale des processus d-altération de la surface dentaire par friction non abrasive et non adhésive: application à l'étude de alimentation de L'Homme fossile. Comptes Rendus de l'Academie Sciences Paris 293, 729–734.

Puech, P-F., Prone, A., Kraatz, R., 1980. Microscopie de l'usure dentaire chez l'homme fossile: bol alimnetaire et environnement. Comptes Rendus de l'Academie des Sciences Paris 290, 1413–1416.

Puech, P-F., Prone, A., Roth, H., Cianfarani, F., 1985. Reproduction expérimentale de processus d'usure des surfaces dentaires des Hominides fossiles: conséquences morphoscopiques et exoscopiques avec application à l'Hominidé I de Garusi. Comptes Rendus de l'Academie des Sciences Paris 301, 59–64.

Puech, P-F., Serratrice, C., Leek, F.F., 1983. Tooth wear as observed in ancient Egyptian skulls. Journal of Human Evolution 12, 617–629.

Purnell, M.A., 1995. Microwear in conodont elements and macrophagy in the first vertebrates. Nature 374, 798–800.

Radlanski, R.J., Jäger A., 1989. Zur mikromorphologie der approximalen Kontaktflächen und der okklusalen Schliffacetten menschlicher Zähne Deutsche Zahnärztl Zeitschrift 44, 196–197.

Rafferty, K.L., Teaford, M.F., Jungers, W.L., 2002. Molar microwear of subfossil lemurs: improving the resolution of dietary inferences. Journal of Human Evolution 43, 645–657.

Raphael, K., Marbach, J., Teaford, M.F., 2003. Is bruxism severity a predictor of oral splint efficacy in patients with myofascial face pain? Journal of Oral Rehabilitation 30, 17–29.

Reinhard, K., de Souza, S.M.F., Rodrigues, C., Kimmerle, E., Dorsey-Vinton, S., 1999. Microfossils in dental calculus: a new perspective on diet and dental disease. In: Williams, E. (Ed.), Human Remains. Conservation, Retrieval, and Analysis. Archaeopress, Oxford, England, pp. 113–118.

Rensberger, J.M., 1978. Scanning electron microscopy of wear and occlusal events in some small herbivores. In: Butler, P.M., Joysey, K.A. (Eds.), Development, Function and Evolution of Teeth. Academic Press, New York, pp. 415–438.

Rensberger, J.M., 1982. Patterns of change in two locally persistent successions of fossil Aplodontid rodents. In: Kurtén, B. (Ed.), Teeth: Form, Function, and Evolution. Columbia University Press, New York, pp. 323–349.

Rensberger, J.M., 1986. Early chewing mechanisms in mammalian herbivores. Paleobiology 12, 474–494.

Rensberger, J.M., Krentz, H.B., 1988. Microscopic effects of predator digestion on the surfaces of bone and teeth. Scanning Microscopy 2, 1541–1551.

Rose, K.D., Walker, A., Jacobs, L.L., 1981. Function of the mandibular tooth comb in living and extinct mammals. Nature 289, 583–585.

Ryan, A.S., 1979. Wear striation direction on primate teeth: a scanning electron microscope examination. American Journal of Physical Anthropology 50, 155–168.

Ryan, A.S., 1981. Anterior dental microwear and its relationship to diet and feeding behavior in three African primates (*Pan troglodytes troglodytes, Gorilla gorilla gorilla*, and *Papio hamadryas*). Primates 22, 533–550.

Ryan, A.S., Johanson, D.C., 1989. Anterior dental microwear in *Australopithecus afarensis*: comparisons with human and nonhuman primates. Journal of Human Evolution 18, 235–268.

Schmidt, C.W., 2001. Dental microwear evidence for a dietary shift between two nonmaize-reliant prehistoric human populations from Indiana. American Journal of Physical Anthropology 114, 139–145.

Schubert, B.W., Ungar, P.S., 2005. Wear facets and enamel spalling in tyrannosaurid dinosaurs. Acta Palaeontologia Polonica 50, 93–99.

Scott, R.S., Ungar, P.S., Bergstrom, T.S., Brown, C.A., Grine, F.E., Teaford, M.F., Walker, A., 2005. Dental microwear texture analysis shows within-species diet variability in fossil hominins. Nature 436, 693–695.

Semprebon, G.M., Godfrey, L.R., Solounias, N., Sutherland, M.R., Jungers, W.L., 2004. Can low-magnification stereomicroscopy reveal diet? Journal of Human Evolution 47, 115–144.

Silcox, M.T., Teaford, M.F., 2002. The diet of worms: an analysis of mole dental microwear and its relevance to dietary inference in primates and other mammals. Journal of Mammalogy 83, 804–814.

Solounias, N., Hayek, L.C., 1993. New methods of tooth microwear analysis and application to dietary determination of two extinct antelopes. Journal of Zoology 229, 421–445.

Solounias, N., Moelleken, S.M.C., 1992a. Tooth microwear analyses of *Eotragus sansaniensis* (Mammalia: Ruminantia), one of the oldest known bovids. Journal of Vertebrate Paleontology 12, 113–121.

Solounias, N., Moelleken, S.M.C., 1992b. Dietary adaptations of two goat ancestors and evolutionary considerations. Geobios 25, 797–809.

Solounias, N., Moelleken, S.M.C., 1994. Differences in diet between two archaic ruminant species from Sansan, France. History and Biology 7, 203–220.

Solounias, N., Semprebon, G., 2002. Advances in the reconstruction of ungulate ecomorphology with application to early fossil equids. American Museum Novitates 3366, 1–49.

Solounias, N., Teaford, M.F., Walker, A., 1988. Interpreting the diet of extinct ruminants: the case of a non-browsing giraffid. Paleobiology 14, 287–300.

Strait, S.G., 1993. Molar microwear in extant small-bodied faunivorous mammals: an analysis of feature density and pit frequency. American Journal of Physical Anthropology 92, 63–79.

Sugawara, K., 1979. Sociological study of a wild group of hybrid baboons between *Papio anubis* and *Papio hamadryas* in the Awash Valley, Ethiopia. Primates 20, 21–56.

Szmulewicz, M.N., Andino, L.M., Reategui, E.P., Woolley-Barker, T., Jolly, C.J., Disotell, T.R., Herrera, R.J., 1999. An Alu insertion polymorphism in a baboon hybrid zone. American Journal of Physical Anthropology 109, 1–8.

Teaford, M.F., 1985. Molar microwear and diet in the genus *Cebus*. American Journal of Physical Anthropology 66, 363–370.

Teaford, M.F., 1988a. A review of dental microwear and diet in modern mammals. Scanning Microscopy 2, 1149–1166.

Teaford, M.F., 1988b. Scanning electron microscope diagnosis of wear patterns versus artifacts on fossil teeth. Scanning Microscopy 2, 1167–1175.

Teaford, M.F., 1991. Dental microwear what can it tell us about diet and Dental function? In: Kelley, M.A., Larsen, C.S. (Eds.), Advances in Dental Anthropology. Alan R. Liss, New York, pp. 341–356.

Teaford, M.F., 1993. Dental microwear and diet in extant and extinct *Theropithecus*: preliminary analyses. In: Jablonski, N.G. (Ed.), *Theropithecus*: The Life and Death of a Primate Genus. Cambridge University Press, Cambridge, pp. 331–349.

Teaford, M.F., 1994. Dental microwear and dental function. Evolutionary Anthropology 3, 17–30.

Teaford, M.F., 2002. Dental enamel microwear analysis. In: Hutchinson, D.L. (Ed.), Foraging, Farming and Coastal Biocultural Adaptation in Late Prehistoric North Carolina. University Press of Florida, Gainesville, Florida, pp. 169–177.

Teaford, M.F., Byrd, K.E., 1989. Differences in tooth wear as an indicator of changes in jaw movement in the guinea pig (*Cavia porcellus*). Archives of Oral Biology 34, 929–936.

Teaford, M.F., Glander, K.E., 1991. Dental microwear in live, wild-trapped *Alouatta palliata* from Costa Rica. American Journal of Physical Anthropology 85, 313–319.

Teaford, M.F., Glander, K.E., 1996. Dental microwear and diet in a wild population of mantled howlers (*Alouatta palliata*). In: Norconk, M., Rosenberger, A., Garber, P. (Eds.), Adaptive Radiations of Neotropical Primates. Plenum Press, New York, pp. 433–449.

Teaford, M.F., Larsen, C.S., Pastor, R.F., Noble, V.E., 2001. Pits and scratches: microscopic evidence of tooth use and masticatory behavior in La Florida. In: Larsen C.S. (Ed.), Bioarchaeology of Spanish Florida: The Impact of Colonialism. University Press of Florida, Gainesville, Florida, pp. 82–112.

Teaford, M.F., Lucas, P.W., Ungar, P.S., Glander, K.E., 2006. Mechanical defenses in leaves eaten by Costa Rican *Alouatta palliata*. American Journal of Physical Anthropology 129, 99–104.

Teaford, M.F., Lytle, J.D., 1996. Diet-induced changes in rates of human tooth microwear: a case study involving stone-ground maize. American Journal of Physical Anthropology 100, 143–147.

Teaford, M.F., Oyen, O.J., 1989a. Differences in the rate of molar wear between monkeys raised on different diets. Journal of Dental Research 68, 1513–1518.

Teaford, M.F., Oyen, O.J., 1989b. *In vivo* and *in vitro* turnover in dental microwear. American Journal of Physical Anthropology 80, 447–460.

Teaford, M.F., Robinson, J.G., 1989. Seasonal or ecological differences in diet and molar microwear in *Cebus nigrivittatus*. American Journal of Physical Anthropology 80, 391–401.

Teaford, M.F., Runestad, J.A., 1992. Dental microwear and diet in Venezuelan primates. American Journal of Physical Anthropology 88, 347–364.

Teaford, M.F., Tylenda, C.A., 1991. A new approach to the study of tooth wear. Journal of Dental Research 70, 204–207.

Teaford, M.F., Ungar, P.S., 2000. Diet and the evolution of the earliest human ancestors. Proceedings of the National Academy of Sciences of the USA 97, 13506–13511.

Teaford, M.F., Ungar, P.S., Grine, F.E., 2002a. Fossil evidence for the evolution of human diet. In: Ungar, P.S., Teaford, M.F. (Eds.), Human Diet: Its Origins and Evolution. London, Bergen, and Garvey, Westport, CT, pp. 143–166.

Teaford, M.F., Ungar, P.S., Grine, F.E., 2002b. Molar microwear and diet of *Praeanthropus afarensis*: preliminary results from the Denan Dora member, Hadar formation, Ethiopia. American Journal of Physical Anthropology Suppl. 34, 154.

Teaford, M.F., Walker, A., 1983. Dental microwear in adult and still-born guinea pigs (*Cavia porcellus*). Archives of Oral Biology 28, 1077–1081.

Teaford, M.F., Walker, A., 1984. Quantitative differences in dental microwear between primate species with different diets and a comment on the presumed diet of *Sivapithecus*. American Journal of Physical Anthropology 64, 191–200.

Turssi, C.P., Ferracane, J.L., Serra, M.C., 2005. Abrasive wear of resin composites as related to finishing and polishing procedures. Dental Materials 21, 641–648.

Ungar, P.S., 1990. Incisor microwear and feeding behavior in *Alouatta seniculus* and *Cebus olivaceus*. American Journal of Primatology 20, 43–50.

Ungar, P.S., 1994. Incisor microwear of Sumatran anthropoid primates. American Journal of Physical Anthropology 94, 339–363.

Ungar, P.S., 1995. A semiautomated image analysis procedure for the quantification of dental microwear II. Scanning 17, 57–59.

Ungar, P.S., 1996. Dental microwear of European Miocene catarrhines: evidence for diets and tooth use. Journal of Human Evolution 31, 335–366.

Ungar, P.S., 2004. Dental topography and diets of *Australopithecus afarensis* and early *Homo*. Journal of Human Evolution 46, 605–622.

Ungar, P.S., 2007. Dental topography and human evolution. In: Bailey, S.E., Hublin, J-J. (Eds.), Dental Perspectives on Human Evolution: State of the Art Research in Dental Paleoanthropology. Springer, Dordrecht.

Ungar, P.S., Brown, C.A., Bergstrom, T.S., Walker, A., 2003. Quantification of dental microwear by tandem scanning confocal microscopy and scale-sensitive fractal analyses. Scanning 25, 185–193.

Ungar, P.S., Grine, F.E., 1991. Incisor size and wear in *Australopithecus africanus* and *Paranthropus robustus*. Journal of Human Evolution 20, 313–340.

Ungar, P.S., Grine, F.E., Teaford, M.F., El-Zaatari, S., 2006. Dental microwear and diets of African early *Homo*. Journal of Human Evolution 50, 78–95.

Ungar, P.S., Simon, J-C., Cooper, J.W., 1991. A semiautomated image analysis procedure for the quantification of dental microwear. Scanning 13, 31–36.

Ungar, P.S., Teaford, M.F., 1996. Preliminary examination of non-occlusal dental microwear in anthropoids: implications for the study of fossil primates. American Journal of Physical Anthropology 100, 101–113.

Ungar, P.S., Teaford, M.F., Glander, K.E., Pastor, R.F., 1995. Dust accumulation in the canopy: a potential cause of dental microwear in primates. American Journal of Physical Anthropology 97, 93–99.

Ungar, P.S., Teaford, M.F., Grine, F.E., 2001. A preliminary study of molar microwear of early *Homo* from East and South Africa. American Journal of Physical Anthropology Suppl. 32, 153.

Ungar, P.S., Teaford, M.F., Kay, R.F., 2004. Molar microwear and shearing crest development in Miocene catarrhines. Anthropology 42, 21–35.

Van Valkenburgh, B., Teaford, M.F., Walker, A., 1990. Molar microwear and diet in large carnivores: inferences concerning diet in the sabretooth cat, *Smilodon fatalis*. Journal of Zoology 222, 319–340.

Walker, A., 1980. Functional anatomy and taphonomy. In: Behrensmeyer, A.K., Hill, A.P. (Eds.), Fossils in the Making. University of Chicago Press. Chicago, pp. 182–196.

Walker, A., 1981. Diet and teeth, dietary hypotheses and human evolution. Philosophical Transactions of the Royal Society of London 292(B), 57–64.

Walker, A., 1984. Mechanisms of honing in the male baboon canine. American Journal of Physical Anthropology 65, 47–60.

Walker, A., Hoeck, H.N., Perez, L., 1978. Microwear of mammalian teeth as an indicator of diet. Science 201, 908–910.

Walker, P.L., 1976. Wear striations on the incisors of cercopithecoid monkeys as an index of diet and habitat preference. American Journal of Physical Anthropology 45, 299–308.

Wu, S.S., Yap, A.U., Chelvan, S., Tan, E.S., 2005. Effect of prophylaxis regimens on surface roughness of glass ionomer cements. Operative Dentistry. 30, 180–184.

Young, W.G., Brennan, C.K.P., Marshall, R.I., 1990. Occlusal movements of the brushtail possum, *Trichosurus vulpecula*, from microwear on the teeth. Australian Journal of Zoology 38, 41–51.

Young, W.G., Robson, S.K., 1987. Jaw movement from microwear on the molar teeth of the koala *Phascolarctos cinereus*. Journal of Zoology 213, 51–61.

5. Tooth wear and diversity in early hominid molars: A case study

L. ULHAAS
Research Institute Senckenberg
Department of Paleoanthropology and Quaternary Paleontology
Senckenberganlage 25
60325 Frankfurt am Main
Germany
lilian.ulhaas@senckenberg.de

O. KULLMER
Research Institute Senckenberg
Department of Paleoanthropology and Quaternary Paleontology
Senckenberganlage 25
60325 Frankfurt am Main
Germany
ottmar.kullmer@senckenberg.de

F. SCHRENK
JWG University Frankfurt, Vertebrate Paleobiology Institut for
Ecology, Evolution & Diversity
Siesmayerstrasse 70,
60054 Frankfurt am Main, Germany
schrenk@bio.uni-frankfurt.de

Keywords: hominids, australopithecines, molar morphology, occlusal topography, wear pattern, wear facets, 3-D analysis, functional morphology, tooth architecture, dental occlusal compass

Abstract

Functional relationships between diet and tooth morphology form an integral part of paleontological research. The detailed description of occlusal relief and wear patterns of molars provides information about food ingestion and mastication. In early hominids overall molar morphology is fairly similar. Size measurements, such as buccolingual or mesiodistal diameter and 2-D cusp area of hominid molars show considerable overlap. The pioneering works of Butler, Mills, Hiiemae, Kay, Maier and others have shown that the wear pattern on the occlusal surface seems to reflect mastication behavior as an indication of diet. However, most of the interpretations are based on two-dimensional analyses. Occlusal relief measured in 3-D highlights functionally important

S.E. Bailey and J.-J. Hublin (Eds.), Dental Perspectives on Human Evolution, 369–390.

features useful for quantifying the complex wear patterns on hominid teeth. However, until recently they could not be measured because techniques and methods were lacking. Nevertheless the results of 2-D analyses so far demonstrate that the occlusal surface of teeth records a significant part of the life history of an individual. The 3-D analysis of wear patterns on hominid teeth may provide additional information regarding the relationships between diet, chewing behavior and early hominid evolution. In this case study we employ a new 3-D approach to compare details on the occlusal surface of worn molars of *Australopithecus afarensis, Australopithecus africanus* and *Paranthropus robustus* in order to examine possible differences in tooth wear patterns. High resolution optical topometry enables us to measure parameters on 3-D computer models of teeth. Here, we compare various occlusal morphologies of worn lower second molars and attempt to interpret function, taking dental and masticatory principles into account. Our results indicate that diverse modes of occlusal wear in *Australopithecus* and *Paranthropus* are evident. A closer look at the occlusal relief and wear facet pattern shows that an assortment of mechanisms for crushing, shearing and grinding on a single tooth are common, since orientation and inclination of wear facets vary. The fact that *A. afarensis* molars show diverse functional areas with little variation among individuals suggests it had a dental toolkit to cope with a wide range of food qualities and may indicate a species-specific dietary spectrum. *A. africanus* and *P. robustus* molars, with their pronounced and relatively rapid flattening of crown relief and diverse individual wear patterns, point towards hard-object feeding and greater intraspecific variation in diet. *P. robustus*, however, with somewhat higher occlusal relief, can be interpreted as an omnivorous generalist with hard objects as fall-back foods.

Introduction

It is generally agreed that tooth wear in mammals is a result of the relationship between the upper and lower jaws, so that the resulting wear patterns yield information about both the mechanisms of food breakdown and the composition of the diet. Teeth are constructed and positioned for a close interaction between the antagonistic upper and lower dentitions. Thus, biomechanical constraints derived from tooth form and tooth size affect how food is prepared and which dietary items are selected. However, it has been hypothesized that tooth relief is adapted to species-specific diet as a compromise of phylogenetic preconditions and actual functional demands (Maier and Schneck, 1981, 1982).

In addition to tooth shape, jaw movement has an important influence on mastication and tooth wear. Previous studies showed that there are two major types of jaw action during the masticatory cycle (Crompton and Hiiemae, 1970; Hiiemae and Crompton, 1971; Hiiemae and Kay, 1972, 1973; Kay and Hiiemae, 1974). The masticatory cycle generally starts with puncture-crushing, in which tooth-tooth contact rarely occurs. Puncture-crushing generates blunting of cusp tips (Hiiemae and Kay, 1973) and the loss of tooth material during this action is caused mainly by abrasion or contact with the various foodstuffs (Stones, 1948; Kay and Hiiemae, 1974). Puncture-crushing generally results in wear areas with coarse surfaces, rather than wear facets that are clearly defined and circumscribed areas of polished enamel (Stones, 1948; Kay, 1977).

The puncture-crushing phase is followed by the chewing phase (Crompton and Hiiemae, 1970). It is during this second type of action (chewing) that tooth-tooth contact occurs more often. After food is sufficiently softened, the antagonistic teeth come into contact and the direction of jaw movement is guided by the crown morphology. During chewing, complementary edges and areas of the antagonistic lower and upper teeth, which form functional units, shear against each other and food is cut between pairs of edges and ground between facets (Maier and Schneck, 1981, 1982). Thus, foodstuffs are exposed to compression, traction, and shearing, depending on the crown relief.

The original or primary relief of the tooth crown is subsequently reshaped by these masticatory processes. Wear facets develop and disappear during the functional life of a tooth. Food ingestion and the mechanical destruction of food particles lead to the loss of enamel and dentine on the occlusal surface, subsequently forming a secondary relief pattern of complementary faces. Numerous other factors influence the type and rate of dental wear (Dahl et al., 1989). These include, among other things, age, sex, occlusal conditions, hyperfunction, bite force, gastrointestinal disturbances, salivary factors and environmental factors (Kim et al., 2001). While some have focused on using tooth wear for individual age determination (Miles, 1963; Scott, 1979; Brothwell, 1981), others have analyzed wear patterns to assess the functional and morphological relationship between tooth features of upper and lower jaws (e.g., Crompton and Hiiemae, 1970; Crompton, 1971; Butler, 1973; Kay, 1973, 1975, 1981; Kay and Hiiemae, 1974; Maier, 1977a, b, 1978, 1984; Maier and Schneck, 1981, 1982; Teaford, 1983; Janis, 1984, 1990; Strait, 1993a, b; Ungar and Williamson, 2000; M'Kirera and Ungar, 2003; Ungar and M'Kirera, 2003; Ulhaas et al., 2004). The latter is based on the idea that even in worn teeth, functional occlusion must be maintained so that food continues to be efficiently processed.

The creation of the secondary relief as the crown surface is worn is closely related to jaw movements and follows a certain pattern. Starting from the point of maximum intercuspation in the centric position, the central fovea of the lower molars accommodates the protocone of the upper molars. The absolute position varies due to morphologic variation of the group-function cusp-to-fossa relationships known as Angle's Classes I–III (Angle, 1900). Kay and Hiiemae (1974) explained how the complex system of wear facets is produced during phases I and II of the chewing cycle in extant and extinct primates. In phase I, facet pairs 1 to 8 are active, while all facet pairs with numbers higher than 8 come into contact in phase II (Figure 1). In general, facets 1 to 8 face buccally in the lower molars and lingually in the upper molars. Five pairs of phase II surfaces (facets 9 to 13) are described on hominoid molars (Maier and Schneck, 1981). These facets are roughly aligned buccally in the upper molars and lingually in the lower molars. Microscopic striations on the facets indicate the general

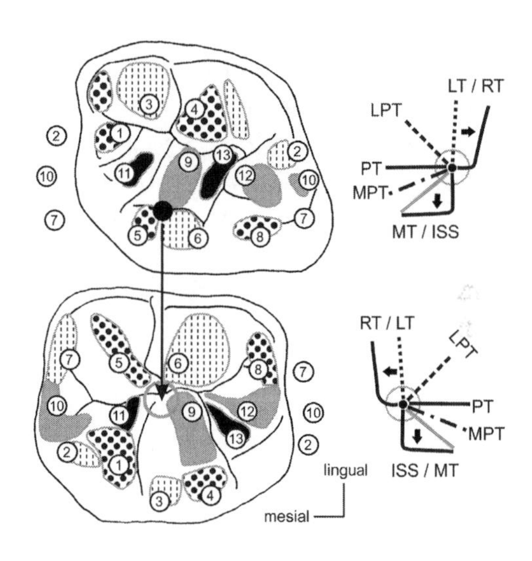

Figure 1. Schematic illustration of complementary wear facets on upper and lower molars; numerical system (1–13) after Maier and Schneck (1981). Arrow indicates occlusal relation in maximum intercuspation of protocone apex of the upper molars and central fovea of the lowers. The circle in the on the lower molar marks the position of the tip of the protocone during maximum intercuspation. Facet pairs 1, 4, 5 and 8 are in contact during lateroretrusion movements (dotted). Facet pairs 2, 3, 6 and 7 correspond to lateroprotrusion (stripped), and facet pairs 9, 10 and 12 are in close contact during mediotrusion and immediate side shift (grey). Facet pairs 11 and 13 correspond to medioprotrusion (black). To the right, the dental occlusal compass (after Schulz, 2003) shows major directions of horizontal occlusal movements for upper molars (above) and lower molars (below), laterotrusion (LT), retrusion (RT), lateroprotrusion (LPT), protrusion (PT), medioprotrusion (MPT), mediotrusion (MT) and immediate side shift (ISS).

orientation of movements (Kay and Hiiemae, 1974; Gordon, 1984). Some wear areas clearly show that movements take place in more than one direction, since their major striation orientations vary (Gingerich, 1972; Kay and Hiiemae, 1974). Kay and Hiiemae (1974, Figure 17) noticed that differences in striae orientation on some facets can also occur as a result of different chewing actions, reflecting the direction of movement of the tooth during phase I of the chewing power stroke and at the same time the direction during the balancing side occlusion in the recovery stroke.

Chewing movements can be reconstructed from dental wear patterns in more detail using the dental occlusal compass (Douglass and DeVreugd, 1997; Schulz, 2003; Schulz and Winzen, 2004). The repertoire of possible chewing movements is restricted by dental relief, which predetermines the pathways on which teeth can be guided against each other. It is also limited by temporomandibular joint morphology, which limits sweeping movements of the lower jaw. Thus, both tooth crown relief and muscle activity play a major role in the biomechanics of occlusion. The dental occlusal compass takes these observations into account and schematically documents the relationships of cusps and possible movement corridors (see Figure 1). The compass depicts the possible directions of antagonistic tooth-on-tooth contact, depending on condylar movements. The fissure pattern on the occlusal surface describes the track of laterotrusion, mediotrusion and protrusion, whereas lateroprotrusion and medioprotrusion are directed towards cusp slopes. There are two threshold areas on the occlusal compass: retrusion and immediate side shift. The retrusion space marks a field for backward movement in which both condyles back up into the articular fossa. The immediate side shift indicates a transverse functional field on the balancing side where the condyle is shifted medially. Jaw motion analyses document that in addition to horizontal movements, upward and downward movements are also important for food breakdown. Thus, mandibular surtrusion (upwards) and detrusion (downwards) act during mastication in combination with horizontal translation to form the occlusal pattern (Douglass and DeVreugd, 1997).

The compass provides information about the possible relative chewing movements at any specific position on the occlusal surface and therefore allows the interpretation of the contact areas, as the contact pattern and the pattern of jaw movement are supposed to be highly correlated during the chewing cycle. For each wear facet there is only one antagonistic area that fits perfectly into the topography of the opposing tooth at a specific moment of the occlusion process.

Our interest in this study is in analyzing the wear patterns of early hominin teeth, based on the above topographical and functional constraints. We try to determine how wear patterns are formed and what movements are responsible for the contacts on the occlusal surface. Given that teeth are abundant in the fossil record, they are a critical source of information for reconstructing jaw movements in fossil hominins. If we look more closely at the wear patterns of teeth and reconstruct the actual progression of wear, we have a good chance of understanding the relationship between tooth wear and diet.

The functional significance of tooth form and size of the australopithecines has been the focus of a great deal of work (e.g., Robinson, 1954, 1956; Tobias, 1967; Sperber, 1973; Grine, 1981). Robinson (1956) mentioned that both, "robust" and "gracile" australopithecines (a.k.a. *Paranthropus* and *Australopithecus*, respectively), wear down their teeth to planar surfaces. He also noted that with advanced wear, the "gracile" australopithecines show a strong buccal inclination in the occlusal plane of the lower molars, while those of the "robust" australopithecines exhibit a strong lingual slope. Later Robinson (1972) claimed that the "robust" postcanines "... tend to wear rapidly to a more or less plane surface..." and that "... these teeth clearly are specialized for crushing and grinding, not slicing and shearing" (p. 228).

Wolpoff (1975, 1976) focused on inter- and intraspecific variation and claimed that the variability of tooth wear patterns in the South African australopithecines has nothing to do with proposed taxonomic schemes, but rather with differences in dietary preferences between populations. In contrast, Wallace (1972, 1973, 1974, 1975, 1978) concluded from his wear pattern analysis that the "gracile" as well as the "robust" dentitions had processed food with similar consistency, although the "robust" dentitions had been adapted more for crushing and grinding than the "gracile" dentitions.

Grine (1981) used deciduous molars in his analyses of macro- and microwear pattern on functional homologous facets in order to infer dietary preferences in South African australopithecines. He wrote that the severe wear of the "robust" australopithecines seemed to indicate the mastication of harder and more fibrous foods than were ingested by the "gracile" australopithecines.

These earlier studies were all undertaken in two dimensions. Given the complexity of occlusion and tooth wear, any dimensionial reduction in describing tooth morphology must result in a loss of information. Therefore, in order to fully understand dental functional morphology it is necessary to apply techniques and methods for acquiring and processing three-dimensional (3-D) data.

The first attempt at this kind of characterization of molar morphology was made by Kay and colleagues (Kay, 1975; Kay, 1978; Kay and Hylander, 1978; Kay and Covert, 1984), who measured the lengths of molar shearing crests in order to calculate a shearing quotient (SQ). Those measurements lack accuracy, however, due to the fact that crests are more concave than straight, and also due to possible orientation problems.

Teaford applied stereophotogrammetric techniques to analyze dental morphology, using x, y and z coordinates (Teaford, 1982, 1983). His was the first attempt at including worn specimens in functional analyses. Strait (1993a) and Reed (1997) were able to quantify dental features in three dimensions using a reflex microscope. Jernvall and Selänne (1999) made use of laser confocal microscopy to obtain 3-D models of teeth for further metrical analysis, but this method is of limited value for fossil hominin teeth because it can only be applied to very small objects ($\leqslant$10 mm). Meanwhile, 3-D scanning techniques like laser scanning (Ungar and Williamson, 2000; Ungar et al., 2002; M'Kirera and Ungar, 2003; Ungar and M'Kirera, 2003; Dennis et al., 2004) and optical 3-D topometry as presented here (see also Kullmer et al., 2002a, b; Ulhaas et al., 2004), enable easy and high-resolution 3-D reproduction of teeth.

This emergence of 3-D technology both facilitates and requires the establishment of new structural parameters. Reed (1997) and Zuccotti et al. (1998) introduced the use of Geographic Information Systems (GIS) to the analysis of occlusal surfaces of molars. These computer tools serve to model and study the surface of the earth and can be used to examine and characterize the occlusal surface of teeth. Treated like landscape data, the dental occlusal surface can be quantified by measuring surface area, relief, and slope of cusps, as well as the whole occlusal surface. With this technique, Ungar and colleagues (Ungar and Williamson, 2000; Ungar et al., 2002; M'Kirera and Ungar, 2003; Ungar and M'Kirera, 2003; Dennis et al., 2004) were able to include worn teeth into functional 3-D analyses that were, until then, restricted to unworn or slightly worn teeth.

In their comparative analysis of worn molars of gorillas and chimpanzees, Ungar and M'Kirera (2003) were able to show that basic functional differences of the molars do not change with wear and that dental function is maintained. The same was observed by Ulhaas et al.. (2004) for cercopithecoids.

In this contribution we introduce the application of portable optical 3-D topometry to occlusal surface analysis of hominid molar crowns. Our approach attempts to determine in more detail the functional morphology of the complex hominid molar occlusal surface. The complex occlusal relief comprises many differently oriented and inclined faces. This arrangement serves to meet different biomechanical demands that are made by a diet composed of foodstuffs with varying physical properties. Analyzing the occlusal surface in more detail should therefore enable a better understanding of the functional capabilities of these molars and a more reliable dietary reconstruction of fossil species. In addition to the quantification of the relief and the overall comparison of tooth morphology, including size, cusp height and surface area, it is of special interest to look at each specimen individually to examine, on the one hand, a general pattern of occlusion and, on the other hand, individual variation. Each tooth crown is unique in terms of size and shape. Regarding wear on the occlusal surface, each tooth shows a specific pattern, representing an "occlusal fingerprint" of the individual.

Our goal is to assess inter- and intraspecific variation in crown topography and wear patterns of fossil hominid molars. Detailed functional analysis of the occlusal morphology, as well as the reconstruction of masticatory movements, will help to assess both differences and similarities among *Australopithecus afarensis, Australopithecus africanus* and *Paranthropus robustus.* This information can then be used to reconstruct the diet of these fossil groups and test traditional dietary hypotheses.

Materials and Methods

Digital Model Generation

A selection of *Australopithecus afarensis* (AL266-1, AL333w-60, LH4), *Australopithecus. africanus* (STW 14, STW 404, MLD 40) and *Paranthropus robustus* (SKW5, SK23, SK34) second lower molars were examined (Figure 2). Each of the three molars in each group represents a different stage of wear: slight wear, progressed wear and advanced wear. No unworn specimens were chosen because our aim was to quantify and compare wear areas. Original specimens of *A. africanus* and *P. robustus* were scanned with the portable optical 3-D measuring system optoTOP (Breuckmann GmbH) at the Transvaal Museum in Pretoria and the University of the Witwatersrand in Johannesburg, South Africa. In the case of the *A. afarensis* specimens, casts from the collection of the Paleoanthroplogy Department of the Research Institute Senckenberg in Frankfurt, Germany were used.

The scanning system is an optical triangulation scanner (optoTOP, Breuckmann GmbH) that is based on pattern projection. The scanner consists of a digital camera and a projector that projects a grid pattern on the object. The light is projected from a known position in a known direction, and the camera and projector are arranged at a fixed distance and fixed angle relative to each other (Figure 3). The camera captures snapshots of a grid pattern projected on the object surface. The pattern used by this system is a coded pattern called "Gray code" that uniquely expresses light directions, combined with a phase-shift approach to increase resolution (Beraldin et al., 2000). If the target surface is flat, the lines of the pattern appear straight and undeformed. Any surface features result in changes of the projected pattern, as the lines of the grid will appear curved, bent, or otherwise deformed. The system calculates

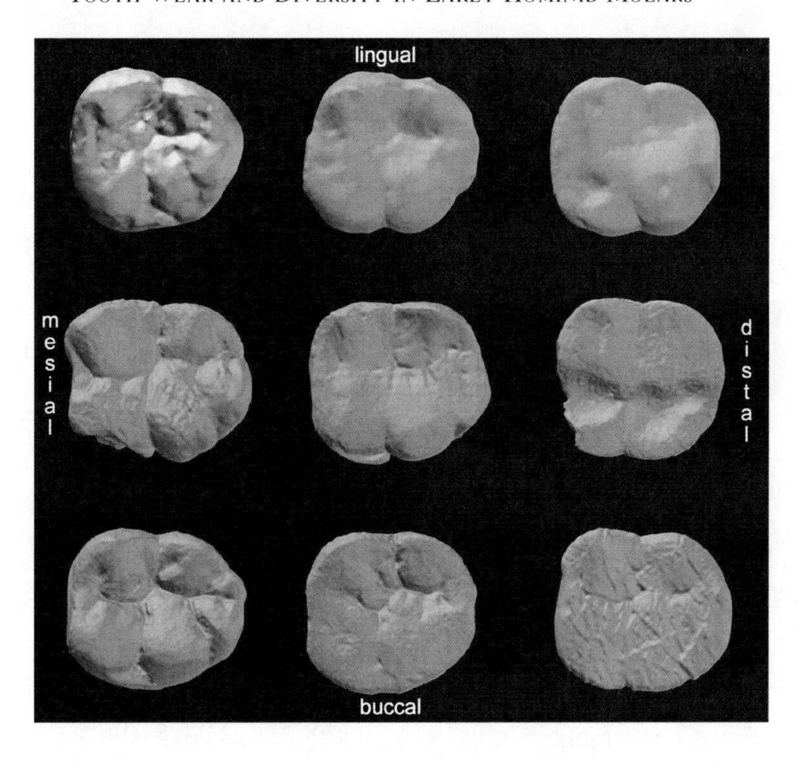

Figure 2. Occlusal aspect of digital models of lower molars included in this case study. All crowns are imaged size independent as left molars, from left to right in the upper row *Australopithecus afarensis* second molars of AL 266-1 (mirrored), AL 333w-60 and LH 4; middle row, *Australopithecus africanus* STW 14 (mirrored), STW 404 and MLD 40 (mirrored) and lower row *P. robustus* SKW 5, SK 23 and SK 34 (mirrored). Wear stage increases from left to right.

the profile of the surface based on known projection angle of the pattern, triangulation formulas and other mathematical techniques (Beraldin et al., 2000; Godin et al., 2002).

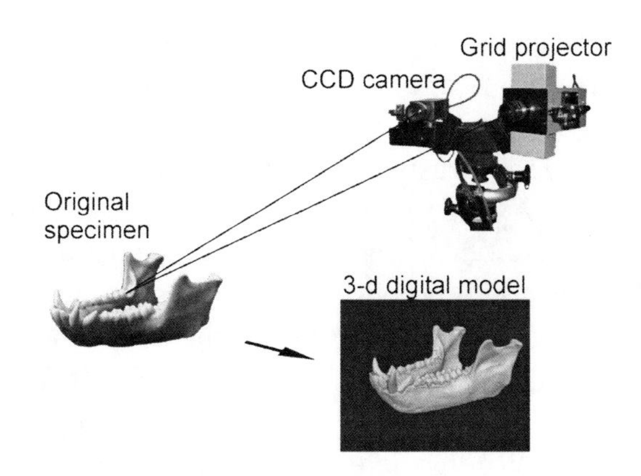

Figure 3. Optical triangulation scanner optoTOP (Breuckmann GmbH) including CCD camera and pattern projector, with the digital model of a *Cercopithecus* mandible.

To obtain data of the whole object, the specimen is scanned from different views. This is achieved by moving the portable scanning unit, which is mounted on a tripod, around the object. The number of views needed depends on surface reflection and complexity of the object. In the case of jaws and teeth we take scans from about 15 different views. Each view results in a 3-D point cloud. To visualize the complete surface, point clouds of the different views are matched and merged. For this purpose, we use the modules IMAlign and IMMerge of the software-package Polyworks™ (InnovMetric Inc.). A triangulated surface model results, with a mean resolution of 0.065mm. Finally, we used 2-D photographs of occlusal, buccal and lingual aspects of each specimen that were taken with a digital camera (Leica DC 300), attached to a binocular microscope (Leica MZ 12) to support the qualitative examination of

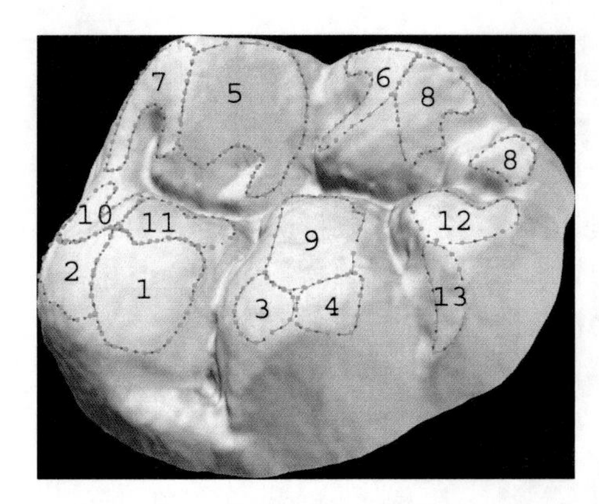

Figure 4. Digital model of SKW 5 lower left second molar with interactively marked wear facets. Facets are numbered according to Crompton (1971), Kay (1977), Maier (1977a, b), and Maier and Schneck (1981).

areas of abrasion, wear facets, exposed dentine and enamel chipping. The images help to mark details like wear facets interactively on the corresponding computerized 3-D model (Figure 4). Interactive marking of the digital object can be carried out in Polyworks™ by drawing a curve onto the model with the help of the mouse, taking the original specimen or the photographs as "draft".

Quantification of Occlusal Relief

For data analysis we apply a method described by Kullmer et al. (2002a) and by Ulhaas et al. (2004), using Polyworks™ modules IMEdit and IMInspect, as well as the CAD software Rhinoceros®.

One major obstacle for comparing tooth crowns of different wear stages is the lack of classical landmarks that can be used as a reference since cusp tips, crests, and fissures are modified with wear. Thus, we chose to define a reference plane that is independent of occlusal wear. According to Wood and Abbott (1983), Wood (1993) and Ungar et al. (2002), the most suitable is a plane that is calculated through the cervical margin of each individual tooth. Therefore, we marked the entire cervical margin of each tooth interactively on the 3-D model. In order to increase accuracy we selected all data points in a distance of 0.5 mm to compensate for uneven patches (e.g., cracks) in the margin. Next, a best fit plane was generated through all selected points. Then, the lowest point of the fissure pattern was marked interactively and the plane was translated parallel to pass through this point. Finally, the model was then sectioned by this plane and the occlusal part of the tooth model was detached and saved separately for surface analysis.

Measurements

In addition to traditional 2-D measurements such as mesio-distal length and bucco-lingual breath, the 3-D surface areas of the entire occlusal surface, as well as of single cusps, was measured using the Polyworks™ modules IMEdit and IMInspect. The contour-lines of the tooth crown and all single cusps are marked interactively on the 3-D model and subsequently extracted and imported in Rhinoceros®. The projection of these contour lines along the z-axis onto the xy-plane allows measuring of the 2-D planar base area of the occlusal surface and the cusps. The 3-D surface area is then divided by the corresponding 2-D base area to get a size-independent index value, representing the complexity of the occlusal relief. Simply put, a low relief is indicated by a low index value. The measurements taken are listed in Table 1. As the relief index was calculated for the occlusal crown and each cusp separately, changes could be observed in more detail.

Additionally, wear areas on the cusp slopes were quantified. We interactively identified areas of abrasion and attrition on the digital models. We labeled the areas of wear by using the numerical system of wear facets according to Crompton (1971), Kay (1977), Maier (1977a, b) and Maier and Schneck (1981). As wear facets can be homologized,

Table 1. Relief index measurements

3-d index	*AL 266-1*	*AL 333w-60*	*LH 4*	*STW 14*	*STW 404*	*MLD 40*	*SKW 5*	*SK 23*	*SK 34*
OA[1]	1.37	1.25	1.19	1.38	1.30	1.41	1.55	1.20	1.38
PC[1]	1.27	1.19	1.13	1.40	1.18	1.28	1.57	1.30	1.17
MC[1]	1.61	1.49	1.28	1.41	1.34	1.68	1.68	1.31	1.54
HC[1]	1.18	1.13	1.13	1.26	1.23	1.23	1.39	1.12	1.19
HCLD[1]	1.29	1.14	1.35	1.45	1.39	1.38	1.59	1.20	1.33
EC[1]	1.39	1.27	1.10	1.39	1.45	1.47	1.55	1.15	1.51
T6[1]	–	–	–	–	–	–	1.52	1.28	1.51

[1]OA: occlusal area PC: Protoconid MC: Metaconid HC: Hypoconid EC: entoconid HCLD: Hypoconulid T6: Tuberculum sextum

this system enables taxonomic as well as functional comparisons of molars.

Kay and Hiiemae (1974) showed that a molar tooth surface consists of a set of planes and features oriented in various angles to each other in space and that these reflect pathways of movement across the surface. Kay and Hiiemae (1974) used a technique widespread in geosciences (Lahee, 1957) to quantify the direction of movement in the jaws. The position of any plane in space can be determined by two angles, "strike" to determine horizontal orientation and "dip" to assess inclination (Figure 5). While the "strike" angle is measured with regard to the longitudinal axis of the tooth crown, the "dip" angle relates to the reference plane, that is, the cervical plane described above. In order to define the longitudinal axis of the tooth crown, we set two tangents to the buccal and lingual aspect, respectively. The bisector of the angle between the tangents gives us the longitudinal axis.

The measurements collected are listed in Table 2. For quantitative analysis, "best-fit" planes were generated to each of the wear facets marked on the digital model with a special software tool. That way each facet was represented by a plane which enabled us to measure its "dip" and "strike" angles with regard to the reference plane and the longitudinal axis of the tooth, respectively. The dip was measured by a software tool that determines the angle between the best-fit plane (representing the wear facet) and the reference plane. The strike (the horizontal orientation of the best-fit plane) is then measured as the angle between the intersection line of the best-fit plane with the reference plane and the longitudinal axis.

To illustrate and interpret the "strike and dip" measurements we used a stereoplot, in which the pattern of curves ("great circles") shows the orientation of the best fit planes. These reflect the average arrangement of contact areas on the occlusal surface (Kullmer et al., 2002a; Ulhaas et al., 2004). Basically, the graph is a projection of a hemisphere, which is intersectioned by one or more planes (Figure 6). The intersections are indicated by the curves. The closer a curve is positioned to the center of the graph, the steeper is the inclination of the plane. A plane parallel to the reference plane, reflecting an absolutely flat area, will be highlighted on the margin of the graph. The position of the endpoints of a great circle indicates the horizontal orientation of the represented plane. With the help of the stereoplot we were able to illustrate and compare the wear patterns of tooth crowns. For teeth of different wear stages, the decrease of relief can be illustrated as well. Detailed comparison of teeth of similar wear stages may indicate differences in relief modification through wear. Generally, the stereoplot illustrates the individual pattern of wear as each tooth is described by an arrangement of great circles that reflects its uniqueness.

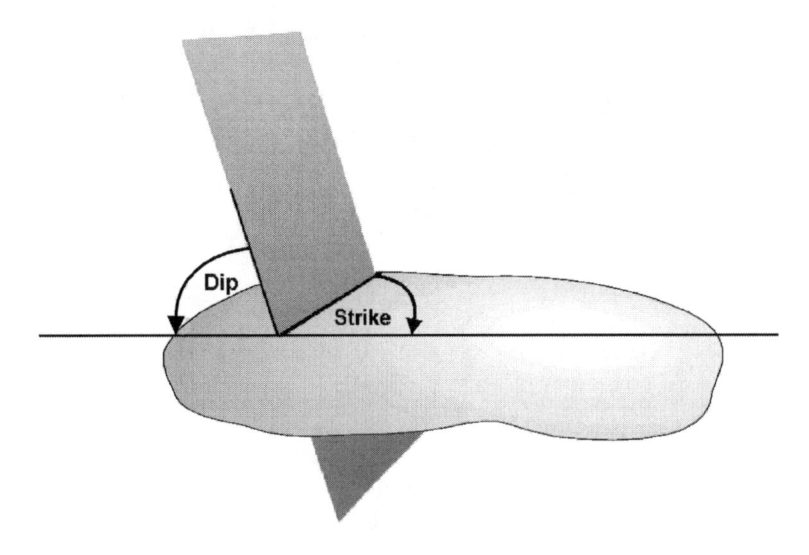

Figure 5. Schematic illustration of "strike" and "dip" measurements.

Table 2. Strike and dip measurements

	STW 14	*STW 404*	*MLD 40*	*AL 266-1*	*AL 333 w-60*	*LH 4*	*SKW 5*	*SK 34*	*SK 23*
Strike facet 1	356.78	79.67	86.48	9.61	12.89	56.02	28.55	56.35	135.01
Dip facet 1	9.84	13.80	9.70	16.06	8.92	10.35	18.80	6.74	5.25
Strike facet 4	337.09	65.69	151.27	44.43	35.44	27.66	0.35	51.96	45.65
Dip facet 4	8.23	24.38	2.35	26.89	29.04	21.89	15.67	10.55	27.96
Strike facet 5	355.58	25.01	18.79	26.53	38.06	17.20	19.00	5.70	27.72
Dip facet 5	26.01	28.29	14.24	36.34	35.99	22.18	43.90	25.96	24.03
Strike facet 8	359.21	79.96	325.62	347.13	53.39	354.03	38.93	327.5	38.19
Dip facet 8	14.67	29.64	10.58	9.22	27.88	13.28	28.71	25.03	13.23
Strike facet 9	244.61	249.58	81.99	230.86	268.84	235.16	274.18	271.28	280.87
Dip facet 9	14.25	13.36	7.39	9.01	7.56	7.27	22.63	6.07	4.92

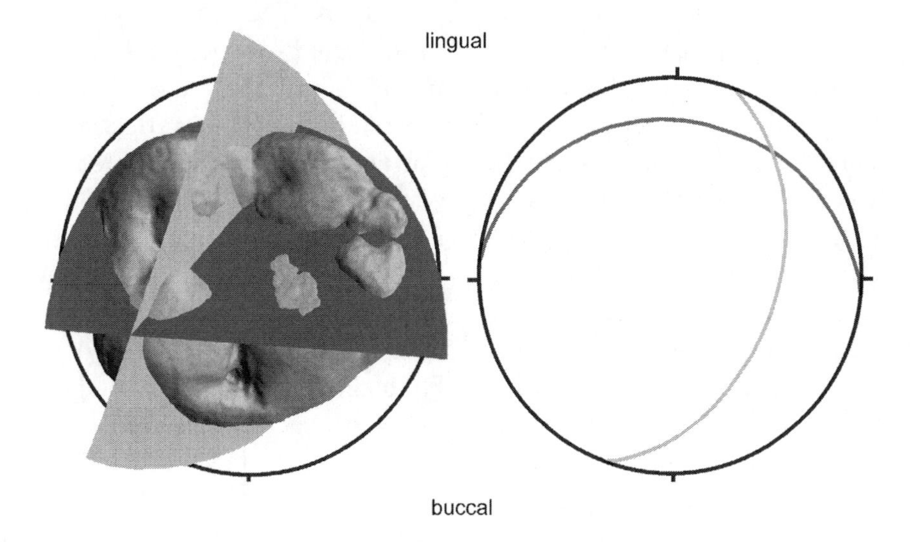

Figure 6. Left: model of SKW 5 lower left M2 with "best-fit" planes of facet 5 (light grey) and facet 9 (dark grey); right: stereoplot diagram of the same tooth with great circles of facet 5 (light grey) and facet 9 (dark grey). To enable a better understanding, the tooth model with planes is depicted as being located on a lower hemisphere.

Dental Occlusal Compass

In order to reconstruct the jaw movements that formed the observed wear patterns, we applied the dental occlusal compass (Douglass and DeVreugd, 1997; Schulz, 2003; Schulz and Winzen, 2004). This compass can be mapped onto any location on the occlusal surface of a tooth and depicts the possible directions of antagonistic tooth-tooth-contact at these contact points, depending on condylar movement (see Figure 1). The central point of the compass where the lines of movement intersect can be positioned at an occlusal contact on a cusp tip, on the slope of a molar or in the central fovea. If applied to wear facets it indicates what kind of jaw movement is responsible for the development of each facet.

Schulz and Winzen (2004), and Schulz (2003) distinguished two major functional directions, three intermediate functional directions, and two threshold fields. On the working side, the main direction is laterotrusion, which is directed slightly distolingually in the lower jaw and mesiobuccally in the upper jaw (Figure 1). On the balancing side, mediotrusion runs distobuccally in the lower jaw and mesiopalatinally in the upper jaw. Whereas the path of the main movements is fan-shaped, the intermediate protrusion is the only functional direction that has fixed coordinates, since in this movement both condyles of the lower jaw move parallel downwards and forwards. Protrusion follows the median plane in the lower jaw distally and in the upper jaw mesially.

On the working side, lateroprotrusion describes pathways between laterotrusion and protrusion, showing a distolingual direction in the lower and a mesiobuccal direction in the upper jaw. Medioprotrusion samples the movements between mediotrusion and protrusion towards mesiopalatinal in the upper jaw and distobuccal in the lower jaw on the balancing side. Retrusion and immediate side shift form the two threshold areas on the occlusal compass. The retrusion space marks a field for backward moves while the immediate side shift indicates a transverse functional field on the balancing side.

Applying the dental occlusal compass to the wear facet pattern on hominoid molars, we can detect functional movements on each of the 13 contact areas (Figure 1). On facet pairs 1, 4, 5 and 8 we find traces of lateroretrusion movement, while facets 2, 3, 6 and 7 are dominated by lateroprotrusion. Higher facet numbers belong to the balancing side with mediotrusion as the leading direction. In facets 9, 10 and 12 the immediate side shift is of greater importance and facets 11 and 13 seem to be in contact during medioprotrusion.

Results

Relief Investigation

Digital models generated with three-dimensional surface scanning provide a first impression of relief height and occlusal wear by qualitative analysis (Figures 2 and 4). The selected M_2s show clear differences in crown morphology. The slightly worn *P. robustus* SKW 5 possesses a relatively high occlusal relief with more or less equally elevated major cusps, while the similarly worn *A. afarensis* AL 266-1 demonstrates distinctly higher lingual cusps compared with the buccal cusps. In AL 266-1 the metaconid is the most prominent cusp, possessing a steep distobuccally directed slope. The *A. africanus* STW 14 specimen has a mesially oriented metaconid tip and a broad distally directing face, while *P. robustus* SKW 5 has a metaconid with a buccally oriented slope. Comparing the M_2s of *A. afarensis* (AL 333w-60), *A. africanus* (STW 404) and *P. robustus* (SK 23) with progressed wear as well as M_2s of *A. afarensis* (LH 4), *A. africanus* (MLD 40) and *P. robustus* (SK 34) with advanced wear, the crown differences described previously are maintained.

A closer look at the buccal margin of the occlusal surface from slight towards advanced wear shows that the *A. afarensis* M_2s, in particular, possess a distinct rounding of the buccal slopes, while *P. robustus* has a sharp, almost right-angled marginal edge. The edge between the occlusal surface and the buccal surface of the molar crown in *A. africanus* is intermediate between these two.

The calculation of the 3-D index produced some surprising results (Figure 7). Regarding the complete occlusal surface, both the *P. robustus* and *A. afarensis* specimens show a clear decrease in 3-D index values from slight to advanced wear. What was unexpected was that while *P. robustus* M_2s have been described as having a low relief relative to those of *A. afarensis* and *A. africanus* (e.g., White et al., 1981; Grine, 1981), in this study the *P. robustus* M_2s have relief index values that are higher than the *A. afarensis* M_2s. Taking the deepest point of the fissures as base for the reference plane and calculating the 3-D index as mentioned above, the measured *P. robustus* M_2s demonstrate the highest relief values in early wear stage of the whole sample. Because *P. robustus* SKW 5 shows comparable wear to *A. afarensis* AL 266-1 and *A. africanus* STW 14, the observed differences in relief index values must be caused by very deep fissures of SKW 5. Whether or not this is typical for *P. robustus* needs to be tested with a larger sample. However, generally, the 3-D index measurements confirm an apparent loss of enamel with extreme flattening of the crown towards advanced wear in *P. robustus* and *A. afarensis*. Our small sample of *A. africanus* specimens, however, did not follow this trend. Although the overall crown relief decreases from slight to progressed wear in this group, it increases considerably in advanced wear. In the *A. africanus* specimen concerned, MLD 40, large areas of dentine are exposed. Deep dentine basins with distinctly stepped margins at the transition from dentine to enamel surface result, as was described by (Grine, 1981). Thus, the 3-D surface area and, consequently, the relief index increases.

More detailed information about the occlusal relief can be extracted by comparing the 3-D index of individual cusps (Figure 7). Prominent metaconid height results in high 3-D index values. While *P. robustus* SKW 5 has the highest value for the slightly worn molars, *A. afarensis* AL 266-1 shows a distinctly higher value than *A. africanus* STW 14. Moderately worn teeth SK 23, AL 333w-60 and STW 404 show the same pattern as the slightly worn teeth with lower 3-D index values. However, in the teeth with advanced wear the 3-D index values give a different picture, with *A. afarensis* LH 4 having the lowest value and *A. africanus* MLD 40 the highest. As mentioned above, the deep dentine basin on the metaconid of MLD 40 is responsible for the high value, since the 3-D surface area is increased. The entoconid gives generally lower results, although SKW 5 has a markedly higher value than AL 266-1 and STW 14. This cusp has a distinctly lower relief index than the metaconid in *A. afarensis* but an almost equal value in the *A. africanus* specimen. With progressing wear, the entoconid relief index in *P. robustus* shows only a slight decrease at first but a stronger decrease in more worn specimens SK 23 and SK34. There is also a remarkable loss of relief in the *A. afarensis* specimens AL 333w-60 and LH 4. In contrast, *A. africanus* specimens STW 404 and MLD 40 exhibit an increase in relief index value. The buccal cusps, protoconid and hypoconid, decrease in relief through wear in all three groups with the above-mentioned exception of MLD 40. Also in the buccal aspect of the occlusal surface, the *P. robustus* specimens seem to possess a relatively high relief, with the exception of the hypoconid in progressed and advanced wear.

All four major cusps of *A. africanus* SKW 5 have high values, although the mesial cusps show slightly higher 3-D index values than

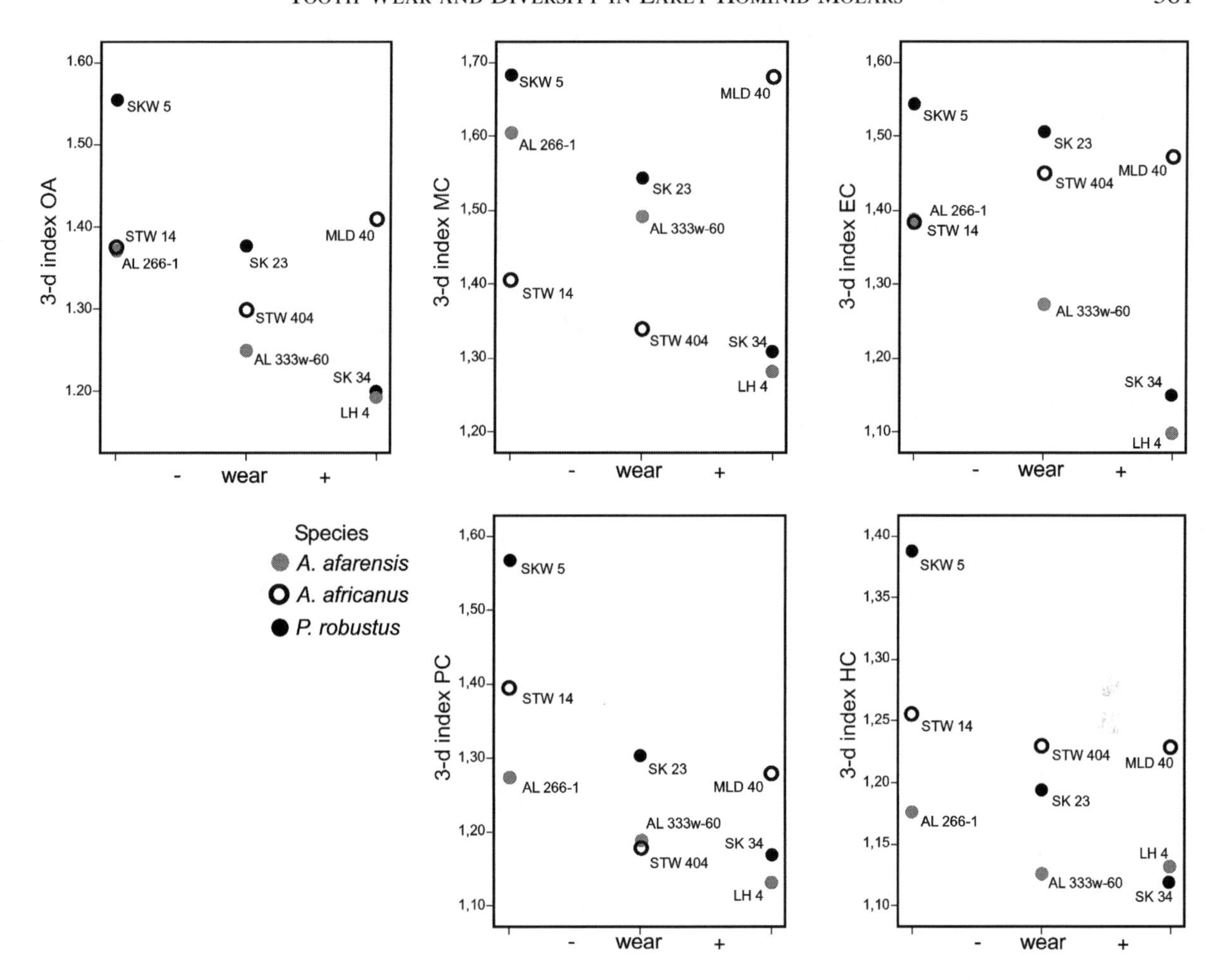

Figure 7. Comparison of 3-d index measurements of studied lower second molars of moderate wear, progressed wear and more advanced wear for occlusal area (OA), metaconid (MC), entoconid (EC), protoconid (PC) and hypoconid (HC).

the distal cusps. *A. africanus* has similar values in all four cusps, which are low-to-intermediate in the specimens with slight and progressed wear, but highest for all cusps in MLD 40. *A. afarensis* generally has the lowest relief index values for all cusps except the metaconid in slight and progressed wear and the hypoconid in advanced wear.

Wear Pattern Analysis

The stereoplot diagrams visualize the "strike" and "dip" of functional areas on the cusps and help to evaluate orientation differences in space (Figure 8). In Figure 8 the graphs are aligned for direct comparison of wear (columns) and species (rows) with an increase in wear stage from left to right. For ease of understanding, we provide an example and have drawn in each stereoplot only the lateroretrusion wear areas, facet 1 on the protoconid, facet 4 on the hypoconid, facet 5 on the metaconid, facet 8 on the entoconid and the medioprotrusion facet 9 on the hypoconid (Figure 1). Each stereoplot represents the pattern of facet orientation for a single molar. Each wear facet or area is represented by a "best fit" plane. This plane's orientation is documented by a "great circle" and a corresponding pole point. The pole

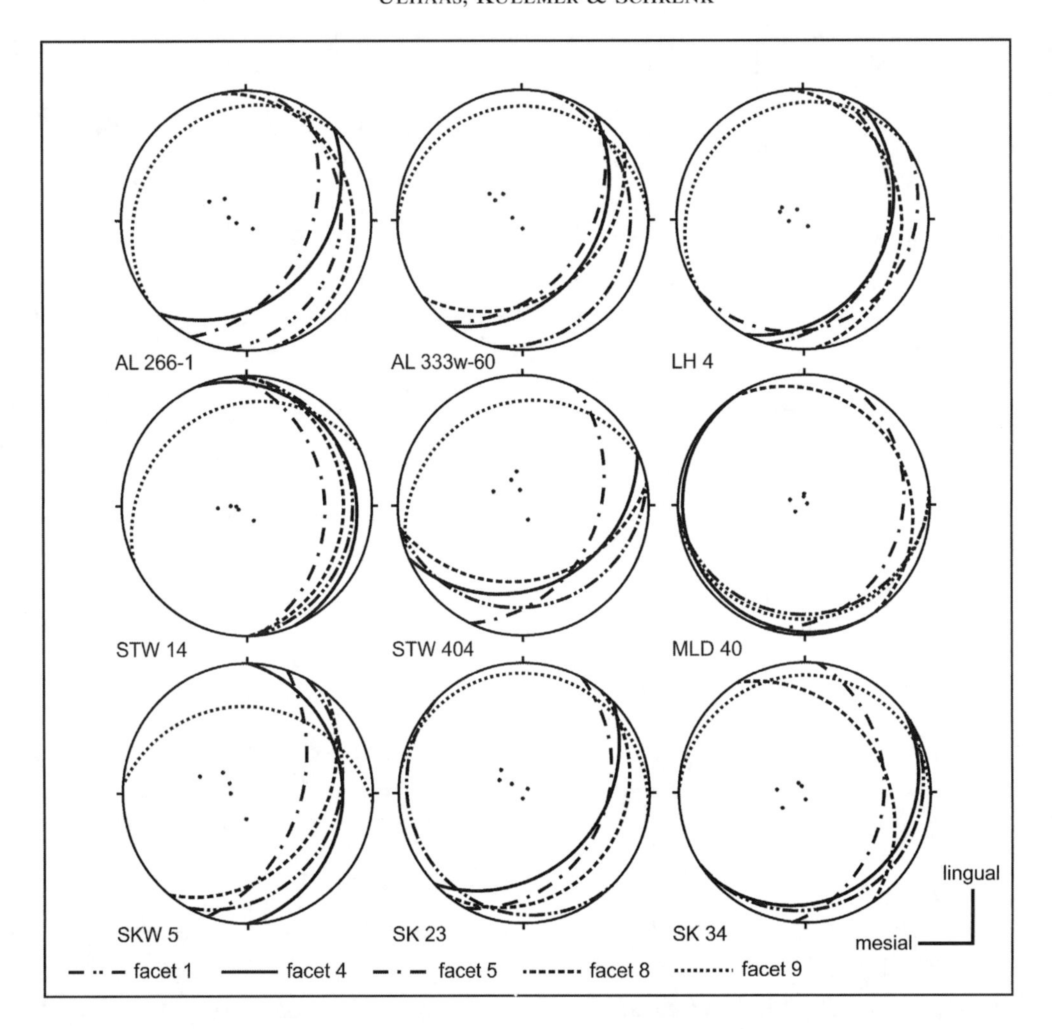

Figure 8. Stereoplot diagrams of lower second molars illustrate diversity in strike and dip angles of lateroretrusion wear facets (1, 4, 5 and 8) and mediotrusion / immediate side shift facet 9 with progressing wear (rows) and between species (columns). Each facet is represented by a great circle, depicting measurements of best fit planes of the facets areas.

point of each plane is marked with a small dot that represents the intersection point of the centered plane normal vector with the surface of a hemisphere. The specific distribution of the pole points in each stereoplot provides information about the similarity of best fit plane orientation on a tooth crown. When the pole points cluster close together we find very similar inclinations on the tooth surface. Alternatively, when they are farther apart the similarity in orientation is low. The pole points can be aligned in a row that illustrates a very similar "strike" orientation and differences in "dip". The point cluster may be positioned close to the center of the graph, indicating rather flat planes, or it may be located eccentric, highlighting steep inclination of the compared areas. Applying the concept of the occlusal compass it is possible to read the stereoplots like a 'guidebook' of occlusion. Facet orientations allow one to ascertain the direction of chewing movements, which provide an estimation of jaw movement based only on the teeth.

We can see in all three slightly worn specimens (AL 266-1, STW 14 and SKW 5) that the lateroretrusion facets point approximately in the same direction, although there is more variation in *A. afarensis* AL 266-1 and *P. robustus* SKW 5 than in *A. africanus* STW 14. The *A. africanus* STW 14 M_2 shows a clear direction of inclination towards the distal end of the crown. *A. afarensis* AL266-1 and *P. robustus* SKW 5 specimens demonstrate a higher variability in both "strike" and "dip". The wear patterns on both specimens reflect a distobuccal orientation of the lateroretrusion areas.

The distally facing facet 1 of the protoconid on *A. afarensis* AL 266-1 indicates a mesial inclination of facet 1 on the mesial edge of the paracone on the antagonistic M^2. In *P. robustus* SKW 5, facet 1 faces distobuccally and consequently on the antagonistic M^2 it must have been mesiolingually oriented. Therefore, the contact area was probably located more lingually. Because the protoconid wears heavily, with progressing wear facet 1 varies most of all lateroretrusion facets in horizontal orientation.

On the metaconid facet 5 shows a rather steep inclination and the "strike" measurements of this facet vary little in all three specimens with slight wear. The corresponding area on the antagonistic M^2 would be located on the mesial edge of the protocone. Facet 5 in both slightly worn *A. afarensis* AL 266-1 and *P. robustus* SKW 5 specimens is directed distobuccally, indicating a position on the mesiolingual slope of the protocone. With progressing wear, facet 5 on the metaconid has relatively stable "strike" angles in the sample. Diverse horizontal orientations of facet 1 compared with the relatively constant "strike" of facet 5 in slightly worn specimens indicate differences in lateroretrusion movements, with the prominent metaconid apparently representing an important cusp for occlusal guidance.

The orientation of facet 8 shows a correspondence with facet 1 in all specimens, mostly in those with slight wear. Both wear areas have distinct similarities in "strike" and "dip" values. Facet 8 is situated on the distal edge of the entoconid and is in occlusal contact with the mesial edge of the hypocone. The position of facet 8 varies due to morphological differences of the hypocone.

Facet 4 has a steep inclination in the slightly worn *A afarensis* AL 266-1 specimen, which is very similar to facet 5. In *A. africanus* STW 14 and *P. robustus* SKW 5 the surface is distinctly flatter, reflecting more or less the inclination of facet 1 and 8. The "strike" orientation varies between distal in STW 14 and SKW 5 and more buccal in AL266-1. In the teeth with advanced wear facet 4 is generally flatter compared to the slightly worn teeth. The strike is comparably oriented, with the exeption of MLD 40, in which it is distinctly different. The orientation and inclination of facet 4 varies more in *A. africanus* than in *A. afarensis* and *P. robustus*. In the *A. afarensis* sample, wear facets 4 and 5 show similar "strike" directions. The "dip" is slightly lower in the heavily worn LH 4 specimen compared to AL 266-1 and AL 333w-60.

The variation in horizontal orientation of the lateroretrusion facets 1, 4, 5 and 8 generally increases with progressing wear, and facets 1, 4 and 8 appear to rotate with progressing wear and decreased cusp height. This can be observed particularly in the *A. africanus* specimens. However, the hypothesis of a rotation of contact areas requires further studies on a larger sample to clarify the development of wear areas from slight to advanced wear.

Facet 9 faces in the opposite direction from areas 1, 4, 5 and 8. Therefore the timing of contact during the chewing cycle must be different. Facet 9 is located on the balancing side, and the contact guides the protocone out of or into the central fovea at a moment in which the corresponding facets 1, 4, 5 and 8 cannot be in close contact. The movement on the surface of area 9 is described as medioprotrusion corresponding to the occlusal compass. In the *A. afarensis* AL 266-1 and *A. africanus* STW 14 specimens, facet 9 is directed mesiolingually with a relatively low "dip". In the *P. robustus* SKW 5 specimen, facet 9 shows a steeper inclination in the lingual direction, indicating differences in occlusal motion. Frequently in advanced wear, facet 9 extends over almost the entire lingual slope of the hypoconid and would normally be in contact with a large facet on the buccal slope of the protocone on the M^2. The "dip" angle of area 9 generally varies only slightly in the studied specimens.

One would expect specimens with progressed and advanced wear to show lower inclination angles of wear areas due to the loss of enamel and decrease in relief. This can be observed clearly in *A. africanus* MLD 40, *A. afarensis* LH 4 and *P. robustus* SK 34 specimens, and to some extent in certain facets of *P. robustus* SK 23. However, the "dip" angles are not as low as expected in the specimens with advanced wear, suggesting that the teeth retain their efficiency by maintaining oblique contact faces for a fairly long time.

The M2s of *Australopithecus. afarensis*, *A. africanus* and *P. robustus* in this study display diverse wear patterns on the occlusal surface. Each molar crown possesses wear areas that indicate capabilities to process a diverse spectrum of foodstuffs. *Australopithecus afarensis* molars show a wide range of occlusal area inclinations but maintain a relatively stable "strike" orientation, whereas *A. africanus* and *P. robustus* have a variable strike pattern. The second molars of SKW 5, SK 34 and SK 23 suggest a more complex occlusional relationship for *P. robustus* dentitions than previously thought and in some ways the characteristics of the wear pattern are very similar to *A. afarensis*.

Discussion

In 1981, Grine noted that to date, all of the observations on relative cuspal height and inclination (intercuspal angulation) have been based upon subjective, qualitative assessments. Although several methods of quantitatively defining intercuspal angulation have been proposed, these techniques cannot account for subtle changes in cuspal slope between the bottom of occlusal fissure and the apex of the cusp (p. 213). At that time, the results of quantitative studies of macroscopic wear patterns and mesowear features were derived from two-dimensional analyses. Today, new possibilities are available with the development of modern digital scanning techniques.

The results of this preliminary study demonstrate that we can detect and measure differences in crown topography and areas of wear on digital models of fossil hominid molar crowns. The sample of second molars of *A. afarensis*, *A. africanus* and *P. robustus* shows clear differences in overall crown and cusp relief height among the groups. The topography of the occlusal surface differs among species and specimens. The 3-D index measurements can be used to highlight these broad topographic differences and also to compare the relief of single cusps. The decrease in 3-D index values with progressing wear is not surprising, and similar results were reported earlier for several primate species (Ungar and M'Kirera, 2003; Ungar and M'Kirera, 2003; Ulhaas et al., 2004; Dennis et al., 2004). This case study confirms what White et al. (1981) observed - that even when worn, *A. afarensis* lower molars maintain prominent lingual cusps, whereas *P. robustus* specimens wear their lower molar crowns down to form a planar occlusal surface. *A. africanus* cheek teeth represent an intermediate state between *A. afarensis* and *P. robustus* molars.

However, we should keep in mind that differences in depth of fissures probably influence the result of 3-D index measurements, as can be seen in the case of the *P. robustus* molar SKW 5. This molar shows distinctly higher relief values than those of *A. afarensis* and *A. africanus*, indicating that

P. robustus molars may in fact have a very high relief at the beginning of their functional life. Additional analyses with a larger sample of *P. robustus* molars will provide further insight.

The analysis of wear facets and areas of occlusal contacts introduced in this study attempts to point out differences in details of the occlusal surfaces that cannot be quantified by measurements such as the 3-D index. The "strike" and "dip" measurements of wear facet orientation present functional characteristics of tooth relief on hominid molars. Early hominids produce an overall wear pattern similar to *Homo sapiens* (Maier and Schneck, 1982). However, the orientation of wear facets, as illustrated in the stereoplot diagrams, varies in all specimens measured in this study. Assuming that the topography of wear areas reflects chewing behavior, the areas of wear in our sample indicate a high variation in jaw movement during lateroretrusion and medioprotrusion. The variation in "strike" measurements increases in teeth with advanced wear. This illustrates that the vertical jaw movement in teeth with lower relief is combined with a greater lateral movement. According to Wallace (1975), low cusped *Paranthropus* molars increase their grinding efficiency through a more horizontal glide path during occlusal movements. In contrast, a relatively high metaconid cusp limits the lateral movement and a steep facet 5 indicates a predominant vertical motion. The variation of "dip" measurements indicates that some teeth exhibit a higher capability for shearing action than others, although all teeth in our study show a set of wear areas with varying inclinations.

The results of this preliminary study indicate a high diversity in wear patterns of certain *Australopithecus* and *Paranthropus* molars. Although the sample is too small to define unambiguous species-specific characteristics, the results demonstrate that *P. robustus* molars do not have a simple plane horizontal occlusal surface, but rather show a complex combination of more and less inclined areas. The relief of the molars of SK 23 and SK 34 demonstrate that even in advanced wear we can find distinct differences in relief orientations on *P. robustus* occlusal surfaces that on first sight give the impression of having a very flat occlusal surface. Therefore, it is necessary to stress that the close topographic examination of the occlusal surface uncovers details that may be important for functional interpretations, as qualitative observations and measuring of overall topography alone cannot detect such particular characteristics.

Both the relief and the orientation of wear facets of the *P. robustus* second molars reflect not only grinding capabilities for processing hard objects but also varying jaw movements that indicate a more generalized diet. Worn molars of *P. robustus* show relatively sharp edges at the occlusal margin, indicating a rapid enamel reduction. This may be caused by hard or tough foods that resulted in a high attrition rate. The relatively high 3-D index of STW 5 may also indicate the high occlusal crown with deep fissures that serves to compensate for this large amount of enamel loss. Our results of the *A. africanus* specimens show a varying wear rate among this group, because facet orientation indicates hard object feeding as well as the capability to process brittle material. The comparison of wear areas on *A. africanus* molars STW 14 and STW 404 illustrates differences in lateroretrusion, which are indicative of idiosyncratic variation in masticatory behavior. It is notable that the specimen STW 14 developed a pattern with buccolingual ridges in slight wear, representing laterotrusion faces that are oriented more distally and mesially. This indicates a dominant lateral motion during chewing, and points to the inclusion of hard and fibrous objects in the daily diet.

Of the three groups examined *A. afarensis* is the most distinct. Its second molars demonstrate the greatest capability for shearing action on lateroretrusion facets 4 and 5, and facet 8. The relatively similar wear pattern of the three specimens suggests similar jaw movements among the individuals and this may indicate comparable food composition. It is hypothesized that the remarkable roundness of buccal cusps in the *A. afarensis* molars is a sign of a relatively low attrition rate, in contrast with the sharp edging on the margin of the occlusal surface as it is observed in the *P. robustus* molars.

Conclusions

This study provides a new perspective on tooth wear using detailed digital models of tooth crowns to document the occlusal wear patterns. The flexibility of a portable scanning system like the optical 3-D topometry system used in this study facilitates the acquisition of a large amount of complex surface data on fossil tooth crowns directly from the original object. The visualization on the computer screen opens the way for new methods of examining teeth and quantifying tooth wear and morphology. A resolution of about 0.065 mm seems to be reasonable for occlusal analysis of hominid molars and enables us to measure wear facet patterns.

One advantage of this method is that all traditional linear measurements can be taken on a digital model without risking damage to the specimen by mechanical calipers. Furthermore, the measurements are reproducible since we can point to the exact measuring points on the polygonal surface mesh from which the data have been collected. Parameters that include surface area, like the 3-D index, facilitate the comparison of crown and cusp height and support the assessment of tooth wear and function.

Our study using a small selection of teeth shows that 3-D analysis of occlusal crown surfaces in different stages of wear can provide new insight into the complexity of tooth surface modification. In our study, the few examples show interspecific differences in the mode of relief decrease resulting in definite differences in wear facet orientation. Since the potential areas of wear on the tooth surface are determined through a general principle of occlusion in hominoid molars, *A. afarensis*, *A. africanus* and *P. robustus* produced a similar overall wear pattern. However in detail, the resulting wear patterns vary among groups, indicating more vertical and short transverse movements in all three *A. afarensis* specimens and wider lateral and mesial excursions during jaw motion in the *P. robustus* molars. Moreover, the *A. afarensis* molars demonstrate the greatest capability for shearing action. This high variability in wear patterns among early hominids supports the hypothesis that they may not have been exposed to a strong interspecific competition for food resources. Instead, they were likely flexible enough to choose their daily diet out of a broad range of resources. Considering our findings, we assume that *P. robustus* probably was an omnivorous generalist rather than a plant-eating specialist. Thus, *P. robustus* was well adapted to feed on various resources. The rapid reduction of enamel of *P. robustus* molars during a relatively long mastication process with wide and variable chewing movements may be caused by freshly unearthed tubers with a portion of soil. However, hard object feeding as fall-back diet cannot be excluded, since mastication power and grinding facets on the occlusal surface are capable of coping with small and brittle foodstuffs (Wood and Strait, 2004).

The diversity of wear patterns among the *A. africanus* specimens may be correlated with individual or population differences in the choice of diverse foods with variable physical properties. We can therefore expect a diverse wear pattern for a larger sample of *A. africanus* teeth.

In *A. afarensis*, the observed inclination and orientation of facets on the occlusal surface indicate a dominating shearing action, which reflects most probably a dietary regime including tough foodstuffs, including some quantity of meat also in addition to fibrous plant material.

Thus, the occlusal surfaces of hominid teeth contain more information than has been previously assumed. The techniques and methods are available to further explore the topography of the occlusal surface of fossil hominid teeth. More research on the fundamentals of occlusion including experimental studies will help to understand the development of wear patterns and their use for the reconstruction of diet. It can be generally concluded that the 3-D inspection of wear patterns on tooth crowns provides a strong potential to gain fundamental insight into functional morphology of occlusal surfaces as well as inter- and intraspecific diversity in early hominids.

Acknowledgments

We want to express our gratitude to Jean-Jacques Hublin and Shara Bailey for organizing this inspiring conference. Many thanks to all staff members of the Department of Human Evolution in the Max-Planck-Institute of Evolutionary Anthropology for the perfect coordination of the meeting and support during our stay in Leipzig. We are very indebted to the Transvaal Museum in Pretoria and the University of the Witwatersrand in Johannesburg for support and access to the original hominid specimens. We are grateful to Olaf Winzen and Dieter Schulz for inspiring discussions and help with the dental occlusal compass. We thank Kerstin Engel and Mathias Huck for scanning of the original specimens in South Africa, and Christine Hemm for acquiring 3-D data of casts of *A. afarensis* specimens, the generation of digital models and help with the figures. Thanks are also due to Shara Bailey, Dick Byer and anonymous reviewers for their helpful comments and revising the English of the manuscript. This research is supported by the Research Institute and Natural Museum Senckenberg in Frankfurt am Main.

References

Angle, E.H., 1900. *Treatment of Malocclusion of the Teeth and Fractures of the Maxillae: Angle's.* 6th Ed. S.S. White Dental Manufacturing Co., Philiadelphia, pp. 37–40.

Beraldin, J.A., Blais, F., Cournoyer, L., Godin, G., Rioux, M. (2000) Active 3D sensing. In: Modelli e Metodi per lo Studio e la Conservazione dell'Architettura Storica. Scuola Normale Superiore Pisa, Quaderni 10, 22–46.

Brothwell, D.R., 1981. *Digging up Bones*, 2nd Ed. British Museum (Natural History) & Oxford University Press, London and Oxford.

Butler, P.M., 1973. Molar wear facets of early tertiary North American Primates. In: Zingeser, M.R. (Ed.), Craniofacial Biology of Primates, Vol. 3 (Symposia of the Fourth International Congress of Primatology). Karger, Basel, pp. 127.

Crompton, A.W., 1971. The origin of the tribosphenic molar. Zoological Journal of the Linnean Society Suppl. 50, 65–87.

Crompton, A.W., Hiiemae, K.M., 1970. Functional occlusion and mandibular movements during occlusion in the American opossum, *Didelphis marsupialis* L. Zoology Journal of the Linnaeus Society 49, 21–47.

Dahl, B.L., Krogstad, B.S., Gaard, B., Eckers-Berg, T., 1989. Differences in functional variables, fillings, and tooth wear in two groups of 19-year-old individuals. Acta Odontologica Scandinavica 47, 35.

Dennis, J.C., Ungar, P.S., Teaford, M.F., Glander, K.E., 2004. Dental topography and molar wear in *Alouatta palliata* from Costa Rica. American Journal of Physical Anthropology 125, 152–161.

Douglass, G.D., DeVreugd, R.T., 1997. The dynamics of occlusal relationships. In: McNeill, C. (Ed.), Science and Practice of Occlusion, Quintessence Publishing Co, Inc., Illinois, pp. 69–78.

Gingerich, P.D., 1972. Molar occlusion and jaw mechanics of the Eocene primate *Adapis*. American Journal of Physical Anthropology 36, 359–368.

Godin, G., Beraldin, J.A., Taylor, J., Cournoyer, L., Rioux, M., El-Hakim, S., Baribeau, R., Blais, F., Boulanger, P., Domey, J., Picard, M., 2002. Active Optical 3D Imaging for Heritage Applications. IEEE Computer Graphics Applications 22, 24–36.

Gordon, K.D., 1984. The assessment of jaw movement direction from dental microwear. American Journal of Physical Anthropology 63, 77–84.

Grine, F.E., 1981. Trophic differences between 'gracile' and 'robust' australopithecines: a scanning electron microscope analysis of occlusal events. South African Journal of Science 77, 203–230.

Hiiemae, K.M., Crompton, A.W., 1971. A cinefluorographic study of feeding in the American opossum, *Didelphis marsupialis*. In: Dahlberg, A.A. (Ed.), Dental Morphology and Evolution. University of Chicago Press, Chicago, pp.299–334.

Hiiemae, K.M., Kay, R.F., 1972. Trends in the evolution of primate mastication. Nature 240, 486–487.

Hiiemae, K.M., Kay, R.F., 1973. Evolutionary trends in the dynamics of primate mastication. In: Symposium of the Fourth International Congress of Primatology, 3. Karger, Basel, pp. 28–64.

Janis, C.M., 1984. Predictions of primate diets from molar wear pattern. In: Chivers, D.J., Wood, B.A., Bilsborough, A. (Eds.), Food Acquisition and Processing in Primates. Plenum Publishing Corporation, New York, pp. 331–340.

Janis, C.M., 1990. The correlation between diet and dental wear in herbivorous mammals, and its relationship to the determination of diets of extinct species. In: Boucot, A.J. (Ed.), Evolutionary Paleobiology of Behaviour and Coevolution. Elsevier Science Publishers B.V., Amsterdam, pp. 241–259.

Jernvall, J., Selänne, L., 1999. Laser confocal microscopy and geographic information systems in the study of dental morphology. Paleontologica Electronica 2, 18 p, 905KB. http://www-odp.tamu.edu/paleo/1999_1/confocal/issue1_99.htm.

Kay, R.F., 1973. Mastication, molar tooth structure and diet in primates. Ph.D. Thesis, Yale University.

Kay, R.F., 1975. The functional adaptations of primate molar teeth. American Journal of Physical Anthropology 43, 195–216.

Kay, R.F., 1977. The evolution of molar occlusion in the Cercopithecoidea and early catarrhines. American Journal of Physical Anthropology 46, 327–352.

Kay, R.F., 1978. Molar structure and diet in extant Cercopithecidae. In: Butler, P.M., Joysey, K.A. (Eds.), Development, Function, and Evolution of teeth. Academic Press, New York, pp. 309–339.

Kay, R.F., 1981. The nut-crackers – a new theory of the adaptations of the Ramapithecinae. American Journal of Physical Anthropology 55, 141–151.

Kay, R.F., Covert, H.H., 1984. Anatomy and behaviour of extinct primates. In: Chivers, D.J., Wood, B.A., Bilsborough, A. (Eds.), Food Acquisition and Processing in Primates. Plenum Press, New York, pp. 467–508.

Kay, R.F., Hiiemae, K.M., 1974. Mastication in *Galago crassicaudatus*, a cinefluorographic occlusal study. In: Martin, R.D., Doyle, G.A., Walker, A. (Eds.), Prosimian Biology. Duckworth, London, pp. 501–503.

Kay, R.F., Hylander, W.L., 1978. The dental structure of mammalian folivores with special reference to primates and Phalangeroidea (Marsupialia). In: Montgomery, G.G. (Ed.), The Ecology of Arboreal Folivores. Smithonian Institute Press, Washington, pp. 173–191.

Kim, S.K., Kim, K.N., Chang, I.T., Heo, S.J., 2001. A study of the effects of chewing patterns on occlusal wear. Journal of Oral Rehabilitation 28, 1048–1055.

Kullmer, O., Huck, M., Engel, K., Schrenk, F., Bromage, T., 2002a. Hominid Tooth Pattern Database (HOTPAD) derived from optical 3D topometry. In: Mafart, B., Delingette, H. (Ed.), Three-dimensional Imaging in Paleoanthropology and Prehistoric Archaeology. Liege, Acts of the XIVth UISPP Congress, BAR International Series 1049, pp. 71–82.

Kullmer, O., Engel, K., Schrenk, F., 2002b. Dreidimensionale Vermessungstechniken erweitern paläontologische Forschung – Virtuelle Zahnmodelle zur Analyse komplexer Zahnoberflächen. Natur und Museum 132, 225–256.

Lahee, F.H., 1957. *Field Geology*. 6th Ed. McGraw Hill, New York.

Maier, W., 1977a. Die bilophodonten Molaren der Indriidae (Primates) – ein evolutionsmorphologischer Modellfall. Zeitschrift Morphologie Anthropologie 68, 307–344.

Maier, W., 1977b. Die evolution der bilophodonten Molaren der Cercopithecoidea. Eine funktionamorphologische Untersuchung. Zeitschrift Morphologie Anthropologie 68, 26–56.

Maier, W., 1978. Zur Evolution des Säugetiergebisses – Typologische und konstruktionsmorphologische Erklärungen. Natur und Museum 108(10), 288–300.

Maier, W., 1984. Tooth morphology and dietary specialization. In: Chivers, D.J., Wood, B.A., Bilsborough, A. (Eds.), Food Acquisition and Processing in Primates. Plenum Press, New York, pp. 303–330.

Maier, W., Schneck, G., 1981. Konstruktionsmorphologische Untersuchungen am Gebiß der hominoiden Primaten. Zeitschrift Morphologie Anthropologie 72, 127–169.

Maier, W., Schneck, G., 1982. Functional morphology of hominoid dentitions. Journal of Human Evolution 11, 693–696.

Miles, A.E.W., 1963. Dentition in the estimation of age. Journal of Dental Research 42, 255–263.

M'Kirera, F., Ungar, P.S., 2003. Occlusal relief change with molar wear in *Pan troglodytes troglodytes* and *Gorilla gorilla gorilla*. American Journal of Primatology 60, 31–41.

Reed, D.N.O., 1997. Contour mapping as a new method for interpreting diet from tooth morphology. American Journal of Physical Anthropology, Suppl. 24, 194.

Robinson, J.T., 1954. Prehominid dentition and hominid evolution. Evolution 8, 324–334.

Robinson, J.T., 1956. The dentition of the Australopithecinae. Transvaal Museum Memoires 9, 1–179.

Robinson, J.T., 1972. *Early hominid posture and locomotion*. University of Chicago Press, Chicago.

Schulz, D., 2003. NAT – Die Naturgemäße Aufwachstechnik. Teil 1: Der anteriore Bereich. Teamwork Media GmbH, Fuchstal.

Schulz, D., Winzen, O., 2004. Basiswissen zur Datenübertragung. Teamwork Media GmbH, Fuchstal.

Scott, E.C., 1979. Dental wear scoring technique. American Journal of Physical Anthropology 51, 213–218.

Stones, H.H., 1948. *Oral and Dental Diseases*. Livingston, Edinburgh, p. 869.

Strait, S.G., 1993a. Differences in occlusal morphology and molar size in frugivores and faunivores. Journal of Human Evolution 25, 471–482.

Strait, S.G., 1993b. Molar morphology and food texture among small-bodied insectivorous mammals. Journal of Mammalogy 74(2), 391–402.

Sperber, G., 1973. Morphology of the cheek teeth of early South African hominids. Ph.D. Dissertation, Witwatersrand University Johannesburg.

Teaford, M.F., 1982. Differences in molar wear gradient between juvenile macaques and langurs. American Journal of Physical Anthropology 57, 323–330.

Teaford, M.F., 1983. The morphology and wear of the lingual notch in macaques and langurs. American Journal of Physical Anthropology 60, 7–14.

Tobias, P.V., 1967. *The Cranium and Maxillary* dentition of *Australopithecus (Zinjanthropus) boisei. Olduvai Gorge vol. 2*. Cambridge University Press, London.

Ulhaas, L., Kullmer, O., Schrenk, F., Henke, W., 2004. A new 3-d approach to determine functional morphology of cercopithecoid molars. Annals of Anatomy 186, 487–493.

Ungar, P.S., 'M'Kirera, F., 2003. A solution to the worn tooth conundrum in primate functional anatomy. Proceedings of the National Academy of Sciences of the USA 100, 3874–3877.

Ungar, P.S., Williamson, M., 2000. Exploring the effects of tooth wear on functional morphology: a preliminary study using dental topographic analysis. Palaeontologia Electronica. (3)1: http://www-odp.tamu.edu/paleo/2000_1/gorilla/main.htm.

Ungar, P.S., Dennis, J.C., Wilson, J., Grine, F., 2002. Quantification of tooth crown shape by dental topographic analysis. American Journal of Physical Anthropology Suppl. 34, 158–159.

Wallace, J.A., 1972. The dentition of South African early hominids: a study of form and function. Ph.D. Dissertation, University of Witwatersrand, Johannesburg.

Wallace, J.A., 1973. Tooth chipping in the australopithecines. Nature 244, 117–118.

Wallace, J.A., 1974. Approximal grooving of teeth. American Journal of Physical Anthropology 40, 285–390.

Wallace, J.A., 1975. Dietary adaptations of *Australopithecus* and early *Homo*. In: Tuttle, R. (Ed.), Paleoanthropology, Morphology and Paleoecology, Mouton, The Hague, pp. 203–223.

Wallace, J.A., 1978. Evolutionary trends in the early hominid dentition. In: Jolly, C. (Ed.), Early Hominids of Africa. Duckworth, London, pp. 285–310.

White, T.D., Johanson, D.C., Kimbel, W.H., 1981. *Australopithecus africanus*: its phyletic position reconsidered. South African Journal of Science 77, 445–470.

Wolpoff, M.H., 1975. Some aspects of human mandibular evolution. In: McNamara, J.A. (Ed.), Determinants of Mandibular Form and Growth. Center of Human Growth and Development, Ann Arbor, pp. 1–64.

Wolpoff, M.H., 1976. Primate models for australopithecine sexual dimorphism. American Journal of Physical Anthropology 45, 497–510.

Wood, B.A., 1993. Early *Homo*: How many species? In: Kimbel, W.H., Martin, L.B. (Eds.), Species, Species Concepts, and Primate Evolution. Plenum Press. New York, pp. 485–522.

Wood, B.A., Abbott, S.A., 1983. Analysis of the dental morphology of Plio-Pleistocene hominids. I. Mandibular molars: crown area measurements and morphological traits. Journal of Anatomy 136, 197–219.

Wood, B.A., Strait, D., 2004. Patterns of resource use in early *Homo* and *Paranthropus*. Journal of Human Evolution 46, 119–162.

Zuccotti, L.F., Williamson, M.D., Limp, W.F., Ungar, P.S., 1998. Modeling primate occlusal topography using geographic information systems technology. American Journal of Physical Anthropology 107, 137–142.

6. 3-D interferometric microscopy applied to the study of buccal enamel microwear

F. ESTEBARANZ
Secc. Antropologia, Dept. Biologia Animal
Fac. Biologia, Universitat de Barcelona
Avgda. Diagonal 645, 08028 Barcelona. Spain
estebaranz@ub.edu

J. GALBANY
Secc. Antropologia, Dept. Biologia Animal
Fac. Biologia, Universitat de Barcelona
Avgda. Diagonal 645, 08028 Barcelona. Spain
jgalbany@ub.edu

L.M. MARTÍNEZ
Secc. Antropologia, Dept. Biologia Animal
Fac. Biologia, Universitat de Barcelona
Avgda. Diagonal 645, 08028 Barcelona. Spain
lmartinez@ub.edu

A. PÉREZ-PÉREZ
Secc. Antropologia, Dept. Biologia Animal
Fac. Biologia, Universitat de Barcelona
Avgda. Diagonal 645, 08028 Barcelona. Spain
martinez.perez-perez@ub.edu

Keywords: SEM, interferometry, microwear, enamel, hominoid

Abstract

Dental microwear analysis is based on the assumption that a correlation exists between ingested diet and microwear patterns on the enamel surface of teeth, such that diet can be reconstructed by quantifying enamel microwear. Abrasive particles, such as plant phytoliths or silica-based sands incorporated into food items, along with food processing techniques and tooth morphology, are responsible for the microwear features observed. Dental microwear has been extensively studied in both extant and extinct primates, including human populations. The dietary and ecological information that can be derived from dental microwear analyses makes it a technique useful for analyzing non-primate species, such as muskrats, sheep, bats, moles, antelopes, pigs and even dinosaurs. In the attempt to reconstruct species' ecology and diet, microwear research has become a successful procedure. The proliferation and persistence of different methods to quantify microwear patterns

S.E. Bailey and J.-J. Hublin (Eds.), Dental Perspectives on Human Evolution, 391–403.

require very accurate definitions of microwear variables, since inter-observer error rates cannot be neglected. The use of semiautomatic methods to quantify microwear features does not guarantee low inter-observer error affecting dental microwear results. Error can be caused by taphonomy, microscopy drawbacks of back-scattered electrons, or differences in SEM reproducibility depending on sample shape and orientation. However, fully automatic procedures lack discrimination between ante-mortem and post-mortem wear processes that affect tooth enamel at various degrees, and their application requires experienced control and evaluation.

Introduction

Plant foods contain significant amounts of phytoliths in their tissues, such that dental microwear is directly related to ecological conditions and diet composition (Teaford, 1994; Ungar and Teaford, 1996; Ungar, 1998). Seeds, shoots, and inflorescence (Ball et al., 1996) are among the main food items that have an effect on enamel microwear (Danielson and Reinhard, 1998; Gügel et al., 2001), although food processing techniques can also influence enamel microwear by incorporating dust and ashes into the ingested foods (Teaford and Glander, 1991, 1996; Ungar, 1995; Daegling and Grine, 1999). In addition, tool technology plays also an important role (Teaford, 1991; Pérez-Pérez et al., 1994). It should be noted that diet-related variables do not affect dental enamel in isolation, but that gnathic morphology must also be taken into account (Gordon, 1982).

Dental microwear of both extinct and extant primates has been widely studied (Ryan, 1979; Gordon, 1982; Kay, 1987; Teaford and Runestad, 1992; Ungar, 1992, 1994, 1996; Ungar et al., 1995; King et al., 1999b; Galbany, 2004; Godfrey et al., 2004; Nystrom et al., 2004; Galbany et al., 2005b), including human populations (Grine, 1986; Lalueza and Pérez-Pérez, 1993; Lalueza et al., 1996; Martínez et al., 2004; Pérez-Pérez et al., 1999, 2003). Enamel microwear analysis has proven to be highly informative regarding dietary habits and paleoecology, and has been applied to a wide range of taxa, including muskrats (Lewis et al., 2000), sheep (Mainland, 2003), bats (Strait, 1993), moles (Silcox and Teaford, 2002), antelopes (Solounias and Hayek, 1993), pigs (Ward and Mainland, 1999), and suids (Hunter and Fortelius, 1994).

When attempting to reconstruct species' ecology and diet, dental microwear research has thus become a successful line of research. The first dental microwear papers were published in the 1950's (Butler, 1952; Mills, 1955), although no quantitative results were given. It was not until the 1980s that several authors proposed alternative methods to quantify microwear features (Gordon, 1982, 1984; Grine, 1986; Ungar et al., 1991, 1995). However, the persistence of an abundance of different methods to quantify microwear patterns greatly limits the comparison of results among researchers (Grine et al., 2002; Galbany, 2005a). Due to high inter-observer error rates, Grine et al. (2002) proposed the adoption of *Microware* 4.0 (by P. Ungar) as standard software for semi-automatic microwear analysis. Nevertheless, Galbany (2005a), using *Sigma Scan Pro* 5.0 (by SPSS) have shown that error rates are independent of the software used, but are highly dependent on how variables are defined and the researcher's expertise (Pérez-Pérez et al., 1999; Galbany et al., 2004b).

The use of semiautomatic methods does not guarantee reliability of results because various sources of measurement error persist, such as using back-scattered or secondary electrons in SEM observation (Pérez-Pérez et al., 2001; Galbany, et al., 2004b), varying the working distance, or any surface tilt. SEM images, which depend on the shape and orientation of the sample (Gordon, 1982; King et al., 1999b; Ungar, 2003), do not generally reproduce the exact tooth surface. In addition, recent studies have demonstrated the existence of important inter-observer and intra-observer error rates due to subjectivity of criteria used to measure microwear features (Grine et al., 2002; Galbany, 2005a). Finally, taphonomic analyses on dental microwear show that post-mortem wear produces a generalized polish of dental enamel, rather than the addition of microwear features, both in occlusal (King et al., 1999a) and buccal tooth surfaces (Martínez and Pérez-Pérez, 2004).

Although the generalized use of a fully automatic quantification procedure might be useful to prevent subjectivity and inter-observer error (Ungar, 2003; Galbany, 2005a), an understanding of the influence of post-depositional processes on enamel microwear is required. The aim of the present contribution is to test whether or not post-mortem enamel microwear can be distinguished from diet-related microwear patterns on buccal tooth surfaces. Dental microwear on the buccal enamel surface is not influenced by tooth-to-tooth contact during food chewing and, thus, reflects only diet (Galbany, 2004; Pérez-Pérez, 2004). In addition, on the buccal surfaces pits are absent, and researchers have to deal only with scratches. Pits are frequently very difficult to characterize due to overlap and their high variability in shape and size (Pérez-Pérez, 2004). Automatic measuring procedures must demonstrate that clear topographic differences exist between post-mortem eroded surfaces and well-preserved enamel. Topographic 3-D techniques have been successfully applied to the analysis of tooth morphology and wear (Mayhall and Kageyama, 1997; Reed, 1997; Zuccotti et al., 1998; Jernvall and Selänne, 1999; Ungar and Williamson, 2000; Kaiser and Katterwe, 2001; Ungar and M'Keirera, 2003; Dennis et al., 2004; Ungar, 2004). However, all methodological procedures based on topographic analysis, at least for tooth surfaces, are bound to be highly sensitive to enamel preservation. An indiscriminate analysis of enamel surfaces without considering preservation is fated to be meaningless in terms of interpretation of dietary adaptations and ecology of fossil populations.

Materials and Methods

The study teeth came from a wide collection of tooth casts curated at the University of Barcelona. The molds were obtained during the course of an international collaborative project on dental microwear (Galbany et al., 2004b). Tooth crown molds were obtained with President MicroSystem Regular Body (Coltène ®) polyvinylsiloxane. This impression material is widely used in dental microwear research (Ungar, 1996; Ungar and Spencer, 1999). It reproduces features with resolutions to a fraction of a micron (Teaford and Oyen, 1989), maintains the resolution for many years (Beynon, 1987), and shows an excellent dimensional stability and reproduction detail (Andritsakis and Vlamis, 1986). Resin positive replicas were obtained from the molds using epoxy resin Epo-Tek #301 as well as polyurethane Feropur PR-55, both showing the same resolution as the original tooth (Galbany et al., 2004b). Once the resin or polyurethane casts were dry, they were mounted on aluminum stubs with term fusible gum. An argent belt (Electrodag 1415M-Acheson Colloiden) was applied between the plastic cast and the aluminum stub, and all casts were sputter-coated with a 40A gold layer (see Galbany et al, 2004b for a more detailed description of these methodological procedures).

The sample included 57 teeth (one tooth per individual was studied) from three hominoid genera (*Gorilla* N = 13, *Pan* N = 4; *Pongo* N = 8) and three hominid species (*Australopithecus anamensis* N = 2, *A. afarensis* n = 22, and *A. africanus* N = 8). Whenever available, the left M_2 was selected. Otherwise, the right M_2 was chosen. If neither was available, then the left M_1 or, in its absence, the right M_1 was studied. Despite some studies in which upper teeth were chosen when the lower ones were was missing (King et al., 1999b), in the present analysis only the lower dentition was studied. The lower M_2 has been extensively studied and the literature on the M_2 is quite rich (Gordon, 1982; King et al, 1999b; Nystrom et al., 2004; Ungar, 2004). No differences were expected in enamel roughness and microwear between M1 and M2 teeth, although no evidence for this is currently available for the sample studied.

For each tooth, a SEM micrograph of the enamel surface was obtained with a Cambridge Stereoscan-120 scanning electron microscope following usual procedures for microwear research (Lalueza et al., 1996; Ungar and Teaford, 1996; Pérez-Pérez et al., 1999; Pérez-Pérez et al., 2003; Pérez-Pérez, 2004). SEM working parameters included acceleration voltage of 15 kV, with working distance ranging between 18 and 25 mm, and 100X magnification. All images were taken at the medial third of the buccal surface, avoiding both cervical and occlusal thirds (Pérez-Pérez et al., 1999). Images were then processed with Adobe ® Photoshop ® 6.0. A semiautomatic measure of striation density (total number of striations over the analysed surface of $0.56\,mm^2$) was obtained using SigmaScan Pro ® 5.0 (SPSS). Striation density on the buccal surface of the human dentition has been shown to depend on the abrasiveness of food, and it appears to be a reliable measure of dietary habits in both ancient and modern human hunter-gatherer populations (Pérez-Pérez et al., 1994; Lalueza et al., 1996; Pérez-Pérez et al., 1999, 2003; Pérez-Pérez, 2004).

At the same time, 3-D topographic images of the same buccal tooth surfaces were obtained using a Veeco NT 1100 interpherometric microscope housed at the Nanotechnology Platform at the "Parc Científic" of the University of Barcelona. Each sample was placed perpendicular to the objective axis. All of the 3-D surfaces were analysed at 50X magnification in VSI mode to control z-axis movements, and with the quality level set to 'full' in order to achieve the highest resolution possible in pixels per area. A total of 199 3-D topographies (*Gorilla* N = 47, *Pan* N = 13, *Pongo* N = 20, *A. anamensis* N = 8, *A. afarensis* n = 79, and *A. africanus* N = 32) were analyzed on the 57 teeth in the study sample. All roughness measures were obtained for a $124.4 \times 94.6\,\mu m$ surface topography. Some image treatment was performed before an automatic measure of surface roughness was derived for each image. First, a histogram of height measures (z score) was drawn to assess the distribution of individual surface parameters. Background noise was reduced by applying a mask filter that eliminated low frequency z scores corresponding to false peaks or valleys. A median smoothing pass filter was then applied to filter out 'noisy and spiky' data (as indicated in the Veeco reference package). This median filter is particularly effective for preserving edges and steps in the data. Finally, several measures of surface roughness were calculated (Rp, Rv, Rt, Rq, and Ra). Rp is the maximum profile peak height, measured as the height difference between the mean line and the highest point over the evaluation length; Rv is the maximum profile valley depth, measured as the height difference between the mean line and the lowest point over the evaluation length; Rt is the peak-to-valley difference calculated over the entire measured array; Rq is the root-mean-squared roughness calculated over the entire measured array; and Ra is the average roughness calculated over the entire measured array, calculated according to the ANSI B46.1 standard.

The independent measures of striation density and surface roughness obtained on the buccal enamel surfaces of the same teeth were analyzed to test the alternative hypotheses that: (1) a significant correlation between microwear striation density and surface roughness can be observed (indicative of a significant association between automatic measures and dietary related habits), and that (2) automatic measures of surface roughness can be used to discriminate between different types of enamel surfaces, including well preserved enamel, post-mortem eroded surfaces, or even inter-proximal enamel facets.

In order to test the first hypothesis, SEM pictures were classified into one of four categories of degree of post-mortem damage (Figure 1): nil, no clear signs of erosion, numerous dietary striations present (1); moderate, some evidence of erosion, though numerous striations still visible (2); marked, clear signs of erosion, low number of striations visible (3); and intense, generalized erosion, striations hardly visible (4).

Results

All measures of surface roughness were highly correlated. In fact, they are all estimations of the overall difference between minimum and maximum z scores. High peak-to-valley distances are indicative of an uneven process of enamel wear affecting certain patches of enamel more intensively than others. Homogeneous wear processes would erode away the enamel without increasing the roughness values, whereas uneven wear might increase roughness throughout the analyzed enamel surfaces. The overall paired linear correlation between the roughness measures and the degree of post-mortem wear was negative and highly significant. In all species, high post-mortem enamel damage was correlated with low surface roughness (Figure 2). Consequently, taphonomic, post-depositional processes may be significant factors to consider when recording automatic measures

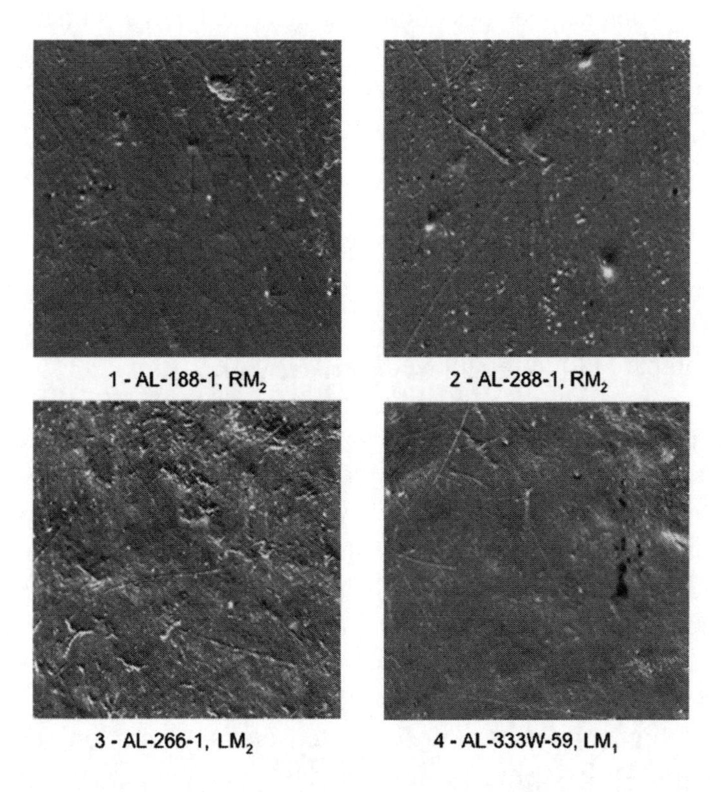

Figure 1. Sample images showing the four categories post-mortem damage: (1) nil, no post-mortem abrasion, (2) low, (3) moderate and (4) high.

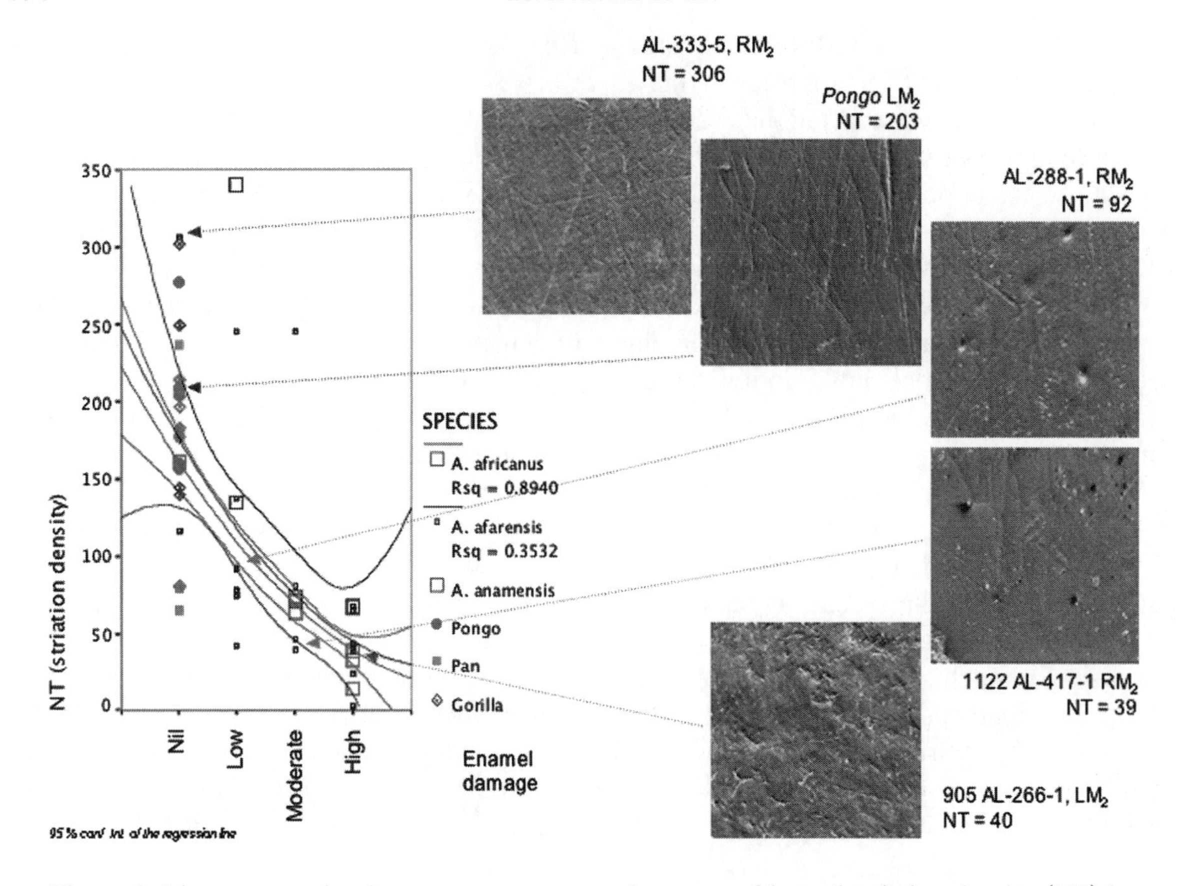

Figure 2. Linear regression between post-mortem damage and buccal striation density (NT) by species. A clear reduction of striation density with increasing post-mortem damage is seen. Sample SEM images of some analyzed teeth are shown in the plot.

of surface relief. However, there is no clear association between striation densities and surface roughness. While striation density tends to decrease with increasing enamel damage, roughness increases in slightly damaged surfaces (compared to non-damaged enamel), and decreases in medium and highly damaged enamel (Figure 3).

The relationship between striation density and enamel roughness depends on post-mortem damage of the enamel surface. When we controlled for this factor, significant quadratic regression coefficients were obtained, demonstrating that roughness tends to increase if enamel damage is slight, but sharply decreases with higher degrees of post-mortem damage (Figure 4). Even by species, those teeth heavily affected by enamel erosion showed a clear tendency toward reduced numbers of striations, whereas moderate rates of erosion showed higher roughness values, and highly striated surfaces showed reduced roughness (Figure 5).

Discussion

Despite the initial goal of this research – to infer the usefulness of automatic procedures to characterize diet-related behavior by comparing roughness and striation density – it became apparent that taphonomic processes are of major concern to topographic analyses. As post-mortem abrasion smoothes the tooth, the microwear density pattern decreases. Intensive erosion on enamel surfaces is certain to cause a progressive obliteration of microwear features that would imply an initial increase in surface roughness, followed

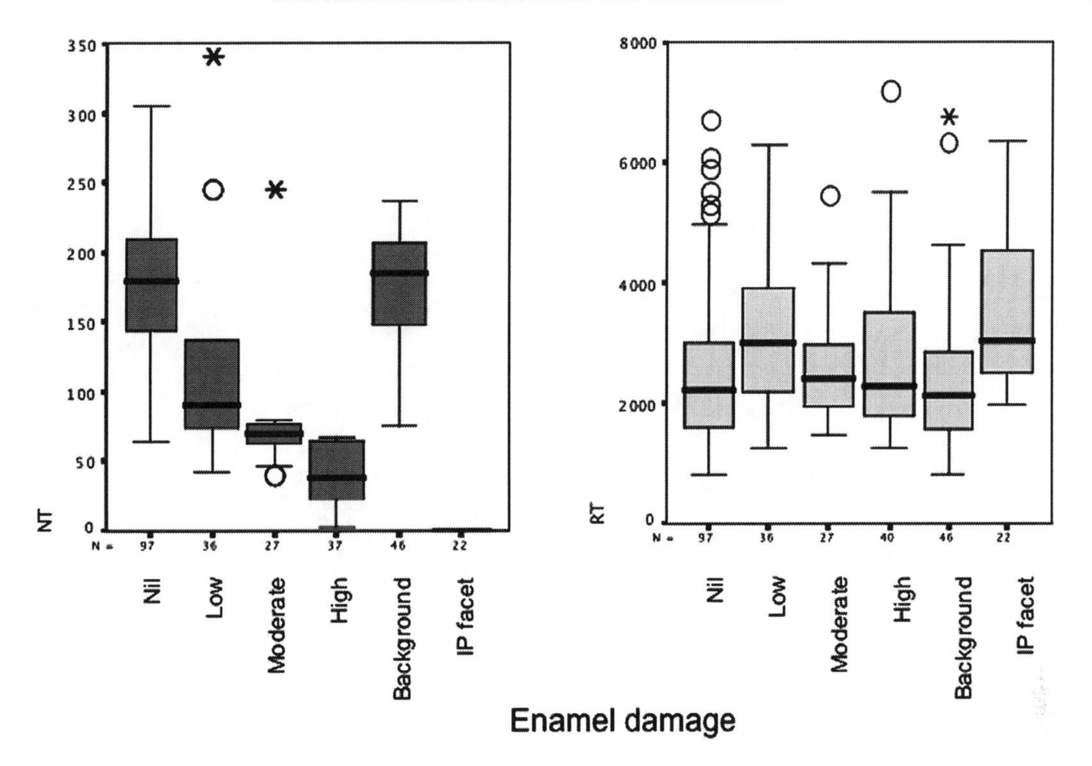

Figure 3. Box plots of striation density NT (left) and roughness values RT (right) for the four degrees of enamel damage (nil, low, moderate and high). The background category includes casts of enamel surfaces that could not be classified into one of the previous groups because of bad preservation or poor replication. The IP group included well-preserved surfaces of inter-proximal facets. No common pattern of variation of NT and RT by enamel damage categories was observed.

by complete polishing of the enamel, with a reduction in overall roughness.

Although it is clear that highly scratched enamel surfaces show higher values of enamel roughness than do smooth or polished surfaces, this relationship is not indicative of a meaningful association between automatic measures of roughness and dietary habits. The primary reason is that roughness is highly sensitive to post-mortem damage. At very low levels of erosion, an increase in roughness could be indicative of higher microwear feature densities, which is a variable clearly dependent on dietary habits. However, further effects by erosive factors rapidly even surface topography, dramatically reducing roughness.

The second hypothesis to be tested is more complicated. In an optimal situation, automatic measures of 3-D topographic relief are required to distinguish not only well preserved from eroded surfaces, but also from other surfaces not informative for inferring diet-related behavior. Roughness measures of surface topography do not appear to be useful for this purpose, as similar measures of roughness can be derived from completely different enamel surfaces, such as well preserved surfaces, enamel with exposed enamel prisms or inter-proximal enamel facets not exposed to food items during food chewing (Figure 6). Automatic procedures still require a thorough process of image selection by the researcher prior to analysis, but a certain degree of inter-observer subjectivity still remains in determining the frontier between no erosion and some incipient erosion.

However, surface analysis is not limited to roughness variables. Measures of anisotropy

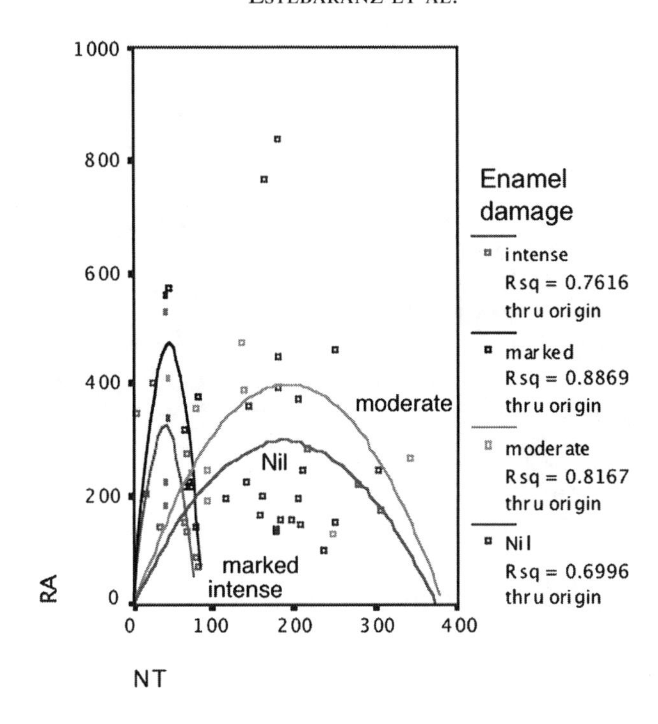

Figure 4. Quadratic regressions between striation density (NT) and average roughness (RA) for all the post-mortem damage categories. Two distinct patterns were noted: nil (1) and low (moderate in the plot) damaged enamel surfaces showed higher variability of striation densities (ranging from 100 to 300), whereas moderate (marked in the plot) and marked (intense in the plot) showed reduced values of striation densities.

over entire enamel surfaces are capable of characterizing heterogeneity of surface relief on an array of orientations, and scale-sensitive fractal analyses of surface topography can easily depict differences between smooth and rough enamel surfaces. However, anisotropy and fractal measurements, as well as roughness, are quantifications of the same underlying phenomenon: the heterogeneity of a peak-to-valley score throughout a three dimensional space. Ideally, unscratched and uneroded enamel surfaces, corrected for spatial curvature, would show no orientation anisotropy, no fractal variations to scale, and no surface roughness. In contrast, enamel surfaces highly affected by uneven processes of either erosion, scratching, pitting, cracking, plucking or prism exposure would show high anisotropy, fractal sensitivity and roughness. Automatic procedures themselves do not account for all the possible causes of surface heterogeneity. Yet, the researcher's expertise is required, and measures of inter-observer error in deciding which surfaces might be indicative of dietary habits and which are not still relevant. Semiautomatic procedures of microwear analysis may be highly subjective. However, automatic procedures require the same type of subjective decision levels during the analysis as the semiautomatic methods do. Further research is still needed to demonstrate that automatic image processing procedures can act as a substitute for the human eye-to-brain decision capabilities. For instance, it is interesting to note that the patterns observed for enamel roughness (Figure 5) vary depending on the species studied: *A. afarensis* and *Pan sp.* share the same patterns, as do *Pongo* and *Gorilla*, whereas *A. africanus* and *A. anamensis* conform to somewhat different models. Whether these differences among species have a biological explanation is something that further investigations will elucidate.

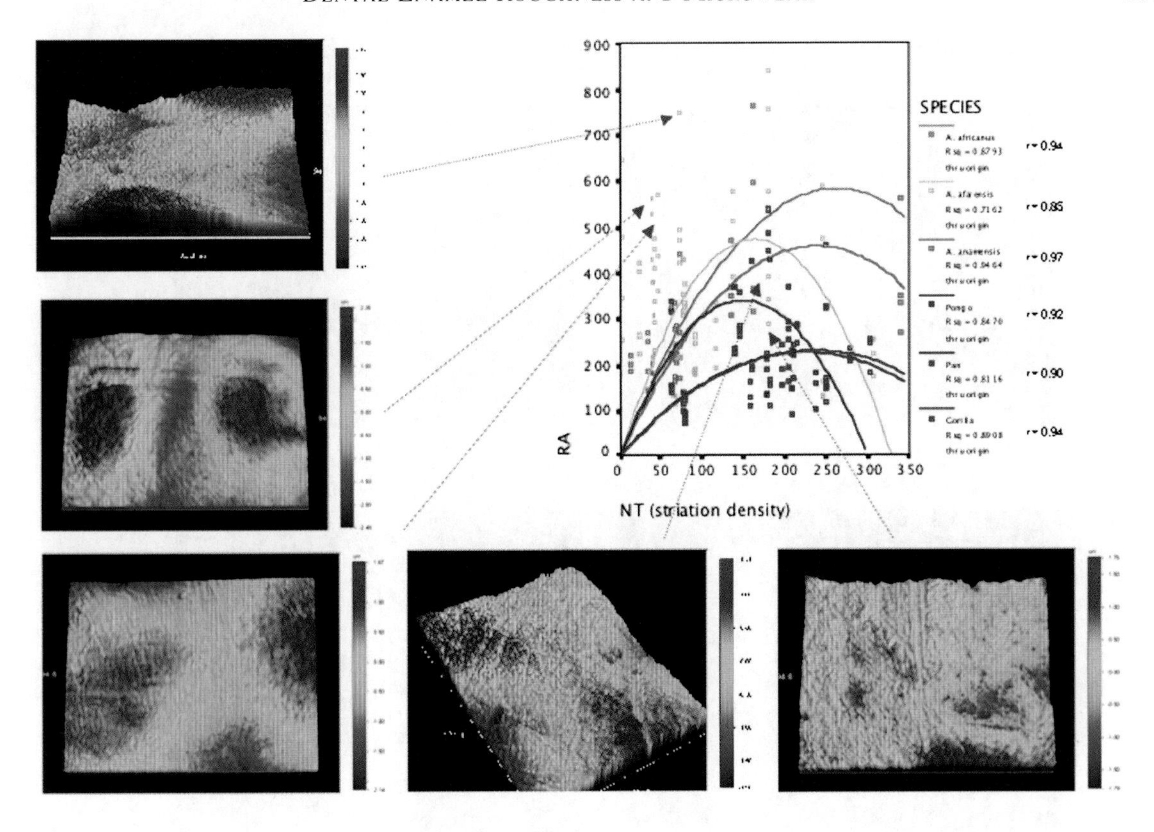

Figure 5. Quadratic regressions between striation density (NT) and average roughness (RA) by species. R^2 values for each species are shown with sample topography images. The regression pattern of *Australopithecus afarensis* and *Pan troglodytes* differed in the variation range of NT from those of *Gorilla, Pongo* and *A. africanus*.

Conclusion

The present study assessed the ability of fully automatic procedures to analyze microwear patterns on primate tooth surfaces through estimation of roughness. Our aim was to determine if roughness is a reliable indicator of dietary related habits rather than a measure of enamel preservation. The results indicate that, in fact, it is true that roughness is a highly sensitive parameter of enamel preservation, and it was not impossible to distinguish different types of enamel preservation and relief. Low degrees of post-mortem damage were characterized by an increase in surface roughness, but as erosion increases roughness tends to decrease, as does striation density. The density of diet-related scratches was highly sensitive to abrasion, as was the roughness variable. However, teeth affected by abrasion are not informative of dietary habits and, in the same way, roughness appears to be more informative of taphonomic processes than of diet. Future research requires that only well-preserved enamel surfaces be studied. The results obtained indicate that there exists a clear and statistically significant correlation between some measures of enamel surface roughness and the microwear pattern. However, this relationship varies and is dependant on the species studied. In addition, the analysis of roughness seems to be scale-sensitive and, thus, the magnification needs to be considered (Estebaranz et al., 2005). Finally, the application of automatic procedures for microwear research will still require a subjective interpretation of what factors are relevant to interpretations of diet and what factors are not.

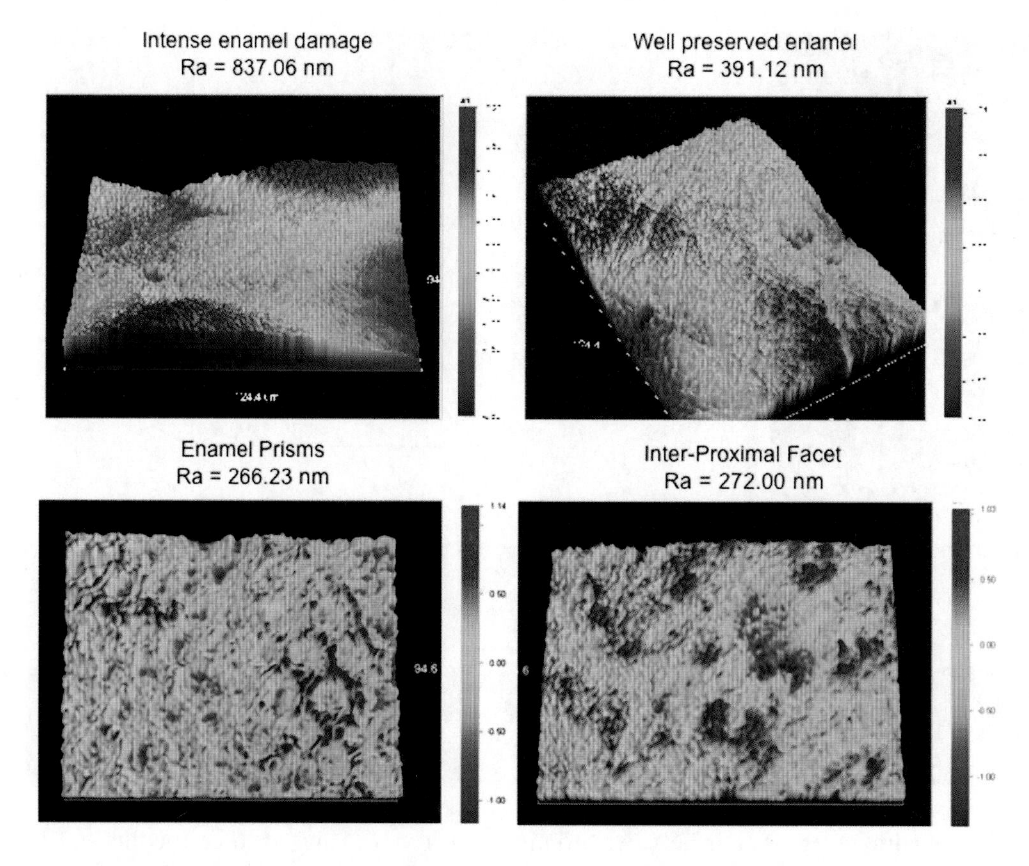

Figure 6. Topographic images of four distinct types of enamel surfaces: intensive enamel damage, well-preserved enamel surface, enamel prisms on a severely damaged surface, and an inter-proximal facet. Although the intensively damaged surface showed a high roughness value, the other surfaces had similar Ra values despite clearly different types of surfaces.

Acknowledgments

The Spanish project CGL2004-00775/BTE of the MEC funded this work. All SEM micrographs were obtained at the *Serveis CientíficoTècnics* (SCT) of the Universitat de Barcelona, and the interferometric images were performed at the *Plataforma de Nanotecnologia* of the *Parc Científic de Barcelona* (PCB).

References

Andritsakis, D.P., Vlamis, K.F., 1986. A new generation of the elastomeric impression materials. Odontostomatol ogike Proodos 40(3), 133–142.

Ball, T., Gardner, J.S., Brotherson, J.D., 1996. Identifying phytoliths produced by the inflorescence bracts of three species of wheat (*Tricutum monococcum* L., *T. dicoccon*, Schrank., and *T. aestivum* L.) using computer-assisted image and statistical analyses. Journal of Archaeological Science 23, 619–632.

Beynon, A.D., 1987. Replication technique for studying microstructure in fossil enamel. Scanning Microscopy 1, 663–669.

Butler, P.M., 1952. The milk molars of perissodactyla with remarks on molar occlusion. Proceedings of the Zoological Society of London 121, 777–817.

Daegling, D.J., Grine, F.E., 1999. Terrestrial foraging and dental microwear in *Papio ursinus*. Primates 40(4), 559–572.

Danielson, D.R., Reinhard K.J., 1998. Human dental microwear caused by calcium oxalate phytoliths in prehistoric diet of the lower Los Pecos Region, Texas. American Journal of Physical Anthropology 107, 297–304.

Dennis, J.C., Ungar, P.S., Teaford, M.F., Glander, K.E., 2004. Dental topography and molar wear in *Alouatta palliata* from Costa Rica. American Journal of Physical Anthropology 125, 152–161.

Estebaranz, F., Losada, M.J., Galbany, J., Martínez, L.M., Pérez-Pérez, A., 2005. Tafonomía y microdesgaste: análisis topográfico de superficies de esmalte dentario. Revista española de antropología 25, 75.

Galbany, J., Martínez, L.M., Hiraldo, O., Espurz, V., Estebaranz, F., Sousa, M., Martínez-López-Amor, H., Medina, A.M., Farrés, M., Bonnin, A., Bernis, C., Turbon, D., Pérez-Pérez, A., 2004a. Teeth: catálogo de los moldes de dientes de homínidos de la Universitat de Barcelona. Universitat de Barcelona, Barcelona.

Galbany J., Martinez, L.M., López-Amor, H.M., Espurz, V., Romero, A., De Juan, J., Pérez-Pérez, A., 2005a. Error rates in buccal-dental mircrowear quantification using scanning electron microscopy. Scanning 27, 23–29.

Galbany, J., Martínez, L.M., Pérez-Pérez, A., 2004b. Tooth replication techniques, SEM imaging and microwear analysis in primates: methodological obstacles. Anthropologie XLII/1, 5–12.

Galbany, J., Pérez-Pérez, A., 2004. Buccal enamel microwear variability in Cercopithecoidea primates as a reflection of dietary habits in forested and open savanna environments. Anthropologie XLII/1, 13–19.

Galbany, J., Pérez-Pérez, A., Moyà-Solà, S., 2005b. Dental microwear variability on buccal tooth enamel surfaces of extant Catarrhini and the Miocene fossil *Dryopithecus laietanus* (Hominoidea). Folia Primatologica 76, 325–341.

Godfrey, L.R., Semprebon, G.M., Jungers, W.L., Sutherland, M.R., Simons, E.L., Solounias, N., 2004. Dental use wear in extinct lemurs: evidence of diet and niche differentiation. Journal of Human Evolution 47, 145–167.

Gordon, K.D., 1982. A study of microwear on chimpanzee molars: implications of dental microwear analysis. American Journal of Physical Anthropology 59, 195–215.

Gordon, K.D., 1984. Hominoid dental microwear: complications in the use of microwear analysis to detect diet. Journal of Dental Research 63, 1043–1046.

Grine, F.E., 1986. Dental evidence for dietary differences in *Australopithecus* and *Paranthropus*. Journal of Human Evolution 15, 783–822.

Grine, F.E., Ungar, P.S., Teaford, M.F., 2002. Error rates in dental microwear quantification using scanning electron microscopy. Scanning 24, 144–153.

Gügel, I.L., Grupe, G., Kunzelmann, K-H., 2001. Simulation of dental microwear: characteristics traces by opal phytoliths give clue to ancient dietary behavior. American Journal of Physical Anthropology 114, 124—138.

Hunter, J.P, Fortelius, M., 1994. Comparative dental occlusal morphology, facet development, and microwear in two sympatric species of *Listridon* (Mammalia: Suidae) from the Middle Miocene of Western Anatolia (Turkey). Journal of Vertebrate Paleontology 14, 105–126.

Jernvall, J., Selänne, L., 1999. Laser confocal microscopy and geographic information systems in the study of dental anthropology. Palaeontologia Electronica 2(1), 1–17.

Kaiser, T.M., Katterwe, H., 2001. The application of 3D-microprofilometry as a tool in the surface diagnosis of fossil and sub-fossil vertebrate hard tissue. An example from the Pliocene Upper Laetoli Beds, Tanzania. International Journal of Osteoarchaeology 11, 350–356.

Kay, R.F., 1987. Analysis of primate dental microwear using image processing techniques. Scanning Microscopy 1(2), 657–662.

King, T., Aiello, L.A., Andrews, P., 1999b. Dental microwear of *Griphopithecus alpani*. Journal of Human Evolution 36, 3–31.

King, T., Andrews, P., Boz, B., 1999a. Effect of taphonomic processes on dental microwear. American Journal of Physical Anthropology 108, 359–373.

Lalueza, C., Pérez-Pérez, A., 1993. The diet of the Neanderthal Child Gibraltar 2 (Devil's Tower) through the study of the vestibular striation pattern. Journal of Human Evolution 24, 29–41.

Lalueza, C., Pérez-Pérez, A., Turbón, D.M., 1996. Dietary inferences through buccal microwear analysis of Middle and Upper Pleistocene human fossils. American Journal of Physical Anthropology 100, 367–387.

Lewis, P.J., Gutierrez, M., Johnson, E., 2000. *Ondatra zibethicus* (Arvicolinae, Rodentia) dental microwear patterns as a potential tool for palaeoenvironmental reconstruction. Journal of Archaeological Research 27, 789–798.

Mainland, I.L., 2003. Dental microwear in grazing and browsing Gotland sheep *Ovis aries* and its implications for dietary reconstruction. Journal of Archaeological Science 30, 1513–1527.

Martínez, L.M., Galbany, J., Pérez-Pérez, A., 2004. Paleodemography and dental microwear of *Homo habilis* from East Africa. Anthropologie XLII/1, 53–58.

Martínez, L.M., Pérez-Pérez, A. 2004. Post-mortem wear as indicator of taphonomic processes affecting enamel surfaces of hominin teeth from Laetoli and Olduvai (Tanzania): implications to dietary interpretations. Anthropologie XLII/1, 37–42.

Mayhall, J.T., Kageyama, I., 1997. A new three-dimensional method for determining tooth wear. American Journal of Physical Anthropology 103, 463–469.

Mills, J.R.E., 1955. Ideal dental occlusion in primates. Dental Practitioner 6, 47–51.

Nystrom, P., Phillips-Conroy, J.E., Jolly, C.J., 2004. Dental microwear in anubis and hybrid baboons (*Papio hamdryas*, Sensu Lato) living in Awash National Park, Ethiopia. American Journal of Physical Anthropology 125, 279–291.

Reed, D.N.O., 1997. Contour mapping as a new method for interpreting diet from tooth morphology. American Journal of Physical Anthropology Suppl. 24, 194.

Pérez-Pérez, A., 2004. Why buccal microwear? Anthropologie XLII/1, 1–3.

Pérez-Pérez, A., Bermúdez de Castro, J.M., Arsuaga, J.L., 1999. Nonocclusal dental microwear analysis of 300,000-year-old *Homo heidelbergensis* teeth from Sima de los Huesos (Sierra de Atapuerca, Spain). American Journal of Physical Anthropology 108(4), 433–457.

Pérez-Pérez, A., Espurz, V., Bermúdez de Castro, J.M., de Lumley, M.A., Turbón, D., 2003. Non-occlusal dental microwear variability in a sample of Middle and Late Pleistocene human populations from Europe and the Near East. Journal of Human Evolution 44, 497–513.

Pérez-Pérez, A., Galbany, J., Fontarnau R., 2001. Feature extinction in back-scattered SEM. In: Universitat de Barcelona (Eds.), Abstracts Microscopy. Universitat de Barcelona. Barcelona, pp. 41–42

Pérez-Pérez, A., Lalueza, C., Turbón, D., 1994. Intraindividual and intragroup variability of buccal tooth striation pattern. American Journal of Physical Anthropology 94, 175–187.

Ryan, A.S., 1979. A preliminary scanning electron microscope examination of wear striation direction on primate teeth. Journal of Dental Research 58, 525–530.

Silcox, M., Teaford, M.F., 2002. The diet of worms: an analysis of mole dental microwear and its relevance to dietary inference in fossil mammals. Journal of Mammalogy 83, 804–814

Solounias, N., Hayek, L.A.C., 1993. New methods of tooth microwear analysis and application to dietary determination of two extinct antelopes. Journal Zoology London 229, 421–445.

Strait, D.S., 1993. Differences in occlusal morphology and molar size in frugivores and faunivores. Journal of Human Evolution 25, 471–484.

Teaford, M.F., 1991. Dental microwear: what can it tell us about diet and dental function? In: Kelley, M.A., Larsen, C.S. (Eds.), Advances in Dental Anthropology. Wiley-Liss, Inc., New York, pp. 341–356.

Teaford, M.F., 1994. Dental microwear and dental function. Evolutionary Anthropology 3(1),17–30

Teaford, M.F., Glander, K.E., 1991. Dental microwear in wild-trapped *Alouata pallaia* from Costa Rica. American Journal of Physical Anthropology 85(3), 313–320.

Teaford, M.R., Glander, K.E., 1996. Dental microwear and diet in a wild population of mantled howlers (*Alouatta palliata*). In: Norconk, M.A., Rosenberger, A.L., Garber, P.A. (Eds.), Adaptive Radiations of Neotropical Primates. Plenum Press, New York, pp. 433–449.

Teaford, M.F., Oyen, O.J., 1989. Live primates and dental replication: new problems and new techniques. American Journal of Physical Anthropology 80, 73–81.

Teaford, M.F., Runestad, J.A., 1992. Dental microwear and diet in Venezuelan primates. American Journal of Physical Anthropology 94, 339–363.

Ungar, P.S., 1992. Feeding behaviour and dental microwear in Sumatran anthropoids. American Journal of Physical Anthropology 88, 347–364.

Ungar, P.S., 1994. Incisor behaviour and dental microwear of Sumatran anthropoid primates. American Journal of Physical Anthropology 94, 339–363.

Ungar, P.S., 1995. A semiautomated image analysis procedure for the quantification of dental microwear II. Scanning 17, 57–59.

Ungar, P.S., 1996. Dental microwear of European Miocene catarrhines: evidence for diets and tooth use. Journal of Human Evolution 31, 335–366.

Ungar, P.S., 1998. Dental allometry, morphology, and wear as evidence for diet in fossil primates. Evolutionary Anthropology 6(6), 205–217.

Ungar, P.S., 2004. Dental topography and diets of *Australopithecus afarensis* and early *Homo*. Journal of Human Evolution 46, 605–622.

Ungar, P.S., Brown, C.A., Bergstrom, T.S., Walker, A., 2003. Quantification of dental microwear by tandem scanning confocal microscopy and scale-sensitive fractal analyses. Scanning 25, 185–193.

Ungar, P.S., M'Kirera, F., 2003. A solution to the worn tooth conundrum in primate functional anatomy. Proceedings of the National Academy of Sciences of the USA 10(7), 3874–3877.

Ungar, P.S., Simon, J.C., Cooper, J.W., 1991. A semiautomated image analysis procedure for the quantification of dental microwear. Scanning 13, 31–36.

Ungar, P.S., Spencer, M.A., 1999. Incisor microwear, diet, and tooth use in three Amerindian populations. American Journal of Physical Anthropology 109, 387–396.

Ungar, P.S., Teaford, M.F., 1996. Preliminary examination of non-occlusal dental micro-wear in anthropoids: implications for the study of fossil primates. American Journal of Physical Anthropology 100, 101–113.

Ungar, P.S., Teaford, M.F., Glander, K.E., Pastor, R.F., 1995. Dust accumulation in the canopy: a potential cause of dental microwear in primates. American Journal of Physical Anthropology 97, 93–99.

Ungar, P.S., Williamson, M., 2000. Exploring the effects of tooth wear on functional morphology: a preliminary study using dental topographic analyses. Palaeontologia Electronica 3(1), 1–18. http://palaeo-electronica.org/2000_1/gorilla/issue1_00.htm

Ungar, P.S., 2001. Microware Software, Version 4.0 A semiautomated image analysis system for the quantification of dental microwear. Fayetteville, AR, U.S.A.

Ward, J., Mainland, I.L., 1999. Microwear in modern free-ranging and stalled pigs. The potential of dental microwear analysis for exploring pig diet and management in the past. Environmental Archaeology 4, 25–32.

Zuccotti, L.F., Williamson, M.D., Limp, F.E., Ungar, P.S., 1998. Technical note: modelling primate occlusal topography using geographical information systems technology. American Journal of Physical Anthropology 107, 137–142.

Index

VERTEBRATE PALEOBIOLOGY AND PALEOANTHROPOLOGY

PUBLISHED AND FORTHCOMING TITLES:

Neanderthals Revisited: New approaches and perspectives
Edited by K. Harvati and T. Harrison
ISBN: 978-1-4020-5120-3, 2006
A Volume in the Max-Planck Institute Sub-series in Human Evolution

The Evolution and Diversity of Humans in South Asia: Interdisciplinary studies in archaeology, biological anthropology, linguistics and genetics
Edited by M. Petraglia and B. Allchin
ISBN: 978-1-4020-5561-4, 2007

Dental Perspectives on Human Evolution: State of the art research in dental paleoanthropology
Edited by S.E. Bailey and J-J. Hublin
ISBN: 978-1-4020-5844-8, 2007
A Volume in the Max-Planck Institute Subseries in Human Evolution

Hominin Environment in the East African Pliocene: An Assessment of the Faunal Evidence
Edited by R. Bobe, Z. Alemseged and A.K. Behrensmeyer
ISBN: 978-1-4020-3097-0, *forthcoming 2007*

Deconstructing Olduvai: A Taphonomic Study of the Bed I Sites
By M. Domínguez-Rodrigo, C.P. Egeland and R. Barba Egido
ISBN: 978-1-4020-6150-9, *forthcoming 2007*

Printed by Publishers' Graphics LLC